D0538550

WIDEBAND WIRELESS DIGITAL COMMUNICATIONS

Andreas F. Molisch
Editor

Prentice Hall PTR
Upper Saddle River, New Jersey 07458
www.phptr.com

Library of Congress Cataloging-in-Publication Data

Wideband wireless digital communications / Andreas F. Molisch.
p. cm.
Includes bibliographical references and index.
ISBN 0-13-022333-6
1. Wireless communication systems. I. Molisch, Andreas F.

TK5103.2.W53 2000
621.382—dc21 00 063703

Editorial/production supervision: *Jane Bonnell*
Cover design director: *Jerry Votta*
Cover design: *Alamini Design*
Manufacturing manager: *Alexis R. Heydt*
Acquisitions editor: *Bernard M. Goodwin*
Editorial assistant: *Michelle Vincenti*
Marketing manager: *Dan DePasquale*

© 2001 by Prentice Hall PTR
Prentice-Hall, Inc.
Upper Saddle River, New Jersey 07458

Prentice Hall books are widely used by corporations and government agencies for training, marketing, and resale.
The publisher offers discounts on this book when ordered in bulk quantities. For more information, contact Corporate Sales Department, Phone: 800-382-3419; FAX: 201-236-7141;
E-mail: corpsales@prenhall.com
Or write: Prentice Hall PTR, Corporate Sales Dept., One Lake Street, Upper Saddle River, NJ 07458.

Disclaimer: The editor, authors, and publisher make no representations or warranties, express or implied, with respect to the merchantability and fitness for a particular purpose of the information in this book.

Product or company names mentioned herein are the trademarks or registered trademarks of their respective owners.

All rights reserved. No part of this book may be reproduced, in any form or by any means, without permission in writing from the publisher.

Printed in the United States of America
10 9 8 7 6 5 4 3 2 1

ISBN 0-13-022333-6

Prentice-Hall International (UK) Limited, *London*
Prentice-Hall of Australia Pty. Limited, *Sydney*
Prentice-Hall Canada Inc., *Toronto*
Prentice-Hall Hispanoamericana, S.A., *Mexico*
Prentice-Hall of India Private Limited, *New Delhi*
Prentice-Hall of Japan, Inc., *Tokyo*
Pearson Education Asia Pte. Ltd.
Editora Prentice-Hall do Brasil, Ltda., *Rio de Janeiro*

CONTENTS

PREFACE

Wireless systems are called wideband if the system bandwidth is comparable to or larger than the bandwidth over which the channel transfer function can be considered constant—typically, on the order of 100 kHz. Equivalently, one can state that time dispersion and thus intersymbol interference play an important role in the system performance. In practice, all *digital* mobile radio systems are wideband to a certain degree. In the early and mid-1990s, this development was mainly driven by the use of multiple-access formats that increase the transmission bandwidth. Recently, the requirements for the data rates of single users have increased so much that transmission bandwidths must be increased still further. After all, the aim of any mobile communications system is to provide the same services and quality as fixed, wired systems. Especially fast wireless internet connections will play a vital role in the future development of mobile communications.

There are four basic approaches to dealing with the "widebandedness" of mobile radio systems. The first one is the use of *unequalized systems* to design the system like a narrowband system and treat the intersymbol interference resulting from the time dispersion; this is a very simple approach that results in a system performance worse than a narrowband system. Alternatively, one can try to combat the intersymbol interference by *equalizers*. This approach exploits the inherent delay diversity of time-dispersive channels and thus gives better performance than narrowband systems. Equalization is usually employed in single-carrier, TDMA/FDMA systems. An alternative is *Code Division Multiple Access* using Rake receivers. This approach will gain special importance in the next years as it has been selected for UMTS (IMT-2000), the third-generation standard that will allow data rates of up to 2 Mbit/s. For even higher data rates, *Orthogonal Frequency Division Multiplexing* is envisioned.

PURPOSE OF THE BOOK: Today, researchers and engineers in the field are faced with different tasks. When designing new systems, they have to determine which approach is the most suitable for the boundary conditions at hand, e.g., the desired data rate, the available processor power in base station and mobile station, etc. The designers of transmitters/receivers for standardized systems are faced with the task of fulfilling certain performance specifications (which are required for type approval), while usually having the freedom to choose the cheapest and most efficient technical realization. Academic researchers usually also strive for optimum or efficient solutions of signal-processing problems within boundary conditions set by the mobile radio channel.

This book gives all these people an overview of the state-of-the-art. It assumes a familiarity with standard digital communications, as elaborated in the classical textbooks, e.g., Proakis' *Digital Communications.* The book aims to provide a path through the minefield of the different methods and results described in the primary literature. It covers the essence of the methods and arranges them in a systematic way. Furthermore, extensive referencing to the literature should enable the reader to access all necessary details of any method he becomes especially interested in.

Apart from serving the practicing researcher, the book can also serve as a text for an advanced graduate course in wireless digital communications. Due to the high information density, an instructor can pick and choose the covered subjects if the course lasts for one semester only. In our experience, it is also effective to introduce students to a particular method by going over the basic description given in the book, and then assigning one or two related papers for self-study or homework.

ORGANIZATION: The book starts with a general part that describes the most popular wideband systems and the wideband mobile radio channel, then gives an overview of the remainder of the book. Each of the subsequent parts II—V describes one of the wideband approaches, namely, (i) unequalized systems, (ii) equalized systems, (iii) multicarrier methods, and (iv) CDMA; a more detailed description of the contents is given in Chapter 4. Each of these parts can be read independently.

We hope that this book will be useful to the researcher in the exciting field of digital wireless communications. We would be happy to hear any comments or suggestions, e.g., by email to Andreas.Molisch@tuwien.ac.at. You will also be able to find some updates and comments at www.nt.tuwien.ac.at/mobile/staff/Andreas.Molisch/wideband_book.htm.

ACKNOWLEDGMENTS: We would like to thank our editor at Prentice Hall, Bernard Goodwin, and the production editor, Jane Bonnell, for their expert guidance through the different stages of the manuscript preparation as well as their patience. Our profound gratitude also to Mrs. Manuela Hueber-Heigl for her help in typing and formatting, which she did not only expertly but also always cheerfully. The authors would also like to thank all the people with whom they have collaborated on the subjects treated in the book as well as the reviewer of the book proposal, who made several helpful suggestions.

Andreas F. Molisch
Vienna, 2000

LIST OF CONTRIBUTORS

George Chrisikos
Dot Wireless, Inc.
6825 Flanders Drive
San Diego, CA 92121
USA

Savo Glisic
Telecommunications Laboratory
University of Oulu
Oulu, 90401 FIN
FINLAND

Alois M. J. Goiser
University of Technology Vienna
Gusshausstrasse 27-29
A-1040 Vienna
AUSTRIA

Brian D. Hart, Ph.D.
Telecommunications Engineering Group
Research School of Information Sciences and Engineering
Institute of Advanced Studies
The Australian National University
Canberra ACT 0200
AUSTRALIA

Aarne Mämmelä, Research Prof.
VTT Electronics
Kaitoväylä 1
P.O. Box 1100
FIN-90571 Oulu
FINLAND

Thomas May
Institute of Telecommunications
Technische Universitaet Braunschweig
D-38092, Braunschweig
GERMANY

Andreas F. Molisch
Institut fuer Nachrichtentechnik und Hochfrequenztechnik
Technische Universitaet Wien
Gusshausstrasse 25/389
A-1040 Vienna
AUSTRIA

Hermann Rohling, Prof.
Institut fuer Nachrichtentechnik
Technical University of Hamburg-Harburg
Eißendorfer Straße 40
21073 Hamburg
GERMANY

Desmond P. Taylor, Tait Professor of Communications
Electrical and Electronic Engineering Dept.
University of Canterbury
Christchurch
NEW ZEALAND

Giorgio M. Vitetta, Prof.
Department of Engineering Sciences
Via Vignolese 905
41100 Modena
ITALY

Moe Z. Win
Wireless Systems Research Department
Newman Springs Laboratory, AT&T Labs-Research
100 Schulz Drive
Red Bank, NJ 07701-7033
USA

Part I

Introduction to Wideband Systems

Andreas F. Molisch

Chapter 1

BASICS

1.1 WHAT ARE WIDEBAND SYSTEMS?

Wideband is a word with many meanings. In the articles of popular science and the mainstream newspapers, it is a buzzword applied to almost any new communication system. In that context, it is abused almost as much as words like "multimedia" and "information society." In the more serious technical literature, it also has different connotations. In analog FM modulation, "wideband" describes a property of the modulation format alone, namely, that the modulation bandwidth is considerably larger than the bandwidth of the modulating signal. In RF engineering, it describes the relative bandwidth of a certain quantity (e.g., antenna bandwidth) compared to the carrier frequency—if this ratio is not much smaller than unity, the system is called wideband. In digital mobile radio, which is the focus of this book, "wideband" has even more complicated implications: it compares certain properties of the system with properties of the propagation environment.

Due to the multipath propagation from the transmitter to the receiver, the received signal is not just a single echo (i.e., an attenuated and phase-shifted version) of the transmitted signal, but consists of several echoes with different runtimes. In other words, the impulse response of the mobile radio channel is not a single delta pulse, but a whole sequence of pulses—the channel is said to be time-dispersive. If now the maximum excess delay (i.e., the delay between the first and the last significant echo) is much smaller than the duration of a digital symbol,[1] then the time dispersion has no appreciable effect on the shape of the received signal, and the system is defined to be *narrowband.* If, on the other hand, this ratio is *not* much smaller than unity, then the system is called wideband. We thus see that from the knowledge of a system alone, we can never judge whether it is wideband or not: we also need the properties of the channel in which it operates. This has led to considerable confusion of terms and obfuscated many discussions. We can make further distinction by checking whether the maximum excess delay is slightly smaller (but not *much* smaller) than the symbol length, or larger. In the former case, the system is usually built like a narrowband system and suffers a few additional

[1]Note that in a spread spectrum system, the relevant quantity is not the information symbol duration, but the chip duration.

errors introduced by the time-dispersion effects (intersymbol interference). If, on the other hand, the maximum excess delay is larger than the symbol length, then we absolutely *havc* to build a system with time-dispersion countermeasures. Such systems are more complicated and expensive but often exhibit better performance than true narrowband systems.

Another way to judge whether a system is wideband or not is to consider it in the frequency domain. There, we compare the system bandwidth to the transfer function of the channel. If this transfer function varies significantly over the system bandwidth, then again we have a wideband system.[2] In a first approximation, the bandwidth over which the transfer function is constant is the inverse of the delay spread. A relation of the various definitions of "widebandedness" is shown in Figure 1.1.

Despite the dependence on transmission channels, it is possible to give some reasonable first estimates as to whether a mobile radio system is wideband or not. In outdoor channels, the maximum excess delay is typically on the order of 5–20 μs, whereas in indoor environments, it is typically on the order of 0.1–1 μs or less. Assuming binary modulation formats, we thus find that outdoor systems with data rates in excess of 10 kbit/s, and indoor systems with 200 kbit/s can be"slightly" wideband in the sense discussed above. Systems with data rates in excess of 50 kbit/s (outdoor) or 1 Mbit/s (indoor) are truly wideband.

1.2 HISTORY

Until about 1990, the interest of service providers lay mainly in the field of speech transmission by analog modulation with frequency division multiple access (FDMA), and channel spacings of 25–30 kHz. Even data transmission was done with a similar channel spacing.

After 1990, however, came the evolution to digital speech transmission. In the United States, it was decided to introduce systems that were compatible with existing analog systems, i.e., using the same channel spacings. Since speech coders require only about 10 kbit/s for the transmission of digital speech, it is possible to put three speech channels within one 30 kHz channel (D-AMPS) by time-division multiple access. Similarly, the Japanese Digital Cellular (JDC) standard puts three speech channels into one 25 kHz radio channels. Both these systems thus are clearly slightly wideband systems as defined above. Time-dispersion countermeasures (equalizers) were foreseen in the D-AMPS standard and spurred research in adaptive equalizers (see Part III). This work could draw on the extensive literature in adaptive filter theory, which had been developed since the advent of digital signal processing in the 1960s (for an overview, see e.g., [1]). While, of course, the theoretical work of the D-AMPS researchers was usually generally valid, the

[2] The exact measures for "delay between first and relevant last echo" and "appreciable variation of the transfer function" is given in Chapter 3. It is also common not to consider a single realization of the channel, but rather an ensemble, which leads us to consider the frequency coherence function instead of the instantaneous transfer function of the channel.

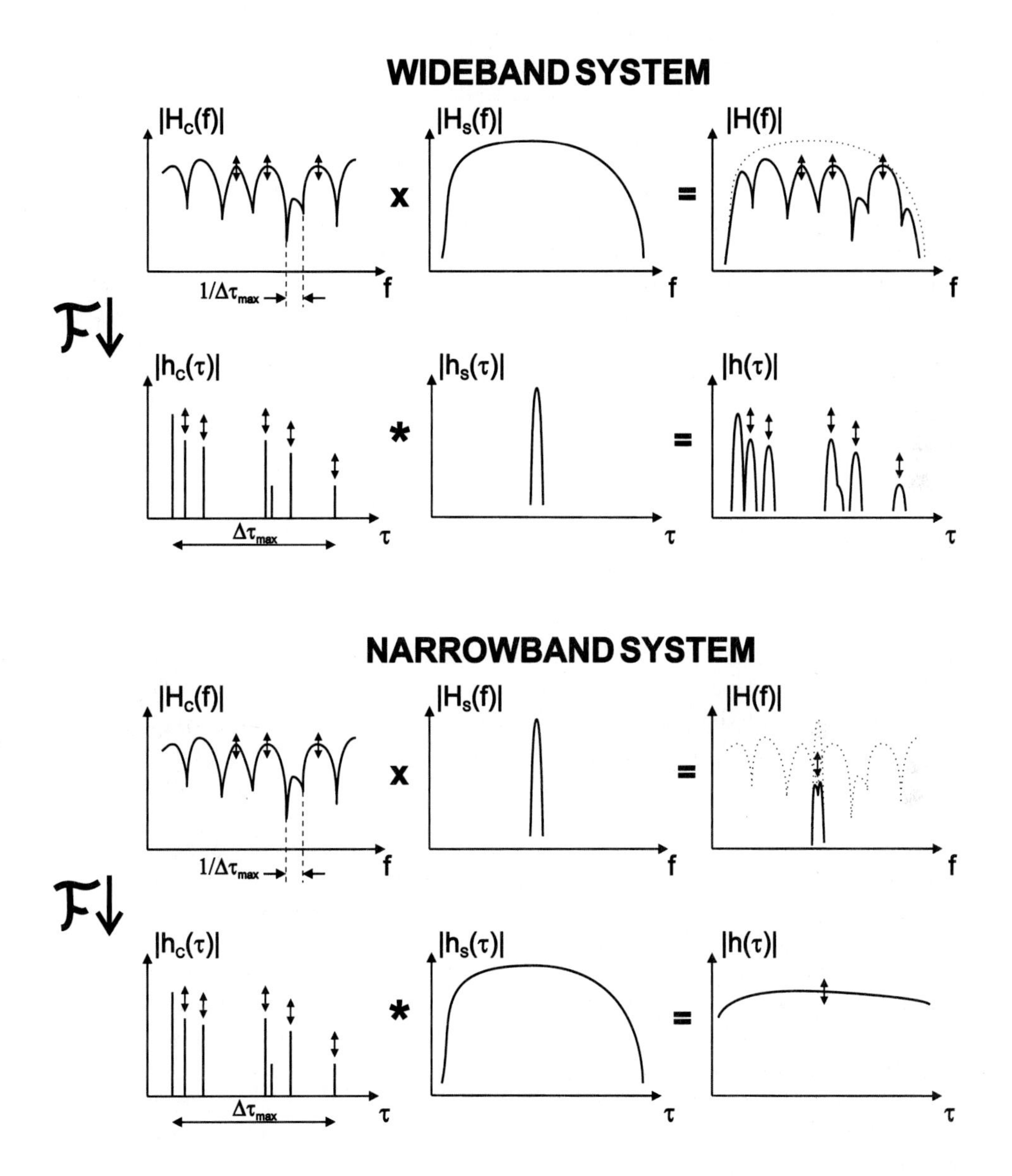

Figure 1.1. Definition of wideband and narrowband systems. $H_C(\omega)$: transfer function of the channel. $H_S(\omega)$: spectrum of transmit signal. $h_C(t)$: impulse response of the channel. $h_S(t)$: waveform for transmission of one bit/chip.

main focus was on equalizers with only a few taps (since the time dispersion was almost never longer than two symbol intervals) and fast time variations (the channel usually changes appreciably during one TDMA time slot).

The JDC standard, on the other hand, did not foresee a compulsory equalizer. Thus, considerable work was performed to find the bit-error rate (BER) for an unequalized system including noise, intersymbol interference due to the channel time dispersion, and also temporal variations of the channel (random FM). Not surprisingly, this work was mainly pushed by Japanese scientists (see, e.g., [2], [3]). This work could also build on previous investigations of time-dispersion effects for ionospheric propagation and directional transmission, which dated as far back as the 1950s, with [4] and [5] as the most important references. For JDC, it turned out that diversity is also quite effective for suppressing intersymbol interference as long as the delay spread stays smaller than a symbol duration. The diversity is also required to mitigate the effects of noise; in contrast to true wideband systems, which have inherent frequency diversity, JDC cannot overcome fading dips without such measures.

For reasons more commercial and regulatory than technical, the JDC system never became a big business success. However, the work on wideband unequalized systems was continued into the middle and late 1990s. At that time, the DECT system, a standard for cordless telephones originally emanating from Europe, was the driving force. DECT is also a TDMA system with a channel spacing of 1.8 MHz. Since it was originally intended for home indoor applications with extremely low delay spreads, no equalizer was foreseen. However, some time after the specifications, it turned out that intersymbol interference could influence the performance. This became especially relevant when DECT became the system of choice for many private (office) networks, wireless local loop applications, and low mobility PCS networks.

GSM has been the biggest success story in mobile radio, and, quite naturally, an enormous amount of research has been done for designing and simulating components or complete systems. GSM is a TDMA system with a channel spacing of 200 kHz. It specifies an equalizer that must be able to cope with echos that exhibit delays of up to four symbol durations. The complexity of the equalizer is thus higher than in the D-AMPS system. The time slot in GSM is sufficiently short that usually the channel does not change significantly during one frame. This simplifies the adaptation algorithms for the equalizer.

In the early 1990s, code-division multiple access (CDMA), which was previously used for military applications, was adapted also to civilian applications, most notably, wireless systems. Qualcomm, a U.S. company, was the driving force in those days and had its system standardized as U.S. interim standard IS-95. In that standard, the information is spread onto a 1.25 MHz broad spectrum by multiplication of each bit with a whole sequence of chips, where each chip is 0.814 μs long. This implies that IS-95 has a bandwidth comparable to a DECT system; however, since it is intended for outdoor applications, it is more wideband. In a CDMA system, a "Rake" receiver is used to first detect all distinguishable echoes and then combine them prior to final detection (for details see Chapter 2 and Part V). The effect is similar to an equalizer, but the implementation is quite different. The CDMA technology was later also adopted by Japanese and European standardization bodies for the definition of third-generation wireless systems, code-worded W-CDMA (wideband-CDMA). This system has a bandwidth of 5 MHz and thus will also be wideband in many indoor applications, as opposed to IS-95. On a systems level, the research until the middle of the 1990s was dominated by analysis of the differences between TDMA and CDMA in general, giving comparatively little attention to the effect of wideband propagation or Rake vs. equalizers. In the last years, there has been much more interest in the effect of bandwidth, since this is one of the key features distinguishing W-CDMA from IS-95. Performance gains due to these aspects are thus of considerable interest. On a more hardware-oriented level, there has always been great interest in the question of how to efficiently realize a Rake receiver, since this is key requirement of any CDMA system. This problem becomes of special importance in combination with multiuser detection and spatial signal processing.

The latest development is the emergence of a new modulation/multiple-access technique, namely, OFDM (orthogonal frequency division multiplexing). Essentially, OFDM distributes the information onto many subcarriers, so that the bits on each subcarrier are much longer, drastically reducing ISI. While OFDM has been known since the 1960s and was first suggested for wireless applications in the 1980s [6], it was for a long time too computationally intensive to be considered for real-time applications, especially in portable equipment. Progress in digital signal processer hardware has rendered it feasible by the middle of the 1990s, as evidenced by its adoption as standard for digital audio and television broadcasting. Judging from conference contributions, the research in this field has increased by a factor of 10 between 1996 and 1999 and by 2000 has already resulted in a first wireless standard (HIPERLAN 2) that applies it. Since it is capable of dealing with higher data rates than either TDMA or CDMA, it is also envisioned for fourth-generation systems.

Up to the late 1990s, mobile radio was used almost exclusively to transmit speech; the large bandwidth required for some systems stemmed from the multiple access format, not from user demand for exclusive use of a large bandwidth. This situation is currently changing as data transmission and multimedia applications enter the mobile radio market. Third-generation systems will allow fast Internet access, video telephony, and many other applications that require hundreds or thousands of kilobits per second for each user. Standards for wireless computer links even demand 20 Mbit/s. Generally, a wireless system will have to provide the same quality and speed as a wired connection, and considering the advances in xDSL broadband connections for the home market, this requirement is indeed a challenge that will need corresponding progress in the broadband wireless field.

Bibliography

[1] S. Haykin, *Adaptive Filter Theory.* Englewood Cliffs, N. J.: Prentice Hall, 1986.

[2] F. Adachi and J. Parsons, "Unified analysis of postdetection diversity for binary digital FM mobile radio," *IEEE Trans. Vehicular Techn.*, vol. 37, pp. 189–198, 1988.

[3] F. Adachi and J. Parsons, "Error rate performance of digital FM mobile radio with postdetection diversity," *IEEE Trans. Comm.*, vol. 37, pp. 200–210, 1989.

[4] P. Bello and B. D. Nelin, "The effect of frequency selective fading on the binary error probabilities of incoherent and differentially coherent matched filter receivers," *IEEE Trans. Comm.*, vol. 11, pp. 170–186, 1963.

[5] J. Proakis, "On the probability of error for multichannel reception of binary signals," *IEEE Trans. Comm.*, vol. 16, pp. 68–71, 1968.

[6] L. J. Cimini, "Analysis and simulation of a digital mobile channel using orthogonal frequency division multiplexing," *IEEE Trans. Comm.*, vol. 33, pp. 665–675, July 1985.

Chapter 2

CURRENT AND FUTURE WIDEBAND SYSTEMS

In this section, we briefly introduce the air-interface specifications of the most important wireless wideband systems. This introduction should help to more easily apply the results of the latter parts of this book, which concentrate more on the basic approaches than on specific systems.

This section is restricted to the most widespread systems. Numerous other standards have been developed for special applications or have a mainly regional importance.

2.1 DECT AND PHS

DECT (Digital Enhanced Cordless Telecommunications) is a standard for cordless telecommunications proposed by ETSI (European Telecommunication Standards Institute) in the early 1990s ([1], see also www.etsi.org/technicalactiv/dect.htm). DECT uses a combination of TDMA and FDMA for multiple access. The 20 MHz band that has been allocated to DECT (1880–1900 MHz) is divided into 10 frequency bands; the carrier frequencies are separated by 1.728 MHz. On each carrier, 12 users can be accommodated in a TDMA scheme: the time axis is divided into frames, which last 10 ms each, and the frames are divided into 24 timeslots. Each user is assigned two timeslots that are separated by 5 ms: the first timeslot is used for the downlink, the second slot for the uplink. This implies that time-domain duplexing is used for separating uplink and downlink. The timeslots are separated from each other by guard times, which are necessary since the runtimes from the different terminals to the base station may be different. The 52 μs guard time prescribed in DECT can theoretically deal with 15 km distance between base station and mobile terminal. This distance is far more than actually needed; even when directional antennas are used, the "cell radius" is usually less than 3 km. The rest of the guard time is used to give the transmitter amplifiers time to power up; the faster these amplifiers have to be, the more expensive they become.

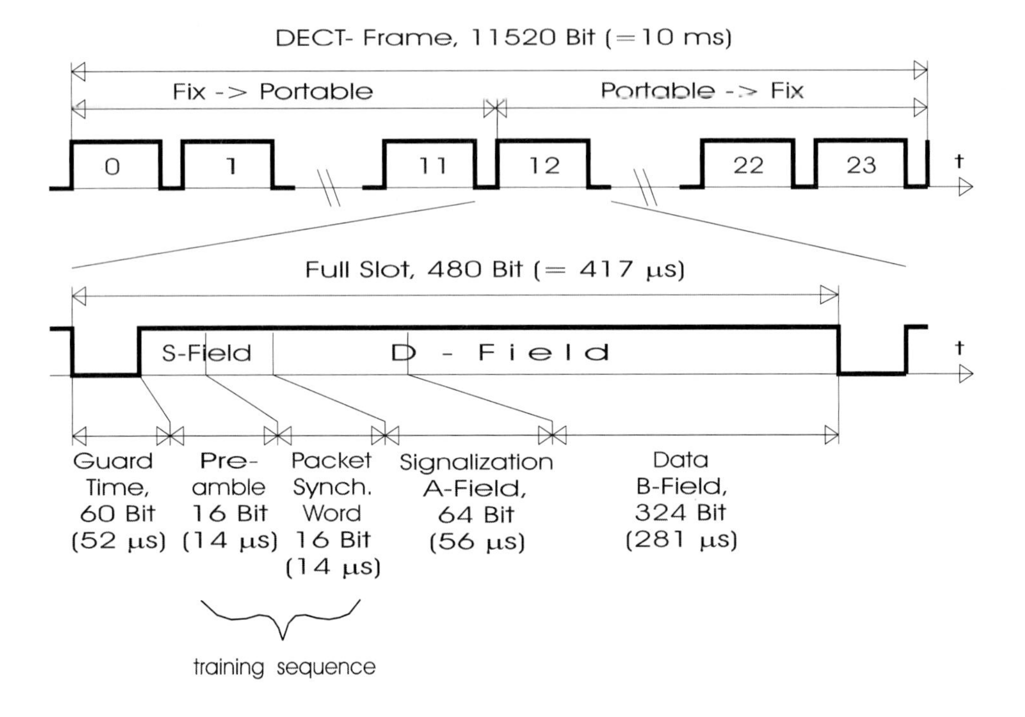

Figure 2.1. Frame structure of DECT

Figure 2.1 illustrates the frame structure for DECT.

Each timeslot begins with a 16-bit preamble and a 16-bit packet synchronization word (the 32-bit S-field). This is followed by the D-field. This D-field contains 64 bits of signalling (A-field) and 324 bits of data. The A-field includes a Cyclic Redundancy Check (CRC) for the detection of packet reception failures; it allows transmission of signalling information with a net rate of 64 kbit/s. Finally, the B-field contains 324 bits of data, enabling the transmission of 32 kBit/s ADPCM coded speech per user and direction. The overall transmission data rate is therefore 1.152 Mbit/s.

This frame/slot structure suggests that DECT terminals can be designed to be far less complex, and thus less expensive, than GSM. DECT does not use a complicated speech coder, which explains the fact that it requires 32 kbit/s instead of the 6.5 or 13 kbit/s prescribed by GSM. It also contains no channel coding for the speech data, so that the bit error probability (BER) for tolerable transmission quality has the rather low value of 10^{-3} (as compared to 10^{-1} for GSM). By combining two or several timeslots, DECT can provide much higher bit rates, up to 552 kbit/s, and carry asymmetric traffic, making DECT attractive for data transmission. The modulation format is GFSK (Gaussian Frequency Shift Keying) with nominal mod-

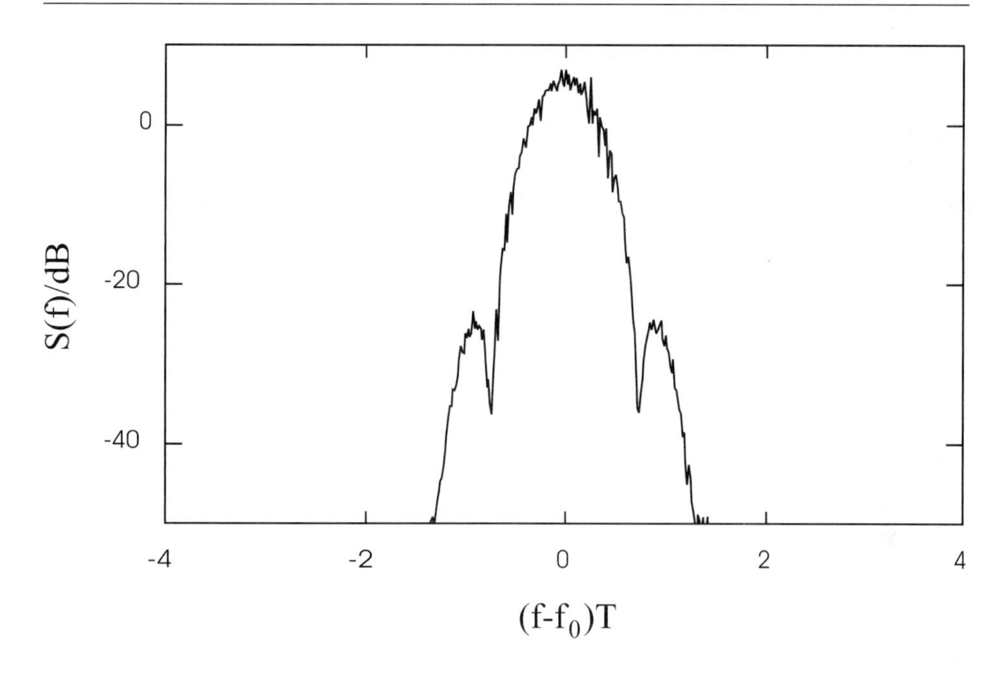

Figure 2.2. Power spectrum of a DECT signal

ulation index $h_{mod} = 0.5$, which is equivalent to Gaussian Minimum Shift Keying (GMSK). However, the tolerances in the modulation index are quite large.

The time-bandwidth product $B_G T$ is 0.5, where B_G is the bandwidth of the Gaussian filter, and T is the symbol period. This value provides a compromise between emitted spectrum and intersymbol interference (see Figure 2.2). The resulting RF bandwidth is about 1.56 MHz (for a 99% power criterion). With the specified carrier spacing of 1.728 MHz, adjacent channel emission is less than -40 dBc.

While DECT has found wide acceptance for cordless communications in many countries, Japan has devised its own rival standard, which also has achieved considerable success. This standard, called PHS (Personal Handyphone System), is mostly identical to DECT in its system design. It also is an FDMA/TDMA/TDD system and uses 32 kbit/s ADPCM without error correction. However, there are some important differences in the radio link: the modulation format is $\pi/4$-DQPSK, so that two bits are sent with each symbol. The pulse shapes used are pulses with a square-root–raised-cosine spectrum with roll-off factor $\alpha = 0.5$. There are only eight timeslots (four users) per carrier frequency; each frame lasts 5 ms. The carrier spacing is only 300 kHz; the bit rate is 384 kbit/s.

In Japan, the frequency band from 1985–1918 MHz has been assigned to PHS [1], giving 77 carriers. These have been divided into 37 lower carriers for home

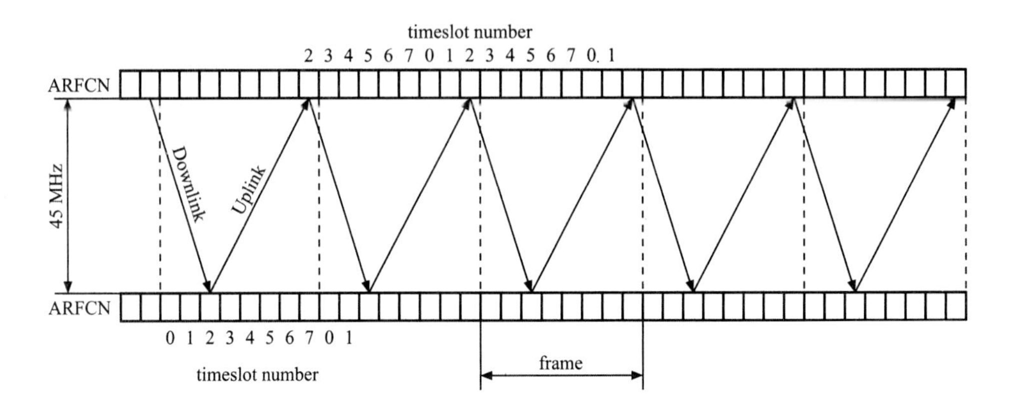

Figure 2.3. Timeslot structure for uplink and downlink in GSM/DCS-1900

and office use (i.e., real cordless and PBX applications) and 40 upper carriers, for a low-mobility PCS system. This system has been very successful in Japan, having several million users. On the other hand, this system requires an enormous investment. The metropolitan area of Tokyo is now covered by several hundred thousand base stations.

We finally note that in the United States, there is the so-called PWT system that combines the system specifications of DECT (including multiple-access specifications) with the $\pi/4$-DQPSK modulation format.

2.2 GSM/DCS-1900

GSM is by far the most successful mobile radio system ever devised and closely tied to the global popularity of mobile radio in general. It was specified in the late 1980s by ETSI and has gained worldwide popularity. It uses a mixture of TDMA and FDMA for multiple access, and FDD for duplexing.

GSM-900 occupies the frequency band from 890–915 MHz for the uplink and 935–960 MHz for the downlink; the duplexing distance is fixed at 45 MHz. In each band, 124 carriers are available (the 100 kHz closest to the lower and upper limits are used as guard bands). In each carrier, up to eight users can be served. This is done by time-division multiple access. The time axis is divided into timeslots, each of which lasts 576.92 μs. A group of eight timeslots is called one "frame" and lasts 4.615 ms. One user occupies one slot within each frame both in the uplink and the downlink frequency band. To facilitate hardware implementation, the timeslots of uplink and downlink are offset (see Figure 2.3).

In each timeslot, 148 active bits are transmitted; the rest of the slot (corresponding to 8.25 bits) is used as guard interval. Of those 148 bits, two blocks of 57 bits each contain useful information. In the middle of the burst, a 26-bit midamble is

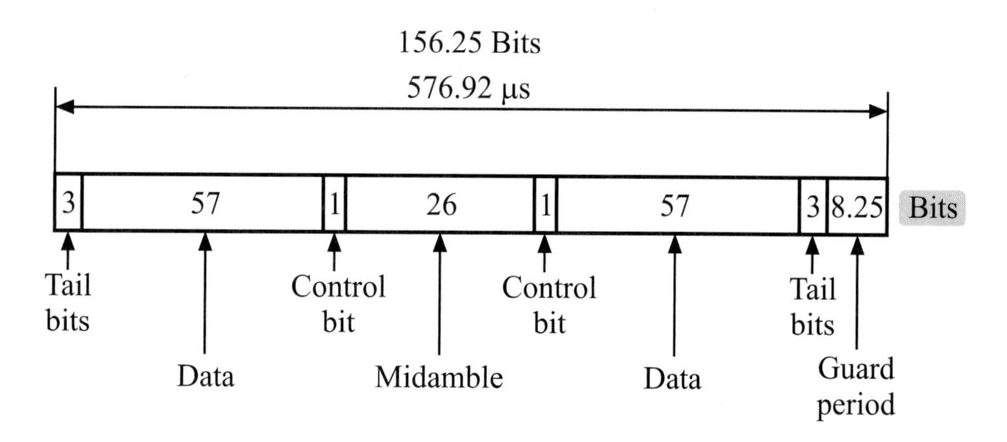

Figure 2.4. Use of bits within a timeslot in GSM

transmitted. This midamble is a known data sequence that is used for synchronization and channel estimation. Furthermore, one burst also contains two control bits and three tail bits at the beginning and end of each frame, which can be used for initialization of a Viterbi detector. The exact slot structure is sketched in Figure 2.4.

The modulation format used in GSM is GMSK, where the phase pulses are rectangular pulses filtered by a Gaussian filter with time-bandwidth product $B_G T = 0.3$. The standard does not prescribe the type of detection. The data are coded with a convolutional code, protected by an error-detecting CRC, and interleaved. The standard GSM speech coder is a Regular-Pulse-Excited Long-Term-Prediction (RPE-LTP) coder that represents the human speech with a data rate of 13 kbit/s. When channel coding is used, the data rate is increased to 22.8 kbit/s. The difference between the 8$\cdot$ 22.8 kbit/s and the actually transmitted 271 kbit/s is used up by the midamble, guard period, control information, etc. Recently, half-rate speech coders have been proposed that can represent speech with only 6.5 kbit/s. This rate allows one to double the number of users in each carrier.

For the actual operation of a GSM system, it is important to note the existence of a variety of logical channels, which are mapped onto the physical channels by a complicated procedure. Frames are combined to give multiframes, superframes, and hyperframes. According to its position in these multiframes, a bit can mean different things—it can contain speech data for a certain user, data of common control channels, or data of dedicated control channels. For some special channels, even the burst structure might differ from the one described above. However, all this has little influence on the wideband transmission properties and equalizer design, and thus is not discussed any further. The interested reader is referred to the extensive literature on this subject, e.g., [2].

GSM 1800 is a practically identical system that uses a carrier frequency in the

1800 MHz band. For the uplink, the frequency band 1710–1785 MHz is used; for the downlink 1805–1880 MHz. This frequency allocation is mainly used in Europe. In the United States, GSM is known as DCS-1900, using the frequency band 1850–1910 MHz for the uplink and 1930–1990 MHz for the downlink.

2.3 IS-136

Like GSM, IS-136 (and its predecessor IS-54) is a TDMA/FDD system [3], [4]. It uses the frequency band from 1850–1919 MHz; IS-54 uses 824–894 MHz with 45 MHz duplexing distance. Both IS-54 and IS-136 are also known as DAMPS. For historical reasons, DAMPS uses a carrier spacing of 30 kHz like in the AMPS analog system.

The speech coder is a Vector-Sum Excited Linear Predictive coder (VSELP). It codes voice language with a data rate of 7.95 kbit/s; half-rate coders with 3.975 kbit/s have been proposed. The speech is divided into 20 ms frames, so that each speech block contains 159 bits. These bits are divided into two classes: class-1 (important) bits that are protected with a rate 1/2–convolutional code and a CRC and class-2 (unimportant) bits that are not protected. All in all, the channel coder produces 260 bits during each speech block. Data from two adjacent speech blocks are then interleaved.

DAMPS is a TDMA system, where one frame lasts 40 ms. In one frame, six slots, each lasting 6.67 ms, fit. For full-rate speech coders, one user occupies two of these slots, while for half-rate coders, he uses only one. During one slot, the 260 bits produced by one speech block, as well as control information, are transmitted. The frame structure in Figure 2.5 shows the exact usage of each symbol. The modulation format is $\pi/4-$ DQPSK with raised-cosine filtering (root-raised-cosine at both transmitter and receiver) with a roll-off factor of 0.35. Apart from the slow associated control channels (SACCH), synchronization symbols and a Coded Digital Verification Color Code are transmitted; G and R denote guard period and ramp-up time. Further control information can be sent instead of speech data (fast associated control channel) or on separate control channels. These separate control channels use either $\pi/4$-DQPSK (in IS-136) or binary FSK (in IS-54) as modulation format.

From the above it follows that the channel symbol rate is $24.3 \cdot 10^3$ symbols/s and the symbol duration is 41.1523 μs. There are thus many environments where the maximum excess delay is not negligible compared to the symbol duration, so that an equalizer has to be used. The exact nature of this equalizer is not prescribed. Due to the long duration of a time slot, the channel might change during the slot, which has implications for the adaptation algorithms.

2.4 IS-95

IS-95 is a CDMA system [5], [6], [7], i.e., a direct-sequence spread spectrum system, with an additional FDMA component. The available frequency range is divided

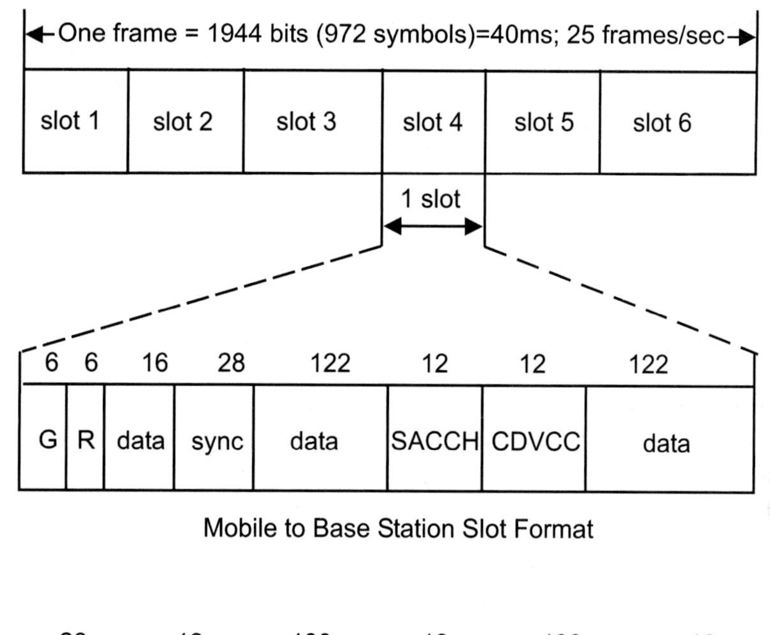

Figure 2.5. Slot and frame structure of IS-136. Adapted from [4], copyright IEEE.

into frequency bands of 1.25 MHz; duplexing is done in the frequency domain. In the United States, the frequencies from 1850–1910 MHz are used for the uplink, and 1930–1990 MHz are used for the downlink band. Within each band, traffic channels, control channels, and pilot channels are separated by the different codes (chip sequences) with which they are spread. The chip rate is 1.2288 Mchip/s. The (downlink) signals generated by one base station are orthogonal to each other. This puts an upper limit of 64 channels per carrier. In the uplink, orthogonality cannot be strictly maintained. Furthermore, interference from other cells reduces the signal quality at base station and mobile station. Thus, 12–18 users can typically be accommodated within one carrier at one time [8].

The pilot channels use a specific PN sequence that is 32,768 chips long, so that it is repeated every 26.67 ms. Pilots from different base stations are distinguished by their offset from an absolute time reference. The offset is a multiple of 64 chips, corresponding to 52.08 μs; thus, there are 512 different pilots. The pilot channels are used for channel estimation (which is vital for Rake receiver configuration) as well as for handover. By monitoring the pilot tones of neighboring cells, the mobile station acquires information as to whether a handover is feasible. During handover, one mobile station can keep connection with two or even more base stations (soft handover). This approach requires an assignment of at least one rake finger to each base station. Due to the different runtimes, the signals arriving from the different base stations have different delays. The structure of a transmitter is sketched in Figure 2.6.

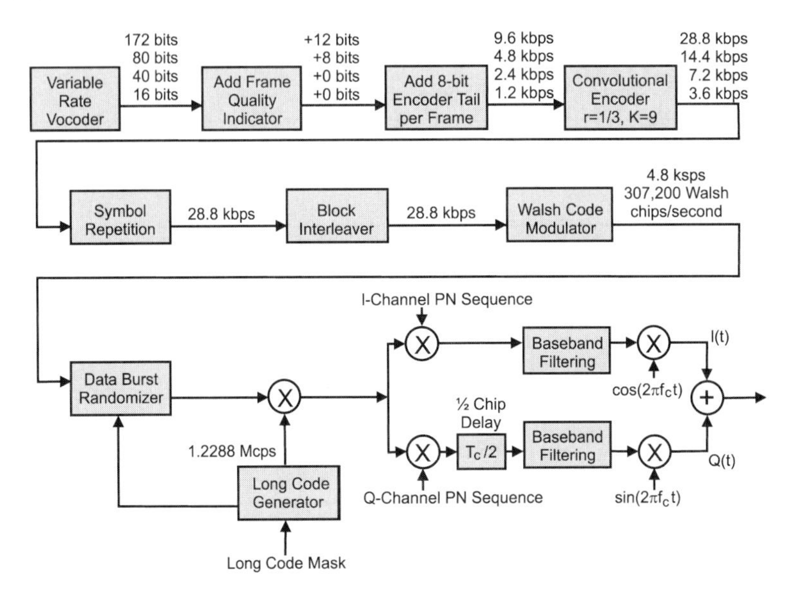

Figure 2.6. IS-95 uplink system sketch. Adapted from [8], reprinted by permission of Prentice-Hall, Inc.

IS-95 specifies two possible speech coder rates: 13.3 or 8.6 kbit/s. In both cases, the output of the vocoder is divided into 20 ms pieces, to which a frame quality indicator (CRC check) and encoder tail bits are then added. Subsequently, the data are sent to a rate-1/3 convolutional coder (for a 8.6 kbit/s speech coder) or a rate-1/2 convolutional coder (for a 13.3 kbit/s speech coder). The output of the encoder has a rate of 28.8 kbit/s in any case. This data sequence is then sent to a Walsh code modulator, where each group of 6 modulation symbols is replaced by one of 64 possible Walsh functions of length 64 chips, which are all orthogonal to each other. Consequently, the output chip rate of the Walsh coder is 307 kchip/s. In the uplink, the next building block is a data burst randomizer. That reduces the average power when the user is not speaking. Subsequently, the data are spread by a long sequence (periodicity of $2^{42} - 1$ chips, i.e., about 4 days) to a chip rate of 1.2288 Mchips/s. This long code is used both for channelization, i.e. distinction between the different traffic and control channels and for encryption. Finally, the chip sequence is split

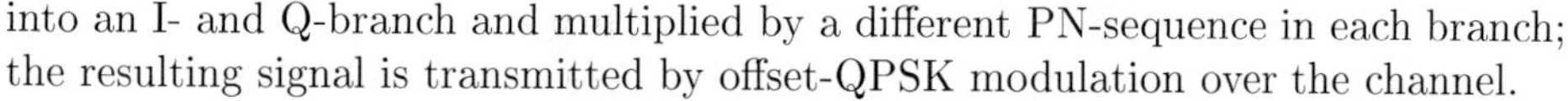

into an I- and Q-branch and multiplied by a different PN-sequence in each branch; the resulting signal is transmitted by offset-QPSK modulation over the channel.

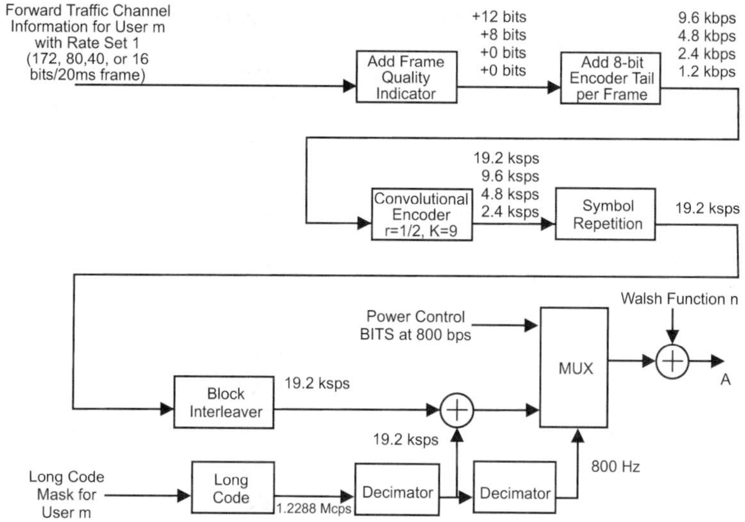

Figure 2.7. IS-95 downlink system sketch. Adapted from [8] reprinted by permission of Prentice-Hall, Inc.

The downlink uses essentially the same building blocks as the uplink; however, they are applied differently. A system sketch is given in Figure 2.7. The first components are identical to the uplink. The first difference is the convolutional encoder: for both rates of the speech coder, a rate 1/2-convolutional coder is used. For the 13.3 kbit/s coder, puncturing is used to obtain a data rate of 19.2 kbit/s; for the other speech coder, this rate is achieved automatically. Certain bits of the long code are then used for encryption purposes, but not for spreading. Finally, the data are spread by Walsh functions with a spreading factor of 64, resulting again in a chip rate of 1.2288 Mchip/s.

2.5 W-CDMA

W-CDMA will be the most important mode of the third-generation cellular standard IMT-2000 [9]. First versions were developed by ETSI and the Japanese ARIB standardization bodies and later reconciled in the 3GPP group. Like IS-95, W-CDMA is a CDMA multiple-access scheme but shows some important differences in the air interface.[1] The most recent version of the standard can be found at www.3gpp.org.

[1] The standard has not yet been officially finalized, so some changes to the description given here are still possible.

The air interface is quite complicated because it is designed to be very flexible. The bandwidth is 5 MHz per carrier, corresponding to a chip rate of 3.84 Mchip/s. The time axis is divided into frames of length 10 ms, with each frame being divided into 15 slots lasting 0.667 ms each. Seventy-two frames form a superframe. The frame and superframe structure are related to control aspects and are not detailed here. Within one time slot, bits have different meanings and possibly different spreading factors, as explained below. The situation is further muddied the fact that uplink and downlink use different spreading/modulation, and that there are so-called spreading codes that provide the spectrum spreading, and channelization codes that distinguish between the different users and cells.

In the *uplink*, user data and control information are transmitted in parallel by quadrature modulation: data on the I-branch and control information on the Q-branch. The useful data can have a spreading factor between 4 and 256, depending on the data rate requirement of the user. The control information always has a spreading factor of 256. The spreading codes are orthogonal, variable-spreading factor (OVSF) codes. The channelization code is a complex code based on VL-Kasami sequences (for short spreading codes) or complex sequences of Gold codes (for long spreading codes).

In the *downlink*, the user data and the control information are first multiplexed to give a single serial bitstream. Since all users in a cell are synchronous, distinction between the users can be done by the OVSF codes, with spreading factors between 4 and 256 possible. The scrambling codes are used only to distinguish between different cells. The output of the coder is then quadrature modulated.

The modulation format used in W-CDMA is QAM, filtered with a raised-cosine filter with roll-off factor $\alpha = 0.22$, which allows for (theoretically) zero emission outside the 5 MHz band. Part of the control data are also pilot bits for channel estimation.

W-CDMA also uses a power control with a very high dynamic (80 dB open loop and 20 dB closed loop), whose values can be updated every 0.667 ms. The use of adaptive antennas is supported by user-specific pilot bits. Multiuser detection is also supported by specification of short codes, which are more suitable for multiuser detection than are long codes.

2.6 HIPERLAN-II

HIPERLAN-II is a standard for short-range mobile radio systems, designed for high capacity in both public and private hot-spot environments. It should allow data rates of up to 54 Mbit/s. It will operate in the 5 GHz band. In the bands from 5.15–5.25 GHz and 5.3–5.35 GHz, transmission with 200 mW mean EIRP (maximum) is allowed in all CEPT countries. The band from 5.25–5.3 GHz will be assigned on a national basis. The band from 5.47–5.725 GHz still is subject to discussion because of necessary sharing with ground-based radar systems. For details of the standard, see www.etsi.org.

HIPERLAN is an OFDM/FDMA system with a carrier spacing of 20 MHz. For each carrier frequency, there are 64 subcarriers spaced 312.5 kHz apart. Due to out-of-band radiation considerations, only 52 of those subcarriers are actually used, 4 of them as pilots. Local oscillator accuracy should be better than 20 ppm. One OFDM symbol lasts for 4 μs, consisting of 3.2 μs data and 0.8 μs cyclic prefix. The data are divided into a preamble with a number of fixed short symbols and the actual payload. The data transmission uses adaptive modulation and coding. Possible modulation methods are BPSK, QPSK, 16QAM, and 64QAM. Possible codes are convolutional codes with rates 1/2 (for BPSK or QPSK), 9/16 (for 16QAM), and 3/4 (for BPSK, QPSK, 16QAM, or 64QAM). The 9/16 and 3/4 codes are derived from a rate 1/2 mother code by puncturing. The combinations of modulation method and codes are chosen according to the transmission quality, allowing data rates of $6, 9, 12, 18, 27, 36$, or possibly 54 Mbit/s.

Bibliography

[1] W. H. W. Tuttlebee, *Cordless Telecommunications Worldwide: The Evolution of Unlicensed PCs.* London, Springer, 1997.

[2] M. Mouly and M. Pautet, "The GSM system for mobile communications," *Cell&Sys, ISBN 2-9507190-0-7*, 1992.

[3] J. D. Gibson, *The Mobile Communications Handbook.* London, CRC Press, 1999.

[4] T. S. Rappaport, *Wireless Communications Principles and Practice.* Piscataway, NJ: IEEE Press, 1996.

[5] J. Lee and L. E. Miller, *CDMA Systems Engineering Handbook.* London, Artech House, 1998.

[6] S. Glisic and B. Vucetic, *Spread Spectrum CDMA Systems for Wireless Communications.* London, Artech House, 1997.

[7] A. J. Viterbi, *CDMA—Principles of Spread Spectrum Communication.* Addison-Wesley Wireless Communications Series, 1995.

[8] J. C. Liberti and T. S. Rappaport, *Smart Antennas for Wireless Communications.* Upper Saddle River, NJ: Prentice Hall, 1999.

[9] T. Ojanpera and R. Prasad, *Wideband CDMA for Third Generation Mobile Communications.* Artech House Publishers, 1998.

Chapter 3

MOBILE RADIO CHANNELS

The performance of wideband systems is determined by the channel over which the transmission takes place. It is thus indispensable to have a good knowledge of wideband (time-dispersive) mobile radio channels. Research in that area has been going on for some 25 years and has concentrated mainly on two areas: (i) Collection of results from extensive measurement campaigns. Since wideband measurements are much more complicated than simple field strength (i.e., narrowband) measurements, the number of measurement campaigns is considerably smaller than for the narrowband case. (ii) From these measurements, channel models are derived, which should fulfill the following two criteria:

- They must be simple enough to allow an analytical computation of basic system performance.
- They must be very close to the physical reality; in other words, the performance computed by these models must be close to the performance measured in actually existing mobile radio channels.

These requirements are contradictory, so models of different complexity and accuracy have been developed. Furthermore, some models might be more suitable for certain systems than others: as an example, we see later that a so-called two-delay model is well suited for unequalized systems, but it is usually too crude for CDMA systems.

A further field of research, which is an intermediate step between measurement and modelling, is *information condensation.* A wideband measurement campaign gives a huge amount of data. For further interpretation and processing, some characteristic "condensed" parameters have to be extracted, which should reduce the amount of data while keeping the loss of information as small as possible.

In the following sections, we summarize the most common methods for describing wideband channels and the channel models that are in widespread use. We also give a brief overview of spatially resolved channel models, since these are of great importance for intelligent antenna systems, which are currently a hot topic, and can be viewed as a generalization of wideband systems.

3.1 FLAT-FADING CHANNELS

Before going into the details of wideband channels, let us briefly recapitulate the properties of flat-fading channels. In a flat-fading channel, the received signal $y(t)$ is given by the transmitted signal $x(t)$, multiplied by a time-varying attenuation $\alpha(t)$, and a noise contribution $n(t)$

$$y(t) = \alpha(t)x(t) + n(t). \tag{3.1.1}$$

The time-variation of the attenuation is known as *fading*. It has been shown experimentally and also argued theoretically that it usually follows a Rayleigh distribution

$$\text{pdf}(\alpha) = \frac{\alpha}{\sigma^2} \cdot \exp\left[-\frac{\alpha^2}{2\sigma^2}\right] \qquad 0 \le \alpha < \infty, \tag{3.1.2}$$

where σ is the variance of the underlying Gauss process. The conditions for this to be valid are that there are many statistically indendent scatterers and no single scatterer makes a dominant contribution. If there is one dominant contribution (usually a line-of-sight LOS component), the distribution of α is a Rice distributed variable characterized by

$$\text{pdf}(\alpha) = \frac{\alpha}{\sigma^2} \cdot \exp\left[-\frac{\alpha^2 + A^2}{2\sigma^2}\right] \cdot I_0\left(\frac{\alpha A}{\sigma^2}\right) \qquad 0 \le \alpha < \infty \tag{3.1.3}$$

where $I_0(x)$ is the modified Bessel function first kind, zero order [1]. The parameter A is the amplitude of the dominant component; the Rice parameter K_r is defined as $A^2/(2\sigma^2)$. The multipath propagation also introduces a phase shift. If the amplitude is Rayleigh fading, the phase shift is statistically independent of the amplitude distribution and is uniformly distributed between 0 and 2π. If the amplitude fading is Rician, the joint pdf of amplitude and phase ψ is [2]

$$\text{pdf}_{\alpha,\psi} = \frac{\alpha}{2\pi\sigma^2} \exp\left(-\frac{\alpha^2 + A^2 - 2\alpha A\cos(\psi)}{2\sigma^2}\right). \tag{3.1.4}$$

An alternative amplitude distribution, which has gained popularity especially for the evaluation of measurements, is the Nakagami m-distribution given by

$$\text{pdf}_\alpha(\alpha) = \frac{2}{\Gamma(m)}\left(\frac{m}{\Omega}\right)^m \alpha^{2m-1} \exp\left(-\frac{m}{\Omega}\alpha^2\right) \tag{3.1.5}$$

for $\alpha \ge 0$ and $m \ge 1/2$. For $m = 1$, this distribution reduces to the Rayleigh distribution. The parameter Ω is the mean square value $\Omega = \overline{\alpha^2}$, and the parameter m is

$$m = \frac{\Omega^2}{\overline{\left(\alpha^2 - \Omega\right)^2}}. \tag{3.1.6}$$

Nakagami and Rice distributions are quite similar and each can be approximately converted to the other for $m \geq 1$:

$$m = \frac{(K_r + 1)^2}{(2K_r + 1)} \tag{3.1.7}$$

and

$$K_r = \frac{\sqrt{m^2 - m}}{m - \sqrt{m^2 - m}}. \tag{3.1.8}$$

The movement of the mobile station leads to a frequency shift of the arriving waves (Doppler effect). If a sinusoidal wave of frequency f_0 is transmitted, the spectrum of the received signal is

$$Y(f) \propto [\mathrm{pdf}_\gamma(\gamma)G(\gamma) + \mathrm{pdf}_\gamma(-\gamma)G(-\gamma)] \frac{1}{\sqrt{\left(f_0 \frac{\mathrm{v}}{c}\right)^2 - (f - f_0)^2}} \tag{3.1.9}$$

in the range $-f_0\mathrm{v}/c < f < f_0\,\mathrm{v}/c$, and 0 elsewhere, where v is the speed of movement, γ is the angle between the direction of incidence of the move and the direction of movement, and $G(\gamma)$ is the antenna pattern. For the case that the waves are all incident horizontally, are uniformly distributed in azimuth, and the antenna has a uniform pattern in azimuth, we get

$$Y(f) \propto \frac{1}{\sqrt{\left(f_0 \frac{\mathrm{v}}{c}\right)^2 - (f - f_0)^2}}. \tag{3.1.10}$$

The Rayleigh- or Rice-fading is also known as *small-scale fading*, since it describes the variation of the amplitude within an area of about ten wavelengths. Over a larger scale, it has been shown experimentally that the small-scale-averaged amplitude F obeys a log-normal distribution

$$\mathrm{pdf}_F(F) = \frac{1}{F\sigma_F\sqrt{2\pi}} \cdot \exp\left[-\frac{(\ln(F) - \mu_F)^2}{2 \cdot \sigma_F^2}\right] \tag{3.1.11}$$

where μ_F is the mean of $\ln(F)$ and σ_F is the variance of $\ln(F)$. The mean is determined mainly by the path loss between base station and mobile station; the variance is typically in the range of 4-8 dB.

3.2 TIME-DISPERSIVE CHANNELS: INTUITIVE DESCRIPTION

Figure 3.1 shows the origin of the fading of the arriving signal. Signal components (echoes) reflected from different obstacles are added up with more or less random phase shifts. For the explanation of flat fading, it was assumed that all those echoes are arriving at effectively the same time and can thus be added up to give $y(t)$. Strictly speaking, however, the runtimes of the echoes are different: only those that are reflected by obstacles on one ellipse in Figure 3.2 arrive at the receiver at exactly the same instant.

In practice, we can relax that requirement somewhat: Since a receiver with bandwidth W cannot distinguish between echoes arriving at τ and $\tau + \Delta\tau$, where $\Delta\tau \ll 1/W$, we can add up all echoes that are reflected in the doughnut-shaped region corresponding to runtimes between $c_0\tau$ and $c_0(\tau + \Delta\tau)$, arguing that they arrive at "effectively" the same time. We can thus divide the impulse response into bins of width $\Delta\tau$ and then compute the sum of the echoes within each bin. If enough non-dominant scatterers are in each doughnut-shaped region, the amplitude distribution of each bin is Rayleigh. We furthermore define the minimum delay as the runtime of the direct path between BS and MS $c_0 \cdot d_{\text{MS-BS}}$, and we define the maximum delay as the runtime from BS to MS via the farthest significant scatterer, i.e., the farthest scatterer that gives a measureable contribution to the impulse response.[1] The maximum excess delay $\tau_{\max}$ is then defined as the difference between minimum and maximum delay.

This definition also gives us a mathematical formulation for narrowband and wideband from a time-domain point of view: a system is narrowband if the inverse of the system bandwidth $1/W$ is much larger than the maximum excess delay $\tau_{\max}$. In that case, all echoes fall into a single delay bin, and the amplitude of this delay bin is $\alpha(t)$. A system is wideband in all other cases, i.e., if more than one amplitude bin has a significant amplitude. This definition also explains the expression *time-dispersive* for a wideband channel. In a wideband system, the *shape* of the arriving signal is different from the shape of the transmitted signal; in a narrowband system, it stays the same (apart from noise).

If the impulse response has a finite extent in the delay domain, it follows from the theory of Fourier transforms that the transfer function $\mathcal{F}\{h(\tau)\} = H(f)$ is frequency dependent. This is an alternative definition of a *wideband channel.* Also, here it is clear that the description of such a channel cannot be done by a simple attenuation coefficient, but that the details of the transfer function must be modelled.

3.3 TIME-DISPERSIVE CHANNELS: SYSTEM-THEORETIC DESCRIPTION

If the BS, MS, and scatterers are stationary, then the channel can be interpreted as a time-invariant filter with the impulse response $h(\tau)$. In that case, the well-known

[1] We see from this definition that the maximum delay is a quantity that is extremely difficult to measure and that it depends on the measurement system.

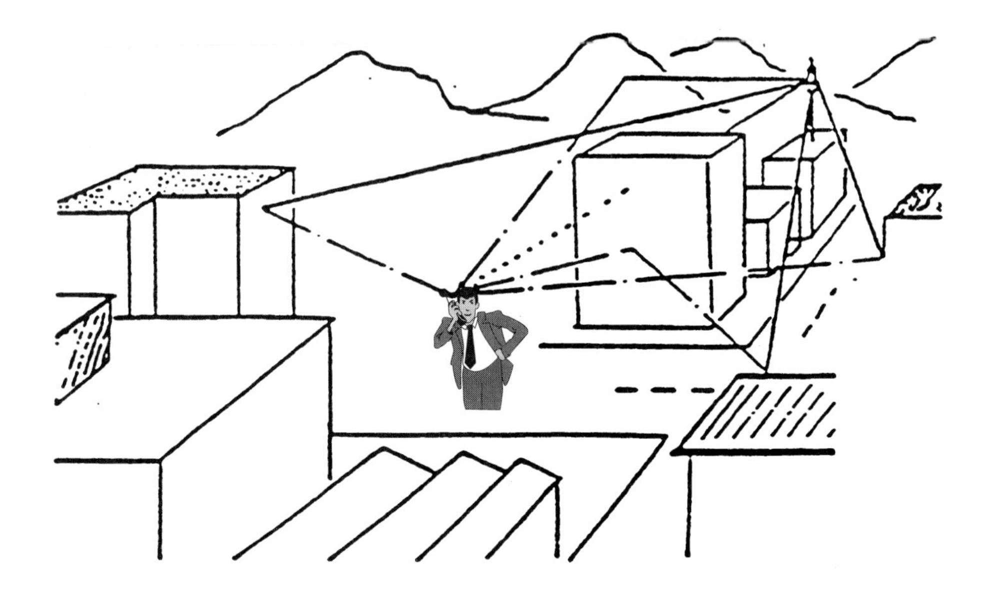

Figure 3.1. Principle of multipath propagation

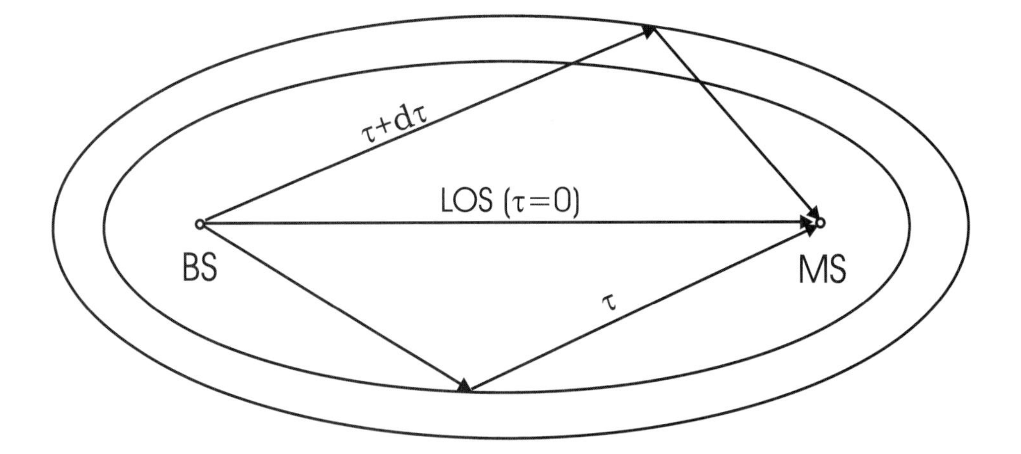

Figure 3.2. Scatterer ellipses

theory of LTI-systems (linear, time-invariant systems) is applicable. However, real-world channels are time-variant, so that the theory of linear time-*variant* systems (LTV) must be used. This theory was founded by Zadeh and extended by Bello in his classical article [3]. It is noteworthy that for slowly time-variant systems, most of the interpretations of the impulse response and the transfer function (as known from LTI-systems) can be retained; we are just using "snapshots" of the channel as an LTI-filter for a limited amount of time. However, if the channel is rapidly varying, these interpretations break down [4], [5]. Also the usual transfer function calculus (multiplication of transfer functions) is no longer applicable [6].

3.3.1 Deterministic interpretation

As for the flat-fading channel, there are essentially two interpretations of the channel: we can view it as a deterministic system, whose impulse response is given once we know the location and electromagnetic property of every relevant scatterer, BS, and MS. On the other hand, it can be considered as a stochastic system, where the impulse response and its temporal development are chosen from the appropriate statistical distributions. In this subsection, we deal with the deterministic interpretation.

The most straightforward way of describing a general linear system is through the kernel function, which links the input and the output of the system. Input and output signals can be described either in the time domain or in the frequency domain, so there are four equivalent kernels $K_1 K_2, K_3, K_4$ defined by

$$y(t) = \int_{-\infty}^{+\infty} x(\tilde{t}) K_1(t, \tilde{t}) d\tilde{t} \tag{3.3.1}$$

$$Y(f) = \int_{-\infty}^{+\infty} X(\tilde{f}) K_2(f, \tilde{f}) d\tilde{f}$$

$$y(t) = \int_{-\infty}^{+\infty} X(f) K_3(t, f) df$$

$$Y(f) = \int_{-\infty}^{+\infty} x(t) K_4(f, t) dt$$

Here, $X(f)$ is the Fourier transform of the input signal $x(t)$, and $Y(f)$ the Fourier transform of the output signal $y(t)$. Also the kernel functions K_1, K_2, K_3, and K_4 are related to each other by Fourier transforms.

From a mathematical point of view, any of the kernel functions is a complete description of the system. However, it does not have a straightforward intuitive interpretation, which is why other system functions based on a "tapped delay line"

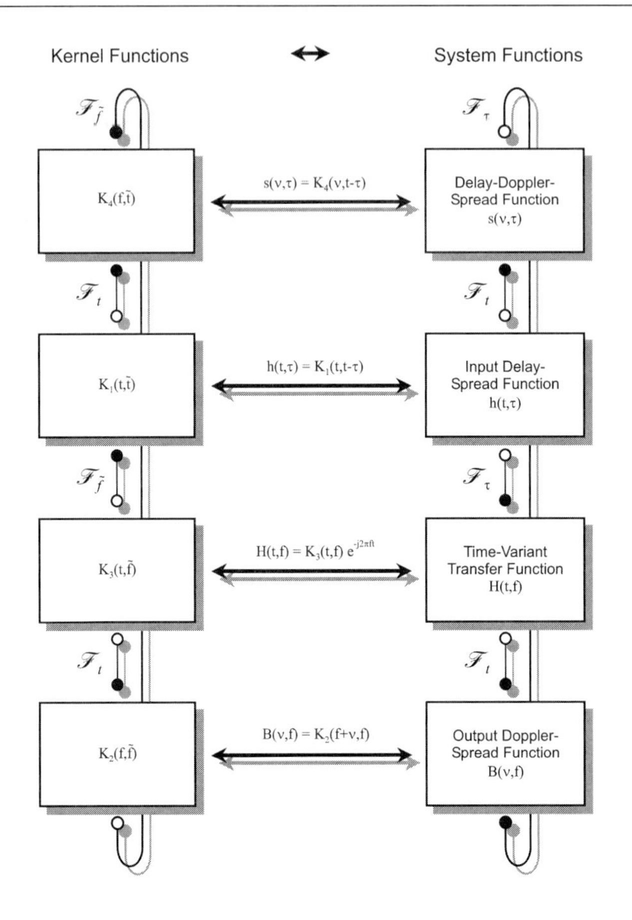

Figure 3.3. Relations between Bello's kernel and system functions. From [7], copyright Shaker Verlag. Modified translation provided by R. Kattenbach.

model or a "frequency conversion chain" model of the channel have gained larger popularity, see Fig. 3.3.

The members of one such possible set of system functions are denoted by Bello as "input delay-spread function" $h(t,\tau)$, "time-variant transfer function" $H(t,f)$, "output Doppler-spread function" $B(\nu,f)$, and "delay-Doppler-spread function" $s(\nu,\tau)$.[2] The tapped-delay line model for the input delay-spread function is sketched in Fig. 3.4a.

The relation between the kernel functions and the tapped-delay-line functions is given by (see also Figure 3.3)

[2]Note that Bello used different symbols for denoting those functions in his paper.

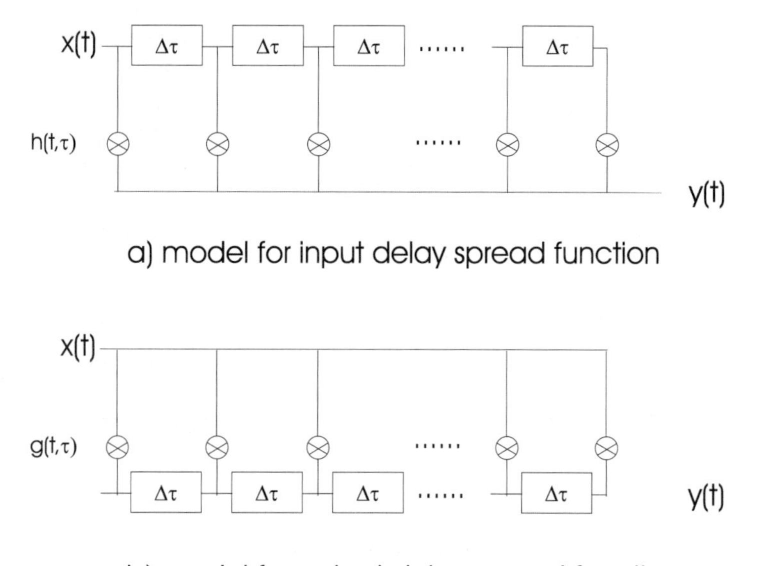

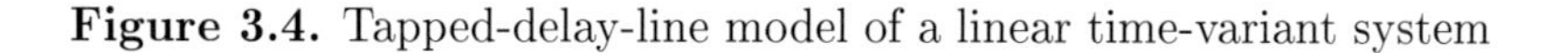

Figure 3.4. Tapped-delay-line model of a linear time-variant system

$$h(t,\tau) = K_1(t, t-\tau) \tag{3.3.2}$$
$$B(\nu, f) = K_2(f+\nu, f) \tag{3.3.3}$$
$$H(t, f) = K_3(t, f)\exp(-j2\pi f t) \tag{3.3.4}$$
$$s(\nu,\tau) = K_4(\nu, t-\tau). \tag{3.3.5}$$

There is also a dual set of system functions, whose members are denoted as "input Doppler-spread function," "frequency-dependent modulation function," "output delay-spread function," and "Doppler-delay-spread function.". They can also be derived from the kernel functions [3]. The model for the output delay spread function is sketched in Fig. 3.4b.

More generally, [4] defined a generalized spreading function $s^{(\alpha)}(\nu,\tau)$

$$s^{(\alpha)}(\nu,\tau) = \int_{-\infty}^{+\infty} K_1\left(t + (1/2-\alpha)\,\tau, t-(1/2+\alpha)\tau\right) e^{-j2\pi\nu t} dt, \qquad |\alpha| \le 1/2. \tag{3.3.6}$$

This function contains Bello's delay-Doppler-spread function and the Doppler-delay-spread function as special cases for $\alpha = 1/2$ and $\alpha = -1/2$, respectively. For the other system functions, similar generalized definitions could be obtained. However, for underspread channels (product of maximum excess delay and maximum Doppler spread much smaller than unity), the system functions are approximately α-invariant. Since most mobile radio channels can be classified in good approximation as underspread, the distinction between functions with different α-parameters is not necessary for wireless applications, and we use in the following only the more popular system functions with $\alpha = 1/2$, i.e., as depicted in Fig. 3.3. The two models are illustrated in Figure 3.4.

Having summarized the possible ways of formally writing the input-output relation of LTV systems, let us now turn to the interpretations used in mobile radio applications. First, we write down the relation between input and output signal, using the delay spread function

$$y(t) = \int_{-\infty}^{+\infty} x(t-\tau)h(t,\tau)d\tau. \tag{3.3.7}$$

This suggests an interpretation of $h(t,\tau)$ as a time-variant impulse response, analogous to the relation for LTI systems. Actually, the denotation "time-variant impulse response" has become far more popular than "input (or output) delay spread function" and we use it henceforth.

Just as for LTI systems, the impulse response can be Fourier-transformed with respect to the delay variable τ to give the time-variant transfer function $H(t,f)$. The output signal $y(t)$ can be calculated, analogously to LTI systems, as the inverse Fourier transform of the product of the input spectrum $X(f)$ and $H(t,f)$:

$$y(t) = \int_{-\infty}^{+\infty} X(f)H(t,f)e^{j2\pi ft}df. \tag{3.3.8}$$

However, this does *not* mean in general that the output spectrum is

$$Y(t,f) = X(f) \cdot H(t,f), \tag{3.3.9}$$

or (more exactly)

$$Y(t,f) = X(t,f) \cdot H(t,f), \tag{3.3.10}$$

with $X(t,f)$ and $Y(t,f)$ being the short-term spectra of input and output signals. Equation (3.3.10) is valid *only* if the channel is slowly time varying [6].

The time-variant impulse response $h(t,\tau)$ and time-variant transfer function $H(t,f)$ are related by a Fourier transform $\tau \leftrightarrow f$. A second possible transform pair is $t \leftrightarrow \nu$, and the new transform variable ν can be interpreted as Doppler frequency ν. A Fourier transform of $h(t,\tau)$ with respect to t yields a function $s(\nu,\tau)$ which is denoted as the *delay-Doppler-spread function*, or simply *spreading function*. Using this function for the input-output relation results in

$$y(t) = \int_{-\infty}^{+\infty}\int_{-\infty}^{+\infty} s(\nu,\tau)x(t-\tau)e^{j2\pi\nu t}d\nu d\tau. \tag{3.3.11}$$

Equation (3.3.11) gives an intuitive interpretation: the output signal $y(t)$ is the sum of delayed and Doppler-shifted replicas of the input signal $x(t)$ weighted by $s(\nu,\tau)$, which can thus be interpreted as the gain of infinitesimal point scatterers.

Finally, we can also Fourier-transform the transfer function with respect to the absolute time t. The resulting function $B(\nu, f)$ has been denoted as *output Doppler-spread function*. The input-output relation reads

$$Y(f) = \int_{-\infty}^{+\infty} X(f-\nu)B(\nu, f-\nu)d\nu. \tag{3.3.12}$$

3.3.2 Stochastic interpretation

For a stochastic interpretation, we first have to clarify what we mean by *ensemble* in the context of mobile radio channels. For a given time (i.e., for a given location and velocity vector of BS, MS, and scatterers), the channel is completely deterministic and has only a single realization. A different realization would imply a different location, e.g., of the scatterers *at the same time instant*, which is physically impossible. It has thus become common in mobile radio to use impulse responses measured at different times as different realizations of the channel. Note that this implies ergodicity of the channel—a fact that will become important when we turn to the WSSUS assumption. Further discussion of this aspect can be found in [7] and [8].

After these preliminaries, let us turn to the stochastic interpretation of the channel. For a complete description, we would need a multidimensional joint probability density function of the impulse response (or an equivalent system function). Since this is too complicated in practice, [3] has suggested that we characterize the channel just by the autocorrelation function (ACF) of the impulse response or one of its Fourier transforms:

$$\begin{aligned} R_s(\nu,\nu';\tau,\tau') &= E\left\{s^*(\nu,\tau)\, s(\nu',\tau')\right\} \\ R_h(t,t';\tau,\tau') &= E\left\{h^*(t,\tau)\, h(t',\tau')\right\} \\ R_H(t,t';f,f') &= E\left\{H^*(t,f)\, H(t',f')\right\} \\ R_B(\nu,\nu';f,f') &= E\left\{B^*(\nu,f)\, B(\nu',f')\right\}. \end{aligned} \tag{3.3.13}$$

If the statistics of the channel are zero-mean Gaussian, then the ACF (i.e., the second-order statistics) are sufficient to completely characterize the channel. If they are not zero-mean, we also need the mean value. Only for non-Gaussian statistics do we need higher-order statistics. We note, however, one possible drawback of this representation: When the system bandwidth becomes very large, the number of scatterers per delay bin becomes smaller and smaller, so that an application of the central limit theorem is no longer possible [7]. Reference [9] has given the pdf for a small number of scatterers; however, this distribution is not in widespread use for system evaluations.

3.4 THE WSSUS ASSUMPTION

Even the autocorrelation function is usually too complicated for a practical description, because it depends on four variables τ, τ', t, t' (or its equivalents). A further simplification can be achieved by the WSSUS (wide-sense stationary uncorrelated scatterers) assumption.

3.4.1 Wide-sense stationarity (WSS)

Mathematically, a channel is wide-sense stationary if the autocorrelation function does not depend on t and t', but only on the difference $\Delta t = t' - t$. Thus we can write

$$R_h(t, t', \tau, \tau') = R_h(t, t + \Delta t, \tau, \tau') = R_h(\Delta t, \tau, \tau'). \tag{3.4.1}$$

Physically, we can interpret the WSS assumption in such a way that the second-order *statistics* of the channel remain the same over time (strict-sense stationarity would also mean that higher-order statistics do not change over time). This interpretation does *not* imply that the impulse response remains the same over time. The equivalent narrowband case would be a Rayleigh-fading amplitude where the σ^2 stays the same over time, but of course the instantaneous attenuation α could change with time. Mathematically, the stationarity would have to be fulfilled over an infinitely long time; practically, we usually assume that it is fulfilled, for example, over several tens of the inverse Doppler frequency.

A further interpretation of the WSS condition can be given by the spreading function $s(\nu, \tau)$. If we insert (3.4.1) into the equation for R_s

$$R_s(\nu, \nu', \tau, \tau') = \int_{-\infty}^{\infty} \int_{-\infty}^{\infty} R_h(t, t', \tau, \tau') \exp[2\pi j(\nu t - \nu' t')] dt dt' \tag{3.4.2}$$

we get

$$\begin{aligned} R_s(\nu, \nu', \tau, \tau') = &\int_{-\infty}^{\infty} \exp[2\pi j t(\nu - \nu')] dt \\ &\cdot \int_{-\infty}^{\infty} R_h(\Delta t, \tau, \tau') \exp[-2\pi j \nu' \Delta t] d\Delta t. \end{aligned} \tag{3.4.3}$$

The first integral is just a Dirac function $\delta(\nu - \nu')$, so that

$$R_s(\nu, \nu', \tau, \tau') = \tilde{P}_s(\nu, \tau, \tau')\delta(\nu - \nu'). \tag{3.4.4}$$

Thus, the WSS assumption implies that the contributions of the different scatterers are uncorrelated if they give different Doppler shifts. Analogously, we get for R_B

$$R_B(\nu, \nu', f, f') = P_B(\nu, f, f')\delta(\nu - \nu'). \tag{3.4.5}$$

3.4.2 Uncorrelated scattering

The uncorrelated scattering (US) assumption is defined as the contributions of scatterers with different delays being uncorrelated. This implies

$$R_h(t, t', \tau, \tau') = P_h(t, t', \tau)\delta(\tau - \tau'). \tag{3.4.6}$$

For R_s the US condition means

$$R_s(\nu, \nu', \tau, \tau') = \tilde{\tilde{P}}_s(\nu, \nu', \tau)\delta(\tau - \tau'). \tag{3.4.7}$$

Physically, the US assumption implies that one echo gives no information about another echo with different delay.

For the transfer function, the US condition implies

$$R_H(t, t', f, f + \Delta f) = R_H(t, t', \Delta f). \tag{3.4.8}$$

Thus, R_H depends only on the frequency difference, but not on the absolute frequency.

3.4.3 WSSUS

It is an obvious step to combine the WSS and the US condition with the WSSUS assumption, which means

$$R_h(t, t + \Delta t, \tau, \tau') = \delta(\tau - \tau')P_h(\Delta t, \tau) \tag{3.4.9}$$

$$R_H(t, t + \Delta t, f, f + \Delta f) = R_H(\Delta t, \Delta f) \tag{3.4.10}$$

$$R_s(\nu, \nu', \tau, \tau') = \delta(\nu - \nu')\delta(\tau - \tau')P_s(\nu, \tau) \tag{3.4.11}$$

$$R_B(\nu, \nu', f, f + \Delta f) = \delta(\nu - \nu')P_B(\nu, \Delta f). \tag{3.4.12}$$

The functions on the right-hand side depend now on two variables, and that simplifies formulation and computations. Because of their great practical importance, the variables have been given names of their own

$$\begin{aligned} P_h(\Delta t, \tau) &\ldots \text{ delay cross-power spectral density} \\ R_H(\Delta t, \Delta f) &\ldots \text{ time-frequency correlation function} \\ P_s(\nu, \tau) &\ldots \text{ scattering function} \\ P_B(\nu, \Delta f) &\ldots \text{ Doppler cross-power spectral density.} \end{aligned} \tag{3.4.13}$$

Analogously to the spreading function, the scattering function has a special importance among these because it can be easily interpreted physically. If only single reflections occur, then each differential element of the scattering function corresponds to a physically existing scatterer. From the Doppler shift we can determine the direction of arrival; from the delay, the radii of the ellipse on which the scatterer lies. Figure 3.5 shows an example of a measured impulse response; Figure 3.6 shows the scattering function computed from it.

The original WSSUS model was intended for ionospherical scattering, so there was no consideration about line-of-sight components or dominant scatterers. These can, however, occur in mobile radio channels, so that we have to consider not just the correlation (or covariance) function, but also the mean values. For the signal envelope, this means that instead of Rayleigh fading, we have Rician fading.

The validity of WSSUS in mobile radio environments has been the subject of considerable discussion in the past. If we assume ergodicity of the channel, as discussed above, we automatically fulfill the WSS assumption. The US assumption, on the other hand, is well fulfilled only in macrocells. In indoor environments or certain special other environments (street corridors, tunnels), it is obvious that the scatterers are not uncorrelated. Still, WSSUS is assumed for most system computations because non-WSSUS models are too complicated.

3.4.4 Special cases of WSSUS system functions

When we set both $\Delta t = 0$ and $\Delta f = 0$, we obtain the power of the signal. The function $R_H(\Delta f) = R_H(0, \Delta f)$ is the *frequency correlation function* while $R_H(\Delta t) = R_H(\Delta t, 0)$ is the *time correlation function.* From the Schwartz inequality, it follows that $R_H(\Delta t)$ and $R_H(\Delta f)$ for $\Delta t \neq 0$ and $\Delta f \neq 0$ are always smaller than or equal than the values occuring for $\Delta t = 0$ and $\Delta f = 0$, respectively. Thus, the correlation functions are decaying functions (but not necessarily monotonically decaying).

The function $P_h(0, \tau)$ was denoted by Bello as *delay power density spectrum,* but is nowadays usually called *power delay profile (PDP).* The integral of $P(\tau)$ over τ yields the power

$$\int_{-\infty}^{+\infty} P(\tau) d\tau = P. \tag{3.4.14}$$

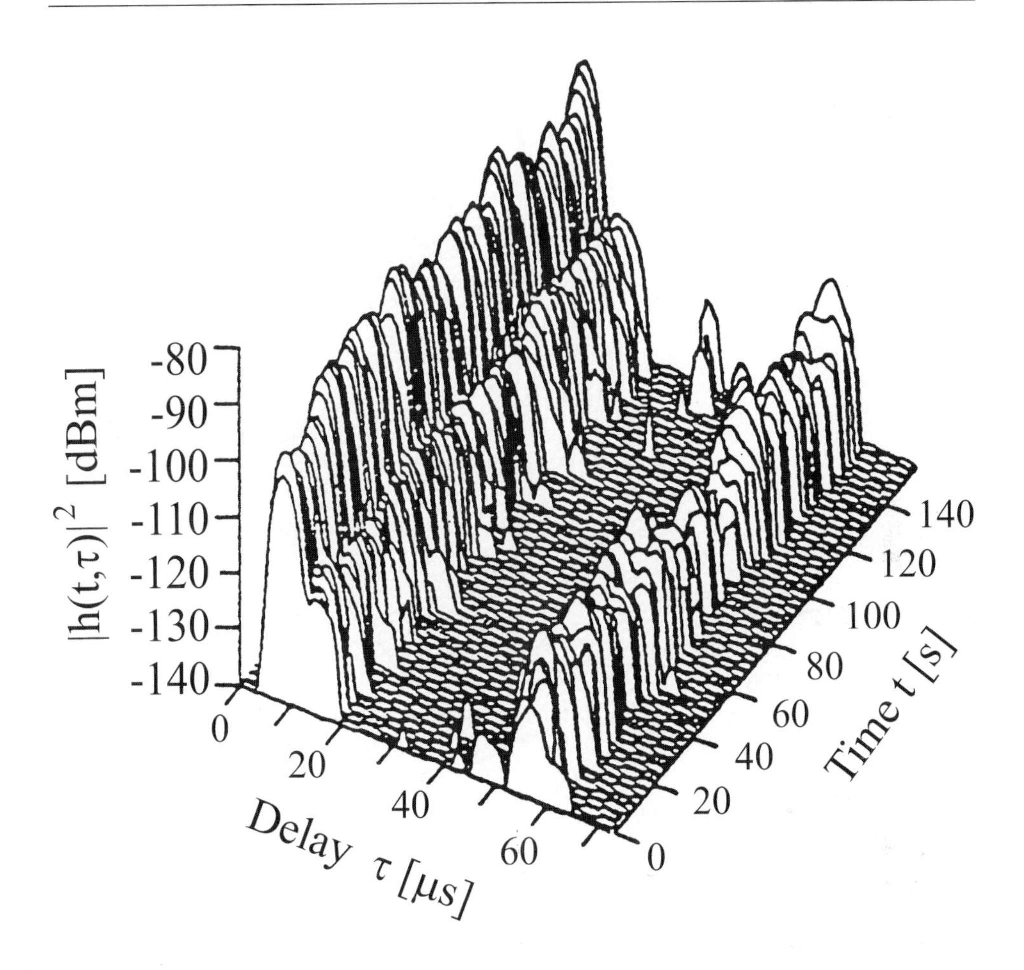

Figure 3.5. $|h(t,\tau)|^2$ measured in the hilly terrain near Darmstadt (Germany). Measurement duration: 140s. Carrier frequency 900 MHz. Adapted from [10] with permission of the authors.

Alternatively, the PDP can be computed from the scattering function as

$$P(\tau) = \int_{-\infty}^{+\infty} P_s(\nu, \tau) d\nu. \tag{3.4.15}$$

The PDP can also be computed directly from the measured values. Assuming ergodicity, we have

$$P(\tau) = E_t\{|h(t, \tau)|^2\} \tag{3.4.16}$$

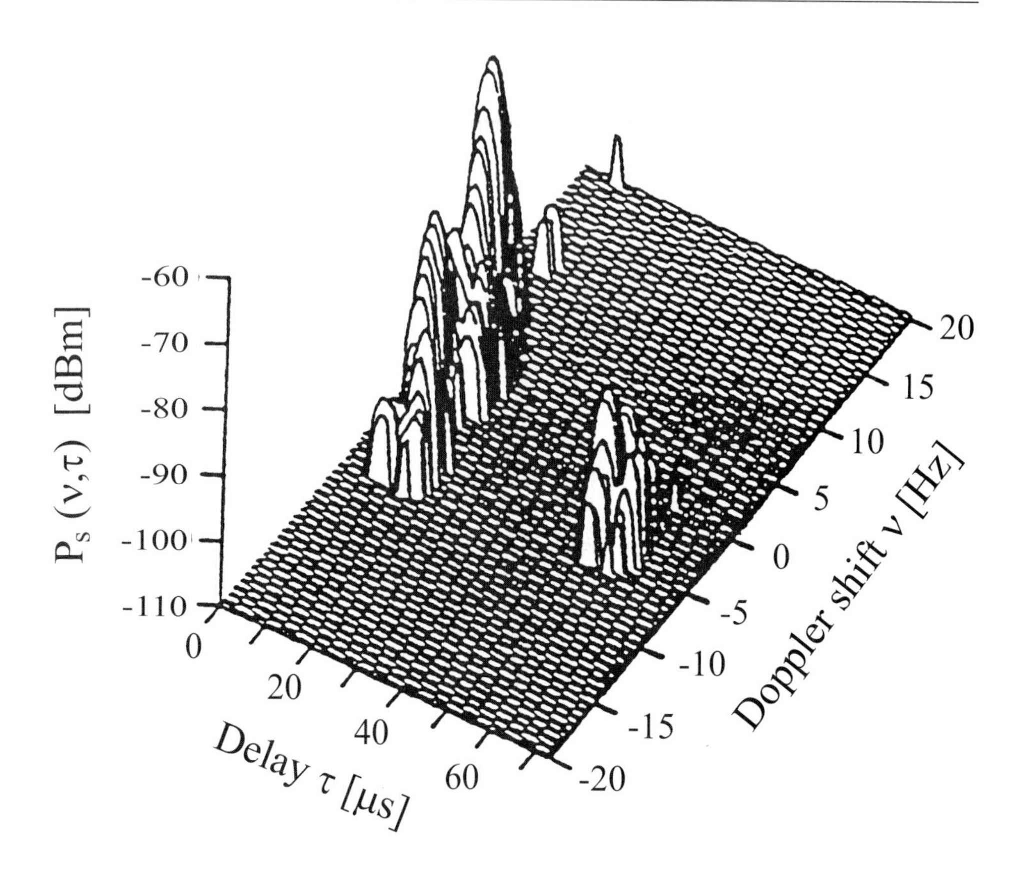

Figure 3.6. Scattering function computed from Figure 3.5. Adapted from [10] with permission of the authors.

where E_t denotes the expectation with respect to absolute time. This interpretation also suggest the definition of an "instantaneous PDP"

$$P(t,\tau) = |h(t,\tau)|^2. \tag{3.4.17}$$

This function is purely deterministic and cannot be interpreted in the stochastic sense.

3.5 PARAMETER FOR CHARACTERIZING TIME-DISPERSIVE CHANNELS

The scattering function or one of its equivalents usually gives a good compromise between the necessity of having a simple description and the requirement of accuracy. Still, for a quick overview of measurement campaigns, an even more condensed description is required: ideally, we would like to have one or two parameters that characterize each type of measured environment. It is obvious that such an information condensation must lead to a considerable loss of information. Furthermore, there are a large number of possible parameters. Still, there are some that have developed into a quasi-standard. The most important of these are the delay spread, the coherence bandwidth, and the delay window.

3.5.1 Delay spread and coherence bandwidth

Rms delay spread: Originally, the average rms delay spread was introduced in a classical paper [3] as a parameter that determines the frequency-selectivity of fading. Since this frequency-selectivity is strongly related to the time dispersion and thus to intersymbol interference (ISI), later papers related the rms delay spread to the irreducible bit error probability in unequalized systems (see Part II). Recently, it has become common to use not only the average (i.e., small-scale averaged) parameters of the channel, but also the instantaneous parameters.

We thus can distinguish two kinds of the rms delay spread [11]: the *average* rms delay spread S_τ

$$S_\tau = \sqrt{\frac{\int_{-\infty}^{\infty} P(\tau)\,\tau^2\,d\tau}{\int_{-\infty}^{\infty} P(\tau)\,d\tau} - \left(\frac{\int_{-\infty}^{\infty} P(\tau)\tau d\tau}{\int_{-\infty}^{\infty} P(\tau)\,d\tau}\right)^2}, \tag{3.5.1}$$

which is computed with the average PDP $P(\tau)$ and the *instantaneous* rms delay spread $S_\tau(t)$

$$S_\tau(t) = \sqrt{\frac{\int^{\infty} P(t,\tau)\,\tau^2\,d\tau}{\int_{-\infty}^{\infty} P(t,\tau)\,d\tau} - \left(\frac{\int_{-\infty}^{\infty} P(t,\tau)\tau d\tau}{\int_{-\infty}^{\infty} P(t,\tau) d\tau}\right)^2}. \tag{3.5.2}$$

The expected value of $S_\tau(t)$ usually differs from S_τ [12].

We see that the delay spread is a second central moment, so that contributions at large excess delays have a large influence. Delay spread also causes problems for measured data. Contributions at late times, which are often artifacts caused by noise, are enhanced by the quadratic weighting of the delay. Appropriate noise reduction techniques and restriction to the useful CIR are thus essential for a meaningful evaluation of the delay spread. Also, filtering of the received signal has a large influence on the measured delay spread.

Coherence bandwidth: The coherence bandwidth is related to the frequency correlation function. Let

$$\widetilde{R}_H(\Delta f) = \frac{E\{H^*(t,f)H(t,f+\Delta f)\}}{E\{|H(t,f)|^2\}} \tag{3.5.3}$$

denote the normalized autocorrelation function of the transfer function. Let us furthermore define the normalized PDP $\widetilde{P}(\tau) = \frac{P(\tau)}{\int_{-\infty}^{\infty} P(\tau) d\tau}$. The normalized frequency correlation function is the Fourier transform of the normalized PDP

$$\widetilde{R}_H(\Delta f) = \int_{-\infty}^{\infty} \exp\left(-j2\pi(\Delta f)\tau\right) \widetilde{P}(\tau)\, d\tau. \tag{3.5.4}$$

The coherence bandwidth B_c of level k is defined in [13] as the smallest number so that $|\widetilde{R}_H(B_c)| < k$. Other authors define the coherence bandwidth not generally for a level k but insert specific values for k (e.g., [14] uses $k = 0.75$, and [15], [16], and [17] use $k = 0.5$).

Quite generally, we have to distinguish between the coherence bandwidth for envelopes (i.e., the absolute value of the transfer function) and for the phase of the transfer function, or for the complex transfer function. The coherence bandwidth for the envelope is the most important for the treatment of noise-limited systems—it makes a statement about the possible inherent diversity gain when an equalizer or Rake receiver is used. Coherence bandwidths for the phase have been computed theoretically [15] but are rarely used in practice.

Relation between the coherence bandwidth and the rms delay spread: Since the PDP and the correlation function are related through a Fourier transform, we can anticipate a certain relationship between the two functions. The best-known relationship is that of Jakes [15], who derived the shape of the frequency correlation function as

$$|\widetilde{R}_H(\Delta f)| = \frac{1}{1 + (2\pi(\Delta f)S_\tau)^2}. \tag{3.5.5}$$

From this he concluded that

$$B_{0.5} = \frac{1}{2\pi S_\tau}. \tag{3.5.6}$$

Jakes' derivation assumes a (one-cluster) *exponential* PDP. He also stated that "the relationship does not change dramatically when other shapes of the delay power profile are used." This statement was (mis)interpreted by many authors into assuming that (3.5.6) is a strictly valid relationship for arbitrary delay power profiles; several authors even define the coherence bandwidth by relation (3.5.6). Similarly, [14] got the empirical relationship

$$B_{0.75} = \frac{1}{8S_\tau}. \tag{3.5.7}$$

The true relationship between coherence bandwidth and delay spread is, however, an uncertainty relationship, which was derived by Fleury [13]:

$$B_k \geq \frac{\arccos(k)}{2\pi} \frac{1}{S_\tau}. \tag{3.5.8}$$

The equality sign is valid only for a two-delay model, where both paths have equal power (in other words, for the model that gives the largest delay spread for a given maximum excess delay). This relationship was also confirmed experimentally.

3.5.2 Delay window and interference ratio

Window parameters are an attempt to distinguish between "useful" and "harmful" signal energy, where long-delayed echoes are deemed harmful. They reflect the self-induced quasi co-channel interference caused by delayed echoes of the received signal [17]. They are directly calculated from the time-variant complex impulse response $h(t,\tau)$ or the Doppler-variant transfer function $B(\nu, f)$. Also in that case, we can have either instantaneous/frequency-dependent or averaged parameters. In the following, we give just the instantaneous/frequency-dependent definitions; the averaged ones are completely analogous. Figure 3.7 illustrates the interference ratio and delay window.

Signal to self-interference ratio: The *signal-to-self-interference ratio* $Q_\tau^{(\Delta\tau)}(t,\tau_0)$, also known as *interference ratio*, is the ratio of the signal power within a window of length $\Delta\tau$ compared to the power outside this window,

$$Q_\tau^{(\Delta\tau)}(t,\tau_0) = 10\log_{10}\left[\frac{\int_{\tau_0-\Delta\tau/2}^{\tau_0+\Delta\tau/2} P(t,\tau)\,d\tau}{\int_{-\infty}^{\infty} P(t,\tau)\,d\tau - \int_{\tau_0-\Delta\tau/2}^{\tau_0+\Delta\tau/2} P(t,\tau)\,d\tau}\right]. \qquad (3.5.9)$$

We see that this ratio depends not only on the length of the time window $\Delta\tau$, but also on the starting time τ_0. This dependence can be eliminated in several ways: (i) by letting the window start at the minimum excess delay $\tau_{\min}$, (ii) by shifting τ_0 to the mean delay of the PDP, (iii) searching for the τ_0 that maximizes the signal-to-self-interference ratio SSIR. Method (i) is appropriate to estimate the performance of receivers that synchronize by detection of a threshold excess of the received signal. Choice (ii) can be used in the same manner, though for receivers that synchronize to the mean delay. Approach (iii) is used by receiver systems that try to put the sampling time and synchronization (and thus the effective window) in such a way that the SSIR is maximized. Most of today's receivers are assumed to operate this way.

Physically, the SSIR describes the self-interference of a system: everything that lies outside the considered window is a (statistically almost independent) interference that acts approximately in the same way as does noise. However, this is only an approximate description.

It is furthermore obvious that the SSIR is closely related to specific systems, since each system might be completely different in its judgement of what is useful. To give but one example: In GSM, the equalizer is specified to deal with echoes that are delayed up to 16 μs, while all echoes that are delayed longer are deemed harmful. A window parameter for GSM thus judges how much signal energy arrives within 16 μs.

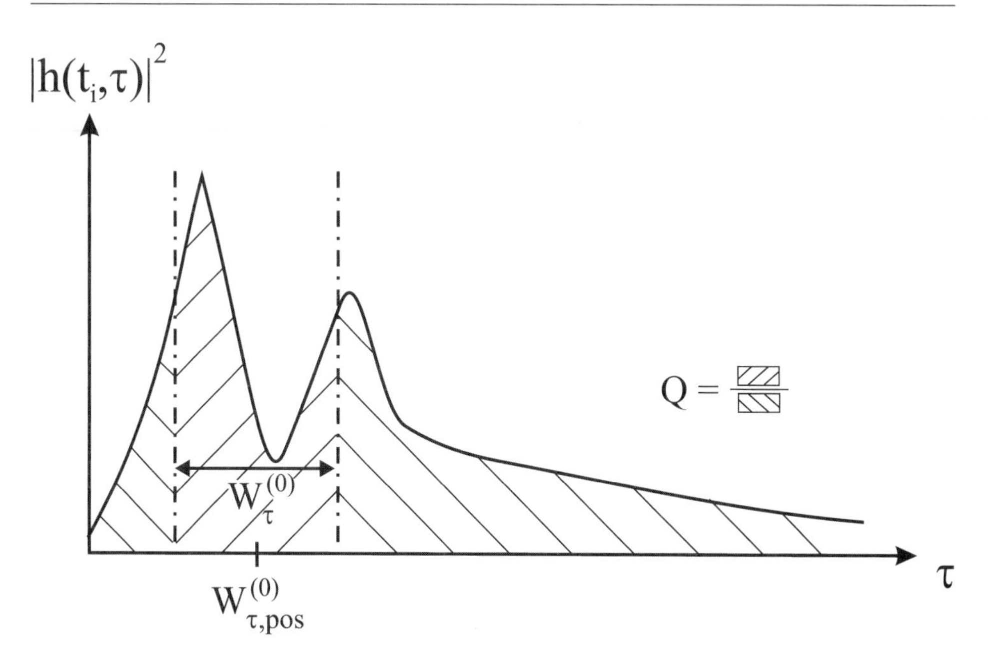

Figure 3.7. Interference ratio and delay window

On the other hand, the DAMPS system specifies an equalizer that must cope with echoes that are delayed by as much as 40 μs, so the window parameter for this system must of course take this into account. A quasi-system-independent characterization is thus only possible if manifold windows, with window sizes relevant for practical systems, are evaluated [8].

Delay Window: The delay window $W_\tau^{(q)}(t)$ is defined as the size of the minimum time window where the ratio of power inside and outside the window is equal to q dB. It says how long the window must be in order that the signal-to-self-interference ratio reaches a certain value.

$$W_\tau^{(q)}(t) = \min_{\tau_0}\{\Delta\tau\}\Big|_{q=Q_\tau^{(\Delta\tau)}(t,\tau_0)}. \tag{3.5.10}$$

The SSIR values $Q_\tau^{(\Delta\tau)}(t)$ are strongly related to the delay windows $W_\tau^{(q)}(t)$; for $W_\tau^{(q)}(t)$, the SSIR is fixed, and the duration of the window is the parameter, and for $Q_\tau^{(\Delta\tau)}(t)$, it is the other way round. For network planning, the signal-to-self-interference ratios are preferred [18], whereas for the task of channel characterization the delay windows are chosen. Figure 3.7 shows the relation between delay windows and SSIRs.

Support in the delay time domain: A quantity that is intimately related to the delay window is the *support* of the channel. The support in the delay domain is defined as the maximum excess delay, in other words, as the delay window that contains *all* of the energy of the impulse response ($q \to \infty$). It is usual to shift the impulse response in such a way that the support is symmetrical with respect to the origin. In other words, the impulse response lasts from $[-\Delta\tau_{\max}/2, \Delta\tau_{\max}/2]$. We will see later that this quantity is important for the basic identifiability of the mobile radio channel. The delay windows can be interpreted mathematically as ε-approximation of the support, i.e., only a percentage $\varepsilon = \frac{1}{1+10^{(q/10)}}$ of the energy is outside the window.

The maximum excess delay is difficult to measure because of the finite signal-to-noise ratio. Contributions with a large excess delay usually have a small amplitude, and very late contributions usually get lost in the noise floor. If we now have a very sensitive measurement system, the maximum excess delay we can measure is larger than for an insensitive system. The measurement problem is thus similar to that of the delay spread (there, the disproportionate weight of the late contributions constituted the main problem), but even more pronounced. In general, the lower the q-value, the less the sensitivity of the delay windows to the SNR.

3.5.3 Summary of relationships

Let us summarize the relation between the system functions, correlation functions, cross-power functions, and condensed parameters. Figure 3.8 shows the relation between all those parameters.

3.6 TIME-DISPERSIVE CHANNEL MODELS

A wealth of models for time-dispersive channels has been developed in the last 40 years. In the following, we give an overview of the most popular ones.

3.6.1 Tapped delay-line model

As mentioned in Section 3.4, a WSSUS channel can be realized as a tapped delay line, where the coefficients multiplying the output from each tap vary with time. The impulse response is then written as

$$h(t, \tau) = \sum_{i=1}^{N} a_i(t) \exp(j\varphi_i(t))\delta(\tau - \tau_i), \tag{3.6.1}$$

where N is the number of taps, $a_i \exp(j\varphi_i)$ are the time-dependent coefficients for the taps (usually complex Gaussian), and τ_i is the delay between input and the ith tap. For each tap, a Doppler spectrum is prescribed in order to describe the changes of the coefficients with time [19]. It is common to assume $\tau_1 = 0$; this is no restriction of generality.

For the choice of the τ_i, there are essentially two possibilities: (i) they are chosen to coincide with the arrival times of the physical paths (scatterer clusters); (ii) they are spaced equidistantly, $\tau_i = i \cdot \Delta\tilde{\tau}$, where the distance between the taps

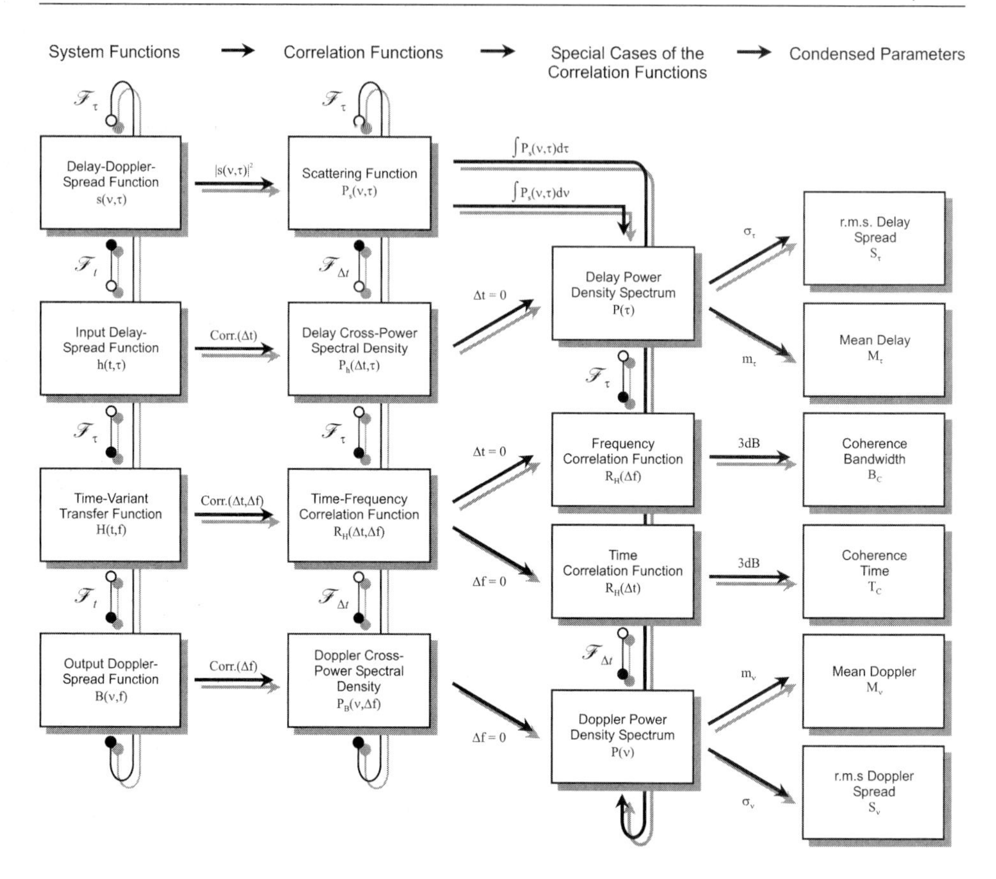

Figure 3.8. Relations between system and correlation functions and their special cases for WSSUS channels with ergodicity with respect to t and f. From [7], copyright Shaker-Verlag. Modified translation provided by R. Kattenbach.

$\Delta\tilde{\tau}$ is determined by the sampling theorem. In the latter case, the tap coefficients are generally *not* statistically independent, even though the WSSUS assumption (statistical independence of the scatterers) is valid.

A further simplification can be achieved if the taps are spaced equidistantly and the coefficients are *defined* to be statistically independent. This model is then known as the *N-delay* (or N-path, or N-beam) *Rice-fading* model if we have a LOS component, or *N-delay Rayleigh-fading* model, if we have no LOS component.[3] For this model, it is furthermore usually assumed that the coefficients change only during a time that is much larger than a symbol length of the data that are trans-

[3]A variant of this model is the Braun-Dersch model, which uses a Nakagami m-distribution for the amplitude variations due to each scatterer clusters and also includes additionally lognormal fading and pathloss attenuation.

mitted; in other words, the channel stays unchanged (has one fixed realization) for the transmission of a few bits. Such a channel is known as a *slow-fading* channel or *quasistatic* channel.

The channel model can be further simplified by restriction of the number of allowed taps N to some small number. The two-delay model has $N = 2$. The two-delay Rayleigh-fading model is popular because it is the simplest possible model that includes time-dispersion effects of the channel; it has also been suggested for standardization [20].

An even simpler model is the satellite channel. In this model, there is just one Rayleigh-fading component and one deterministic LOS component. This arrangement corresponds to the physical situation in a mobile-to-satellite connection, where one part of the signal arrives directly from (or to) the satellite, and one part is reflected by the surroundings of the mobile.

An excellent overview of the models and their implementation methods can be found in [21].

3.6.2 COST 207 model

A special case of the tapped delay-line model is the COST 207 model, which specifies the PDPs or tap weights and Doppler spectra for four typical environments [22]. These PDPs have been evaluated by numerous measurement campaigns performed in France, United Kingdom, Netherlands, Sweden, and Switzerland. The following classes have been specified (see Figure 3.9):

- Rural Area (RA)

$$P(\tau) = \begin{cases} \exp(-9.2\frac{\tau}{\mu s}) & \text{for } 0 < \tau < 0.7\mu s \\ 0 & \text{elsewhere} \end{cases} \tag{3.6.2}$$

- Typical Urban area (TU)

$$P(\tau) = \begin{cases} \exp(-\frac{\tau}{\mu s}) & \text{for } 0 < \tau < 7\mu s \\ 0 & \text{elsewhere} \end{cases} \tag{3.6.3}$$

- Bad Urban area (BU)

$$P(\tau) = \begin{cases} \exp(-\frac{\tau}{\mu s}) & \text{for } 0 < \tau < 5\mu s \\ 0.5\exp(5 - \frac{\tau}{\mu s}) & \text{for } 5 < \tau < 10\mu s \\ 0 & \text{elsewhere} \end{cases} \tag{3.6.4}$$

- Hilly Terrain (HT)

$$P(\tau) = \begin{cases} \exp(-3.5\frac{\tau}{\mu s}) & \text{for } 0 < \tau < 2\mu s \\ 0.1\exp(15 - \frac{\tau}{\mu s}) & \text{for } 15 < \tau < 20\mu s \\ 0 & \text{elsewhere} \end{cases} \tag{3.6.5}$$

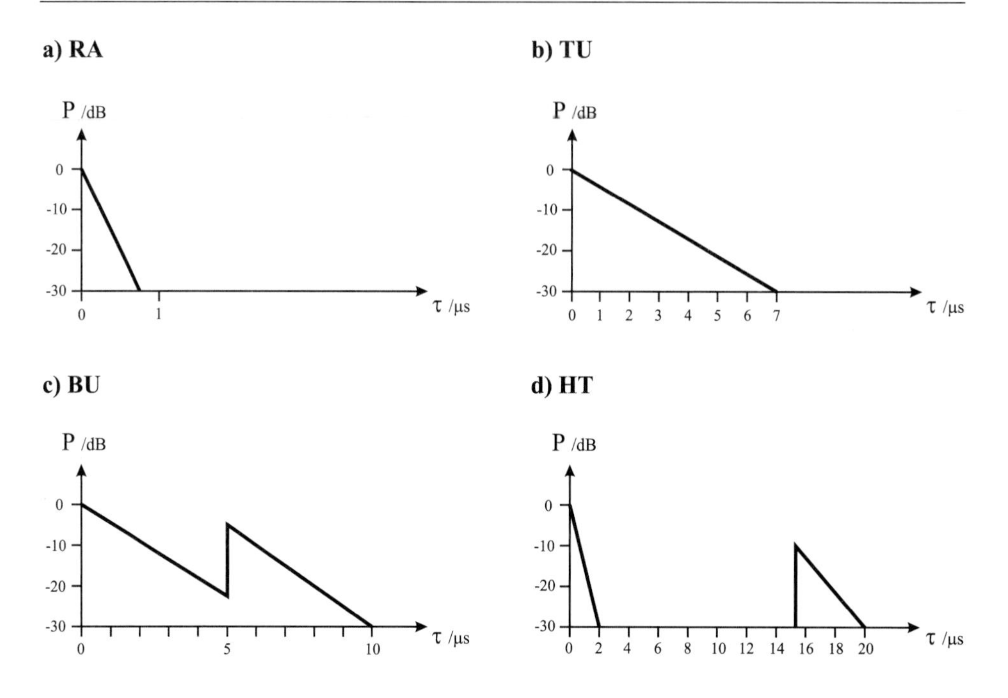

Figure 3.9. COST 207 power delay profiles

In a tapped delay-line realization, each channel has a Rayleigh distributed amplitude and a Doppler spectrum $P_s(\nu, \tau_i)$, where i is the index of the tap. The following four types of Doppler spectra are defined in the COST 207 model, where $\nu_{\max}$ represents the maximum Doppler shift and $G(A, \nu_1, \nu_2)$ is the Gaussian function

$$G(A, \nu_1, \nu_2) = A \exp\left(-\frac{(\nu - \nu_1)^2}{2\nu_2^2}\right) \tag{3.6.6}$$

and A is a normalization constant so that $\int P_s(\nu, \tau_i) d\nu = 1$.

a) *CLASS* is the classical (Jakes) Doppler spectrum and is used for paths with delays not in excess of 500 ns ($\tau_i \leq 0.5\ \mu$s):

$$P_s(\nu, \tau_i) = \frac{A}{\sqrt{1 - \left(\frac{\nu}{\nu_{\max}}\right)^2}} \tag{3.6.7}$$

for $\nu \in] - \nu_{\max}, \nu_{\max}[$;

b) *GAUS1* is the sum of two Gaussian functions and is used for excess delay times in the range of 500 ns to 2 μs ($0.5\ \mu\text{s} \leq \tau_i \leq 2\ \mu\text{s}$):

$$P_s(\nu, \tau_i) = G(A, -0.8\nu_{\max}, 0.05\nu_{\max}) + G(A_1, 0.4\nu_{\max}, 0.1\nu_{\max}) \qquad (3.6.8)$$

where A_1 is 10 dB below A;

c) *GAUS2* is also the sum of two Gaussian functions and is used for paths with delays in excess of 2 μs ($\tau_i \geq 2\ \mu\text{s}$):

$$P_s(\nu, \tau_i) = G(B, 0.7\nu_{\max}, 0.1\nu_{\max}) + G(B_1, -0.4\nu_{\max}, 0.15\nu_{\max}) \qquad (3.6.9)$$

where B_1 is 15 dB below B;

d) *RICE* is the sum of a classical Doppler spectrum and one direct path, such that the total multipath contribution is equal to that of the direct path. This spectrum is used for the shortest path of the model for propagation in rural areas:

$$P_s(\nu, \tau_i) = \frac{0.41}{2\pi\nu_{\max}\sqrt{1 - \left(\frac{\nu}{\nu_{\max}}\right)^2}} + 0.91\delta(\nu - 0.7\nu_{\max}) \qquad (3.6.10)$$

for $\nu \in]-\nu_{\max}, \nu_{\max}[$;

Four ways of applying these Doppler spectra to the four propagation classes suggested by COST 207 are shown in Tables 3.1–3.4.

3.6.3 Hashemi-Suzuki-Turin model

In its original version this model was developed by Turin et al. [23] to investigate urban multipath channels. The original model has been improved and extended by Suzuki [24] and Hashemi [25]. Turin et al. [23] assumed a priori that the carrier phases φ_i of the various paths are mutually independent random variables uniformly distributed over $[0, 2\pi]$. After performing various measurement runs, they claimed that the amplitudes r_i are distributed following a lognormal distribution. Suzuki [24] found another distribution function, the so-called Suzuki distribution. Because the extraction of the parameters for this distribution from measurements was quite difficult, he used the Nakagami-distribution (see Section 3.1), which showed a good fit to measured data.

As a first-order approximation, Turin et al. [23] assumed that objects that cause reflections in an urban area are located randomly in space, giving rise to a *Poisson distribution* for the excess delays. However, a second-order model [24], gives much better agreement with measurements. This model is a modified Poisson process (the $\Delta - K$ model), which takes into account the possibility that paths may arrive in groups. This model has two states: S_{-1}, where the mean arrival rate is $\lambda_0(t)$, and S_{-2}, where the mean arrival rate is $K\lambda_0(t)$. The process starts in S_{-1}. If a path

Tap#	Delay [μs]	Power [dB]	Doppler category
1	0	0	RICE
2	0.2	-2	CLASS
3	0.4	-10	CLASS
4	0.6	-20	CLASS

Table 3.1. Parameters for rural (non-hilly) area (RA)

Tap#	Delay [μs]	Power [dB]	Doppler category
1	0	-3	CLASS
2	0.2	0	CLASS
3	0.6	-2	GAUS1
4	1.6	-6	GAUS1
5	2.4	-8	GAUS2
6	5.0	-10	GAUS2

Table 3.2. Parameters for urban (non-hilly) area (TU)

Tap#	Delay [μs]	Power [dB]	Doppler category
1	0	-3	CLASS
2	0.4	0	CLASS
3	1.0	-3	GAUS1
4	1.6	-5	GAUS1
5	5.0	-2	GAUS2
6	6.6	-4	GAUS2

Table 3.3. Parameters for hilly urban area (BU)

Tap#	Delay [μs]	Power [dB]	Doppler category
1	0	0	CLASS
2	0.2	-2	CLASS
3	0.4	-4	CLASS
4	0.6	-7	CLASS
5	15	-6	GAUS2
6	17.2	-12	GAUS2

Table 3.4. Parameters for hilly terrain (HT)

arrives at time t, a transition is made to S_{-2} for the interval $[t, t+\Delta]$. If no further paths arrive in this interval, a transition is made back to S_{-1} at the end of the interval. Note that for $K = 1$ or $\Delta = 0$, the above-mentioned process reverts to a standard Poisson process. Hashemi [25] additionally introduced two exponentially decreasing functions that which take into account the spatial correlation on the arrival times.

3.7 MODELS INCLUDING ANGULAR DISPERSION

Research in intelligent antennas has exploded in the last five years. As we explain in Chapter 5, an intelligent antenna can be interpreted as a spatial equalizer or Rake receiver, which combines phase-corrected versions of the echoes incident from different directions. For these applications, we also need channel models that include the directional components. The standard wideband models described in the previous section are unsuitable because the relation between Doppler spectrum and directions of arrival at the MS is not unique: On one hand, the $\cos(\gamma)$ enters the Doppler spectrum, so that at least two angles result in the same Doppler shift. On the other hand, not only the MS, but also the scatterers (and even the BS) might be moving, which would change the Doppler shift but not the angle of arrival. Even more importantly, the Doppler shift is not related to the directions of arrival *at the BS*.

There are mainly two types of spatial channel models: purely stochastic and semistochastic models:

- In the purely stochastic approach, the angle-resolved impulse response $h(t, \tau, \varphi, \theta)$ is modelled stochastically, where φ and θ are the azimuth and elevation at the receiver. Commonly, the ADPS (Angular Delay Power Spectrum) $E_t\{|h(t, \tau, \varphi, \theta)|^2\}$ is prescribed, and the instantaneous realizations are chosen accordingly. The amplitude statistics are assumed to be Rayleigh, Rice, or Nakagami. Measurement campaigns indicate that in most environments, there are one or two scatterer clusters, where each cluster gives rise to an ADPS of the form

$$\text{ADPS}(\tau, \varphi, \theta) \propto \exp(-(\tau - \tau_{\min})/S_\tau) \exp(-|\varphi - \varphi_0|/\sigma_\varphi)\delta(\theta) \qquad (3.7.1)$$

  i.e., an exponential distribution in time, a Laplacian distribution in azimuth, and a delta function in elevation (i.e., all radiation is incident in the horizontal plane).

- In the semistochastic model, also known as the geometry-based stochastic channel model (GSCM) [26], not the ADPS, but the location of the scatterers is prescribed. Furthermore, it is usually assumed that only single scattering occurs on the way from the BS to the MS. Thus, the GSCM proceeds in two steps: (i) the location of the scatterers is chosen at random from the given

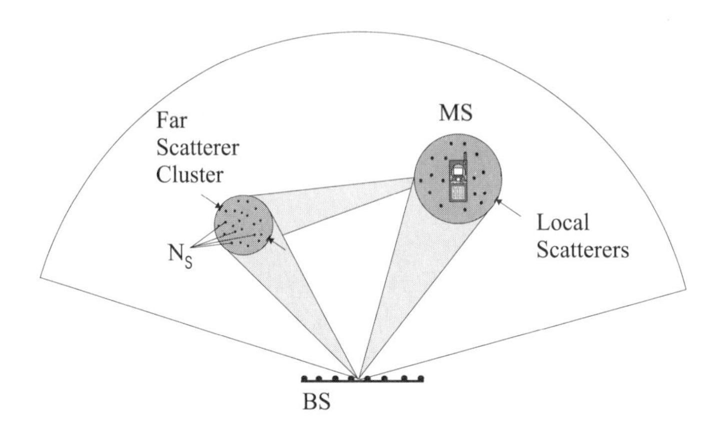

Figure 3.10. Principle of the GSCM spatial channel model

PDF; (ii) the angle-resolved impulse response is computed by a simple ray tracing. In most cases, the scatterers are situated around the MS. However, far scatterers, like mountains or high-rise buildings, can also be included (see Figure 3.10).

In principle, the two modelling approaches are equivalent. The purely stochastic approach is usually faster, whereas the GSCM is better suited for the inclusion of long-term variations of the channel and can be more easily parametrized from simple geometrical considerations. An extensive overview of the literature in the field is given in the review papers [27] and [28].

Bibliography

[1] M. Abramowitz and I. A. Stegun, *Handbook of Mathematical Functions.* New York, Dover, 1965.

[2] S. O. Rice, "Statistical properties of a sine wave plus random noise," *Bell System Techn. J.*, vol. 27, pp. 109–157, 1947.

[3] P. Bello, "Characterization of Randomly Time-Variant Linear Channels," *IEEE Trans. Comm.*, vol. 11, pp. 360–393, 1963.

[4] W. Kozek, "On the transfer function calculus for underspread LTV channels," *IEEE Trans. Signal Proc.*, vol. 45, pp. 219–223, 1997.

[5] W. Kozek and A. F. Molisch, "On the eigenstructure of underspread wssus channels," in *Proc. First Signal Processing Workshop Advances in Wireless Communications (SPAWC'97)*, pp. 325–328, 1997.

[6] G. Matz and F. Hlawatsch, "Time-frequency transfer function calculus (symbolic calculus) of linear time-varying systems (linear operators) based on a generalized underspread theory," *J. of Math. Phys. (Special Issue on Wavelet and Time-Frequency Analysis)*, vol. 39, no. 8, pp. 4041–4070, 1998.

[7] R. Kattenbach, *Characterization of Time-Variant Indoor Radio Channels by Means of Their System- and Correlation Functions (in German).* PhD thesis at University GhK Kassel, published by Shaker-Verlag, Aachen, 1997.

[8] A. F. Molisch and M. Steinbauer, "Condensed parameters for characterizing wideband mobile radio channels," *Int. J. Wireless Information Networks*, vol. 6, 1999.

[9] D. S. Polydorou and C. N. Capsalis, "A new theoretical model for the prediction of rapid fading variations in indoor environment," *IEEE Trans. Vehicular Techn.*, vol. 46, pp. 748–755, 1997.

[10] U. Liebenov and P. Kuhlmann, "Determination of scattering surfaces in hilly terrain," *COST231 TD(93)119*, 1993.

[11] J. B. Andersen, "A note on definitions of terms for impulse responses," *EURO-COST 231*, vol. TD(89)060, 1989.

[12] A. F. Molisch, "Statistical properties of the rms delay spread of mobile radio channels with independent Rayleigh-fading paths," *IEEE Trans. Vehicular Techn.*, vol. 45, pp. 201–205, 1996.

[13] B. H. Fleury, "An uncertainty relation for WSS processes and its application to WSSUS systems," *IEEE Trans. Comm*, vol. 44, pp. 1632–1634, 1996.

[14] M. J. Gans, "A power-spectral theory of propagation in the mobile radio environment," *IEEE Trans. Vehicular Techn.*, vol. 21, pp. 27–38, 1972.

[15] W. C. Jakes, *Microwave Mobile Communications.* Piscataway, NJ: IEEE Press, 1974.

[16] R. S. Kennedy, *Fading Dispersive Communication Channels.* New York: Wiley-Interscience, 1969.

[17] J. P. DeWeck, *Real-Time Characterization of Wideband Mobile Radio Channels.* PhD thesis, TU Wien, 1992.

[18] H. Buehler, *Estimation of Radio Channel Time Dispersion for Mobile Radio Network Planning.* PhD thesis, TU Wien, 1994.

[19] J. D. Parsons, *The Mobile Radio Propagation Channel.* Halstead Press, 1992.

[20] Electronics Industry Association, *Dual-Mode Subscriber Equipment Compatibility Specification*, 1989.

[21] M. Paetzold, *Mobilfunkkanaele (in German).* Vieweg, 1999.

[22] C. of the European Communities, *Digital Land Mobile Radio Communications—COST 207.* ECSC-EEC-EAEC, 1989.

[23] G. L. Turin, F. Clapp, T. Johnston, S. B. Fine, and D. Lavry, "A statistical model of urban multipath propagation," *IEEE Trans. Vehicular Techn.*, vol. 21, pp. 1–9, 1972.

[24] H. Suzuki, "A statistical model for urban radio propagation," *IEEE Trans. Comm.*, vol. 25, pp. 673–680, 1977.

[25] H. Hashemi, "Simulation of the urban radio propagation channel," *IEEE Trans. Vehicular Techn.*, vol. 28, pp. 213–225, 1979.

[26] J. Fuhl, A. F. Molisch, and E. Bonek, "Unified channel model for mobile radio systems with smart antennas," *IEE Proc. Radar, Sonar and Navigation*, vol. 145, pp. 32–41, 1998.

[27] R. B. Ertel, P. Cardieri, K. W. Sowerby, T. S. Rappaport, and J. H. Reed, "Overview of spatial channel models for antenna array communications systems," *IEEE Personal Comm. Mag.*, pp. 10–22, Feb. 1998.

[28] U. Martin, J. Fuhl, I. Gaspard, M. Haardt, A. Kuchar, C. Math, A. F. Molisch, and R. Thomä, "Model scenarios for intelligent antennas in cellular mobile communication systems—scanning the literature," *Wireless Personal Comm., Special Issue on Space Division Multiple Access*, vol. 11, pp. 109–129, 1999.

Chapter 4

OUTLINE

After the introductory focus in this part, the rest of this book covers (i) unequalized systems, (ii) equalizers, (iii) OFDM, and (iv) Rake receivers. All these four parts are self-contained, so they can be read in arbitrary sequence.

PART II - UNEQUALIZED SYSTEMS by *Molisch* discusses those systems at the boundary between wideband and narrowband where the time dispersion is smaller, but not *much* smaller, than the symbol duration (slightly time-dispersive systems) and do not contain equalizers. Typically, those systems were originally designed as true narrowband systems, where it later turned out that under some circumstances time dispersion can play a role, or where equalizers were simply too expensive.

After a brief introduction, Chapter 7 gives a generic system description for both transmitter and receiver. The most important modulation formats, namely, phase-shift keying, quadrature amplitude modulation, and continuous-phase frequency shift keying are described mathematically, together with the standard detection methods (coherent, differentially coherent, and incoherent).

Chapter 8 then describes various mathematical methods for computing the bit error probability (BER) of unequalized systems under the influence of time-dispersive signals. Although most textbooks on mobile radio describe the standard approach for flat-fading, namely, first computing the BER for AWGN channels and then averaging over the distribution of channel attenuation, this approach is no longer possible for time-dispersive channels. Methods that circumvent these problems are, e.g., based on certain properties of quadratic forms of Gaussian variables, probability density functions of the angles between Gauss-distributed vectors, or the group delay distributions of the channel. Unfortunately, none of these can be considered to be a master approach that is optimum for all possible cases. Methods based on Gaussian variables are usually the simplest but always require the channel to be Rayleigh- or Rice-fading, and encounter problems, e.g., in Nakagami channels. Within this group, the method based on quadratic forms can most easily include arbitrary time dispersion but cannot easily be extended to multilevel modulation formats, e.g., M-ary PSK, which can be better treated by pdfs of Gaussian vectors or the two-path equivalent matrix method. Chapter 8 describes the different methods in sufficient detail for the reader to judge what is needed to solve a specific problem, but details are covered only in a few examples.

Closed-form equations and figures of the performance of standard systems are the contents of Chapter 9. The influence of modulation format, delay spread, shape of the power delay profile, sampling instant, and other parameters are presented.

In Chapter 10, the author describes special modulation formats and receiver structures that are used for the reduction of ISI-induced errors (error floor). Some modulation formats try to actually exploit the time dispersion and thus approach the performance of an equalized system in a time-dispersive environment. The penalty paid for this approach lies in a higher bandwidth requirement, which precludes these formats from application in cellular systems, but the technique might be interesting for wireless LANs and cordless applications.

Some receiver structures, on the other hand, try just to reduce the error floor by eliminating the ISI before detection, thus making the performance comparable to a true flat-fading channel. These stay within the framework of standard modulation formats and thus usually have only a penalty in the form of a slightly more complex receiver structure (but still much simpler than an equalizer). A technique that also has these properties is *adaptive sampling*, more exactly, the adaptive choice of the sampling instant according to the channel constellation. Adaptive sampling is described in Chapter 11.

Chapter 12 describes antenna diversity. Although antenna diversity is usually known to reduce noise-induced errors, it was also shown to be an effective countermeasure for ISI-induced errors. After a description of the different ways to combine the diversity signals, the mathematical methods of Chapter 8 are modified to include the diversity effects. Those methods are useful not only for conventional multidimensional diversity, but also for the performance of Rake receivers, since these use the time-delayed echoes of the original signal as diversity signals that are combined with maximum-ratio combining. Subsequently, the performance with the diversity antennas is presented in some examples. A summary (Chapter 13) concludes this part.

PART III - EQUALIZERS by *Vitetta, Hart, Mammela, and Taylor* focuses on the equalization techniques for single-carrier, unspread digital signals transmitted over multipath fading channels and is organized as described below.

Some preliminary topics are presented in Chapter 14, where the mathematical models of the transmitted signal and of the wireless channel are illustrated. An optimal receiver must not discard useful information present in the continuous time received signal; digital processing, however, is an inevitable requirement of any modern receiver. Therefore, filtering and discretizing of the received signal are discussed. Notations for the various scalar and matrix quantities needed in the remainder of the Part are given; in particular, matrix representations for the received signal samples are provided to simplify the derivation of equalization algorithms. Finally, reduced complexity channel models, i.e., parsimonious representations or parameterizations of the channel impulse response, are introduced to simplify the channel estimation problem and the equalizer design. In particular, the authors focus on the Karhunen-Lóeve expansion, on the complex exponential parameterization, and on the power series models.

Any equalization algorithm processes the received signal, producing a set of real quantities, known as *metrics*, that are evaluated by the receiver to make decisions on the transmitted data. In Chapter 15 we derive, interpret and analyze the performance of the metrics computed by *optimal* detectors under the assumption that the CIR (or some equivalent quantity) is *known*, *estimated*, or *averaged over* for the doubly selective wireless channel and its special cases: the frequency-flat and frequency-selective channels. Both the *maximum likelihood* (ML) and *maximum a posteriori* (MAP) methods are discussed as optimality criteria, and their application to bits, symbols and sequences of symbols is illustrated. In particular MAP *bit detectors* (MAPBDs), MAP *symbol detectors* (MAPSDs) and ML *sequence detectors* (MLSDs) are considered. In addition, performance bounds for both the ML and MAP detectors are provided; they represent usefuls tool to assess their error performance in some situations.

In Chapter 16, we describe various *equalization algorithms* corresponding to different ways of implementing the computation of the derived metrics, ranging from optimal to highly suboptimal. Again the primary division into sections is related to how the CIR is treated. Within each section, the structure, complexity, and performance measures of each type of equalizer are described. In particular, in Section 16.1 we begin by assuming that the CIR is known exactly *a priori*, and we derive five important classes of equalizers: MLSDs, MAPSDs and MAPBDs, reduced complexity sequence detectors, decision feedback equalizers (DFEs), and linear equalizers (LEs). The aim of Section 16.2 is twofold. First, we show some mathematical tools for channel estimation and we discuss blind equalization techniques. Secondly, we illustrate equalization techniques incorporating channel estimation strategies, like adaptive MLSDs and adaptive MAPBD/MAPSDs. Adaptive LEs and DFEs and pilot-based detection techniques for frequency-flat fading channels are also considered. Equalization when the CIR is averaged-over is investigated in Section 16.3. MLSDs and MAPSDs are developed and are related to their adaptive counterparts. Finally, various equalization/detection strategies for FF fading channels are illustrated.

PART IV - OFDM by *May and Rohling* describes orthogonal frequency division multiple access (OFDM), also known as multicarrier modulation. Chapter 17 introduces OFDM and describes its historical evolution. Chapter 18 describes the basic transmission/reception technique, both in a time-continuous formulation that is intuitively appealing and in a time-discrete formulation that illuminates the important relation to discrete Fourier transforms. The shaping of the basic pulses is discussed. In a conventional system, rectangular pulses are used as a basis for constructing the total signal, but this might not be optimum in dispersive channels. Chapter 19 then treats frame, timing, and carrier frequency synchronization.

Chapters 20 and 21 are devoted to modulation and demodulation. Many aspects are identical to those of single-carrier systems, but there are also some specific points that are unique to OFDM systems. When, for example, using differential modulation/demodulation, we have a choice whether we take the difference with the previous symbol (in time) or the adjacent symbol (in frequency). For coherent demodulation, the design of the pilot tones also plays a vital role. Trade-offs between the quality of the channel estimation and the spectral efficiency loss because of the insertion of non-information-carrying symbols must be considered.

An OFDM system can really exploit its advantages in wideband channels only when appropriate coding (discussed in Chapter 22) is used. Transmission on carriers in fading dips will always result in a high bit-error rate, and only appropriate interleaving and coding can recover the information in such a way that the frequency diversity of the wideband transmission is really exploited. Soft-decision demodulation is also one major aspect and realizes large gains compared to hard decisions. Convolutional codes, possibly concatenated, and trellis-coded modulation are additional performance-enhancing methods.

In contrast to conventional modulation methods, which are usually constant-envelope, OFDM can exhibit a high Crest factor. The reason is that an OFDM signal is a sum of many (hundreds or thousands) partial signals whose phase is determined by the data sequence to be transmitted. For some data, these partial signals can add up in such a way that the output signal has a huge instantaneous amplitude. Amplifiers for the transmission of such signals would have to have a large dynamic range, which is difficult to implement. There are thus intensive investigations, described in Chapter 23, on how to reduce the Crest factor. Solution proposals range from insertion of redundancy, application of correcting functions, to special coding.

Nonlinearities of the transmission amplifier are also one reason for out-of-band emissions. Also, the shaping of the basis pulses, as discussed for Chapter 18, plays a role. In any case, these out-of-band transmissions must be eliminated as much as possible, since especially in a cellular system, adjacent channel interference requirements are very strict. This reduction is usually achieved by filtering, which in turn increases intersymbol interference. The trade-off between these two considerations is the focus of Chapter 24.

Chapter 25 relates conventional OFDM systems to multicarrier OFDM, where information symbols are spread over several carriers, and to single-carrier transmission with frequency-domain equalization. All of these systems can be viewed as special cases of a generic OFDM system; this viewpoint facilitates comparisons of the various advantages and disadvantages.

PART V - CDMA by *Goiser (Chapters 26–28), Win and Chrisikos (Chapter 29), and Glisic (Chapters 30–33)* describes how CDMA systems exploit the time-delay diversity that is inherent in multipath propagation. This exploitation is achieved by the so-called Rake receiver, which with its several "fingers" detects several time-delay replicas of the original signal and then adds them coherently.

Chapter 26 introduces direct-sequence spread spectrum, which forms the basis for CDMA systems. It describes how the information is spread by multiplication with a spreading sequence from a bandwidth $1/T_{\text{symbol}}$ to a bandwidth $P/T_{\text{symbol}} = 1/T_{\text{chip}}$, where P is the processing gain and T_{chip} is the chip duration of a chip in a spreading sequence. This chapter also explains some basic trade-offs in the design of a CDMA network, including trade-offs between number of users, spreading gain, achievable BER, etc.

Since the chip duration is quite short, a lot of echoes can be resolved by a receiver. The correlation between the received signals and the spreading sequence thus exhibits several peaks, each corresponding to one echo. The Rake receiver now adds up the largest echoes with the correct phases. This technique exploits the path diversity inherent in the wideband channel.

Chapter 27 compares the performance of such a Rake receiver to a simpler MUW receiver that receives only one echo and suppresses the other echoes. Simulation curves show the gain that is possible by the Rake receiver in different channels for a single-user environment.

Chapter 28 describes a receiver concept known as D-Rake (decorrelating Rake receiver) that is especially suitable for a multiuser environment, as always occurs in a multiuser environment. This receiver not only exploits the multiple echos but also suppresses the signals from other users very efficiently. Interference from other users is always present because the used codes are never orthogonal in a multipath environment and the power control has only a finite accuracy and dynamic. Thus, interference from users in the same cell is usually the capacity limit.

Chapter 29 explores the theoretical performance of a Rake receiver with a finite number of fingers. Such a receiver selects the echoes that have the highest instantaneous energy. The authors introduce a technique that allows a closed-form computation of the resulting SNR and BER.

One of the most important practical aspects of Rake receivers is code synchronization, which consists of acquisition and tracking. Code acquisition is treated in Chapter 30. After the description of performance analysis tools, especially the signal flow graph method, the performance of acquisition methods in quasi-synchronous and asynchronous networks are computed for different multipath channel models.

Chapter 31 then treats code tracking. First, three different types of code tracking loops are explained: the *baseband full-time early-late tracking loop*, the *full time early-late noncoherent tracking loop*, and the *tau-dither early-late non-coherent tracking loop*. After this, the influence of a wideband channel on the tracking loop performance is computed.

Chapter 32 treats the problem of channel estimation, especially with a subspace-based method, which is explained in detail and shown to give good results in channels with near/far effect.

Chapter 33 treats the capacity of CDMA networks. This is an especially important subject since high capacity is one of the key arguments for the use of CDMA in mobile radio. After describing the system model, some general equations for the capacity are given. Then, the effect of power control, and the influence of the Rake receiver and interference-cancelling receivers are shown. In general, Rake receivers with many fingers should give higher capacity than those with few fingers. However, due to channel estimation errors, there can actually be also a decrease.

An appendix with mathematical background material concludes Part V.

Chapter 5

OUTLOOK

5.1 COMPARISON OF APPROACHES

Since four distinct approaches for wideband transmission are treated in this book, the question naturally arises which of them is the best. This question does not allow an easy answer. Each approach has its advantages and drawbacks, making it preferable under specific circumstances. The following gives qualitative guidelines for comparisons. More quantitative approaches require exact mathematical analyses according to the methods described in the latter parts of this book.

Unequalized systems are by definition suboptimum. The best one can usually hope for is to approximate the performance of narrowband systems. However, the frequency diversity inherent in a wideband transmission is not exploited. The main advantage of this approach lies in its simplicity, allowing for cheap, compact, low power consumption receivers. With digital signal processing becoming cheaper every year, the production costs of a simple equalizer become less significant compared to other parts of the receiver. It is thus to be expected that emphasis for unequalized systems in the future will shift from cordless telephones to even simpler applications, like wireless local loop systems, telemetric applications, etc.

Both TDMA and CDMA make use of the frequency diversity of a wideband system. The visualization of the operation of an equalizer (in a TDMA system) is easiest in the frequency domain: in the simplest form (zero-forcing equalizer), it tries to invert the transfer function of the channel, so that the combination of channel and equalizer transfer function is flat. A more sophisticated form minimizes the mean-square error between transmitted and received signal. This statement is used below to compare equalizers to Rake receivers, since mean-square error is a quantity that is easily transformed between time and frequency domains.[1] An equalizer can be realized by a tapped delay line. Then, the number of computer operations is proportional to the equalizer length L (i.e., the number of taps L) during data detection, and proportional to either L, L^2, or L^3 during the training phase (determination of the weights of the tapped delay line), according to the algorithm used for training. Another form of equalizer is the Viterbi algorithm, where the computational effort scales with S^L, where S is the number of states

[1] More exactly, the L_2-norm of a quantity is invariant with respect to a Fourier transform.

in the trellis. The length of the equalizer in turn is determined by the ratio of the maximum excess delay of the channel compared to the symbol duration. It plays no role whether the different echoes of the channel are all closely spaced or whether there are long periods in the delay power profile where no significant echo occurs.[2] Due to these restrictions, TDMA is only feasible for a restricted range of wideband systems. An equalizer length of L=4–5 seems currently the maximum that is practically realizable.

The operation of the Rake receiver in a CDMA system can most easily be visualized in the delay domain. Each finger of the Rake receives and detects one echo of the transmitted signal, where all echoes are assumed to have a sufficiently different delay (closely spaced echoes are smeared together by the receiver filtering). The outputs of the Rake fingers are then coherently added, so that maximum-ratio combing of the different echoes is achieved. Since maximum ratio combining also achieves minimum mean-square error, the effect of this operation can be seen to be quite similar to an equalizer.

There are, however, important practical differences. The number of required Rake fingers is determined by the number of significant echoes. The distance between two Rake fingers should be at least one chip length; echoes that are more closely spaced in the delay domain are strongly correlated because of the receiver filtering. The ratio of maximum excess delay to the chip length is thus an upper bound to the number of Rake fingers, in contrast to the equalizer, where it was the minimum number of equalizer taps. Even if fewer Rake fingers L are used than there are significant echoes N_{sign}, performance still is good, since (i) the Rake will select the L strongest echoes, and (ii) the correlation receiver will suppress all other echoes, so that they cannot act as interference. If an equalizer is shorter than the maximum excess delay, all echoes between LT_{symbol} and $N_{\text{sign}}T_{\text{symbol}}$ act as interference. Furthermore, these echoes are not suppressed by the equalizer.

The decision between a TDMA and a CDMA system is, however, usually dictated by other considerations than the tradeoff equalizer versus Rake receiver. CDMA systems are more complicated and require more digital signal processing; however, they offer slightly higher capacity in wireless systems and are more flexible in accomodating different data rates and allow easier cell planning. Up to 2000, TDMA was the dominant multiple access method because of its use in GSM and IS-136. For the next 10 years, however, CDMA seems to gain the upper hand, since it was adopted for the IMT-2000 proposals by European, American, and Asian standards organizations.

Even CDMA systems have upper limits of the possible data rate: the maximum number of Rake fingers is also on the order of 5, so that, for example, in a system with 20 significant echoes, they are losing too much of the available energy. A system that is able to deal with almost arbitrary data rates is OFDM. In this system, the information is distributed over a large number of narrowband

[2]This statement is not exact. Recently, there has been some effort in devising algorithms with reduced complexity for the case when there are only a few relevant echoes in the power delay profile.

subcarriers. In contrast to TDMA and CDMA, this system does not inherently exploit frequency diversity for each separate bit: bits that are transmitted on a subcarrier in a fading dip will be lost. OFDM thus requires a sophisticated coding and interleaving strategy, possibly combined with adaptive modulation. If these measures are included, the achievable bit error rates are comparable to those of equivalent TDMA or CDMA systems. The computational effort for one OFDM block is essentially $N_{\text{sub}} \log_2(N_{\text{sub}})$, where N_{sub} is the number of subcarriers. A further system requirement is that the distance between subcarriers is considerably (about a factor 10) smaller than the inverse of the maximum excess delay. Since one OFDM block contains N_{sub} data symbols, the computational effort *per data symbol* is proportional to $\log_2(\tau_{\text{max}}/T_{\text{symbol}})$, compared to approximately $\tau_{\text{max}}/T_{\text{symbol}}$ for the other approaches.

5.2 FUTURE DEVELOPMENTS

For third- and fourth-generation systems, wideband transmission techniques will often be combined with two key innovations of the last years: adaptive antennas and multiuser detection. These two methods are outside the scope of this book, but we give here a brief introduction to show their potential and possible combination with wideband transmission techniques.

5.2.1 Adaptive Antennas

One exciting possibility for improving wideband system is the addition of adaptive antennas. Adaptive antennas consist of several antenna elements whose signals are combined or otherwise processed by digital signal processing. Conceptually, adaptive antennas show many similarities with wideband systems. They exploit the fact that signals arrive at the receiver not only with different delay, but also from different directions. If a receiver is thus capable of resolving those different echoes, it can add them up in a constructive way.

Various overview papers present introductions to smart antenna technology: [1] gives a short introduction to smart antennas, their benefits and application. More comprehensive studies with extensive references are [2] and [3], where Part II focuses on beamforming. In [4], [5] the focus lies on space–time processing for TDMA and CDMA systems. An introduction of smart antenna arrays for CDMA systems is presented in [6] and, especially, in [7]. Reference [8] presents an overview on parametric DOA (direction of arrival) estimators for array signal processing.

Basically, a smart antenna processor can include the following:

- Spatial processor: In the straightforward approach, an adaptive beamformer realizes a spatial filter (Figure 5.1), i.e., a single-weight vector $\mathbf{w}$ is applied to combine the received signals that results in the user signal $\mathbf{y}$. The smart antenna processor is followed typically by a conventional baseband processor that includes equalizer and detector. Thus, this approach utilizes a spatial filter and a subsequent *independent* temporal filter (equalizer).

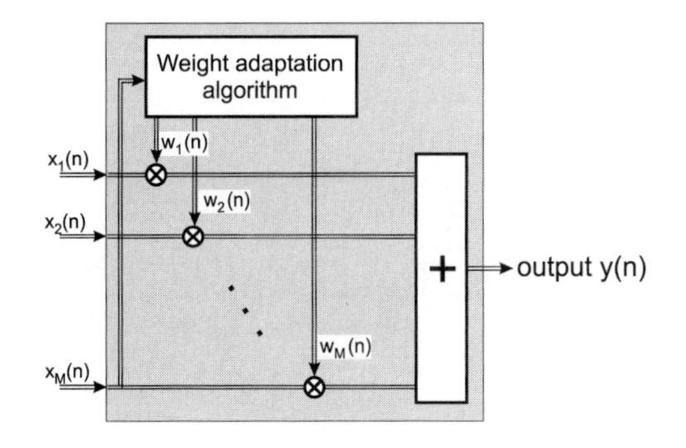

Figure 5.1. Space-only processing. The smart antenna processor includes only a spatial processing structure, which is followed by a temporal equalizer. Reproduced from [9] with permission of the author.

- Space/time equalizer: An extension of the above concept is to combine the spatial filter with the temporal filter (Figure 5.2). The output of the space–time filter is directly put into a detector without further equalization because temporal equalization already has been carried out in the space/time equalizer. Combined spatio–temporal processing will improve performance, especially in systems with large delay spread. It will not boost performance significantly compared to the previous approach when no significant variation of the angular power distribution over delay occurs, because then temporal and spatial equalization can be done serially without loss of performance. In [4] the concept of space-only and space-time processing are discussed, both for TDMA and CDMA systems.

- Space/time detector: In the most elaborate approach the space-time equalization and detection is combined (Figure 5.3). Such systems generally show superior performance, but most are today computationally too demanding for real-time implementation. However, first steps have been taken to reduce the complexity and so allow a real-time implementation [10]. Here, the smart antenna processor already includes the detector, and thus it also includes the baseband processing unit. A drawback is that there is no equivalent concept for transmission.

In the first two concepts—spatial processor and space/time equalizer—we can think of antenna patterns that are implicitly or explicitly formed to optimize reception/transmission. In the case of the space/time detector, *no* weight vector is calculated anymore and thus the concept of beamforming is not applicable here. Actually, the intermediate step of extracting a user-specific, complex baseband signal is omitted. Instead the final user symbol stream is extracted in a single procedure, hence the name *space-time detector.*

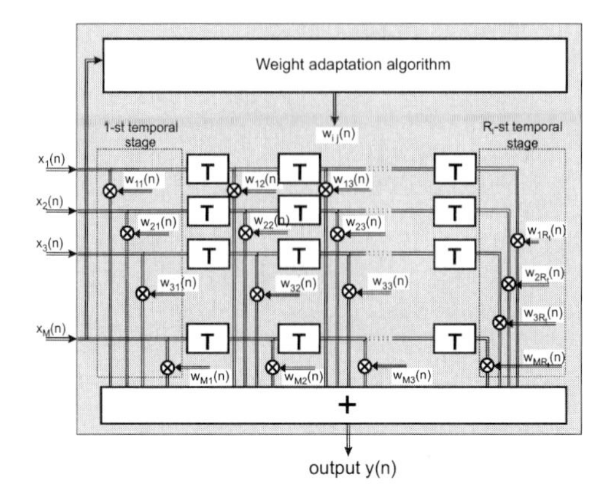

Figure 5.2. Space/time equalization. The smart antenna processor includes a spatio–temporal processing structure. The output of the smart antenna processor can be detected without further equalization. Reproduced from [9] with permission of the author.

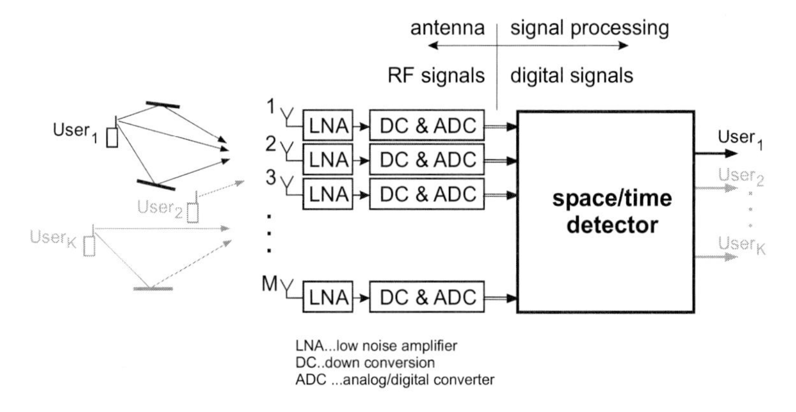

Figure 5.3. A space–time detector combines the spatio-temporal equalization with the detection process. Reproduced from [9] with permission of the author.

The specific choice of the spatio-temporal processing method depends very much on the specific transmission techniques. For unequalized systems, there is only a spatial signal processing, which a priori offers only the possibility of DOA-based or training-sequence-based processing. In both cases, one key duty of the adaptive antenna lies in the reduction of the delay spread. For DOA-based algorithms, the antenna pattern is chosen in such a way that nulls are put in the direction of long-delayed echoes. For diversity combining, the signals from the various antenna elements should be combined in such a way that the minimum mean-square error is minimized, not the signal strength maximized (optimum combining). Diversity combining is described in detail in Chapter 12.

For TDMA and CDMA, the question arises whether to use the (optimum) spatio-temporal processing or to use spatial and temporal processing as separate stages. In the latter case, the techniques described in this book can be applied without modification since a standard equalizer (or Rake receiver) is applied after the spatial processor as described above. This application is independent of whether the adaptive antenna uses DOA-based or training-sequence-based algorithms.

The situation is somewhat more involved when true spatio-temporal reception is used. The generalization of a Rake receiver from purely temporal to spatio-temporal is conceptually the simplest one. Instead of detecting echoes that are distinguished by their separate delays, the Rake now detects echoes that are distinct in delay and direction of arrival. The Rake detects the L strongest echoes and combines them coherently (maximum ratio combining). The higher number of resolved echoes allows us to approximate very closely the performance of an AWGN channel and to null out interferers. The computational effort compared to the separate temporal and spatial processing does not increase dramatically; the most serious problem lies in the channel identification.

Joint space-time equalizers are considerably more complicated than purely temporal equalizers. Abbreviating the number of antenna elements with P, we now require PL taps in the equalizer, and the data matrix used during the training phase is also increased in size by a factor P^2. Reduction of complexity will surely be one major topic of research in the next years.

5.2.2 Multiple-input, multiple-output systems

One exiting development in mobile communications is the advent of systems with smart antennas at both link ends [11], [12], also called MIMO (Multiple-input, multiple-output) systems. They have a much higher channel capacity than do conventional systems; essentially, capacity increases linearly with the number of antenna elements (if there is the same number of elements at transmitter and receiver). By using appropriate coding strategies (so-called space-time codes, see [13]), one can come close to actually realizing these capacities.

Most of the existing investigations are for flat-fading channels; however, recently the performance in frequency-selective channels has drawn more attention. Especially, the combination of OFDM with MIMO systems is under intensive scrutiny, see, e.g., [14], [15], [16], [17]. Although the field is quite new, it is extremely active, and much progress can be anticipated for the next years.

5.2.3 Multiuser detection

An important development of communication theory in the 1980s was the invention of multiuser detection. In contrast to the conventional matched filter, where interference is treated in the same way as noise, the multiuser detector demodulates the signals of the desired user as well as those of the interferers. This allows us to exploit the a priori information about the interfering signal, e.g., constant modulus and finite alphabet. The conceptually simplest multiuser detector is the interference cancellation scheme. The user with the strongest field strength is first detected; this detected signal is then subtracted from the total received signal. Then, the second-strongest user is detected, its signal is subtracted, and so on. This procedure is known as interference cancellation; however, it is suboptimum. An optimum detector is a maximum-likelihood sequence estimation (MLSE, Viterbi algorithm). The trellis for this algorithm has at least $N_{\mathrm{symb}}^{N_{\mathrm{us}}}$ states, where N_{symb} is the alphabet size of the modulation and N_{us} is the number of users. The resulting MSLE thus has to search through a huge number of possible transitions, especially if the time dispersion is large compared to the bit length, and thus leads to a large constraint length of the Viterbi algorithm. For this reason, suboptimum algorithms, especially linear approximations, have become very popular. The reader interested in more details of this subject is referred to the book of Verdu [18].

Bibliography

[1] J. H. Winters, "Smart antennas for wireless systems," *IEEE Personal Comm. Mag.*, vol. Feb. 1998, pp. 23–27, 1998.

[2] L. C. Godara, "Applications of antenna arrays to mobile communications, Part I: Performance improvement, feasibility, and system considerations," *Proc. IEEE*, vol. 85, pp. 1031–1061, 1997.

[3] L. C. Godara, "Applications of antenna arrays to mobile communications, Part II: Beam-forming and direction-of-arrival considerations," *Proc. IEEE*, vol. 85, pp. 1195–1245, 1997.

[4] A. Paulraj and C. Papadias, "Space-time processing for wireless communications," *IEEE Signal Processing Mag.*, vol. 14, no. 6, pp. 49–83, 1997.

[5] A. J. Paulraj and B. C. Ng, "Space-time models for wireless personal communications," *IEEE Signal Processing Mag.*, pp. 49–83, Feb. 1998.

[6] J. S. Thompson, P. M. Grant, and B. Mulgrew, "Smart antennas for CDMA systems," *IEEE Personal Comm. Mag.*, pp. 16–25, Oct. 1996.

[7] J. C. Liberti and T. S. Rappaport, *Smart Antennas for Wireless Communications.* Upper Saddle River, NJ: Prentice Hall, 1999.

[8] H. Krim and M. Viberg, "Two decades of array signal processing research," *IEEE Signal Processing Mag. (Special Issue on Array Processing)*, vol. 13, July 1996.

[9] A. Kuchar, *Real-Time Smart Antenna Processing for GSM1800.* PhD thesis, Vienna, Austria, 2000.

[10] J. Laurila, R. Tschofen, and E. Bonek, "Semi-blind space-time estimation of co-channel signals using least squares projections," in *IEEE-Proceedings of Vehicular Technology Conference*, (Amsterdam), pp. 1310–1315, 1999.

[11] J. H. Winters, "On the capacity of radio communications systems with diversity in rayleigh fading environments," *IEEE J. Selected Areas Comm.*, 1987.

[12] G. J. Foschini and M. J. Gans, "On limits of wireless communications in fading environments when using multiple antennas," *Wireless Personal Comm.*, vol. 6, pp. 311–335, 1998.

[13] V. Tarokh, N. Seshadri, and A. R. Calderbank, "Space-time coding for high data rate wireless communications: Performance criterion and code construction," *IEEE Trans. Information Theory*, vol. 44, pp. 744–765, 1998.

[14] Y. G. Li, N. Seshadri, and S. Ariyavisitakul, "Channel estimation for OFDM systems with transmitter diversity in mobile wireless channels," *IEEE J. Selected Areas Comm.*, vol. 17, pp. 461–471, 1999.

[15] Y. G. Li, J. C. Chuang, and N. R. Sollenberger, "Transmitter diversity for OFDM systems and its impact on high-rate data wireless networks," *IEEE J. Selected Areas Comm.*, vol. 17, pp. 1233–1243, 1999.

[16] D. Agrawal, V. Tarokh, A. Naguib, and N. Seshadri, "Space-time coded OFDM for high data-rate wireless communication over wideband channels," *VTC'98*, vol. 3, pp. 2232–2236, 1998.

[17] H. Boelcskei, D. Gesbert, and A. J. Paulraj, "On the capacity of OFDM-based multi-antenna systems," p. submitted, 2000.

[18] S. Verdu, *Multiuser Detection*. Cambridge, U.K.: Cambridge Univ. Press, 1998.

Part II

Unequalized Systems

Andreas F. Molisch

Chapter 6

WHY UNEQUALIZED SYSTEMS?

In cellular radio, time dispersion (as opposed to flat fading) is usually considered as a positive effect because it can be used for diversity reception: since the multipath components arrive at the receiver at different times, they can be separated and constructively added by a Rake receiver or equalizer. However, in many applications, equalizers are undesireable because of cost and complexity. This part of the book thus concentrates on the *BER* in unequalized systems that use TDMA and/or FDMA as multiple access format.

Such systems have many applications, whose importance have increased dramatically in the past years:

- Cordless telephones: the most important cordless phone systems are PWT (Personal Wireless Telecommunications) and PACS (Personal Access Communications System), the American standards [1], DECT (Digital Enhanced Cordless Telecommunications) [2], the European standard, and PHS (Personal Handyphone System), the Japanese standard. These phones have sold more than 20 million units to date, and the market is still growing. None of these systems were intended to be operated with equalizers.[1]

- Wireless local loop (WLL): cordless phones or similar systems can also be used for wireless local loop applications, obviating the need for expensive cable installation for each home. These applications become of ever greater importance in a liberalized telephone market, where new operators have to establish their links to potential customers. In WLL, the time dispersion encountered is usually larger than in cordless applications because the distances are larger. Field trials based on the DECT system are currently being made in several European cities [3], [4].

- Low-mobility PCS (Personal Communications System) systems: the success of PHS in Japan stems not only from its use as cordless system but also from the

[1] There have been some attempts to perform equalization also in these systems, e.g., by Viterbi equalizers or decision-feedback equalizers. Equalizer principles are covered in Part III of this book.

fact that networks with literally hundreds of thousands of base stations have been built as low-mobility cellular systems. Field trials for similar European systems based on DECT have been made [5]. Since most of the connections are in outdoor environments, the time dispersion is usually larger than in classical cordless applications.

- Cellular systems: the Japanese digital system JDC (Japanese Digital Cellular) also known as PDC Pacific Digital Cellular) does not foresee an equalizer. It is basically a narrowband system, having a channel width of only 25 kHz. However, in some environments, even this system experiences time dispersion whose influence on the BER cannot be neglected.

From this enumeration, it becomes clear that the computation of the BER in time-dispersive environments is of great practical, as well as theoretical, importance. Work has actually been going on since 1963, starting with the classical paper of Bello and Nelin [6]. The main motivation of the work at that time was not cellular systems, but ionospheric communication links. However, the mathematical principles remain the same. One might think that after so many years, the field should long be mature in the sense that all worthwhile information has been found out and published. However, this is not the case. On one hand, the classical problem is analyzed even today, with the aims of finding simpler computation methods, gaining new physical interpretations, and replacing previously used approximations by exact formulations. On the other hand, there is considerable research on how the BER due to the time dispersion of the channel can be reduced by methods other than equalizers and channel coding.

In this Part, we use the following policy in citing conference papers and journal papers in foreign languages: If a paper in an English-language journal has similar contents, we cite only that paper. Furthermore, we cite no unpublished work, inaccessible (or difficult to obtain) documents like internal reports, theses, etc. We also do not cite any work that deals only with (fast or slow) *flat*-fading channels—although this situation was important for older, low-data-rate systems, it is of less interest for the newer systems, and by definition outside the scope of this Part. Finally, the deadline for inclusion in the literature list was set at December 31, 1999, to help in future literature searches.

Bibliography

[1] C. C. Yu, D. Morton, C. Stumpf, R. G. White, J. E. Wilkes, and M. Ulema, "Low-tier wireless local loop radio systems. Part 1: Introduction; Part 2: Comparison of systems," *IEEE Comm. Mag.*, pp. 84–98, March 1997.

[2] E. T. S. I. (ETSI), "Radio equipment and systems: European cordless telecommunications standard interface," *DECT Specifications*, vol. part 1–3, p. version 02.01, 1991.

[3] W. Tuttlebee, *Cordless Telecommunications Worldwide.* London, Springer, 1997.

[4] E. Toivanen, "High quality fixed radio access as a competitive local loop technology," in *Int. Symp. Subscriber Line Services*, pp. 181–186, 1996.

[5] M. Pettersen and R. Raekken, "Dect offering mobility on the local level—experiences from a field trial in a multipath environment," in *Proc. VTC'96*, pp. 829–833, 1996.

[6] P. Bello and B. D. Nelin, "The effect of frequency selective fading on the binary error probabilities of incoherent and differentially coherent matched filter receivers," *IEEE Trans. Comm.*, vol. 11, pp. 170–186, 1963.

Chapter 7

SYSTEM MODEL

In this chapter, we introduce the system model that forms the basis of our considerations in the rest of this part. We outline the mathematical formulation for transmitter, channel, and receiver, in Sections 7.1, 7.2, and 7.3. Finally, we compare co-channel interference to noise.

7.1 TRANSMITTER

7.1.1 Phase Shift Keying

Many important modulation schemes can be described as *pulse amplitude modulation*, where a basis pulse $g(t)$ is multiplied by complex modulation symbols c_m

$$d(t) = \sum_{m=-\infty}^{\infty} c_m g(t - mT) \tag{7.1.1}$$

where T is the symbol duration. Here, $d(t)$ is the signal in the equivalent baseband representation.

For the case of unfiltered M-ary PSK, $g(t)$ is a rectangular pulse of duration T and unit amplitude, and the coefficients are $c_m = \sqrt{2E_s/T}\exp(j\phi_m)$, where E_s is the symbol energy, and $\phi_m = 2\pi m/M$. 4-QAM (often called simply QAM), the most common QAM format, can also be interpreted as 4-PSK. In higher-order QAM schemes, $|c_m|$ can also depend on m and not just on the symbol energy, i.e., in the constellation diagram, the signal points are not all on a circle.

In mobile radio applications, differential encoding and detection of PSK is common, to allow easy elimination of a phase shift by the channel. In that case, we transmit not the absolute phase but the difference between the phase associated with the current and the preceding symbol

$$c_m = \sqrt{2E_s/T}\exp[j(\Delta\phi_m + \beta)] \qquad \Delta\phi_m = \phi_m - \phi_{m-1} \tag{7.1.2}$$

where β is an additional phase shift whose purpose will become clear below.

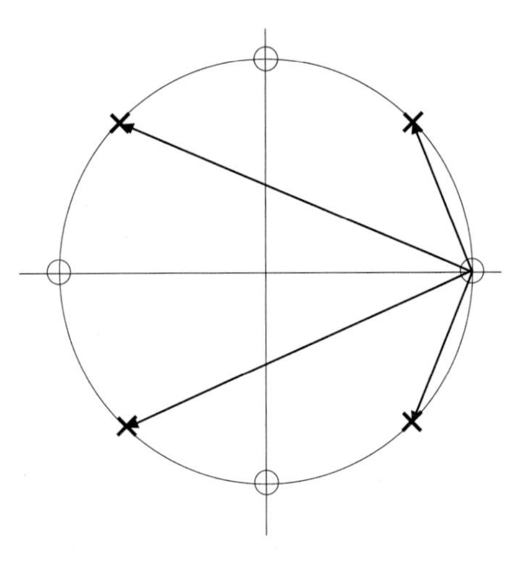

Figure 7.1. Signal space representation of $\pi/4$-DQPSK

The most common differential phase modulation types are DBPSK ($M = 2$, $\beta = 0$), symmetrical DBPSK ($M = 2$, $\beta = \pi/2$), DQPSK ($M = 4$, $\beta = 0$), $\pi/4$-shifted DQPSK ($M = 4$, $\beta = \pi/4$), and 8-DPSK. Among these, $\pi/4$-shifted DQPSK occupies a unique position because it is used in a large number of current systems: IS-54, IS-136 (American cellular standards), PHS, PACS and PWT, and TETRA (Trans-European Trunk RAdio, the European trunk radio standard). Figure 7.1 shows the signal constellation diagram. The major advantage of the $\pi/4$ phase shift is that no signal transition goes through the origin; in other words, the envelope of the signal shows no nulls. This gives advantages in the hardware transmitter implementation.

When the spectrum is restricted, it is preferable not to use rectangular basis pulses, but instead to use pulses that show a steeper decay in the spectrum. For such applications, Nyquist pulses are popular because they avoid ISI in slow flat-fading channels. The most common of the Nyquist filters is the raised-cosine filter, defined by

$$H_N(f) = \begin{cases} T & 0 \le |f| \le (1-\alpha)\cdot\frac{1}{2T} \\ \frac{T}{2}\cdot\left(1-\sin\left(\frac{\pi T}{\alpha}\left(|f|-\frac{1}{2T}\right)\right)\right) & (1-\alpha)\cdot\frac{1}{2T} \le |f| \le (1+\alpha)\cdot\frac{1}{2T} \\ 0 & (1+\alpha)\cdot\frac{1}{2T} \le |f|\,. \end{cases} \tag{7.1.3}$$

The roll-off parameter α determines how fast the decay of the spectrum is. For most practical applications, α lies between 0.2 and 1.0.

When the system uses matched-filter reception, receiver and transmitter filter must have the same transfer function, namely, $\sqrt{H(f)}$. If the overall transfer function is a raised-cosine shape $H_N(f)$, such filters are known as square-root raised cosine filters. The impulse response for the raised-cosine filter is

$$g(t) = \frac{\sin[\pi(1-\alpha)t/T] + 4(\alpha t/T)\cos[\pi(1+\alpha)t/T]}{(\pi t/T)[1-(4\alpha t/T)^2]}. \tag{7.1.4}$$

The impulse response for square-root raised cosine filters is rather involved, but can also be found in Ref. [1].

7.1.2 Frequency Shift Keying

The pulse amplitude modulation (7.1.1) can be generalized to the multipulse transmission scheme [2] where the transmitted signal is

$$d(t) = \sum_{m=-\infty}^{\infty} g_{c_m}(t - mT), \tag{7.1.5}$$

i.e., where we have several different pulses at our disposal and the choice of which of these is transmitted is determined by the symbol c_m.

For M-ary FSK, we have M signals whose frequencies are shifted by an amount $f_{\text{mod}} = m\Delta f/2$ from the carrier, where $m = \pm 1, \pm 3, ... \pm (M-1)$. Usually, Δf is chosen in such a way that all signals are orthogonal to each other. We thus have a multipulse transmission scheme with $g_{c_m}(t) = \exp(j2\pi m\Delta f t/2) \cdot \text{rect}(t)$, where $\text{rect}(t)$ is the rectangular pulse. To make the spectrum decay more quickly, we can also use some other (e.g., Nyquist) pulse; however, doing so influences orthogonality.

The situation is somewhat different for continuous-phase FSK (CPFSK). In this case, the multipulses are put together in such a way that there is no discontinuity in the phase (which results in better spectral efficiency). The amplitude is usually chosen to be constant; for this case, the phase $\Phi(t)$ of the signal $d(t)$ can be described by

$$\Phi(t) = 2\pi h_{\text{mod}} \sum_{m=-\infty}^{\infty} c_m \int_{-\infty}^{t} \tilde{g}(u - iT)du. \tag{7.1.6}$$

h_{mod} is the modulation index ($h_{\text{mod}} = 0.5$ for minimum shift keying MSK), and $\tilde{g}(t)$ is the phase-shaping pulse. For Gaussian MSK (GMSK), $c_m = \pm 1$ and $\tilde{g}(t)$ is [3]

$$\tilde{g}(t) = \frac{1}{4T}\left[erfc\left(\frac{2\pi}{\sqrt{2\ln(2)}}B_G T\left(-\frac{t}{T}\right)\right) - erfc\left(\frac{2\pi}{\sqrt{2\ln(2)}}B_G T\left(1-\frac{t}{T}\right)\right)\right] \tag{7.1.7}$$

where $erfc(\cdot)$ is the complementary error function and B_G is the bandwidth of the Gaussian filter; the pulses are normalized to $\int_{-\infty}^{\infty} \tilde{g}(t)dt = 1/2$. GMSK is the most important FSK modulation format, being used in the GSM/DCS1800/PCS1900 (worldwide cellular standards) and DECT systems (strictly speaking, the DECT standard specifies GFSK with $h_{\text{mod}} = 0.5$ plus/minus tolerances).

7.2 CHANNEL

The transmitted signal passes through the mobile radio channel, which distorts it and adds noise. The relation between the received and the transmitted signal is given by

$$r(t) = \int h(t,\tau)d(t-\tau)\mathrm{d}\tau + n(t). \tag{7.2.1}$$

Note that generally, $h(t,\tau)$ is the impulse response of a time-varying channel, as discussed in Part I, Chapter 3. It can be interpreted in a deterministic or stochastic way. However, for the remainder of this Part, we assume that the channel is slowly time varying if not stated otherwise. In other words, it can be assumed to be static both for the duration of a transmitted symbol and for the length of the impulse response (the latter assumption corresponds to the *underspread* condition). For the stochastic interpretation of the channel, we also assume that WSSUS is valid and that the fading statistics follow a Rayleigh or Rice distribution if not stated otherwise.

The noise added by the channel is additive white Gaussian noise if not stated otherwise. This is the standard assumption in the literature, even though the man-made noise often exhibits non-Gaussian characteristics. For all our considerations, a possible receiver filter will be shifted to the transmitter. This is admissible when the channel varies only during times that are much larger than the symbol duration—a condition that is fulfilled for all high-data-rate systems. Furthermore, the properties of the noise then have to modified when such a shift is done; the noise added by the channel is not white anymore, but colored (with noise spectral density $N_0 \left|H_R(f)\right|^2$, where $H_R(f)$ is the receiver filter transfer function).

7.3 RECEIVER

7.3.1 Coherent and noncoherent demodulation

The optimum coherent demodulation in an AWGN channel for memoryless modulation (i.e., *not* for CPFSK) is, as is well known, the matched filter receiver, see, e.g., [4]. In this method, the received signal is sent through a bank of M filters, each of which has as impulse response $g^*_{c_m}(T-t)$. The symbol with index m corresponding to the filter that generates the largest output is chosen as the most likely symbol to have been sent. An equivalent reception method based on cross-correlators is possible. The matched-filter reception requires complete phase recovery.

A simpler method of reception is the *noncoherent receiver*, where (for orthogonal signals) the output from each filter is sent through an envelope detector and the signal that gives the largest envelope is chosen. For mathematical convenience, often not the envelope, but its square is assumed to be detected.

Coherent detection of CPFSK is useful mainly when the additional information that is contained in the continuity of the phase is exploited in some way. For MSK, e.g., we know that when the previously transmitted bit corresponds to the point $(1,0)$ in a signal constellation diagram, then the next bit must correspond to either $(0,1)$ or $(0,-1)$.

7.3.2 Differential detection of PSK and CPFSK

Differential detection is a simple and effective method to eliminate phase shifts introduced by the channel. In the simplest case of single-bit differential detection,[1] only the phase difference between the signal received at two subsequent sampling instants is analyzed. Since the decision is based on that phase difference, a phase shift by the channel that is constant within one symbol duration is eliminated by that method.[2]

The most straightforward implementation of the differential detector for PSK and CPFSK is the product detector. For the detection of PSK, we compute the quantities

$$q_m = \mathrm{Re}\left\{r(t_s)r^*(t_s - T)\exp\left(-j\Delta\phi_m\right)\right\} \tag{7.3.1}$$

for each possible nominal value $\Delta\phi_m$ and select the one with the largest q_m. Here, $r(t)$ is the received signal and t_s is the sampling instant; a more thorough discussion on how to find this quantity is given at the beginning of Chapter 8.

7.3.3 Frequency-discriminator detection of (CP)FSK

For frequency-discriminator detection, we determine the instantaneous frequency of the received signal.[3] The instantaneous frequency can be computed as

$$f_{\mathrm{inst}} = \frac{\mathrm{Im}\left(r^*(t_s)\frac{\mathrm{d}r(t)}{\mathrm{d}t}|_{t=t_s}\right)}{|r(t_s)|^2}. \tag{7.3.2}$$

If we have binary FSK, then the decision can be based on the *sign* of the instantaneous frequency, so that the decision basis can be also written as

[1]When we speak of differential detection, will henceforth assume single-bit differential detection when not explicitely stated otherwise. Some special techniques for differential detection are also treated in Chapter 10 and 11.

[2]For MSK, [5] has suggested the use of the phase difference of signals separated not by one-bit duration but by two-bit durations, so that essentially $\cos(\phi(t) - \phi(t - 2T))$ is detected; results for this methods are briefly given in Chapter 9. A more general version of that method is multiple-symbol differential detection [6]; however, this topic is be treated further here.

[3]We shift the receiver filter to the transmitter as discussed in Section 7.2. We stress, however, that this is possible only for filters that are situated *in front of the discriminator*. A filter after the discriminator cannot be shifted because the discriminator is a nonlinear device.

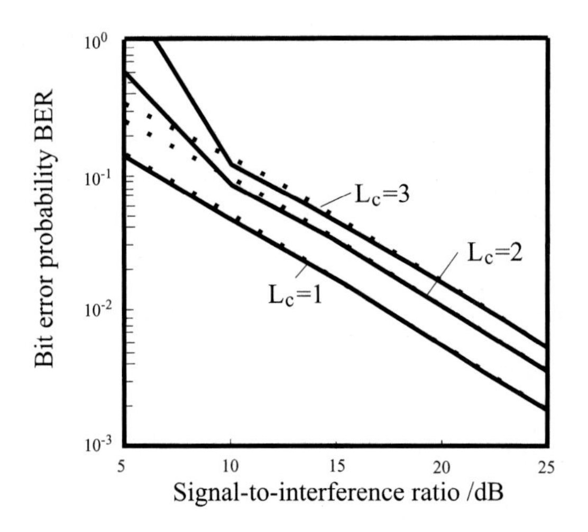

Figure 7.2. Exact (solid) and approximate (dashed) BERs of GMSK with $B_G T = 0.3$ in flat Rayleigh-fading environment with L_c co-channel interferers; limiter-discriminator detection; SNR = 40 dB. Adapted from [7], copyright IEEE.

$$\mathrm{Re}\left(r^*(t_s)\frac{\mathrm{d}r(t)}{\mathrm{d}t}|_{t=t_s}\exp\left(-j\frac{\pi}{2}\right)\right). \tag{7.3.3}$$

This is the same form that occurs for differential detection of PSK.

7.4 TREATMENT OF CO-CHANNEL INTERFERENCE

The performance of mobile radio systems is often influenced more by co-channel interference than by noise, especially in urban and indoor environments. It is common to model the interference as zero-mean Gaussian process, which can then be treated in exactly the same way as noise. This is, however, an approximation in two respects: (i) the interferers might also have line-of-sight to the receiver; this is, however, a rare case in practice; (ii) there must be a large number of interferers with independent bit sequences if the sum signal should show noiselike behavior. Also, co-channel interference is not spatially white (which is important for smart antennas) and might be eliminated by multiuser detectors due to its inherent structure. Still, in most practical cases the Gaussian approximation is quite well fulfilled, as was shown, e.g., in [7]; see also Figure 7.2. In the following chapters, we always use this approximation, as did most other authors in the field, and we no longer differentiate between noise and interference.

Bibliography

[1] S. Chennakeshu and G. Saulnier, "Differential detection of pi/4-shifted DQPSK for digital cellular radio," *IEEE Trans. Vehicular Techn.*, vol. 42, pp. 46–57, 1993.

[2] E. A. Lee and D. G. Messerschmitt, *Digital Communication.* Boston, Kluwer Academic, 1st ed., 1988.

[3] A. Yongacoglu, D. Makrakis, and K. Feher, "Differential detection of GMSK using decision feedback," *IEEE Trans. Comm.*, vol. 36, pp. 641–649, 1988.

[4] J. G. Proakis, *Digital Communications.* New York, McGraw Hill, 3rd ed., 1995.

[5] M. K. Simon and C. C. Wang, "Differential detection of Gaussian MSK in a mobile radio environment," *IEEE Trans. Vehicular Techn.*, vol. 33, pp. 307–320, 1984.

[6] D. Divsalar and M. K. Simon, "Multiple-symbol differential detection of MPSK," *IEEE Trans. Comm.*, vol. 38, pp. 300–308, 1990.

[7] C. Tellambura and V. Bhargava, "Performance of GMSK in frequency selective Rayleigh fading and multiple cochannel interferers," in *Proc. VTC'95*, pp. 211–215, 1995.

Chapter 8

COMPUTATION METHODS FOR FIXED SAMPLING

When we speak of fixed sampling, we define that the sampling happens at time instants $t'_s = kT + t_s$, where $k = 0, 1, 2, \ldots$ and t_s is a quantity that does not change during the interval $-\infty < t < \infty$; in the following sections, we usually do not distinguish between t'_s and t_s and call them both *the sampling time*. When we have coherent, noncoherent, or differential detection, t_s is typically either equal to the minimum excess delay or to the average mean delay (see Chapter 3). When we have frequency-discriminator detection, t_s is shifted by $T/2$ with respect to the minimum excess delay or average mean delay. Theoretically, however, all values of t_s are possible.

8.1 GENERAL CONSIDERATIONS

8.1.1 Averaging over symbol sequences

Due to the filtering and the time dispersion of the channel, the decision $\widehat{b}_k$ for one received bit is not only influenced by the transmitted bit b_k but also by preceding and following bits. Theoretically, the ISI of, e.g., a Gaussian filter lasts until infinity; in practice, however, the overwhelming part of the ISI energy is concentrated within a few symbols. It is thus justified in practice to say that N_{sym} transmitted symbols influence a decision for one received symbol, where N_{sym} is a finite number, usually on the order of 3 to 7. To compute the average BER, we then have to compute the BER for each of the possible $M^{N_{\mathrm{sym}}}$ symbol combinations, and then average over all these combinations. We usually assume that all symbol sequences are equally likely.

There might be some cases (for high-order modulation formats) where averaging over all possible bit combinations is still too computationally expensive. In that case, an evaluation with the help of the pdf of the intersymbol interference might prove helpful; for such methods, see [1] and references therein.

8.1.2 Canonical receiver analysis

Let us now consider a canonical receiver model for the different modulation formats, following [2] and illustrated in Figure 8.1. The received signal is first bandpass-filtered and then split to get the in-phase and quadrature-phase signals. The two signals are then multiplied by a reference signal (for the quadrature component, this reference signal is phase-shifted by $\pi/2$). The resulting signals are then low-pass filtered and serve as input for the phase decision device. The signals $r_I(t)$ and $r_Q(t)$ are zero-mean Gaussian random variables if the channel is Rayleigh fading, and finite-mean-value Gaussian variables if the channel is Rice fading. The above receiver structure describes both coherent detection (CD) and differential phase detection (DPD). In the former case, the reference signal is a local oscillator derived from some carrier recovery circuit; it can be either noisy or non-noisy [3], [4]. For DPD, on the other hand, the reference signal is simply the received signal delayed by one symbol period. Generalizing this result, we also propose that limiter-discriminator detection can be treated in such a way for binary FSK: in that case, the reference signal is the derivative of the received signal.

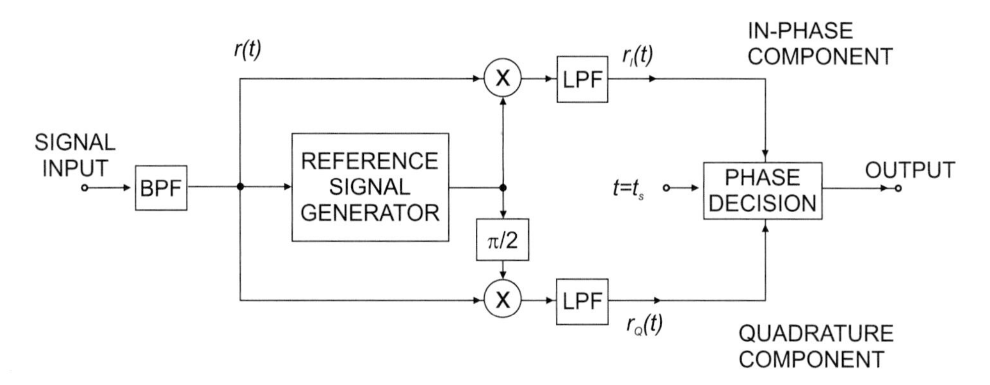

Figure 8.1. Canonical receiver model. Adapted from [2], copyright IEEE.

The next step is the determination of the decision region in the signal constellation diagram. The ideal signal points for PSK are equally spaced at $2\pi i/M$. The symbol error probability is now given as [2]

$$P_s = \begin{cases} P_M & M = 2 \\ 2P_M - P_M^2 & M = 4 \\ 2P_M & M = 8, 16 \end{cases} \tag{8.1.1}$$

where P_M is the probability that the detected signal vector falls into region I or III in Figure 8.2.

The next step is the rotation of the error region by an amount of $(\pi/2) - \phi_{\text{err\_reg}}$, as can be seen from Figure 8.3. This rotation ensures that the boundary between

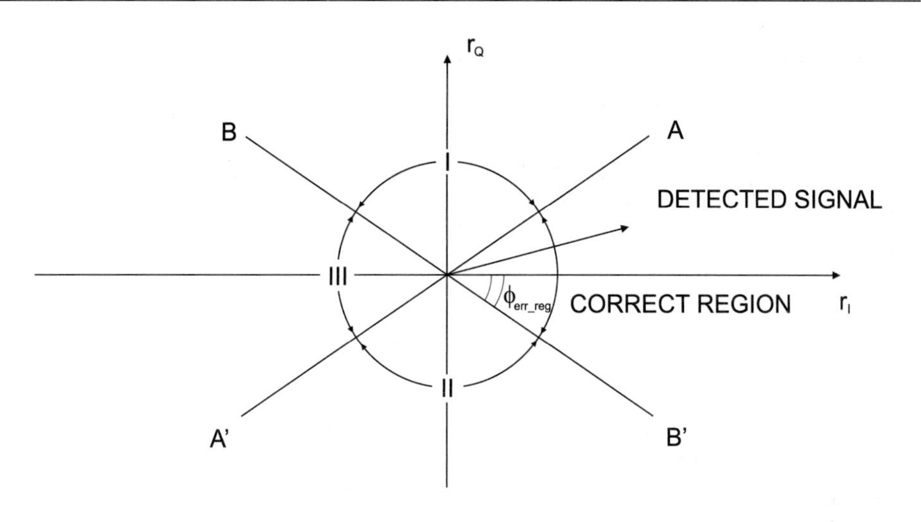

Figure 8.2. Detected signal diagram. Adapted from [2], copyright IEEE.

correct and wrong decisions is the y-axis; we will see later that this boundary is useful for the so-called QFGV technique (see Section 8.3). The conversion from the symbol-error probability to the bit-error probability can be done by the equations that are well known from the AWGN case [5], [6].

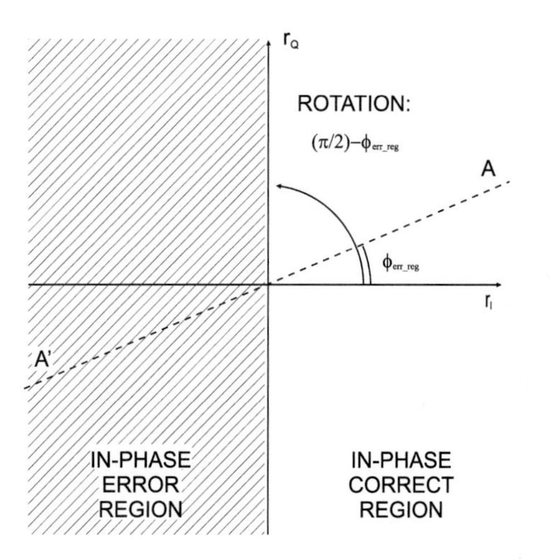

Figure 8.3. Rotation of error region. Adapted from [2], copyright IEEE.

8.1.3 Correlation properties of the received signal

General correlation properties of the received signal [1]

For many of the computation methods, we will need correlation coefficients between distorted, noisy signals in WSSUS channels. For differential detection, for example, we see later that we need $E\{r(t)r^*(t-T)\}$, where $r(t)$ is the received signal. To compute this value, we proceed along the following lines: the received signal can be written as

$$r(t) = \int h(t,\tau)d(t-\tau)\mathrm{d}\tau + n(t). \tag{8.1.2}$$

Replacing the time-variant impulse response by the spreading function $s(\nu,\tau)$ (see Chapter 3), this becomes

$$r(t) = \int\int s(\nu,\tau)\exp(+j2\pi\nu t)d(t-\tau)\mathrm{d}\tau\mathrm{d}\nu + n(t). \tag{8.1.3}$$

The expected value of $r(t)r^*(t-T)$ becomes then

$$\begin{aligned} E\{r(t)r^*(t-T)\} = \int\int\int\int & E\{s(\nu,\tau)s^*(\nu',\tau')\}\exp[+j2\pi\nu t]d(t-\tau) \\ & \exp[-j2\pi\nu(t-T)]d^*(t-T-\tau')\mathrm{d}\tau'\mathrm{d}\nu'\mathrm{d}\tau\mathrm{d}\nu \\ + E\{\int\int & N(f)\exp(j2\pi ft)H_R(f)N^*(f') \\ & \exp(-j2\pi f'(t-T))H_R^*(f')\}\mathrm{d}f\mathrm{d}f'\} \end{aligned} \tag{8.1.4}$$

where $H_R(f)$ is the transfer function of the receiver filter and $N(f)$ is the noise spectrum. The quadruple integral can be simplified because in a WSSUS channel, $E\{s(\nu,\tau)s^*(\nu',\tau')\} = \xi_s(\nu,\tau)\delta(\tau-\tau')\delta(\nu-\nu')$ and $\int \xi_s(\nu,\tau)\exp(j2\pi\nu T)\mathrm{d}\nu = \xi_h(T,\tau)$, so that we finally get

$$E\{r(t)r^*(t-T)\} = \int_{-\infty}^{\infty} \xi_h(T,\tau)d(t-\tau)d^*(t-\tau-T)\mathrm{d}\tau + 2B_nN_0\xi_n(T) \tag{8.1.5}$$

where $\xi_n(t)$ is the noise correlation function

$$\xi_n(t) = \frac{1}{B_n}\int_{-\infty}^{\infty} |H_R(f)|^2\exp(j2\pi ft)\mathrm{d}f \quad \text{and} \quad B_n = \int_{-\infty}^{\infty} |H_R(f)|^2\mathrm{d}f. \tag{8.1.6}$$

N_0 is the single-sided noise power spectral density; note that for white noise [7],

[1] Note, in this subsection, we assume a time-variant, i.e., not necessarily slow-fading, channel.

$$E\{N(f)N^*(f')\} = 2N_0\delta(f - f'). \tag{8.1.7}$$

Similarly, the autocorrelation coefficients, $E\{r(t)r^*(t)\}$, $E\{r(t-T)r^*(t-T)\}$ can be computed.

Correlation coefficients for differential detection. We need the correlation coefficients of the signals at the two sampling times, $X = r(t_s)$, $Y = r(t_s - T)$. It follows from the previous subsection that

$$\begin{aligned}
\mu_{xx} &= \frac{1}{2}\int_{-\infty}^{\infty} \xi_h(0,\tau)|d(t_s-\tau)|^2 \mathrm{d}\tau + B_n N_0 \qquad (8.1.8)\\
\mu_{yy} &= \frac{1}{2}\int_{-\infty}^{\infty} \xi_h(0,\tau)|d(t_s-\tau-T)|^2 \mathrm{d}\tau + B_n N_0\\
\mu_{xy} &= \frac{1}{2}\int_{-\infty}^{\infty} \xi_h(T,\tau)d(t_s-\tau)d^*(t_s-\tau-T) \mathrm{d}\tau + B_n N_0 \xi_n(T)
\end{aligned}$$

where μ_{xx} is the second central moment, $\mu_{xx} = \frac{1}{2}E\{(X-\overline{X})(X-\overline{X})^*\}$, etc. We also define the normalized correlation coefficient

$$\rho = \rho_c + j\rho_s = \frac{\mu_{xy}}{\sqrt{\mu_{xx}\mu_{yy}}}. \tag{8.1.9}$$

If the channel is Rician fading, the channel impulse response is a complex Gaussian process with mean a_{LOS}. It is then often advantageous to introduce a normalized notation.The auto-covariance function is written as

$$\begin{aligned}
&E\left\{[h(t,\tau) - a_{\mathrm{LOS}}]\left[h(t+\Delta t,\tau') - a_{\mathrm{LOS}}\right]^*\right\} \qquad (8.1.10)\\
&\quad = \xi_h(\Delta t,\tau)\delta(\tau-\tau')\\
&\quad = P_d\widetilde{\xi_h}(\Delta t,\tau)\delta(\tau-\tau')
\end{aligned}$$

where P_d gives the attenuation of the power in the diffuse path ($= 2\sigma^2$) and $\widetilde{\xi_h}$ is normalized so that $\int \widetilde{\xi_h}(0,\tau)d\tau = 1$. P_d is related to the Rice-factor K_{Rice} as $K_{\mathrm{Rice}} = a^2_{\mathrm{LOS}}/P_d$. The transmitted signal is written as $d(t) = \sqrt{2E_s/T}\widetilde{d}(t)$. We then have the relations

$$\mu_{xx} = \frac{P_d E_s}{T} \int_{-\infty}^{\infty} \widetilde{\xi}_h(0,\tau) |\widetilde{d}(t_s - \tau)|^2 \mathrm{d}\tau + B_n N_0 \tag{8.1.11}$$

$$\mu_{yy} = \frac{P_d E_s}{T} \int_{-\infty}^{\infty} \widetilde{\xi}_h(0,\tau) |\widetilde{d}(t_s - \tau - T)|^2 \mathrm{d}\tau + B_n N_0$$

$$\mu_{xy} = \frac{P_d E_s}{T} \int_{-\infty}^{\infty} \widetilde{\xi}_h(T,\tau) \widetilde{d}(t_s - \tau) \widetilde{d}^*(t_s - \tau - T) \mathrm{d}\tau + B_n N_0 \xi_n(T)$$

$$\overline{X} = a_{\mathrm{LOS}} \sqrt{\frac{2E_s}{T}} \widetilde{d}(t_s)$$

$$\overline{Y} = a_{\mathrm{LOS}} \sqrt{\frac{2E_s}{T}} \widetilde{d}(t_s - T).$$

For CPFSK, we proceed in a completely analogous way: we rotate the product $r(t)r^*(t-T)$ in the complex plane by an amount that corresponds to ϕ_{rot}, the phase shift that would occur for ideal transmissions. The complex number describing this rotation is henceforth denoted as $C = \exp(-j\phi_{\mathrm{rot}})$; for MSK, for example, $C = \exp(-j\pi/2)$. The equations for the correlation coefficients remain unchanged.

Correlation coefficients for frequency-discriminator detection. For other detection methods, correlations between other signal components are required, but the basic method stays the same. For frequency-discriminator detection, the (correlated) variables are now not the sampling values at two subsequent time instants, but the sample value at one instant and its derivative. The correlation coefficients for these two variables (for Rayleigh fading) are given in [7] as

$$\mu_{xx} = \frac{1}{2} \int_{-\infty}^{\infty} \xi_h(0,\tau) |d(t_s - \tau)|^2 d\tau + B_n N_0, \tag{8.1.12}$$

$$\begin{aligned} \mu_{yy} = \frac{1}{2} \int_{-\infty}^{\infty} \Bigg[& \xi_h(0,\tau) |\frac{\mathrm{d}}{\mathrm{d}\tau} d(t_s - \tau)|^2 \\ & -2\,\mathrm{Im}\left(\frac{\mathrm{d}\xi_h(t,\tau)}{\mathrm{d}t}|_{t=0} \right) \mathrm{Im}\left(d(t_s - \tau) \frac{\mathrm{d}}{\mathrm{d}\tau} d^*(t_s - \tau) \right) \\ & - \frac{\mathrm{d}^2 \xi_h(t,\tau)}{\mathrm{d}t^2}|_{t=0} \cdot |d(t_s - \tau)|^2 \Bigg] \mathrm{d}\tau - B_n N_0 \frac{\mathrm{d}^2 \xi_n(t)}{\mathrm{d}t^2}|_{t=0} \end{aligned} \tag{8.1.13}$$

$$\begin{aligned} \mu_{xy} = \frac{1}{2} \int_{-\infty}^{\infty} \Bigg[& \frac{\mathrm{d}\xi_h(t,\tau)}{\mathrm{d}t}|_{t=0} \cdot |d(t_s - \tau)|^2 \\ & -\xi_h(0,\tau) d(t_s - \tau) \frac{\mathrm{d}}{\mathrm{d}\tau} d^*(t_s - \tau) \Bigg] \mathrm{d}\tau + B_n N_0 \frac{\mathrm{d}\xi_n(t)}{\mathrm{d}t}|_{t=0} \end{aligned} \tag{8.1.14}$$

from which we can again compute the normalized correlation coefficient ρ that we need later for the BER computations.

8.2 MONTE CARLO (MC) SIMULATIONS

8.2.1 Computation aspects

A Monte Carlo simulation is just a computer implementation of the physical process that happens also in the real world. A random bit sequence is chosen, sent through the modulator, and this signal is then put into the channel model. The amplitudes and phases of the signal components created by the various channel echoes are chosen at random from the appropriate pdfs, and LOS components and AWGN noise are added. The resulting signal is sent into the receiver filter and detector, and a decision is made concerning which bits were sent. This bit sequence is then compared to the originally transmitted sequence, and the number of errors is counted.

In most cases, the Doppler spread is much smaller than the symbol rate, in other words, the time during which the channel changes significantly is much longer than the symbol duration (slow fading). In that case, MC simulations can be considerably speeded up by the following procedure. (i) We determine how many symbols can influence one decision. This number of symbols, called N_{symbol}, is determined by the maximum excess delay of the channel and the filtering at transmitter and receiver. (ii) We select at random one channel configuration. (iii) We send all possible symbol sequences of length N_{symbol} over this channel and compute the error probability for each.

If we are just interested in the error floor, computation of the error probability for each sequence is straightforward (it can only be either 0 or 1). If noise is included, then we compute the error probability including noise for the given channel configuration (for coherent detection, this error probability is a Q-function with an appropriately modified distance between the signal points in the signal constellation diagram; for differential detection, this error probability can be computed according to the ABGV; see Section 8.4 and Appendix A of Part II). This procedure considerably reduces the computation time, because the statistical averaging over noise samples is eliminated [8].[2]

When comparing Monte Carlo simulations to analytical solutions, we find the MC simulations are much easier to implement and can easily be adapted to various nonidealities that are practically impossible to include in analytical approaches. The big advantage of the analytical solution is that once the solution is computed, the influences of the various parameters can be clearly seen, and we do not have to start a new, lengthy computation every time a parameter changes. Both approaches thus have philosophies that can be well justified.

[2]Actually, the combination of measured impulse responses with BER computation by the Q-function has no Monte Carlo, i.e., random elements, in it anymore. However, we mention it here because the basic philosophy is similar.

8.2.2 Bibliographical notes

The first paper to use MC simulations for error floor computations is the well-known paper by Chuang [9] and its sequel [10]. He analyzed the influence of small delay spread on the error floor for various modulation formats, various detection methods (coherent, differential, sampling on the instantaneous or the average mean delay). He used two-delay channels, and also applied measured-delay power profiles. For GTFM (Generalized Tamed Frequency Modulation) with discriminator detection, simulations were done in [11]. Reference [12] reported similar investigations for pilot-symbol-assisted modulation.

Fung et al. [13] used measured delay power profiles for MC simulations of error probabilities of $\pi/4$-DQPSK. The most important distinction of the measured channels as compared to the two-delay model is that in the measured situations, the delays of the multipath components are not constant, but vary (slowly) with time. This can have a considerable effect on the BER. Reference [14] compared the error probability of GMSK and $\pi/4$-DQPSK by MC simulations with COST 207 standard channel models, whereas [15] used measured profiles to compare the sensitivity of various modulation formats to delay-spread effects. Lopes [16] used MC simulations to determine the probability of CRC (Cyclic Redundancy Check) failures in DECT systems (also for diversity).

8.3 QUADRATIC FORM OF GAUSSIAN VARIABLES (QFGV)

The most widespread analytical method is based on quadratic forms of Gaussian variables, so we call it the QFGV method. It is based on the classical papers of Bello and Nelin [3], [17] and Proakis [18], which in turn rely on a paper by Turin [19].

8.3.1 Formulations

The essential idea is to reformulate the condition for the occurence of an error as

$$BER = P\{D < 0\} \tag{8.3.1}$$

where D is a quadratic form

$$D = (X^* Y^*) \begin{pmatrix} A & C \\ C^* & B \end{pmatrix} \begin{pmatrix} X \\ Y \end{pmatrix} = A|X|^2 + B|Y|^2 + CXY^* + C^*X^*Y$$

where A and B are real constants, C is a complex constant, and X and Y are complex Gaussian random variables (they need not be zero-mean); solutions for $P\{D < 0\}$ are given in Appendix A. The computation of the error probability is thus practically solved if we can represent the decision problem at the receiver in a form $D < 0$, and we can compute the mean values and correlation coefficients of the Gaussian variables, as we did for several detection methods in Section 8.1.3. Considering the canconical receiver analysis in the previous section, we saw, e.g., that by setting $C = j\exp(-j\phi_{\mathrm{err\_reg}})$, we have reduced the problem to a form that can be treated by the QFGV approach.

Depending on the detection type, we have either $C = 0$ or $A = B = 0$ (see Section 9.2 and Appendix A) [3].

For the case of pure Rayleigh fading and binary FSK, a much simpler representation can be given [20]. The starting point is the conditional error probability (for given $R = |r(t_s)|$), which is

$$BER(R) = \frac{1}{2} erfc\left(\frac{b_0 \operatorname{Im}\{\rho\}}{\sqrt{1-|\rho|^2}} \frac{R}{\sqrt{2\mu_{xx}}}\right) \tag{8.3.2}$$

where b_0 is the bit that should actually be transmitted. This error probability must then just be averaged over the Rayleigh distribution of the received envelope.

The QFGV method can also be seen as an extended version of Stein's "unified analysis," see [4], [21], who considered the problem $P(D < 0)$, but with

$$D = |X|^2 - |Y|^2 \tag{8.3.3}$$

and $\mu_{xy} = 0$. This theory was used in [22], results obtained in this reference are a special case of the results of [20] presented in Chapter 9. For two-bit differential detection, the theory was modified in [23].

Finally, a method by Biglieri et al. [24], [25] in a sense generalizes the QFGV method. Again, the starting point is the fact that we want to compute the probability that some variable D is smaller than zero. We assume furthermore that the Laplace transform of the pdf of D, $\Pi_D(s)$ is known. The error probability can then be computed as inverse Laplace transform

$$P(D < 0) = \frac{1}{2\pi j} \int_{\varkappa - j\infty}^{\varkappa + j\infty} \Pi_D(s) \frac{1}{s} \mathrm{d}s. \tag{8.3.4}$$

This integral can then be evaluated by various function-theoretical methods, e.g., the saddle-point method. Reference [24], however, suggested a variable transform that reduces the infinite integral to an integral over the range $[-1, 1]$, which is then evaluated by application of Gauss-Chebyshev quadrature. The method has the advantage that D need not to be a quadratic form of Gaussian variables; in fact, it can be an arbitrary decision variable. On the other hand, the computation of the Laplace transform of the pdf of D, which is necessary before we can apply (8.3.4), might constitute the biggest part of the problem.

8.3.2 Bibliographical notes

An application of the QFGV method can be found in [26], which combined this computation method with measured scattering functions, and in various papers on diversity (see Chapter 12). An approach similar to Proakis' formulation is computing the pdf of the interference terms and the noise by using their characteristic functions, [27], [28]. Reference [29] used this method to also include co-channel interferers. Reference [30] used a closely related method to analyze the BER of 16-DAPSK, and [31] applied it to MPSK. The method was also applied in [32] for MSK and OQPSK; for DPSK in [33] (where also the influence of shadowing, co-channel interferers, etc., was included) and [34]; for binary FSK in [35].

Reference [36] developed a method for coherent detection of M-ary PSK that can be considered as generalization of the Proakis' method. This approach reduces to the Proakis method in the special case of BPSK. It is essentially based on computing the conditional density of the phase; however, it requires in the general case the evaluation of multiple integrals. Furthermore, the authors of that reference defer many of the details to a future paper, so currently a detailed assessment of their method is not possible.

References [37] and [38] derived results similar to those of [7] for frequency-discriminator detection and differential phase detection, respectively. There are, however, some important differences in the methodology. References [37] and [38] assume a two-delay model with a single co-channel interferer. They first compute the pdf of the received frequency (for each possible bit combination). They then integrate over this pdf from 0 to ∞ to arrive at the total probability of error. A method that is also similar to that of [7] was developed in [39] for coherent detection. The decision variable conditioned on a certain channel constellation can be written as a form involving Gaussian variables plus some non-Gaussian terms. However, the exact evaluation requires numerical integration.

8.4 ANGLE BETWEEN GAUSSIAN VECTORS (ABGV)

One of the key papers for the computation of error probabilities with differential detection is [40]. In this paper, Pawula et al. solve the following mathematical problem: given two (deterministic) vectors, each of which is perturbed by (possibly correlated) complex Gaussian noise, what is the pdf of the angle between the two resulting vectors? The general solution, following the approach of [40], is given in Appendix A.

The applicability of the ABGV approach to mobile radio problems stems from the fact that the noise components need not represent just noise but can describe Rayleigh-fading signals as well, since they are complex Gaussian processes.[3] The only parameters that have to be determined are the correlation coefficients ρ_c and ρ_s, which can be computed in essentially the same way as in Section 8.1.3. In the

[3]The difference from noise is that they are signal dependent and must thus be computed separately for different signals.

literature, the ABGV formulation has been used mainly for special models of the mobile radio channel, namely the satellite channel, in [41] and a series of papers by Korn [42], [43], [44], [45], [46], [47], [48], [49], [50], [51]. Modifications for an arbitrary Rayleigh-fading channel are given in [52].[4] The derivations of the equations in those papers is rather more complicated than for the QFGV method, and thus is not reproduced here; the complications seems to arise mainly from the fact that the computation of the correlation coefficients and the actual computation of the error probability are intermingled, whereas these steps were done separately for the QFGV. It would seem, however, also possible to use the correlation coefficients computed in Section 8.1 for a more straightforward application of ABGV method; see [55].

Generally, when we compare the QFGV and ABGV methods, we find that from the point of view of the computational effort, the QFGV method has the advantage that it requires one numerical integral that depends on two parameters (the Marcum Q function), whereas the ABGV method requires an integral that depends on *three* parameters. This advantage can be an important aspect when a tabulation of values is to be used. On the other hand, the ABGV method gives more information, namely, the true pdf of the angles and not just the information that a value is smaller or larger than zero. This knowledge is of great importance when we want to compute M-ary PSK in an *exact* way.

8.5 CORRELATION-MATRIX-EIGENVALUE METHOD

References [56] and [57] developed a method for the computation of the error probability with coherent detection when the channel is modelled as an N-delay, Rayleigh-fading channel with possible correlations. We write the Euclidean distance between two signals as

$$d_e^2 = \sum_{i=1}^{N}\sum_{j=1}^{N} a_i a_j^* \int u(\tau-\tau_i)u^*(\tau-\tau_j)\mathrm{d}\tau = \vec{a}^H \underline{W}\,\vec{a} \tag{8.5.1}$$

where $u(\tau) = d_1(\tau) - d_2(\tau)$. This expression is then diagonalized

$$d_e^2 = \vec{q}^H \underline{U}^H \underline{\Lambda}\,\underline{U}\,\vec{q} = \sum_{i=1}^{N} \lambda_i |y_i|^2 \tag{8.5.2}$$

where $\vec{q}$ is a vector of zero-mean Gaussian variables, and $\underline{U}$ is a unitary transformation matrix. $\underline{\Lambda}$ is the matrix of the eigenvalues of $\underline{R}\,\underline{W}$, where $\underline{R}$ is the correlation

[4]Su et al. [53] use a similar approach to compute the BER with multiple co-channel interferers; they assume, however, that the interference is Gaussian (with possible LOS components). Also, [54] computes 16DAPSK with interference by a related method.

matrix of the channel, $R_{i,j} = E\{a_i a_j^*\}$. If there are no multiple eigenvalues in $\underline{\Lambda}$, the error probability is given as

$$BER = \frac{1}{2}\sum_{i=1}^{N}\left(\prod_{j\neq i}\frac{\lambda_i}{\lambda_i-\lambda_j}\right)\left(1-\frac{1}{\sqrt{1+4N_0/\lambda_i}}\right). \tag{8.5.3}$$

When multiple eigenvalues occur, a similar (but more complicated) equation can be given [57].

8.6 GROUP DELAY METHOD

The group delay T_g of the channel transfer function is defined as

$$T_g = -\frac{\partial\Phi_c}{\partial\omega}|_{\omega=0} \tag{8.6.1}$$

where Φ_c is the phase of the transfer function of the channel (in equivalent baseband) and ω is the angular radian frequency. Using a Taylor expansion that is truncated after the second term, the phase shift introduced by the channel can be written as

$$\vartheta(\omega) = \vartheta(0) - \omega T_g. \tag{8.6.2}$$

The first term in this expression corresponds to the average mean delay and can be omitted if we sample on the average mean delay. Omitting the higher terms in the Taylor expansion of the phase shift is admissible only when the delay spread is very small, and even then it is only an approximation.

Let us now explain how the error floor can be computed; we use the example of differentially detected MSK in [58], [59]. Since we are interested in the phase *difference* between the two sampling times, we have to compute the difference in the channel phase shifts at the two sampling times

$$\Delta\vartheta = -\Delta\omega\cdot T_g \tag{8.6.3}$$

where $\Delta\omega$ is the difference in the instantaneous angular frequencies at the two sampling times—in MSK, it is either 0 or π/T. Without distortions by the channel, MSK gives a phase difference equal to $\pm\pi/2$. Errors will occur if the additional phase shift by the channel has an absolute magnitude larger than $\pi/2$.[5] We thus must evaluate the probability that the group delay is so large that it leads to phase shifts (modulo 2π) $|\Delta\vartheta| > \pi/2$. The pdf of the group delay was shown to follow a Student's t distribution [60]

[5]Strictly speaking, phase shifts in the interval $[3\pi/2, 5\pi/2]$, $[7\pi/2, 9\pi/2]$, etc., also do not lead to errors, but they are so improbable that they are not considered further.

$$\mathrm{pdf}_{T_g}(T_g) = \frac{1}{2S} \frac{1}{\left[1 + (T_g/S)^2\right]^{3/2}} \tag{8.6.4}$$

where S is the average rms delay spread. From this, it follows that the error floor is

$$BER = \frac{4}{9}\left(\frac{S}{T}\right)^2 \approx \frac{1}{2}\left(\frac{S}{T}\right)^2. \tag{8.6.5}$$

Bibliographical Notes: The approximate relation given above has also been used in [61] and verified experimentally [62]. The group delay theory was applied to PSK by Bach Andersen [63]. Extensions to include noise, LOS components, and sampling at time instances other than the average mean delay are given in [64]. We also note that an approximate method for estimating the outage probability by Jakes [65] is related to the group-delay method but relies on amplitude distortions instead of phase distortions.

8.7 ERROR REGION METHOD

The error region method allows the computation of the error floor in a two-delay channel. Molisch et al. [66] and independently Karasawa et al. [67] derived a formulation that allows an exact computation of the error floor and gives some new insights in the error mechanism. A two-delay channel is characterized by two parameters: $\tilde{r} = a_2/a_1$, and $\varphi = \varphi_2 - \varphi_1$ (and τ_2, which is assumed to be fixed); the absolute channel amplitude a_1 and the phase $\varphi_1 + \varphi_2$ do not play a role.

The next step is to interpret these two free parameters as polar coordinates in a complex plane. Each realization of the channel impulse response is thus assigned one point in this complex plane; we call the quantity $\tilde{r}\exp[j(\varphi_2 - \varphi_1)]$ the normalized channel phasor. For one bit combination, a certain realization of the impulse response leads either to a correct or a wrong decision (i.e., the system transmitter + channel + receiver can be considered as completely deterministic), so that the instantaneous BER can only take the values 0 or 1.

We then can identify all channel realizations (i.e., all points in the complex plane) that lead to wrong decisions; all such points lie in certain regions of the complex plane, the so-called *error regions*, which are often circles [66], see Figure 8.4. These regions depend on the sampling time, the filtering, and the transmitted bit combination. The coordinates are given in closed form for differentially detected (G)MSK in [68] and Appendix A in this Part, and for BPSK and $\pi/4$-DQPSK in [69]. In all these cases, analytical computation of the error regions is possible. If the decision rules are very complicated, a numerical procedure might be necessary, namely, by stepping through the complex plane on a grid and computing the decision (true/false) for each point numerically. Even in that case, the method is more efficient than Monte Carlo simulations.

Once the error regions are identified, the BER is simply the probability that the normalized channel phasor falls into an error region

$$BER = \int \int_{\text{error-region}} p_{r,\phi}(r,\phi) \mathrm{d}r \mathrm{d}\phi \tag{8.7.1}$$

where $p_{r,\phi}(r,\phi)$ is the joint pdf of magnitude and phase of the normalized channel phasor. It can be easily computed from the statistics of the two phasors [70], and is given explicitly in Appendix A and [66].[6] Since the error regions are usually near the point $(-1, 0)$, the BER can be approximated very well as

$$BER \approx p_{r,\phi}(1,\pi) \text{Area}_{\text{error-region}}. \tag{8.7.2}$$

The error region method is related to a method that was developed in the 1970s mainly to compute outage probability in fading microwave radio links [71], [72], [73]. For these computations, the noise had to be taken into consideration; this implies that the absolute channel amplitude a_1 also has to be taken into account. Furthermore, when noise is present, the instantaneous BER can take on *all* values in the range $[0, 1]$. For computing the average BER, we again average the instantaneous BER over the probability distribution of the channel parameters

$$BER = \int \int \int p_{a_1,r,\phi}(a_1,r,\phi) BER(a_1,r,\phi) \mathrm{d}a_1 \mathrm{d}r \mathrm{d}\phi \tag{8.7.3}$$

where $p_{a_1,r,\phi}(a_1,r,\phi)$ is the joint pdf of the channel parameters. While this shows only a small *formal* change compared to (8.7.1), the difference in the actual implementation is considerable: for one value of a_1, the integration is now not over a small error region, but over the whole complex plane.

These problems are circumvented when we are interested not in the average BER but only in the outage probability, i.e., how large the probability is that the instantaneous BER exceeds a certain threshold. For a fixed a_1, we can identify *outage regions*, i.e., closed regions where the instantaneous BER exceeds a certain value. Computation of these outage regions is usually done in an approximate way; although an exact computation, e.g., by the ABGV method, would be feasible, this computation is too complicated. Finally, the outage probability is averaged over the distribution of a_1. Computations based on this method were also done by Andrisano et al. [74], [75], [76].

[6] Note a misprint in reference [66]: c in that reference denotes σ_1/σ_2, and not σ_2/σ_1, as written there; $b = a_{\text{LOS}}/(\sqrt{2}\sigma_1)$.

8.8 EQUIVALENT CHANNEL MODELS

The error region method allows a straightforward derivation of the BER, but only for a restricted channel model. It seems thus advantageous to convert the general channel models to equivalent two-delay channels and compute the error probability there. There are essentially two approaches to computing these equivalent channels: (i) phenomenological, i.e., to prescribe from physical reasoning the parameters for the two-delay channel, and (ii) mathematically enforced equivalence. The first approach is called the ETP (Equivalent Transmission Path) model in [67], [77]. Since the error floor is proportional to the delay spread of the diffuse (i.e., non-LOS) components, the equivalent channel should give the same diffuse delay spread as the real channel. Furthermore, the LOS contribution should remain unchanged. The advantage of this approach is that it is intuitively appealing; the drawback is that the basic premise (error floor is proportional only to the delay spread) is only an approximation and does not consider the effects of long-delayed echoes (see discussion in Chapter 9). Furthermore, inclusion of noise effects is not simple.

An exact approach for finding the equivalent channel is called the TPEM (Two-Path Equivalent Matrix) method [78].[7] It is based on the following fact: Since the contributions from the diffuse components and the noise are all zero-mean Gaussian processes, the received signals (i.e., the signal at the two sampling times) are statistically completely characterized by a 4x4 real correlation matrix (or equivalently by the μ_{xx}, μ_{yy}, and μ_{xy} mentioned in Section 8.1). Our next step is then to find an equivalent *noise-free*, two-delay channel that results in the same correlation matrix as the original channel. For differential detection, this can be achieved by solving the following system of equations for the α, Φ, and σ

$$\begin{aligned}
\mu_{xx} &= \alpha_{1,k+1}^2\sigma_1^2 + \alpha_{2,k+1}^2\sigma_2^2 \\
\mu_{yy} &= \alpha_{1,k}^2\sigma_1^2 + \alpha_{2,k}^2\sigma_2^2 \\
\mathrm{Re}\{\mu_{xy}\} &= \alpha_{1,k+1}\alpha_{1,k}\sigma_1^2\cos(\Phi_{1,k+1} - \Phi_{1,k}) + \\
&\quad \alpha_{2,k+1}\alpha_{2,k}\sigma_2^2\cos(\Phi_{2,k+1} - \Phi_{2,k}) \\
\mathrm{Im}\{\mu_{xy}\} &= -\alpha_{1,k+1}\alpha_{1,k}\sigma_1^2\sin(\Phi_{1,k+1} - \Phi_{1,k}) - \\
&\quad \alpha_{2,k+1}\alpha_{2,k}\sigma_2^2\sin(\Phi_{2,k+1} - \Phi_{2,k})
\end{aligned} \tag{8.8.1}$$

where the $\alpha_{i,k}$ and $\Phi_{i,k}$ are the amplitude and phase of the received signal in the ith path ($i = 1, 2$) at the kth sampling time. In other words, we take the correlation matrix that we get from our actual channel and find (arbitrary) values for the α, Φ, and σ that result in the same correlation matrix. When a LOS component is present, $\Phi_{1,k}$, $\Phi_{1,k+1}$, $\alpha_{1,k}$, $\alpha_{1,k+1}$ must agree with the phase and amplitude of the signals from the LOS component of the real channel. In any case, we have enough free variables to find an equivalent channel also in that case. Since the signals transmitted through the equivalent channel have the same statistical properties as

[7] The TPEM was derived for differentially detected modulation formats, but extension to limiter-discriminator detection is straightforward.

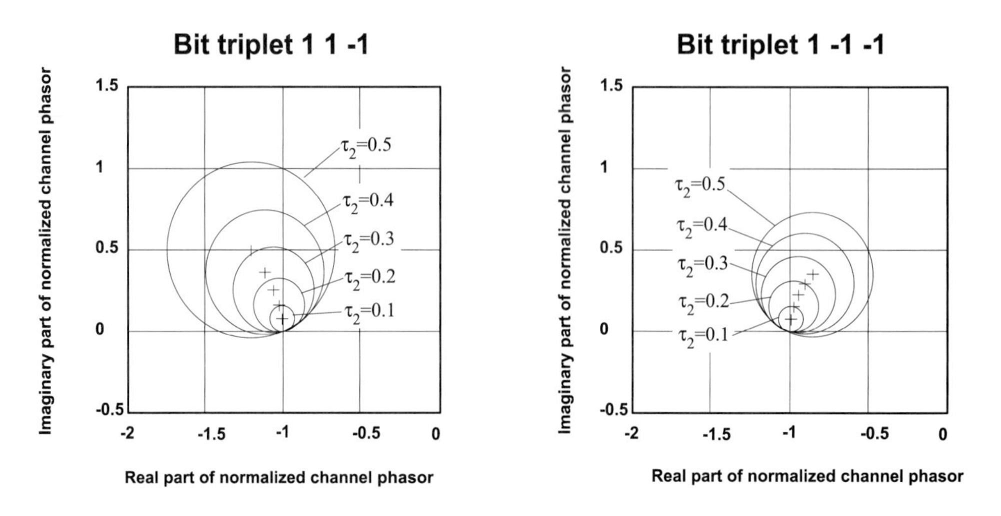

Figure 8.4. Error regions for pure MSK. Adapted from [66], copyright IEEE.

the signals transmitted through the real channel, the error probability must remain the same when the equivalent channel is used.

Compared to the ETP, this method has the big advantage that it is exact; moreover, long-delayed echoes, noise, etc., can be included in a straightforward manner. When compared to the QFGV method, we find that any postprocessing of the received (sampled) signals, like quantization, can be taken into account in a straightforward way: we just have to include this processing in our computation of the error regions, but no changes in the computation of the equivalent channel are required.

8.9 FURTHER METHODS: BIBLIOGRAPHICAL NOTES

References [79], [80] computed the BER by first determining the pdf of the received signal: they multiplied the characteristic functions of the pdfs of each contributing echo (and similarily for the noise). However, their results are not in agreement with other computations in the literature; for example, they get an error floor that is lower for GMSK than for MSK.

Reference [81] computed the error floor of MSK in an N-path, Rayleigh-fading channel with small maximum excess delay. It proceeded by converting the phase distortions of the channel into equivalent amplitude distortions and averaging directly over the resulting Rayleigh statistics. It neglects, however, the error probabilities of the symmetric bit combinations; this was remedied in a later paper [82]. Reference [83] computes the probability of eye closure without noise (i.e., error floor) and averages over the channel parameters; using this technique, it compares various PSK and QAM modulation schemes.

For $\pi/4$-DQPSK in a two-delay channel, [84] derived approximate expressions that are somewhat similar to the ABGV method. It approximates the ISI and the co-channel interference as Gaussian process. There are thus four Gaussian processes: the first echo of the desired signal, the second echo, the noise, and the co-channel interference. The correlation matrix of these components is then computed. It is used to compute the pdf of the angle ψ between the signals at two subsequent sampling times, and thus the probability for wrong decisions. The resulting equations are an approximation to the results of [85] described in Chapter 9. The method was also used for pilot-symbol-aided detection of coherent PSK [86].

Reference [87] developed a very intricate channel model for wideband line-of-sight links; it assumes, however, that the outage is induced also for the frequency-selective case only by low field strength and does not take phase distortions into account.

Reference [88] represents the received signal for a noncoherent FSK receiver by a finite series according to the sampling theorem. Computing the pdfs of the signals by using their characteristic functions, it finally arrives at equations for the BER in the form of infinite series. The same problem was analyzed by a Fourier series expansion of the fading process [89].

Reference [90] derives an upper bound for the BER for 16-QAM by first identifying the bit combination that leads to the largest BER and then computing the BER for it. This is an upper bound for the BER for all bit combinations.

Reference [91] first computed the signature curves[8] and then averaged over the channel statistics to arrive at the outage probability.

Reference [92] derived equations for frequency-discriminator detection of generalized tamed frequency modulation and some related modulation formats by series expansions.

Foschini and Salz [93] give an upper bound for the error probability in a (non-fading) AWGN channel derived from an information-theoretical model. For QAM, for example, it reads

$$BER \leq \exp\left[-\frac{1 - MSE/\sigma_c^2}{MSE}\right] \tag{8.9.1}$$

where MSE is the mean-square error and σ_c^2 is the expected value of the squared absolute value of the transmission alphabet symbols.

An exciting new development is a method of Simon and Alouini ([94], [95], and references therein). This method is based on an alternative representation of the Q-function

[8]The signature curve is a characteristic in a two-delay channel. Given a notch frequency, one must compute the notch depth that leads to a desired outage probability. The curve *notch depth* versus *notch frequency* is called the signature curve.

$$Q(x) = \frac{1}{\pi} \int_0^{\pi/2} \exp\left(-\frac{x^2}{2\sin^2 u}\right) du \tag{8.9.2}$$

and similarily for Marcum's Q-function. In contrast to the conventional representation of the Q-function, the limits of the integral are definite (and also finite), allowing interchanging order of integration, for example, when averaging over different channel constellations. This method offers a wealth of new possibilities for finding closed-form equations for average error probabilities, even if the fading statistics are neither a Rayleigh nor a Rice distribution. Simon and Alouini have also applied their results to time-dispersive channels with ideal Rake reception (see Chapter 12 and Part V) but not to time-dispersive unequalized systems. This application might become an interesting aspect in the future.

Bibliography

[1] M. Reuter, "Numerically efficient fourier-based technique for calculating error probabilities with intersymbol interference," *IEEE Trans. Comm.*, vol. 45, pp. 629–632, 1997.

[2] H. Suzuki, "Canonic receiver analysis for m-ary angle modulations in Rayleigh fading environment," *IEEE Trans. Vehicular Techn.*, vol. 31, pp. 7–14, 1982.

[3] P. Bello and B. D. Nelin, "The effect of frequency selective fading on the binary error probabilities of incoherent and differentially coherent matched filter receivers," *IEEE Trans. Comm.*, vol. 11, pp. 170–186, 1963.

[4] M. Fitz, "Further results in the unified analysis of digital communication systems," *IEEE Trans. Comm.*, vol. 40, pp. 521–532, 1992.

[5] J. G. Proakis, *Digital Communications.* New York, McGraw Hill, 3rd ed., 1995.

[6] P. J. Lee, "Computation of the bit error rate of coherent M-ary PSK with gray code bit mapping," *IEEE Trans. Comm.*, vol. 34, pp. 488–491, 1986.

[7] F. Adachi and J. Parsons, "Error rate performance of digital FM mobile radio with postdetection diversity," *IEEE Trans. Comm.*, vol. 37, pp. 200–210, 1989.

[8] G. Janssen, P. Stigter, and R. Prasad, "Wideband indoor channel measurements and BER analysis of frequency selective multipath channels at 2.4, 4.75, and 11.5 GHz," *IEEE Trans. Comm.*, vol. 44, pp. 1272–1288, 1996.

[9] J. Chuang, "The effects of time delay spread on portable radio communications channels with digital modulation," *IEEE Selected Areas Comm.*, vol. 5, pp. 879–888, 1987.

[10] J. Chuang, "The effect of delay spread on 2-PSK, 4-PSK, 8-PSK and 16-QAM in a portable radio environment," *IEEE Trans. Vehicular Techn.*, vol. 38, pp. 43–45, 1989.

[11] C. P. Donoghue, C. J. Burkley, and M. O'Droma, "The performance of GTFM in a frequency-selective Rayleigh fading channel," in *Proc. VTC'89*, pp. 878–883, 1989.

[12] H. Lau and S. Cheung, "Performance of a pilot symbol-aided technique in frequency-selective Rayleigh fading channels corrupted by co-channel interference and Gaussian noise," in *Proc. VTC'96*, pp. 1008–1012, 1996.

[13] V. Fung, T. Rappaport, and B. Thoma, "Bit error simulation for pi/4 DQPSK mobile radio communications using two-ray and measurment-based impulse response models," *IEEE Selected Areas Comm.*, vol. 11, pp. 393–405, 1993.

[14] M. Wittmann, J. Marti, and T. Krner, "Impact of the power delay profile shape on the bit error rate in mobile radio systems," *IEEE Trans. Vehicular Comm.*, vol. 46, pp. 329–339, 1997.

[15] A. Johnson, "Simulation of digital transmission over mobile channels at 300 kb/s," *IEEE Trans. Comm.*, vol. 39, pp. 319–327, 1991.

[16] L. B. Lopes, "Performance of the DECT system in fading dispersive channels," *Electronics Letters*, vol. 26, pp. 1416–1417, 1990.

[17] P. Bello, "Binary error probabilities over selectively fading channels containing specular components," *IEEE Trans. Comm.*, vol. 14, pp. 400–406, 1966.

[18] J. Proakis, "On the probability of error for multichannel reception of binary signals," *IEEE Trans. Comm.*, vol. 16, pp. 68–71, 1968.

[19] G. L. Turin, "The characteristic function of Hermitian quadratic forms in complex normal variables," *Biometrika*, vol. 47, pp. 199–201, 1960.

[20] F. Adachi and J. Parsons, "Unified analysis of postdetection diversity for binary digital FM mobile radio," *IEEE Trans. Vehicular Techn.*, vol. 37, pp. 189–198, 1988.

[21] S. Stein, "Unified analysis of certain coherent and noncoherent binary communications systems," *IEEE Trans. Inform. Theory*, vol. 10, pp. 43–51, 1964.

[22] K. Hirade, M. Ishizuka, F. Adachi, and K. Ohtani, "Error-rate performance of digital FM with differential detection in land mobile radio channels," *IEEE Trans. Vehicular Techn.*, vol. 28, pp. 204–212, 1979.

[23] J. Horikoshi and S. Shimura, "Multipath distorsion of differentially encoded GMSK with 2-b differential detection in bandlimited frequency-selective mobile radio channel," *IEEE Trans. Vehicular Techn.*, vol. 39, pp. 308–315, 1990.

[24] E. Biglieri, G. Caire, G. Taricco, and J. Ventura-Traveset, "Simple method for evaluating error probabilities," *Electronics Letters*, vol. 43, pp. 191–192, 1996.

[25] E. Biglieri, G. Caire, G. Taricco, and G. Ventura-Traveset, "Computing error probabilities over fading channels: A unified approach," *European Trans. Telecomm.*, vol. 9, pp. 15–25, 1998.

[26] R. Bultitude and A. Leslie, "Propagation measurement-based probability of error predictions for digital land-mobile radio," *IEEE Trans. Vehicular Techn.*, vol. 46, pp. 717–729, 1997.

[27] K. Wu, N. Morinaga, and T. Namekawa, "Error rate perfromance of binary DPSK system with multiple co-channel interference in land mobile radio channels," *IEEE Trans. Vehicular Techn.*, vol. 33, pp. 23–31, 1984.

[28] H. Ma, N. Shehadeh, and J. Vanelli, "Effect of intersymbol interference on a Rayleigh fast-fading channel," *IEEE Trans. Comm.*, vol. 28, pp. 128–131, 1980.

[29] C. Tellambura and V. Bhargava, "Performance of GMSK in frequency selective Rayleigh fading and multiple cochannel interferers," in *Proc. VTC'95*, pp. 211–215, 1995.

[30] Y. Chow, A. Nix, and J. McGeehan, "Error performance of circular 16-DAPSK in frequency-selective Rayleigh fading channels with diversity reception," in *Proc. VTC'95*, pp. 419–423, 1995.

[31] C. C. Chui, C. S. Ng, and T. T. Tjhung, "Performance of pi/M—MPSK in frequency-selective Rician fading with non-zero fading bandwidth," in *Proc. Int. Conf. Inform., Comm. and Signal Processing*, pp. 138–142, 1997.

[32] D. Hummels and F. Ratcliffe, "Calculation of error probability for MSK and OQPSK systems operating in a fading multipath environment," *IEEE Trans. Vehicular Techn.*, vol. 30, pp. 112–120, 1981.

[33] B. Glance and L. Greenstein, "Frequency-selective fading effects in digital mobile radio with diversity combining," *IEEE Trans. Comm.*, vol. 31, pp. 1085–1094, 1983.

[34] F. Garber and M. Pursley, "Performance of differentially coherent digital communications over frequency-selective fading channels," *IEEE Trans. Comm.*, vol. 36, pp. 21–31, 1988.

[35] F. Garber and M. Pursley, "Performance of binary FSK communications over frequency-selective Rayleigh fading channels," *IEEE Trans. Comm.*, vol. 37, pp. 83–89, 1989.

[36] T. Staley, R. North, W. Ku, and J. Zeidler, "Performance of coherent MPSK on frequency selective slowly fading channels," in *Proc. VTC'96*, vol. 5, pp. 784–788, 1996.

[37] P. Varshney and S. Kumar, "Performance of GMSK in a land mobile radio channel," *IEEE Trans. Vehicular Techn.*, vol. 40, pp. 607–615, 1991.

[38] P. Varshney, J. Salt, and S. Kumar, "BER analysis of GMSK with differential detection in a land mobile channel," *IEEE Trans. Vehicular Techn.*, vol. 42, pp. 683–689, 1993.

[39] M. Dechambre and A. J. Levy, "Limits of data rates in the urban mobile channel," in *Proc. VTC'87*, pp. 541–546, 1987.

[40] R. F. Pawula, S. O. Rice, and J. H. Roberts, "Distribution of the phase angle between two vectors perturbed by Gaussian noise," *IEEE Trans. Comm.*, vol. 30, pp. 1828–1841, 1982.

[41] L. J. Mason, "Error Probability Evaluation for Systems Employing Differential Detection in a Rician Fast Fading Environment and Gaussian Noise," *IEEE Trans. Comm.*, vol. 35, pp. 39–46, 1987.

[42] I. Korn, "Error probability of M-ary FSK with differential phase detection in satellite mobile channel," *IEEE Trans. Vehicular Techn.*, vol. 38, pp. 76–85, May 1989.

[43] I. Korn, "GMSK with differential phase detection in the satellite mobile channel," *IEEE Trans. Comm.*, vol. 38, pp. 1980–1986, Nov. 1990.

[44] I. Korn, "M-ary Frequency Shift Keying with Limiter-Discriminator-Integrator Detector in Satellite Mobile Channel with Narrow-Band Receiver Filter," *IEEE Trans. Comm.*, vol. 38, pp. 1771–1778, 1990.

[45] I. Korn, "GMSK with limiter discriminator detection in satellite mobile channel," *IEEE Trans. Comm.*, vol. 39, pp. 94–101, Jan. 1991.

[46] I. Korn, "Differential phase shift keying in two-path Rayleigh channel with adjacent channel interference," *IEEE Trans. Vehicular Techn.*, vol. 40, pp. 461–471, 1991.

[47] I. Korn, "Error floors in the satellite and land mobile channels," *IEEE Trans. Comm.*, vol. 39, pp. 833–837, 1991.

[48] I. Korn, "M-ary frequency shift keying with limiter discriminator detection in mobile channels with narrowband receiver filter," *Archiv Elektronik Uebertragungstechn.*, vol. 47, pp. 69–76, 1993.

[49] I. Korn, "The effect of pulse shaping and transmitter filter on the performance of FSK-DPD and CPC-DPD in satellite mobile channel," *IEEE J. Selected Areas Comm.*, vol. 13, pp. 245–249, 1995.

[50] I. Korn, "Effect of transmitter filter on the performance of FSK-DPD and DPSK-DPD in satellite mobile channel," *European Trans. Telecomm.*, vol. 6, pp. 581–585, 1995.

[51] I. Korn, "Binary CPM-DPD with diversity in Rician fading channels," *Electronics Letters*, vol. 31, pp. 519–521, 1995.

[52] I. Korn, "GMSK with frequency-selective Rayleigh fading and cochannel interference," *IEEE Selected Areas Comm.*, vol. 10, pp. 506–515, 1992.

[53] Y. T. Su, W. C. Kao, and J. S. Li, "The effects of Rician fading and multiple CCI on differentially detected GMSK signals," in *Proc. PIMRC'97*, pp. 959–963, 1997.

[54] J. Y. Lee, Y. M. Chung, and S. U. Lee, "On the bit error probability of 16DAPSK in a frequency-selective fast Rayleigh fading channel with cochannel interference," *IEICE Trans. Comm.*, vol. E82-B, pp. 532–541, 1999.

[55] F. Adachi, "Bit error rate analysis of M-ary DPSK in frequency-selective Rician fading," *Electronics Letters*, vol. 30, pp. 1734–1736, 1994.

[56] D. Dzung and W. Braun, "Performance of coherent data transmission in frequency-selective Rayleigh fading channels," *IEEE Trans. Comm.*, vol. 41, pp. 1335–1341, 1993.

[57] C. Schlegel, "Error probability calculation for multibeam Rayleigh channels," *IEEE Trans. Comm.*, vol. 44, pp. 290–293, 1996.

[58] I. Crohn and G. Schultes, "Error performance of differentially-detected MSK in small delay-spread Rayleigh channel," *Electronics Letters*, vol. 28, pp. 300–301, 1992.

[59] I. Crohn, G. Schultes, R. Gahleitner, and E. Bonek, "Irreducible error performance of a digital portable communication system in a controlled time-dispersion indoor channel," *IEEE Selected Areas Comm.*, vol. 11, pp. 1024–1033, 1993.

[60] J. B. Andersen, S. L. Lauritzen, and C. Thommesen, "Distribution of phase derivatives in mobile commnications," *Proc. IEE, Part H*, vol. 137, pp. 197–204, 1990.

[61] A. Kukushkin and I. Sharp, "Evaluation of DECT receiver performance," in *Proc. VTC '96*, pp. 820–823, 1996.

[62] G. Schultes and I. Crohn, "Measured performance of DECT transmission in low despersive indoor radio channel," *Electronics Letters*, vol. 28, pp. 1625–1627, 1992.

[63] J. B. Andersen, "Propagation parameters and bit errors for a fading channel," in *Proc. Commsphere '91*, p. 8.1, 1991.

[64] A. F. Molisch, M. Paier, and E. Bonek, "Analytical computation of the error probability of (g)msk with adaptive sampling in mobile radio channels," *European Transactions on Telecommunications*, vol. 9, pp. 551–559, 1998.

[65] W. C. Jakes, "An approximate method to estimate an upper bound on the effect of multipath delay distortion on digital transmission," *IEEE Trans. Comm.*, vol. 27, pp. 76–81, 1979.

[66] A. F. Molisch, J. Fuhl, and P. Proksch, "Error floor of MSK modulation in a mobile-radio channel with two independently-fading paths," *IEEE Trans. Vehicular Techn.*, vol. 45, pp. 303–309, 1996.

[67] Y. Karasawa, T. Kuroda, and H. Iwai, "The equivalent transmission-path model—a tool for analysing error floor characteristics due to intersymbol interference in Nakagami-Rice fading environments," *IEEE Trans. Vehicular Techn.*, vol. 46, pp. 194–202, 1997.

[68] A. F. Molisch, L. Lopes, M. Paier, J. Fuhl, and E. Bonek, "Error floor of unequalized wireless personal communications systems with MSK modulation and training-sequence-based adaptive sampling," *IEEE Trans. Comm.*, vol. 45, pp. 554–562, 1997.

[69] A. F. Molisch and E. Bonek, "Reduction of error floor of differential PSK in mobile radio channels by adaptive sampling," *IEEE Trans. Vehicular Techn.*, vol. 47, pp. 1276–1280, 1998.

[70] A. Papoulis, *Probability, Random Variables, and Stochastic Processes.* Tokyo, McGraw Hill, 1965.

[71] L. J. Greenstein and V. K. Prabhu, "Analysis of multipath outage with applications to 90MBit/s PSK systems at 6 and 11 GHz," *IEEE Trans. Comm.*, vol. 27, pp. 68–75, 1979.

[72] W. Lundgren and W. D. Rummler, "Digital radio outage due to selective fading - observation vs prediction from laboratory simulations," *Bell System Techn. J.*, vol. 58, pp. 1073–1100, 1979.

[73] A. J. Giger and W. T. Barnett, "Effects of multipath propagation on digital radio," *IEEE Trans. Comm.*, vol. 29, pp. 1345–1352, 1981.

[74] O. Andrisano and V. Tralli, "Analytical outage evaluation of TDMA local radio system with coding and diversity," *IEEE Trans. Comm.*, vol. 40, pp. 1725–1736, 1992.

[75] O. Andrisano, "The combined effects of noise and multipath propagation in multilevel PSK radio links," *IEEE Trans. Comm.*, vol. 32, pp. 411–418, 1984.

[76] O. Adrisano, G. Corazza, and G. Immovilli, "On the availability of multilevel CPFSK systems with modulation pulse shaping during multipath propagation," *IEEE Trans. Comm.*, vol. 33, pp. 975–985, 1985.

[77] J. I. Takada, H. Mochida, and K. Araki, "A BER floor simulation for the indoor multipath environment," in *Proc. PIMRC'97*, pp. 145–148, 1997.

[78] A. F. Molisch, "A new method for the computation of the error probability of differentially detected FSK and PSK in mobile radio channels—the case of minimum shift keying," *Wireless Personal Comm.*, vol. 9, pp. 165–178, 1999.

[79] T. Kuerner and W. Wiesbeck, "Einfluss der Mehrwegeausbreitung auf die Bitfehlerrate," *Frequenz*, vol. 48, pp. 270–278, 1994.

[80] F. Kuechen, T. Zwick, and W. Wiesbeck, "Symbol error rate prediction for mobile receivers in urban single frequency networks," in *Proc. VTC'97*, pp. 1952–1956, 1997.

[81] V. Lipovac and A. F. Molisch, "On the performance of MSK signal transmission over a multipath channel with small time dispersion," in *Proc. VTC'95*, pp. 25–29, 1995.

[82] A. F. Molisch, J. Fuhl, and V. Lipovac, "An improved equation for the computation of the bit error probability of MSK in mobile radio channels with small delay spread," in *COST 231*, p. TD95(62), 1995.

[83] L. Greenstein and B. Czekaj-Augun, "Performance comparisons among digital radio techniques subjected to multipath fading," *IEEE Trans. Comm.*, vol. 30, pp. 1184–1197, 1982.

[84] C. L. Liu and K. Feher, "Bit error performance of pi/4-DQPSK in a frequency-selective fast Rayleigh fading channel," *IEEE Trans. Vehicular Techn.*, vol. 40, pp. 558–568, 1991.

[85] F. Adachi and K. Ohno, "BER performance of QDPSK with postdetection diversity reception in mobile radio channels," *IEEE Trans. Vehicular Techn.*, vol. 40, pp. 237–249, 1991.

[86] C.-L. Liu and K. Feher, "Pilot-symbol aided coherent M-ary PSK in frequency-selective fast Rayleigh fading channels," *IEEE Trans. Comm.*, vol. 42, pp. 54–62, 1994.

[87] J. Lavergnat, M. Sylvain, and J.-C. Bic, "A method to predict multipath effects on a line-of-sight link," *IEEE. Trans. Comm.*, vol. 38, pp. 1810–1822, 1990.

[88] H. Chadwick, "The error probability of a wide-band FSK receiver in the presence of multipath fading," *IEEE Trans. Comm.*, pp. 699–707, 1971.

[89] S. Kwon and N. Shehadeh, "Noncoherent detection of FSK signals in the presence of multipath fading," *IEEE Trans. Comm.*, vol. 26, pp. 164–168, 1978.

[90] L. B. Milstein, D. Schilling, and R. Pickholtz, "Comparison of performance of 16-ary QASK and MSK over a frequency selective Rician fading channel," *IEEE Trans. Comm.*, vol. 29, pp. 1622–1633, 1981.

[91] K. Metzger and R. Valentin, "An analysis of the sensitivity of digital modulation techniques to frequency-selective fading," *IEEE Trans. Comm.*, vol. 33, pp. 986–993, 1985.

[92] M. Quacchia and V. Zingarelli, "An analytical evaluation of bit error probability in mobile radio systems with 12pm3 modulations," *IEEE Trans. Vehicular Techn.*, vol. 37, pp. 135–151, 1988.

[93] G. Foschini and J. Salz, "Digital communications over fading radio channels," *Bell Syst. Techn. Jour.*, vol. 62, pp. 429–456, 1983.

[94] M. K. Simon and M. S. Alouini, "A unified approach to the performance analysis of digital communications over generalized fading channels.," *Proc. IEEE*, vol. 86, pp. 1860–1877, 1998.

[95] M. K. Simon and M. S. Alouini, *Digital Communications over Generalized Fading Channels: A Unified Approach to Performance Analysis.* New York: Wiley, 2000.

Chapter 9

RESULTS FOR FIXED SAMPLING

9.1 INFLUENCE OF MODULATION, CHANNEL, AND RECEIVER

One of the most important decisions when designing a system is to determine the modulation format that is best suited for the environments in which the system should operate. While many facts, like envelope fluctuations, spectral efficiency, etc., influence the decision, we address here only the subject of the BER. It has been found repeatedly (e.g., [1], [2], [3]) that the error floor is approximately proportional to the square of the normalized delay spread, $BER = K \cdot (S/T)^2$, where the proportionality constant K depends on the modulation format, the filtering at transmitter and receiver, and the sampling time. The proportionality between BER and delay spread is true, however, only under some special assumptions, the most important of which are [4] the following ones: (i) the maximum excess delay of the channel must be much smaller than a symbol length, and (ii) the channel must be Rayleigh-fading, i.e., must not have a LOS component. Furthermore, it is only an approximation, because the shape of the delay power profile actually does influence the BER; see Figure 9.1.

Figure 9.2 shows the error floor for various modulation formats as a function of the delay spread. The figure gives the impression that high-order modulation formats are a bad choice in a time-dispersive environment; this conclusion can be intuited from the fact that the signal points in the constellation diagram are closer together than those of lower-order formats so that a smaller distortion by the channel is sufficient to induce errors. However, we note that for this figure, the delay spread is normalized to the *symbol* length. If we normalized to the *bit* length (which is often a meaningful measure for system design), then the higher-order modulation formats would fare much better.

Another important parameter that determines the error floor is the filtering at the transmitter and receiver. Strong filtering leads to intersymbol interference even in a flat-fading channel (though the filtering has to be *very* strong, e.g., BT

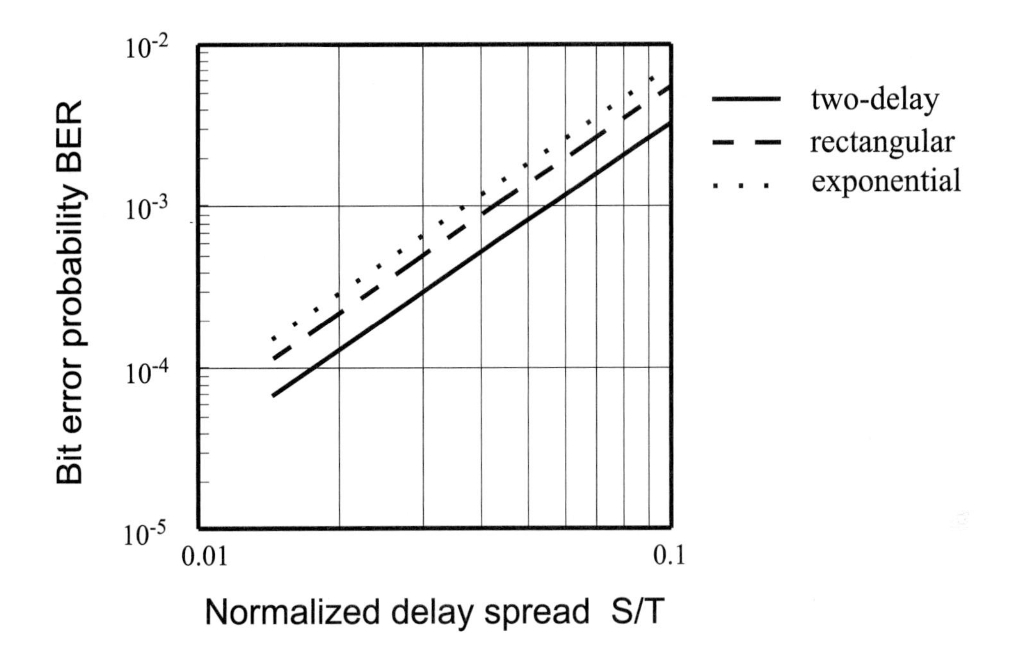

Figure 9.1. Error floor of unfiltered MSK with sampling on the average mean delay in a slowly fading Rayleigh time-dispersive channel. Power delay profiles: two-delay, rectangular, exponential.

products on the order of 0.1 or less, in order to cause an error floor in a flat-fading channel). This ISI then also makes the system more sensitive to errors due to channel distortions. Figure 9.3 shows the error floor for raised-cosine-filtered QPSK as a function of the roll-off parameter. We see that the error floor changes strongly. The influence of the filtering is much weaker for GMSK, where changing the Gaussian filter bandwidth B_GT from 0.125 to 2 has practically no effect on the error floor [2]; for the receiver filtering, a similar effect can be observed [4]. The situation is, however, quite different when also noise is present: due to the ISI and the closer spacing of the signal points in the constellation diagram, the system becomes also more sensitive to noise when the filtering becomes tighter (on the other hand, more of the noise is filtered out by the receiver filter). There is thus a quite pronounced optimum for the receiver filter bandwidth, depending on the SNR. Figure 9.4 shows an example for this trade-off.

Finally, the sampling instant plays an important role. Although it has been believed for a long time that the average mean delay is the optimum sampling instant, recent investigations have shown that this is not true for MSK [4], [6], [4]. These papers give equations for the optimization of the sampling time for a given power delay profile (PDP) and show that the average mean delay is optimum only for some

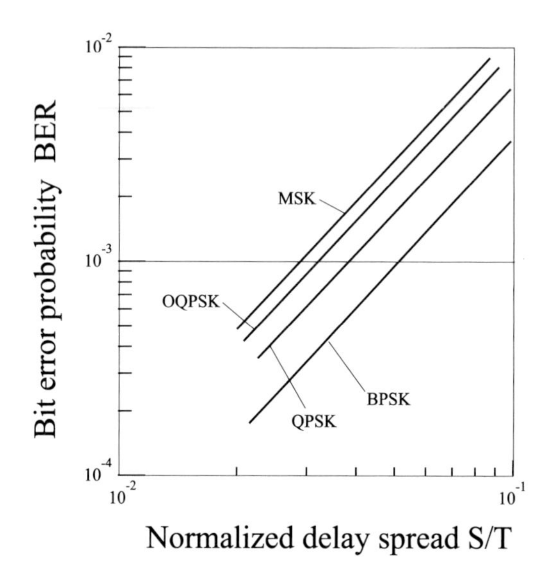

Figure 9.2. Error floor for different modulations with coherent detection for a channel with Gaussian PDP. Adapted from [2], [5], copyright IEEE.

PDPs (two-delay channel with equal average power in the two paths; rectangular delay power profile). Figure 9.5 shows the BER as a function of sampling time for MSK. This suggests also that similar results might hold true for other modulation formats.

Further results for fixed sampling are also given in Chapter 12 on diversity; "normal" reception is, after all, just a special case of diversity reception, with the number of diversity signals equal to unity, $L = 1$.

9.2 CPFSK

The case of binary FSK with differential detection in a purely Rayleigh-fading channel is a special case of the canonic analysis that we described in Chapter 8. However, because of its great practical importance, we give the equations for the BER explicitly. Following [7] we get

$$BER = \frac{1}{2} - \frac{1}{2} \frac{b_0 \operatorname{Im}\{\rho\}}{\sqrt{\operatorname{Im}\{\rho\}^2 + (1 - |\rho|^2)}} \tag{9.2.1}$$

where ρ is the correlation coefficient from Section 8.1.3. For the flat-fading case, this becomes

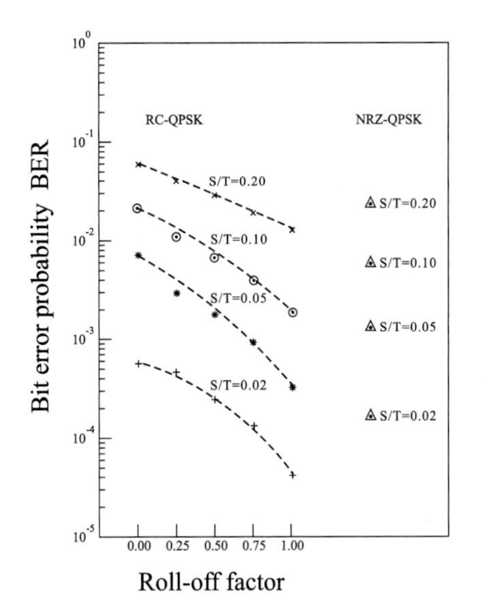

Figure 9.3. Error floor for QPSK with coherent detection and raised-cosine filtering. For comparison, results without raised-cosine filtering are also shown. Reprinted from [2], copyright IEEE.

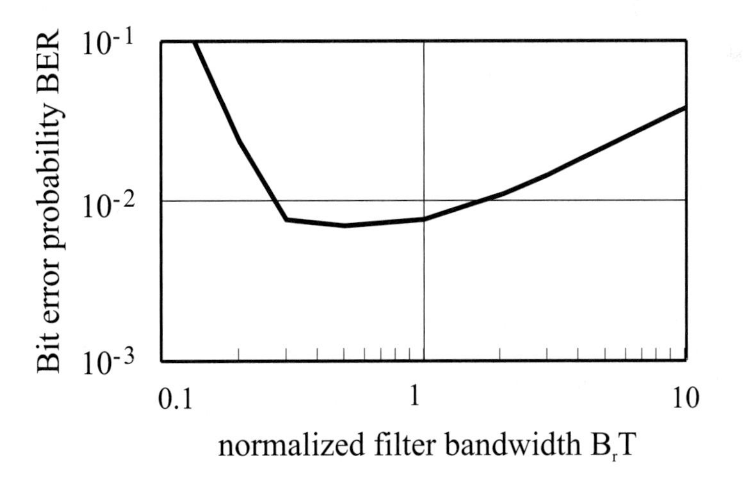

Figure 9.4. Error probability BER of filtered MSK as a function of receiver filter bandwidth B_r; $H(f) = B_r^2/(B_r^2 + f^2)$. Channel has 20 independently fading paths ($\sigma_i = 1$); delay spread $S = 0.1$; sampling on the average mean delay. SNR=12 dB.

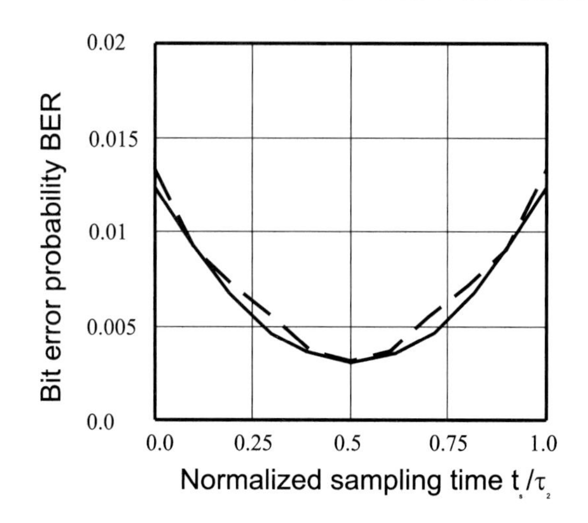

Figure 9.5. Error floor of MSK for a symmetrical two-delay delay power profile: analytical computations (solid), MC simulations (dashed).

$$BER_{\text{flat}} = \frac{1}{2}\left(1 - \frac{\Gamma \sin(h_{\text{mod}}\pi)}{\sqrt{(\Gamma+1)^2 - \Gamma^2 \cos^2(h_{\text{mod}}\pi)}}\right). \tag{9.2.2}$$

where Γ is the mean SNR. The general expression for the error floor can be approximated for small delay spread as

$$BER_{\text{floor}} = \frac{1}{2}\left[T \frac{\left|\frac{\mathrm{d}}{\mathrm{d}t} d(t)|_{t=0} d(-T) - \frac{\mathrm{d}}{\mathrm{d}t} d(t)|_{t=-T} d(0)\right|}{\sqrt{2}\,\mathrm{Im}\left\{d(0)d^*(-T)\right\}}\right]^2 \left(\frac{S}{T}\right)^2. \tag{9.2.3}$$

Previously, it was concluded that in the limit of very small delay spread, the influence of the shape of the PDP vanishes [7]. However, recent investiations [6], [8] showed that for MSK, the exact limiting function is

$$BER_{\text{floor}} = \left[\frac{\pi\sqrt{2}/4}{\int_{-\infty}^{\infty} P(\tau) d\tau} \right]^2 \tag{9.2.4}$$

$$\Bigg[\int_{-\infty}^{\infty} P(\tau)\mathrm{d}\tau \int_{-\infty}^{\infty} P(\tau)(\tau - t_s)^2 \mathrm{d}\tau$$

$$- \left(\int_{-\infty}^{t_s} P(\tau)(t_s - \tau)\mathrm{d}\tau \right)^2$$

$$- \left(\int_{t_s}^{\infty} P(\tau)(\tau - t_s)\mathrm{d}\tau \right)^2$$

$$- \int_{-\infty}^{t_s} P(\tau)(t_s - \tau)\mathrm{d}\tau \int_{t_s}^{\infty} P(\tau)(\tau - t_s)\mathrm{d}\tau \Bigg],$$

where $P(\tau)$ is the PDP. An example for the influence of the shape of the PDP was already shown in Section 9.1.

Bibliographical Notes: Reference [9] analyzed GMSK in a two-delay channel with two-bit differential detection [10]. Evaluations for CPFSK in a two-delay Rice-fading channel are given in [11]. Reference [12] analyzed the error floor in a two-delay channel that suffers not only from Rice-fading but also from shadowing the author found that the small-scale-averaged BER must be averaged over the shadowing and that the use of a large-scaled-averaged power-delay profile in the equations of this section is not admissible.

9.3 FSK

The coherent and noncoherent detection of FSK are special cases of the canonic receiver analysis of Section 8.1.2, so that a more detailed treatment is not necessary. We also note that for noncoherent detection of binary FSK, noncoherent detection reduces to finding whether the expression $X^2 - Y^2$ is larger or smaller than zero (where X and Y denote the outputs of the filters matched to the two-basis waveforms). The decision variable is thus a special case of the general quadratic form used in the QFGV method. Evaluations for the case of Gaussian delay power profiles can be found in [13].

For frequency-discriminator detection of FSK, the equations of the BER have the same functional form as for DPD (differential phase detection). The difference lies in the definition of the correlation coefficient, see Section 8.1.3. This definition also leads to a strong change in the actual *values* of the error probability. For the flat-fading case, we get

$$BER_{\text{flat}} = \frac{1}{2} \left(1 - \frac{\sqrt{3}\Gamma h_{\text{mod}}}{(\Gamma + 1)^{1/2}(1 + 3\Gamma h_{\text{mod}}^2)^{1/2}} \right). \tag{9.3.1}$$

Especially, the ISI-induced error floor strongly depends on the detection type. For frequency discriminator (FD) detection,

$$BER_{\text{floor}} = \left[T\frac{\left|\left[\frac{\mathrm{d}}{\mathrm{d}t}d(t)|_{t=T/2}\right]^2 - \frac{\mathrm{d}^2}{\mathrm{d}t^2}d(t)|_{t=T/2}d(T/2)\right|}{\sqrt{2}\,\mathrm{Im}\left\{\frac{\mathrm{d}}{\mathrm{d}t}d^*(t)|_{t=T/2}d(T/2)|\right\}}\right]^2\left(\frac{S}{T}\right)^2. \tag{9.3.2}$$

In contrast to DPD, receiver filter bandwidth strongly influences the error floor for FD. A physical interpretation of this fact will become obvious in Section 10.1, when we keep in mind that FD detection is the limiting case of fractional-bit detection. Under most circumstances, FD detection is preferable to one-bit DPD and two-bit DPD [14].

Reference [15] performed experiments with FSK with discriminator detection. It confirmed that to a first approximation, the error floor depends only on the delay spread normalized to the bit length. Wedge [16] obtained results for FSK with various modulation indices by MC simulations. He found that increasing the modulation index leads to a strong reduction of the error floor: going from $h_{\text{mod}} = 0.5$ to $h_{\text{mod}} = 1$ decreased the error floor by more than an order of magnitude. Multi-h pulse modulation (MHPM) is analyzed in [17].

9.4 COHERENTLY DETECTED PSK

For the case of an ideal matched filter receiver for binary modulation formats, the decision is based on the variable $XY^* + X^*Y$ and thus also is a special case of the canonic receiver analysis. Also, a noisy reference could in principle be handled by that model. Liu and Feher [18] have analyzed pilot-symbol aided carrier estimation. The fading characteristic of the channel is estimated from a pilot signal, which is one (known) symbol inserted every N_{pilot} data symbols. How often this pilot symbol is necessary depends on the fading rate and on how good the channel estimate has to be: the more pilot symbols we have, the more easily noise can be eliminated from them. The (interpolated) values of the pilot tone at times nT are written as $W(nT) + jZ(nT)$, where W and Z are Gaussian processes. We then define a correlation coefficient between the total received signal $U(nT) + jV(nT)$ and the pilot tone

$$\rho^2_{\text{pilot}} = \frac{\mu^2_{UW}}{\mu_{UU}\mu_{WW}} + \frac{\mu^2_{VW}}{\mu_{VV}\mu_{WW}}. \tag{9.4.1}$$

Making several simplifying assumptions (two-delay channel, delayed component and CCI modelled as Gaussian processes), the error probability for symmetrical BPSK is

$$\frac{1}{2} - \frac{1}{2}\rho_{\text{pilot}} \tag{9.4.2}$$

and for $\pi/4$-QPSK, it is

$$\frac{1}{2} - \frac{\rho_{\text{pilot}}/2}{\sqrt{2-\rho_{\text{pilot}}^2}}. \tag{9.4.3}$$

The results of this analysis show that the error floor is decreased by about a factor of 1.3 compared to differential detection. Pilot-symbol-aided 16 PSK and 16 QAM is analyzed in [19].[1]

9.5 DIFFERENTIALLY DETECTED PSK

For a purely Rayleigh-fading channel, the results for the error probability can be best represented in the notation of [22]. All we need for this are the correlation coefficients of the signals that we already computed in Section 8.1.3. For $\pi/4$-DQPSK we get

$$BER = \frac{1}{2} - \frac{1}{4}\left\{\frac{b_0 \operatorname{Re}\{\rho\}}{\sqrt{(\operatorname{Re}\{\rho\})^2 + (1-|\rho|^2)}} + \frac{b_0' \operatorname{Im}\{\rho\}}{\sqrt{(\operatorname{Im}\{\rho\})^2 + (1-|\rho|^2)}}\right\} \tag{9.5.1}$$

where b_0 and b_0' are the two bits transmitted by one symbol. Let us next analyze the limiting cases: for the flat-fading case, we get

$$BER_{\text{flat}} = \frac{1}{4} B_n N_0 \frac{|d(0)|^2 + |d(-T)|^2 - 2\xi_n(T) \operatorname{Re}\left\{e^{-j\pi/4} d(0) d^*(-T)\right\}}{\operatorname{Re}^2\left\{e^{-j\pi/4} d(0) d^*(-T)\right\}}. \tag{9.5.2}$$

For the error floor due to time dispersion, we get

$$BER_{\text{floor}} = \frac{1}{2}\left[T \frac{\left|\frac{\mathrm{d}}{\mathrm{d}t} d(t)|_{t=0} d(-T) - \frac{\mathrm{d}}{\mathrm{d}t} d(t)|_{t=-T} d(0)\right|}{\sqrt{2} \operatorname{Re}\left\{e^{-j\pi/4} d(0) d^*(-T)\right\}}\right]^2 \left(\frac{S}{T}\right)^2. \tag{9.5.3}$$

Bibliographical Notes: The influence of log-normal fading is discussed in [23]. A discussion is also given in [24]; it analyzed the diversity case, and we discuss results in Chapter 12. Reference [25] analyzed the performance of $\pi/4$-DQPSK with various detection schemes. It found that a coherent open-loop detector gave slightly better performance in the flat-fading case (noise and CCI) than did a differential detector, but that there was practically no difference when ISI-induced errors dominate.

[1] We note in that context that [20] has also considered a pilot-symbol-based MPSK system, but in a flat-fading environment; these results were also used in [21], combined with measured delay power profiles.

Bibliography

[1] P. A. Bello and B. Nelin, "Predetection diversity combining with selectively fading channels," *IEEE Trans. Comm.*, vol. 10, pp. 32–42, 1962.

[2] J. Chuang, "The effects of time delay spread on portable radio communications channels with digital modulation," *IEEE Selected Areas Comm.*, vol. 5, pp. 879–888, 1987.

[3] I. Crohn, G. Schultes, R. Gahleitner, and E. Bonek, "Irreducible error performance of a digital portable communication system in a controlled time-dispersion indoor channel," *IEEE Selected Areas Comm.*, vol. 11, pp. 1024–1033, 1993.

[4] A. F. Molisch, "A new method for the computation of the error probability of differentially detected FSK and PSK in mobile radio channels—the case of minimum shift keying," *Wireless Personal Comm.*, vol. 9, pp. 165–178, 1999.

[5] J. Chuang, "The effect of delay spread on 2-PSK, 4-PSK, 8-PSK and 16-QAM in a portable radio environment," *IEEE Trans. Vehicular Techn.*, vol. 38, pp. 43–45, 1989.

[6] V. Lipovac and A. F. Molisch, "On the performance of MSK signal transmission over a multipath channel with small time dispersion," in *Proc. VTC'95*, pp. 25–29, 1995.

[7] F. Adachi and J. Parsons, "Error rate performance of digital FM mobile radio with postdetection diversity," *IEEE Trans. Comm.*, vol. 37, pp. 200–210, 1989.

[8] A. F. Molisch, J. Fuhl, and V. Lipovac, "An improved equation for the computation of the bit error probability of MSK in mobile radio channels with small delay spread," in *COST 231*, p. TD95(62), 1995.

[9] J. Horikoshi and S. Shimura, "Multipath distorsion of differentially encoded GMSK with 2-b differential detection in bandlimited frequency-selective mobile radio channel," *IEEE Trans. Vehicular Techn.*, vol. 39, pp. 308–315, 1990.

[10] M. K. Simon and C. C. Wang, "Differential detection of Gaussian MSK in a mobile radio environment," *IEEE Trans. Vehicular Techn.*, vol. 33, pp. 307–320, 1984.

[11] I. Korn, "M-ary CPFSK-DPD with L-diversity maximum ratio combining in Rician fast-fading channels," *IEEE Trans. Vehicular Techn.*, vol. 45, pp. 613–621, 1996.

[12] A. F. Molisch, "Error floor of Gaussian minimum shift keying in mobile radio channels with large-scale and small-scale fading," *Archiv f. Elektronik u. Uebertragungstechn.*, vol. 51, pp. 290–295, 1997.

[13] P. Bello and B. D. Nelin, "The effect of frequency selective fading on the binary error probabilities of incoherent and differentially coherent matched filter receivers," *IEEE Trans. Comm.*, vol. 11, pp. 170–186, 1963.

[14] I. Korn, "Error floors in the satellite and land mobile channels," *IEEE Trans. Comm.*, vol. 39, pp. 833–837, 1991.

[15] H. Arnold and W. Bodtmann, "The performance of FSK frequency-selective Rayleigh fading," *IEEE Trans. Comm.*, vol. 31, pp. 568–572, 1983.

[16] D. Wedge, "Error rate analysis of broadband binary FM in an indoor channel," in *Proc. PIMRC'97*, pp. 251–255, 1997.

[17] F. Xiong and S. Bhatmuley, "Performance of MHPM in Rician and Rayleigh fading mobile channels," *IEEE Trans. Comm.*, vol. 45, pp. 279–283, 1997.

[18] C.-L. Liu and K. Feher, "Pilot-symbol aided coherent M-ary PSK in frequency-selective fast Rayleigh fading channels," *IEEE Trans. Comm.*, vol. 42, pp. 54–62, 1994.

[19] H. K. Lau and S. W. Cheung, "Pilot-symbol aided 16PSK and 16QAM for digital land mobile radio systems," *Wireless Personal Comm.*, vol. 8, pp. 37–51, 1998.

[20] J. Proakis, "Probabilities of error for adaptive reception of M-phase signals," *IEEE Trans. Comm.*, vol. 16, pp. 71–81, 1968.

[21] R. Bultitude and A. Leslie, "Propagation measurement-based probability of error predictions for digital land-mobile radio," *IEEE Trans. Vehicular Techn.*, vol. 46, pp. 717–729, 1997.

[22] F. Adachi and K. Ohno, "BER performance of QDPSK with postdetection diversity reception in mobile radio channels," *IEEE Trans. Vehicular Techn.*, vol. 40, pp. 237–249, 1991.

[23] N. Benvenuto and L. Tomba, "Performance comparison of space diversity and equalization techniques for indoor radio systems," *IEEE Trans. Vehicular Techn.*, vol. 46, pp. 358–368, 1997.

[24] A. A. Abu-Dayya and N. Beaulieu, "Micro- and macrodiversity MDPSK on shadowed frequency-selective channels," *IEEE Trans. Vehicular Trans.*, vol. 43, pp. 2334–2343, 1995.

[25] S. Goode, H. Kazecki, and D. Dennis, "A comparison of limiter-discriminator, delay and coherent detection for pi/4 QPSK," in *Proc. VTC'90*, pp. 687–694, 1990.

Chapter 10

MODULATION FORMATS AND RECEIVER STRUCTURES FOR THE REDUCTION OF THE ERROR FLOOR

10.1 FRACTIONAL-BIT DETECTION

Let us now consider a situation where an unfiltered MSK signal is sent through a channel whose maximum excess delay is smaller than one bitlength. Strong distortions are introduced only in the bit transition regions, i.e., $0 < t < \tau_{\max}$,[1] where the contributions from adjacent bits that arrive in different multipath components overlap in a constantly changing way. During the period $\tau_{\max} < t < T$, on the other hand, the interference of the different multipath components stays constant, resulting only in an amplitude scaling and a constant phase shift. Such a phase shift and scaling is, however, easily cancelled out by a differential detector. If thus a differential detector for MSK did not compare the phases of the received signal at $t = 0$ and $t = T$, as is usually done, but instead compared the phases at $t = \tau_{\max}$ and $t = T$, then the distortions in the bit transition regions would be avoided [1] and no error floor would occur. Similar schemes are also possible for PSK; here the integration in the matched filter, i.e., an integrate-and-dump filter, would integrate not from 0 to T, but from $\tau_{\max}$ to T. An analogous scheme should be possible also for ASK but has to the best of our knowledge not yet been proposed in the literature.

The resistance to ISI-induced errors is bought at the price of worsened AWGN performance. Since not all the signal energy is used in the receiver, the effective SNR decreases, and thus the errors due to noise increase. This increase in the *BER*

[1] Remember that we assume $\tau_{\min} = 0$.

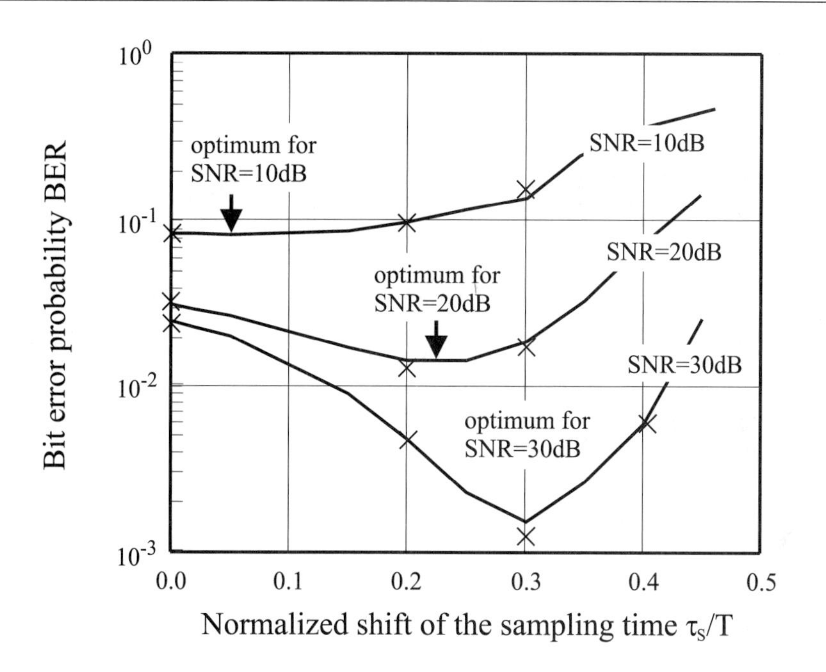

Figure 10.1. *BER* for *MSK* as function of sampling time shift τ_s (normalized to bitlength) for various SNR. Delay power profile: 10-path Rayleigh-fading channel with $S = 0.2$. Theory (solid) and Monte Carlo simulations (crosses). Note that the first sampling instant is shifted by $+\tau_s$ and the second by $-\tau_s$.

is not very pronounced; in the AWGN case, the penalty of using, e.g., 50% partial-bit integration for PSK is only about a factor of two in the *BER* [2]. However, when the maximum excess delay becomes close to one bitlength, then the penalty is too large, and a trade-off between residual time dispersion and noise is necessary. An example for this trade-off is given in Figure 10.1.

A further effect that decreases the efficiency of the fractional-bit detection scheme is the filtering at transmitter and receiver. The filtering "smears" the bit transition region over a larger range. It is thus not possible to completely exclude the bit transitions regions from the decision, so that an error floor will be present in any case; of course, the tighter the filtering, the larger the error floor. Figure 10.2 shows the error floor for various receiver widths for MSK modulation.

There are various ways to compute the *BER* with fractional-bit detection. Most of the methods mentioned in Chapter 8 do not require modifications: they are based on evaluation of Gaussian variables, and it does not matter whether these variables are created by integration over a whole bit period or only over part of it. The QFGV (Section 8.3) method [1] and the TPEM (Section 8.8) method [3] do not require any changes. The group delay method requires one simple modification [4]: we define

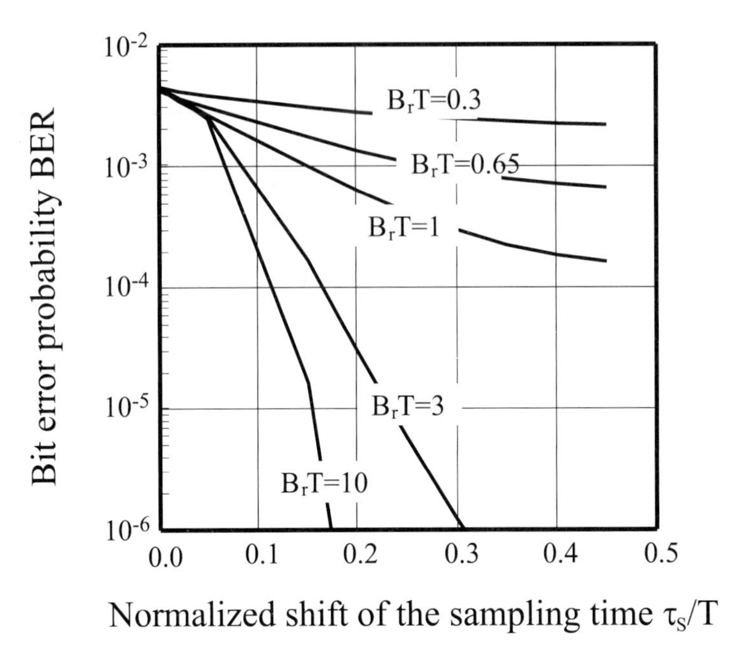

Figure 10.2. Error floor of fractional-bit detection. 10-path Rayleigh-fading channel (rectangular delay power profile; $S/T = 0.1$). $H(f) = B_r^2/(B_r^2 + f^2)$.

an *effective* delay spread that is the square root of the second central moment of that part of the power delay profile that lies within the detection interval. This *effective* delay spread is then inserted into (8.6.5).

10.2 NONLINEAR FREQUENCY DISCRIMINATOR

Another alternative to reduce errors in binary FSK is the use of a receiver structure that consists of a frequency discriminator, a limiter, and a low-pass filter (usually, an integrate-and-dump filter) [5]. This structure differs in one important respect from the usual limiter-discriminator-integrator (LDI) structure: the limiter is placed *after* the frequency discriminator and allows only instantaneous frequencies with $|f| < h_{\text{mod}}/2$. It can be shown that in the usual LDI, time-dispersion errors are caused by bursts in the instantaneous frequency that occur in the bit transition regions. If these bursts are clipped, i.e., their magnitude is limited, the error floor is drastically reduced. For unfiltered MSK and $\tau_{\max} < T/2$, the error floor is completely eliminated; as for the fractional-bit detection, the error floor becomes finite when filtering is applied because the filtering smears the instantaneous-frequency bursts over a longer duration. Again, there is a strong dependence of the error floor on the tightness of the filtering. Figure 10.3 shows the error floor for vari-

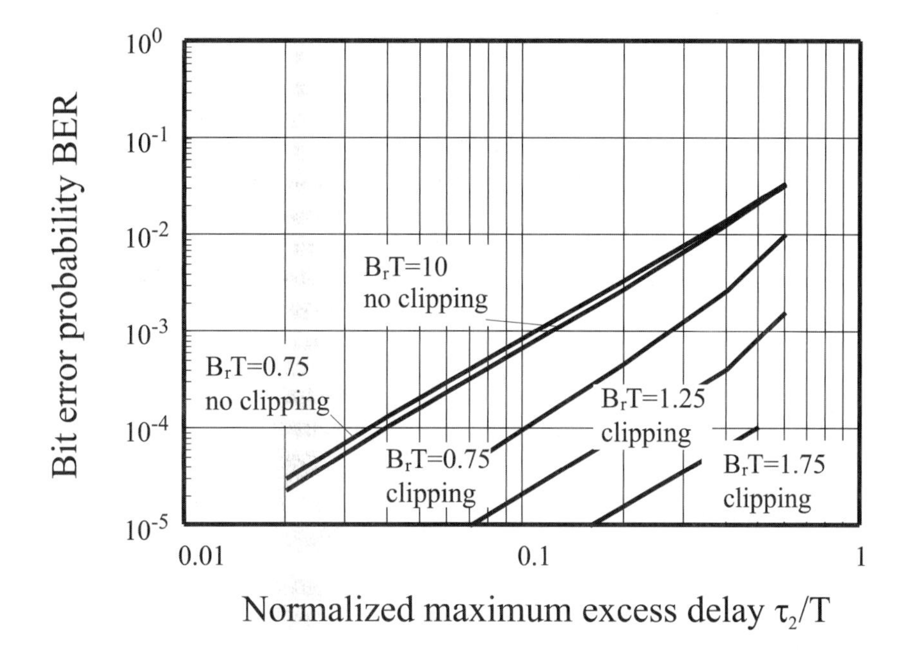

Figure 10.3. Error floor with nonlinear frequency-discriminator-integrator for various widths of the receiver filter. Reproduced from [5], copyright IEEE.

ous filter bandwidths. As a computation method, the error region method is well suited; other methods based on the properties of Gaussian variables are less suitable because the receiver is nonlinear.

10.3 MODULATION FORMATS FOR REDUCING THE ERROR FLOOR

An ingenious scheme to exploit time dispersion by the choice of the modulation scheme has been proposed by Ariyavisitakul et al. [6]. This scheme, called DSK (double phase-shift keying) is based on the principle of shifting the phase not only once during one bit period, as is done at the beginning of each bit for conventional symmetrical BPSK, but twice. The shift in the middle of the bit is always by $+\pi/2$ Assuming a two-delay channel with $\tau_2 < T$, the output signal is in region a (see Figure 10.4):

$$\begin{array}{ll} 1-(\widetilde{r})^2 & 0 \to 1 \\ -\left[1-(\widetilde{r})^2\right] & 1 \to 0 \\ 1+(\widetilde{r})^2+2\,(\widetilde{r})\sin(\varphi) & 1 \to 1 \\ -\left[1+(\widetilde{r})^2-2\,(\widetilde{r})\sin(\varphi)\right] & 0 \to 0 \end{array} \tag{10.3.1}$$

in region b

$$\begin{array}{ll} 1+(\widetilde{r})^2+2\,(\widetilde{r})\cos(\varphi) & x \to 1 \\ -\left[1+(\widetilde{r})^2+2\,(\widetilde{r})\cos(\varphi)\right] & x \to 0 \end{array} \tag{10.3.2}$$

in region c

$$\begin{array}{ll} 1+(\widetilde{r})^2+2\,(\widetilde{r})\sin(\varphi) & x \to 1 \\ -\left[1+(\widetilde{r})^2-2\,(\widetilde{r})\sin(\varphi)\right] & x \to 0 \end{array} \tag{10.3.3}$$

and region d just like region b; x denotes an arbitrary bit, and $\widetilde{r}$ has been defined in Section 8.7. The difference from conventional symmetrical BPSK lies in the presence of region c. The sum of the contributions from region b and c always gives a nonzero value (i.e., there is no eye closure). Furthermore, the contribution from region a never becomes larger than the contribution from the other regions. The optimum sampling time (when adaptive sampling is used; see Chapter 11) always lies between 0 and τ_2, in contrast to the standard case. This is also illustrated in Figure 10.4.

The big advantage of DSK lies, however, not only in the fact that it eliminates the detrimental effects of time dispersion but that it exploits its beneficial effects, namely, the built-in diversity, without the need for an equalizer. While it does not have the full diversity effect, e.g., of a Viterbi equalizer, it leads to a considerable reduction of the BER (more than an order of magnitude) for a two-delay profile, and even more so when more independent multipath components are present. The properties of this scheme were also investigated experimentally [8].

The drawback of this modulation format is its low spectral efficiency. Since there are two phase changes per bitlength, straightforward Fourier analysis shows that the required bandwidth must be larger than for BPSK. There are essentially two ways to remedy that: (i) the phase shift during each bit duration is not made equal to $\pi/2$, but to some smaller angle θ_{DSK}, e.g., $\theta_{\mathrm{DSK}} = \pi/4$ [6]; and (ii) the transmitted signal can be filtered. A combination of these two approaches can make DSK more spectrally efficient than BPSK and comparable to MSK.

In the last ten years, several similar modulation formats have been suggested. References [7], [9] proposed PSK-RZ, where the PSK signal is transmitted only during the first half of each bit, while in the second half, no signal is transmitted. This

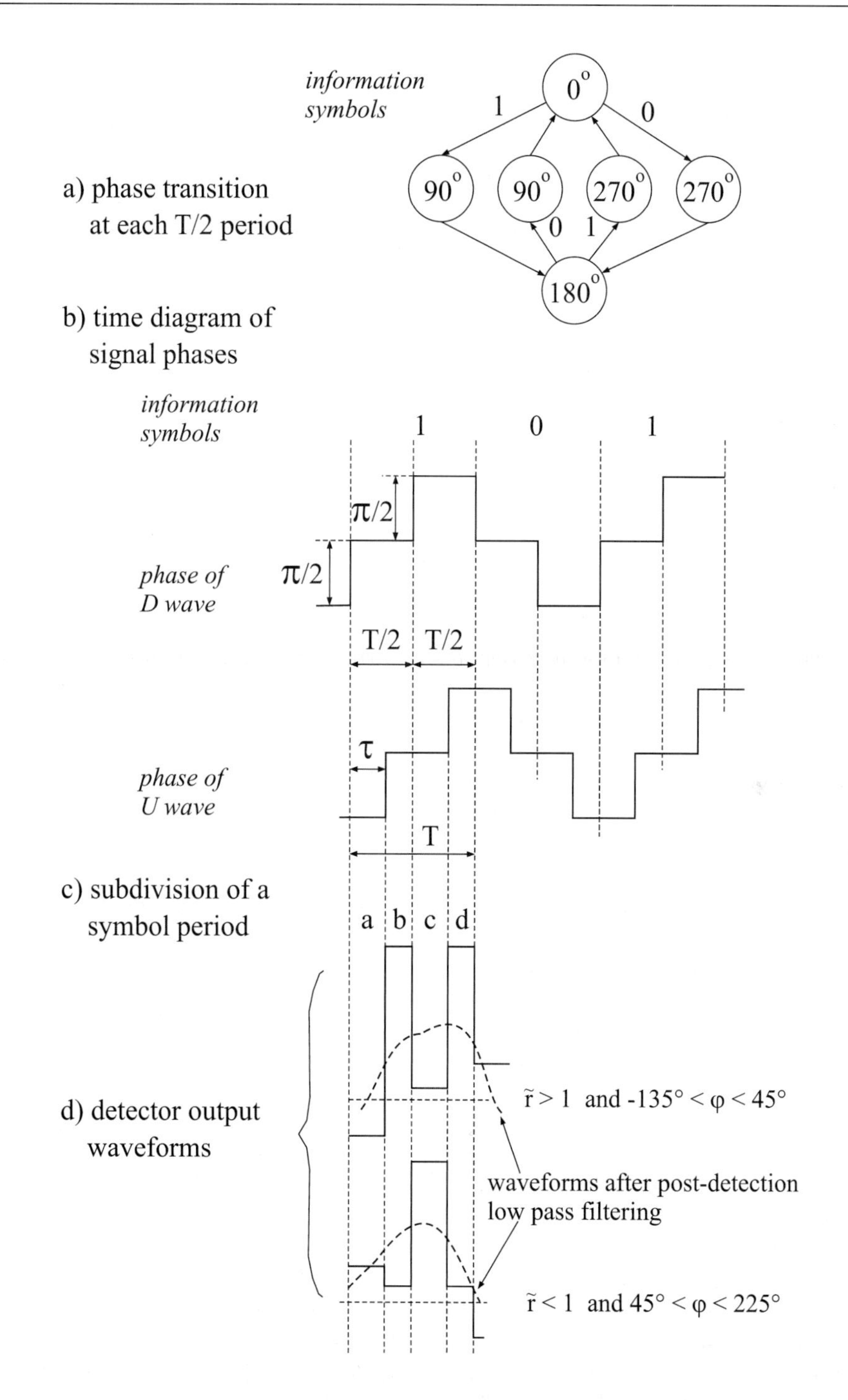

Figure 10.4. Signal format of DSK and computed waveforms for two-delay model. D und U wave are the contributions from the direct and the delayed channel echo, respectively. Adapted from [7], copyright IEEE.

modulation scheme allows excess delays of up to one bitlength. An alternative is to combine the PSK-modulator with a Manchester coder [10]. This can be generalized to a system called PSK-VP (PSK with variable phase), where the phase is not kept constant during one time slot but can vary in some way [11], [12]. Convex forms for the phase during one bitlength give the best performance. Reference [13] compared all these schemes; the authors found that Manchester-coded-PSK showed rather poor performance in comparison to the other schemes, and—not surprisingly—that the filtering had a strong influence on the performance.

All the modulation formats described above have the problem that they give a rather low spectral efficiency because of the additional phase modulation, as explained above for DSK. This decrease in spectral efficiency can be mitigated by introduction of asymmetrical pulse shapes for signalling that show essentially the same spectral efficiency as conventional PSK signals with raised-cosine filters [14]. However, the gain in admissible delay spread is rather small. For FSK, a similar and very promising method, ARC-FSK (asymmetric, raised-cosine, frequency-shift keying), was proposed [15]. For this format, there is hardly any decrease in spectral efficiency compared to GMSK if -20 dB sidelobes are taken as a criterion; however, for -40 dB sidelobes, the decrease in efficiency is larger. The decrease in the error floor is considerable; for a two-path model, delay spreads up to $0.5T$ can be tolerated for a BER of 10^{-3}.

Reference [16] compared MSK to a modulation format called sinusoidal FSK (a continuous-phase modulation format, where the phase transitions between two sampling points are not linear but sinusoidal, leading to a better spectral roll-off).

Pulse amplitude modulation in a slightly time dispersive channel (linearly frequency selective, i.e., the Taylor expansion of the transfer function includes only a linear term) allows an analytical optimization of the basis pulse $g(t)$. According to [17], the convolution of the basis pulse with its matched filter must be $\mathrm{sinc}^2(t/T)$. This choice allows one to practically eliminate the error floor up to maximum excess delays of about $0.1T$ but is less spectrally efficient than other pulse shapes, e.g., raised-cosine pulses.

Another modulation technique is an adaptive precoding of the transmit signal, so that the distortions of the precoding and of the channel cancel each other [18]. Essentially, this means shifting the effort of equalization from the transmitter to the receiver. The theoretical advantage of this technique is that at the transmitter, no error propagation (as in decision-feedback equalizers) can occur. From a practical point of view, the complicated and expensive equalization can be shifted from the mobile station to the base station. Problems can occur because the transmitter needs an excellent knowledge of the channel, which is difficult in FDD systems and in time-varying TDD systems.

Reference [19] proposed a symbol-rate adaptive modulation scheme, where the symbol rate is chosen according to the current SNR and delay spread of the channel. Reference [20] suggested doing frequency hopping and using the channels with good transmission quality to transmit the more significant bits, while using those with bad quality for the less significant bits.

Bibliography

[1] S. Ariyavisitakul, S. Yoshida, F. Ikegami, and T. Takeuchi, "Fractional-bit differential detection of MSK: A scheme to avoid outages due to frequency-selective fading," *IEEE Trans. Vehicular Techn.*, vol. 36, pp. 36–42, 1987.

[2] R. Pawula, "Refinements to the theory of error rates for narrow-band digital FM," *IEEE Trans. Comm.*, vol. 36, pp. 509–513, 1988.

[3] A. F. Molisch and E. Bonek, "Computation of the bit error probability of MSK with fractional-bit detection in time-dispersive awgn fading channels," in *Proc. 1996 SNRV and NUTEK Conf. on Radio Sciences and Telecommunications—RVK 96*, pp. 390–394, 1996.

[4] I. Crohn and E. Bonek, "Suppression of the irreducible errors in a frequency channel by fractional-bit differential detection," *IEEE Trans. Vehicular Techn.*, vol. 43, pp. 1039–1047, 1994.

[5] R. Petrovic and A. F. Molisch, "Reduction of multipath effects for FSK with frequency-discriminator detection," in *Proc. PIMRC'97*, pp. 943–948, 1997.

[6] S. Ariyavisitakul, S. Yoshida, F. Ikegami, and T. Takeuchi, "A novel anti-multipath modulation technique DSK," *IEEE Trans. Comm.*, vol. 35, pp. 1252–1264, 1987.

[7] S. Ariyavisitakul, S. Yoshida, F. Ikegami, K. Tanaka, and T. Takeuchi, "A power-efficient linear digital modulator and its application to an anti-multipath modulation PSK-RZ scheme," in *Proc. VTC'87*, 1987.

[8] T. Takeuchi, F. Ikegami, S. Yoshida, and N. Kikuma, "Comparison of multipath delay characteristics with BER performance of high speed digital mobile transmission," in *Proc. VTC'88*, pp. 119–126, 1988.

[9] S. Yoshida, T. Takeuchi, M. Nakamura, and F. Ikagami, "High bit-rate field transmission of an anti-multipath modulation technique PSK-RZ," in *Proc. VTC'90*, pp. 527–532, 1990.

[10] S. Yoshida and F. Ikegami, "An anti-multipath modulation technique - Manchester-coded PSK (MC-PSK)," in *Proc. ICC'87*, pp. 39.3.1–39.3.5, 1987.

[11] H. Takai, "BER performance of anti-multipath modulation scheme PSK-VP and its optimum phase-waveform," *IEEE Trans. Vehicular Techn.*, vol. 42, pp. 625–639, 1993.

[12] H. Takai, "In-room transmission BER performance of anti-multipath modulation PSK-VP," *IEEE Trans. Vehicular Techn.*, vol. 42, pp. 177–185, 1993.

[13] K. Ju, U. Goni, and A. Turkmani, "Comparative evaluation of the performance of anti-multipath modulation techniques for digital mobile radio systems," in *Proc. VTC'94*, pp. 1575–1561, 1994.

[14] C.-L. Liu and K. Feher, "An asymmetrical pulse shaping technique to combat delay spread," *IEEE Trans. Vehicular Techn.*, vol. 42, pp. 425–433, 1993.

[15] P. Leung, "ARC-FSK—a new anti-multipath modulation for mobile and personal communications systems," in *Proc. VTC'96*, pp. 805–809, 1996.

[16] M. Tahernezhadi and J. Cruz, "Frequency selective fading and adjacent channel interference performance of MSK and SFSK," *IEEE Trans. Comm.*, vol. 38, pp. 102–108, 1990.

[17] W. S. Leon, U. Mengali, and D. P. Taylor, "Equalization of linearly frequency-selective fading channels.," *IEEE Trans. Comm.*, vol. 45, pp. 1501–1503, 1997.

[18] W. Zhuang and W. V. Huang, "Phase precoding for frequency-selective Rayleigh and Rician slowly fading channels.," *IEEE Trans. Vehicular Techn.*, vol. 46, pp. 129–142, 1997.

[19] T. Ue, S. Sampei, and N. Morinaga, "Symbol rate controlled adaptive modulation/TDMA/TDD for wireless personal communication systems," *IEICE Trans. Comm.*, vol. E78-B, pp. 1117–1124, 1995.

[20] T. Kumagai, K. Kobayasi, K. Kawazoe, and S. Kubota, "Bit significance selective frequency diversity transmission," *IEICE Trans. Comm.*, vol. E81-B, pp. 545–552, 1998.

Chapter 11

ADAPTIVE SAMPLING

11.1 BLIND ADAPTIVE SAMPLING

Up to now, we have considered the sampling time t_s to be fixed for all times. With adaptive sampling, we mean that t_s is adjusted to the instantaneous channels configuration.

Yoshida et al. [1], [2] did a pioneering study on the effects of adaptive sampling on the error probability in time-dispersive environments for various modulation formats. Their analysis is based on coherent reception and a two-delay channel model; a Gaussian receiver filter is included. They analyze the error probability when the sampling time is either midway between two zero-crossings of the eye pattern or at the time of maximum eye opening (optimum blind sampling time). Their computation procedure works in two steps. (i) For one set of channel parameters a_1, a_2, ϕ, find the sampling point and determine the minimum eye opening, i.e., the opening caused by the worst bit combination. In the case of optimum blind sampling, determination of the sampling time and of the eye opening is intermingled. (ii) Average over the channel statistics according to (8.7.3). This averaging is essentially the evaluation of a triple integral that can be approximated as [2]

$$BER = \frac{1}{2\pi}\int_0^{\pi}\int_0^1 \{1 - [1 + w(y,\phi)]\exp[-w(y,\phi)]\}\,\mathrm{d}y\mathrm{d}\phi, \tag{11.1.1}$$

with

$$w(y,\phi) = \frac{\varkappa}{\Gamma}\frac{1+\sigma_1^2/\sigma_2^2}{(1-y)[\,G\,(\sqrt{\frac{y}{1-y}\frac{\sigma_1^2}{\sigma_2^2}},\phi,\tau_2)\,]^2} \tag{11.1.2}$$

where Γ is the average SNR, G is the eye height at the specified sampling time, and $\varkappa$ is a threshold so that a step function (with value 0.5 for SNR$<\varkappa$ and zero elsewhere) is a good approximation to the AWGN *BER* curve. Their most important result was that BPSK is superior to QPSK, OQPSK, and MSK because it exhibited no error floor.

A later paper [3] showed microscopic BER measurements and confirmed that the errors are really mainly due to fluctuations in the clock timing. A problem that was little addressed in that paper is how the original word synchronization is achieved and what happens when synchronization is lost. It seems that ideal word synchronization was assumed. Reference [4] analyzed the number of cycle slips per second. A cycle slip happens when the optimum sampling instant jumps from $-T/2$ to $T/2$ (or vice versa). The number of cycle slips per second was determined as

$$\sqrt{2}\pi f_D \frac{S}{T}. \tag{11.1.3}$$

11.2 ADAPTIVE SAMPLING WITH TRAINING SEQUENCE

A somewhat different situation occurs when the optimum sampling instant is determined with the help of a training sequence [5]. Let us assume a TDMA system where the channel stays constant during one data burst. At the beginning of a burst, a training sequence is transmitted. During this training sequence, the received signal is detected with N_{samp}-fold oversampling. We thus can build up the following sequences of demodulator inputs:

$$\begin{array}{cccccccc} s_1 & = & r_{1,1} & r_{2,1} & r_{3,1} & r_{4,1} & .. \\ s_2 & = & r_{1,2} & r_{2,2} & r_{3,2} & r_{4,2} & .. \\ .. & .. & .. & .. & .. & .. & .. \\ s_n & = & r_{1,n} & r_{2,n} & r_{3,n} & r_{4,n} & .. \\ .. & .. & .. & .. & .. & .. & .. \end{array} \tag{11.2.1}$$

where $r_{k,n} = r\left[kT + (n - N_{\text{samp}})/N_{\text{samp}}\right]$. In (11.2.1), we let n run from 1 to $3N_{\text{samp}}$, so that the sampling offset t_s lies between $-T$ and $2T$; it can be shown that the optimum t_s always lies in that interval. Letting t_s vary in that range can also be interpreted as performing a joint bit synchronization (i.e., finding the best sampling offset within one bit length, $0 < t_s < T$) and a word synchronization (i.e., finding whether a certain decided bit $\widehat{b}_k$ corresponds to the transmitted bit b_{k-1}, or b_k,...). We then compare the s_n to the original data of the training sequence and compute the BER during the training sequence for each possible n. For the reception of the data, we use the n that results in the lowest BER during the training sequence, or some similar criterion; see below. For pure MSK with differential detection, this approach completely eliminates the error floor if no filtering is employed. If filtering is introduced in the data sequence, in the transmitter, or the receiver, an error floor occurs that is smaller than in the fixed-sampling case. This error floor is $K(S/T)^2$, where the proportionality constant K depends on how narrow the filtering is. Similar results can be obtained for differential BPSK [6]. However, for $\pi/4$-DQPSK, the use of adaptive sampling reduces the error floor only by a factor of 2 even in the unfiltered case, and even less in the filtered case. This reduction is caused by quadrature component interference [7].

To understand the basic difference between adaptive sampling (with training sequence) and fixed sampling, we have to discuss why, in a time-dispersive channel, the optimum sampling time can be shifted away from the mean delay. Consider Figure 11.1, where we show the decision variable $\Delta\Phi_r$, i.e., the phase difference between $r(kT + t_s)$ and $r((k-1)T + t_s)$, for various bit tuplets as function of the sampling time t_s; this results in a graph similar to an eye pattern. In a well-behaved channel, sampling on the average mean delay, i.e., in the middle of the impulse response, gives good results because the eye at that point is wide open (Figure 11.1a). In a strongly distorting channel (Figure 11.1b), the eye at the average mean delay is completely closed, while it is open near $-T/2$.[1] Note that the eye is seemingly open near $T/2$; however, we cannot make correct decisions when sampling there, because the decision variables for those cases where a $+1$ and a -1 were actually transmitted are intermingled (i.e., the dashed and the solid lines are not separated by the threshold). This shows that adaptation without a training sequence is not possible: without a training sequence, we cannot distinguish between a seemingly open and a truly open eye.

A similar situation occurs in Figure 11.1c; there, the eye is open at $T/2$, closed at 0, and seemingly open at $-T/2$. Note that all those eye patterns were computed for a maximum excess delay of $0.1T$, but the effect is a shift of the optimum sampling time by $\pm T/2$ or more with respect to the average mean delay. This also explains why adaptive sampling on the *instantaneous mean delay* is not effective: the instantaneous mean delay always lies between 0 and the maximum excess delay; in the above figure, we could thus vary the sampling time only between 0 and $0.1T$.

For the computation of the error floor in a system with adaptive sampling and filtering, an extension of the error region method can be used [5]. An error region contains all channel configurations, i.e., points in the complex plane, where we cannot get error-free reception *regardless of the sampling time.* Thus, the sampling time for each point in the complex plane can be different; this is in marked contrast to the fixed-sampling case described in Chapter 8. Another important difference is that now we have to consider all possible bit combinations *jointly.* For a fixed sampling time, we can identify the error regions for each bit combination separately and then add up the errors. For adaptive sampling, we have to determine the optimum sampling time for each channel configuration by averaging the instantaneous BER over all possible bit combinations.

Figure 11.2 shows the error floor for various amounts of filtering. Also for adaptive sampling, it is true that the error floor depends, to a first approximation, only on the delay spread. This is true *only* if the maximum excess delay is smaller than a bitlength T.

The effect of noise can be included by a combination of the group-delay method

[1] There is also a small range near $T/2$ where the eye is truly open. However, even a small jitter in the sampling time will get us outside this range, so this is not considered further. For details, see [8].

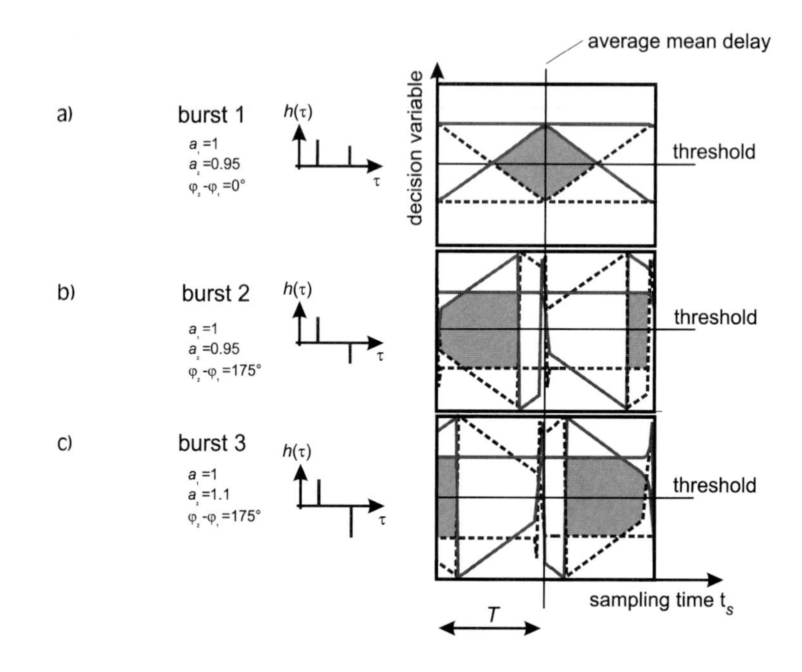

Figure 11.1. Decision variable (phase difference) as a function of t_s for various bit combinations. Solid lines: bit we want to detect is +1; dotted lines: −1. $\tau_{\max} = 0.1T$.

with the ABGV method [11]. As in the usual group delay method, the channel distortions are characterized by a single parameter, the group delay. With this parameter, it is possible to compute $\Delta\Phi_r$, the effective phase difference without the influence of noise. Once we know this phase difference, we can compute the effective error probability for this channel configuration by the ABGV equations. We then choose the sampling time that results in the lowest instantaneous BER. This is done on a discrete grid for all possible group delays. Then, we average the instantaneous BER over the pdf of the group delay.

Finally, we note that for GMSK in a time-dispersive channel, an instantaneous threshold of 0 is not optimum for all situations; rather, the threshold should be chosen adaptively. The most straightforward approach to finding the optimum threshold is an exhaustive search, which, however, is quite CPU-time intensive. A semianalytical algorithm both for single-bit and two-bit differential detection has been proposed [12]. This adaptive choice of the threshold decreases the error floor by an order of magnitude. We note, however, two caveats: (i) the decision in that receiver is not based on the phase difference between two subsequent samples, but on the quantity $\mathrm{Im}\{r(t)r^*(t-T)\}$; and (ii) the synchronization method in that reference is the maximum correlation method (see below), and the error floor achieved with

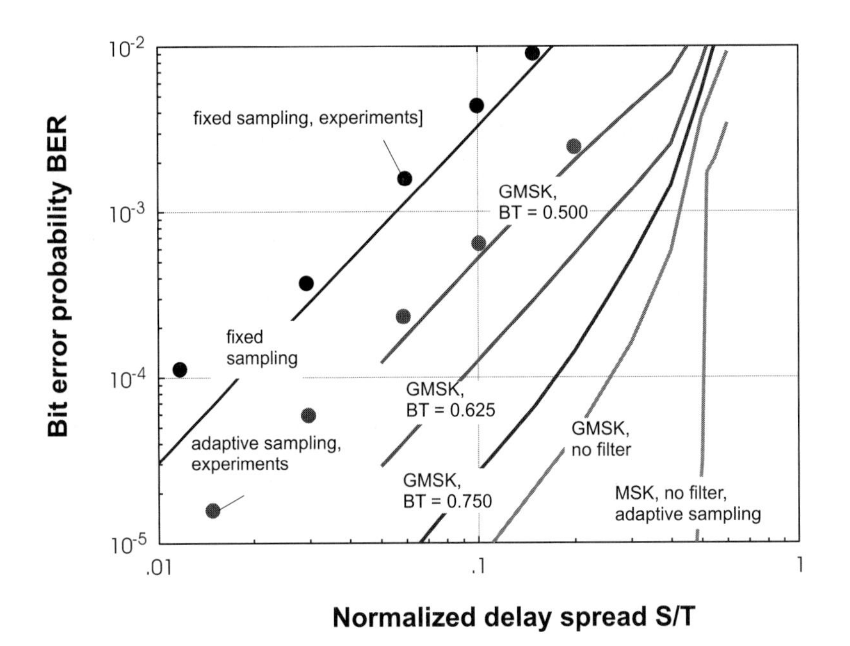

Figure 11.2. Error floor of filterered (G)MSK with adaptive sampling. For GMSK, $B_G T = 0.5$ is used. From [8]. Fixed sampling experiments from [9], adaptive sampling experiments from [10].

that method is quite large. It is uncertain how this adaptive threshold affects other synchronization methods.

The idea of first determining the optimum sampling phase and then using it in connection with symbol-spaced samples for the actual detection process can also be applied to systems with equalizers; this has been proposed and verified by simulations in [13].

11.3 SYNCHRONIZATION WITH TRAINING SEQUENCE

Choosing the sampling time that gives the smallest number of errors during the training sequence is optimum for minimizing the error floor. It is, however, not suitable for noisy channels. In recent years, there has been considerable interest in finding good synchronization mechanisms for time-dispersive fading channels. Usually, the synchronization algorithms try to simultaneously estimate the sampling time and the frequency offset between the received carrier and the local oscillator. To simplify the discussion, we assume here that the frequency offset is zero and we just have to estimate the sampling time. Furthermore, we also neglect the papers that deal with synchronization in an equalized system; for this subject, see, e.g.,

[14] and references therein.

Estimation of channel phase shift: One class of synchronization algorithms tries to estimate the phase shift ϑ introduced by the channel and, by compensating for it, produce flat-fading conditions. One such algorithm is described, e.g., in [15]. The signal is sampled with the Nyquist rate $1/\Delta T$ so that no information is lost. The shift ϑ is determined by minimizing the log-likelihood function

$$\lambda'_s(\vartheta) = \sum_n \mathrm{Re}\left\{ r(n\Delta T) d^*(n\Delta T) \exp(-j\vartheta) \right\} \tag{11.3.1}$$

with respect to ϑ.

Eye pattern analysis: Lopes and coworkers [16], [17] have proposed a metric that is based on an eye-pattern analysis of the received signal. The received signal is oversampled. When there is just one region, $t_1 < t < t_2$, where the eye is open, the actual sampling time is placed near the middle of this eye. When there are several disjoint regions, then the sampling time is put in the middle of the largest region; this technique results in a small sensitivity to timing jitters. The description given above is an intuitive (but only approximate) interpretation of the metric given in [17].

Maximum correlation: Another method lies in finding the time instant that gives maximum correlation between the training sequence and the received signal [18], [12]. This method can be best applied if the training sequence is a PN sequence. For the DECT system, a 12-bit subsection of the training sequence is a PN sequence. This procedure should in principle be capable of giving not only the optimum sampling instant, but also the word synchronization. Simulation results [12] show, however, results that are very close to the *fixed-sampling* results. It is not completely clear from that paper whether this result is due to the principle of the algorithm or whether the word-synchronization was done in a suboptimum way.

Metrics for oversampled signals: An alternative lies in oversampling the received signal during a training sequence and finding the sample that fits best to a given metric, as described in (11.2.1). Our aim is thus to find the n that results in the best performance for the transmission of the actual data. We determine the optimum value of n by finding the extremum (maximum or minimum with respect to n) of the expression

$$\sum_k F(r_{k,n}, b_k) \tag{11.3.2}$$

where $F(.)$ is some function (metric). For the noise-free case, the metric is simply the BER. If the channel is not noise-free and the training sequence is short (this is usually the case), alternative expressions for the metrics should be used. One possibility is to minimize the expected value of the BER during the data sequence with the help of the ABGV formulation; this approach is preferable to just measuring the BER during the training sequence. For constant-envelope signals, we can, e.g., compute the (quadratic) deviation of the received signal phase from the transmitted one. For non-constant-envelope signals, the deviations of the amplitude should also be taken into account. The performance of differentially detected GMSK with these metrics is described in [19].

For QPSK, it has been proposed to use [20]

$$\arctan\left(\frac{Q_k^{\mathrm{id}} + Q_{k,n}^{\mathrm{dev}}}{I_k^{\mathrm{id}} + I_{k,n}^{\mathrm{dev}}}\right) - \arctan\left(\frac{Q_{k-1}^{\mathrm{id}} + Q_{k-1,n}^{\mathrm{dev}}}{I_{k-1}^{id} + I_{k-1,n}^{dev}}\right) \qquad (11.3.3)$$
$$- \arctan\left(\frac{Q_k^{\mathrm{id}}}{I_k^{\mathrm{id}}}\right) + \arctan\left(\frac{Q_{k-1}^{\mathrm{id}}}{I_{k-1}^{\mathrm{id}}}\right)$$

where I^{id} and Q^{id} are the in-phase and quadrature components of the ideal signals and I^{dev} and Q^{dev} denote the in-phase and quadrature component of the deviations. k is the index for the bits, and n is an index that runs only from 0 to N_{samp}. The performance of a system using such a metric in a time-dispersive enviromnment was analyzed by Monte Carlo simulations in [21]. For $\pi/4$-DQPSK, [22] uses a metric that is the square of the phase error; this metric is similar to the phase deviation metric mentioned above.

Bibliography

[1] S. Yoshida and F. Ikegami, "A comparison of multipath distortion characteristics among digital modulation techniques," *IEEE Trans. Vehicular Techn.*, vol. 34, pp. 128–135, 1985.

[2] S. Yoshida, S. Onoe, and F. Ikegami, "The effect of sampling timing on bit error rate performance in a multipath fading channel," *IEEE Trans. Vehicular Techn.*, vol. 35, pp. 168–174, 1986.

[3] S. Yoshida, F. Ikegami, and T. Takeuchi, "Causes of burst in multipath fading channel," *IEEE Trans. Comm.*, vol. 36, pp. 107–113, 1988.

[4] Y. Karasawa, T. Kuroda, and H. Iwai, "Cycle slip in clock recovery on frequency-selective fading channels," *IEEE Trans. Comm.*, vol. 45, pp. 376–383, 1997.

[5] A. F. Molisch, L. Lopes, M. Paier, J. Fuhl, and E. Bonek, "Error floor of unequalized wireless personal communications systems with MSK modulation and training-sequence-based adaptive sampling," *IEEE Trans. Comm.*, vol. 45, pp. 554–562, 1997.

[6] A. F. Molisch and E. Bonek, "Reduction of error floor of differential PSK in mobile radio channels by adaptive sampling," *IEEE Trans. Vehicular Techn.*, vol. 47, pp. 1276–1280, 1998.

[7] A. F. Molisch and H. Bolcskei, "Error floor of pulse amplitude modulation with adaptive sampling phase in time-dispersive fading channels," in *Proc. PIMRC'98*, pp. 884–890, 1998.

[8] A. F. Molisch, M. Paier, and E. Bonek, "Performance of DECT receivers with burst-by-burst adaptive synchronization," in *Proc. IRCTR Colloquium Indoor Comm.*, pp. 27–43, 1997.

[9] I. Crohn, G. Schultes, R. Gahleitner, and E. Bonek, "Irreducible error performance of a digital portable communication system in a controlled time-dispersion indoor channel," *IEEE Selected Areas Comm.*, vol. 11, pp. 1024–1033, 1993.

[10] G. Schultes, J. Radlbauer, H. Egger, D. Walter, K. Pillekamp, and B. Pauli, "Performance of the Siemens DECT-prototype gigaset 95x in a dispersive indoor environment," in *Proc. IEEE VTC'94*, pp. 1079–1084, 1994.

[11] A. F. Molisch, M. Paier, and E. Bonek, "Analytical computation of the error probability of (g)msk with adaptive sampling in mobile radio channels," *European Transactions on Telecommunications*, vol. 9, pp. 551–559, 1998.

[12] N. Benvenuto, A. Salloum, and L. Tomba, "Performance of digital DECT radio links based on semianalytical methods," *IEEE J. Selected Areas Comm.*, vol. 15, pp. 667–675, 1997.

[13] H. Amca, T. Yenal, and K. Hacioglu, "Adaptive equalisation of frequency selective multipath fading channels based on sample selection," *IEE Proc. Comm.*, vol. 146, pp. 55–60, 1999.

[14] U. Lambrette, J. Horstmannshoff, and H. Meyr, "Techniques for frame synchronization on unknown frequency selective channels," in *Proc. VTC 1997*, pp. 1059–1063, 1997.

[15] G. Ascheid, M. Oerder, J. Stahl, and H. Meyr, "An all digital receiver architecture for bandwidth efficient transmission at high data rates," *IEEE Trans. Comm.*, vol. 37, pp. 804–813, 1989.

[16] L. Lopes and S. Safavi, "Relationship between performance and timing recovery mechanisms for a DECT link in dispersive channels," *Electronics Letters*, vol. 29, pp. 2173–2174, 1993.

[17] A. Brandao, L. Lopes, and D. McLernon, "Method for timing recovery in presence of multipath delay and cochannel interference," *Electronics Letters*, vol. 30, pp. 1028–1029, 1994.

[18] S. Safavi and L. Lopes, "Novel techniques for performance optimisation of DECT receivers with non-linear front-end," in *Proc. VTC'96*, vol. 5, pp. 824–828, 1996.

[19] M. Paier, A. F. Molisch, and E. Bonek, "Determination of the optimum sampling time in DECT-like systems," in *Proc. EPMCC'97*, pp. 459–465, 1997.

[20] N. R. Sollenberger and J. C. I. Chuang, "Low-overhead symbol timing and carrier recovery for TDMA portable radio systems," *IEEE Trans. Comm.*, vol. 38, pp. 1886–1892, 1990.

[21] J. C.-I. Chuang and N. Sollenberger, "Burst coherent demodulation with combined symbol timing, frequency offset estimation, and diversity selection," *IEEE Trans. Comm.*, vol. 39, pp. 1157–1164, 1991.

[22] S. Chennakeshu and G. Saulnier, "Differential detection of pi/4-shifted-DQPSK for digital cellular radio," *IEEE Trans. Vehicular Techn.*, vol. 42, pp. 46–57, 1993.

Chapter 12

ANTENNA DIVERSITY

Antenna diversity is one of the most common methods to combat multipath effects in mobile radio. If the antennas are sufficiently separated from each other, the signals arriving at them are approximately statistically independent. It is, for example, improbable that two antennas are simultaneously in fading dips. If we thus combine the signals from the antennas in an appropriate way, the total signal will be less likely to suffer from fading. Similarily, it is not so likely that all antennas simultaneously suffer from high instantaneous delay spread (i.e., higher than the average delay spread) [1].

One other important type of diversity is the path or frequency (i.e., delay) diversity in Rake receivers. The echoes of the signal arriving at the receiver can be interpreted as diversity signals, and the Rake receiver sums them up. Thus, the computation methods described in this section can be applied to the computation of the BER in Rake receivers. However, each of the echoes is *flat* fading, so that the methods described in this chapter (intended for time-dispersive diversity signals) are actually more general than necessary. For more details on Rake receivers, see also Part V.

12.1 TYPES OF ANTENNA DIVERSITY

The most common types of diversity are

- RSSI-driven selection diversity (RSSI-SC): here, the signal with the largest instantaneous power (received signal strength indication, RSSI) is chosen; the signals from the other antennas are thrown away. For RSSI-SC, the envelope R of the total signal is simply

$$R = \max(R_1, R_2, ...R_L) \tag{12.1.1}$$

where L is the number of diversity branches.

- Equal-gain combining (EGC): for this method, the received signals are added with correct phase; the signals from all antennas enter with equal weight. For EGC, the envelope of the total signal is

$$R = \frac{1}{\sqrt{L}} \sum_{l=1}^{L} R_l. \tag{12.1.2}$$

- Maximum-ratio combining (MRC): for this method, the signals from the antennas are also added with their correct phases, but the signals are first weighted with their SNRs. The envelope of the total signal is then

$$R = \sqrt{\sum_{l=1}^{L} R_l^2}. \tag{12.1.3}$$

- BER-driven selection diversity (BER-SC): MRC emphasizes signals with high field strength, but for frequency-selective channels, a signal might be strongly distorted even though it has high field strenth. In situations with large CCI, high field strength can mean high interference. Thus, MRC is not optimum in those cases. Better performance can be achieved by BER-driven selection diversity. For this approach, we need the transmission of a training sequence. The received signals from all diversity branches are demodulated, and the demodulated signals are compared to the original training sequence. Defining a metric (similarily to Chapter 11), we choose the diversity branch that gives the largest (or smallest) metric and use it until the next training sequence is transmitted.

- Minimum mean-square error diversity (MMSE-diversity) also known as *optimum combining* [2]: BER-SC gives good results also in the presence of ISI, but is by definition suboptimum, because it is *selection* diversity. Efficient combination is achieved by MMSE-diversity (minimum mean-square error diversity), i.e., the diversity signals are combined in such a way that the mean-square error is minimized. The general performance of this approach in a time-dispersive channel is discussed in [3], [4], [5]; optimum combining for a DECT-like system is considered in [6]. A wealth of investigations deals with spatio-temporal processing with equalizers or Rake receivers; see also the references cited in Chapter 5.

- Postdetection combining: Adachi and coworkers ([7], [8], [9], [10]) proposed for M-ary PSK a "blind" postdetection scheme where the detected phases, not the received signals, are weighted with the instantaneous received powers in the received branches and then added.[1] This method is simpler but less efficient than MRC.

---

[1] Reference [10] actually replaced the instantaneous received power by a weighting derived from previous decisions to allow cheaper implementation.

Diversity is called macrodiversity when the antennas are so far apart that the lognormal (shadow) fading on the two antennas is approximately uncorrelated. Usually, this requires separate base stations (ports). Except when we explicitly say so, *diversity* henceforth means *microdiversity*, i.e., the type of diversity where the distance between the antennas is on the order of one wavelength.

12.2 THE QFGV METHOD

Adachi and Parsons [11], [12] investigated the error probability of binary FSK with RSSI-SC, EGC, and MRC. The starting point for their analysis is Equation (8.3.2). Next, they derive the statistics of the received signal envelope. Performing the averaging of the conditional BER over these distributions, they get for the average BER for SC

$$BER = \frac{1}{2} - \frac{1}{2}\sum_{l=1}^{L}\binom{L}{l}(-1)^{l+1}\frac{b_0 \operatorname{Im}\{\rho\}}{\sqrt{\operatorname{Im}\{\rho\}^2 + l(1-|\rho|^2)}}; \tag{12.2.1}$$

for EGC

$$BER = \frac{1}{2} - \frac{1}{2}\frac{b_0 \operatorname{Im}\{\rho\}}{\sqrt{\operatorname{Im}\{\rho\}^2 + \varepsilon(1-|\rho|^2)}} \sum_{l=0}^{L-1}\frac{(2l-1)!!}{(2l)!!}\left[\frac{(1-|\rho|^2)}{\operatorname{Im}\{\rho\}^2 + \varepsilon(1-|\rho|^2)}\right]^l \tag{12.2.2}$$

where $\varepsilon = L/\left[(2L-1)!!\right]^{1/L}$ and $(2L-1)!!$ means $1 \cdot 3 \cdot 5 ...... (2L-1)$. For MRC

$$BER = \frac{1}{2} - \frac{1}{2}\frac{b_0 \operatorname{Im}\{\rho\}}{\sqrt{1-\operatorname{Re}\{\rho\}^2}}\sum_{l=0}^{L-1}\frac{(2l-1)!!}{(2l)!!}\left(\frac{(1-|\rho|^2)}{1-\operatorname{Re}\{\rho\}^2}\right)^l. \tag{12.2.3}$$

We note that the equations for SC and MRC are exact, whereas the equation for EGC is an approximation. Similar equations are given for $\pi/4$-DQPSK in [13]. The influence of correlations is considered for FSK in [12], for $\pi/4$-DQPSK in [14]. Some of the results for FSK were confirmed experimentally for QPSK in [15].

The problem of maximum-ratio combining (MRC) has also been solved in a general way by Proakis [16]. We mentioned in Section 8.3 the solutions for the quadratic form $D = A|X|^2 + B|Y|^2 + CXY^* + C^*X^*Y$. The same reference, also gives the results for a form

$$D = \sum_{l=1}^{L} A|X_l|^2 + B|Y_l|^2 + CX_lY_l^* + C^*X_l^*Y_l \tag{12.2.4}$$

where the X_l and Y_l are independent, identically distributed Gaussian random variables. Solutions for this problem are also given in Appendix A.

Bibliographical Notes: Reference [17] evaluated (12.2.4) for M-ary CPFSK with differential phase detection with L-ary diversity in a satellite fading channel. Reference [18] computes bounds for the performance of M-PSK and M-DPSK with possible correlation between the diversity branches.

Reference [19] used a variant of the QFGV method to analyze various diversity schemes for $\pi/4$-DQPSK: it used the QFGV no-diversity equations to construct the expectation over the noise to arrive at error probabilities conditioned on the channel configuration. Then, averaging over the channel constellations gives the total error probability. This method also allows one to treat non-Gaussian channel statistics and various types of diversity combining.

For SC and MRC, [20] analyzed the flat-fading case based on the equations; it seems likely that their analysis can be extended to the frequency-selective case, but no explicit evaluations have been given up to now. The investigations of [21] compared diversity to block coding; Reference [22] computed the error probability with diversity including co-channel interference. In the latter case, both the Gaussian approximation for the interferers and an "exact" model were included—however, the exact model assumed that the timing offset between the desired signal and each interferer is known and fixed. Offset receiver diversity, where the signals of the diversity branches are mixed down to different intermediate frequencies and then combined, was analyzed for GMSK in [23] by a method that is similar to that of [12]; see also Section 8.3.

An extensive investigation of diversity was also done in [24], [25]; however, this dealt with diversity combined with equalization and is thus beyond the purview of this chapter. Similarly, a series of papers, [26], [27], [28], covered diversity combined with trellis-coded modulation. Selection diversity was analyzed by Monte Carlo simulations in [29]. Refererence [3] computed upper bounds of the BER of MMSE-diversity by the method of [30]. Also, the method of [31] as described in Chapter 8 can be used to treat diversity in flat-fading channels (and equivalently, Rake reception). Similarly, a large number of other papers cover diversity in flat-fading channels, see, e.g., the references cited in [31].

12.3 THE ERROR REGION METHOD

The error region method can be applied to the computation of the error floor of selection diversity in two-delay channels [32]. As usual in the error region method, the first step is the computation of the channel constellations that lead to errors; this step is the same as in the no-diversity case. The next step is the computation of the pdf of the normalized phasor of the received signal (normalized to the phasor of the first echo in the first diversity branch). This joint pdf is

$$\mathrm{pdf}(a_A, a_B, \varphi_A, \varphi_B) = \frac{1}{(2\pi)^2} \frac{1}{\sigma_1^2\sigma_2^2\sigma_{\mathrm{tot}}^2} \frac{a_A a_B}{\left(\frac{1}{2\sigma_1^2} + \frac{1-2a_A\cos(\varphi_A)+a_A^2}{2\sigma_2^2} + \frac{a_B^2}{2\sigma_{\mathrm{tot}}^2}\right)^3} \quad (12.3.1)$$

where indices A and B denote the diversity branches and indices 1 and 2 the two paths. a_A and φ_A denote the magnitude and phase of the phasor $1 + (a_{2,A}/a_{1,A})\cos(\varphi_{2,A} - \varphi_{1,A})$, and a_B and φ_B denote the magnitude of $a_{1,B}/a_{1,A}\cos(\varphi_{1,B} - \varphi_{1,A}) + (a_{2,B}/a_{1,A})\cos(\varphi_{2,B} - \varphi_{1,A})$. σ_1 and σ_2 are the variances in the first and second path (assumed to be identical in the two branches), and $\sigma_{\mathrm{tot}}^2 = \sigma_1^2 + \sigma_2^2$. Note that since here the *total* phasor is considered, the appropriate error circle has the origin $(1 + x_0/y_0)$, with x_0, y_0 given in Appendix A.3 of Part II.

When we have the usual (RSSI-driven) selection diversity, then the phasor with the largest magnitude is selected. The error probability is then

$$BER = 2 \iint_{\substack{\text{error}\\\text{region}}} BER(a_A, \phi_A) \int_0^{a_a}\int_0^{2\pi} \mathrm{pdf}(a_A, a_B, \phi_A, \phi_B)\mathrm{d}a_B\mathrm{d}\phi_B\mathrm{d}a_A\mathrm{d}\phi_A \quad (12.3.2)$$

where $BER(a_A, \phi_A)$ is the instantaneous BER, averaged over the bit combinations, for a channel characterised by a_A, ϕ_A. For BER-driven selection diversity, the diversity branch with the lowest BER is selected. We essentially have to compute the probability that the two phasors from the two diversity branches fall into error regions at the same time.

12.4 DIVERSITY IN SHADOWED CHANNELS

The starting point of the analysis of M-DPSK macrodiversity in a shadowed channel [33] is the BER-equation for the Rayleigh-fading channel as given in Chapter 9. The diversity antenna with the largest (long-time-averaged) power is selected. The pdf for the arriving power is then

$$\mathrm{pdf}_x(x) = \sum_{l=1}^{L_P} \frac{1}{\sqrt{2\pi}\sigma_l} \exp\left(-\frac{(x-u_l)^2}{2\sigma_l^2}\right) \prod_{m\neq l}\left[1 - Q\left(\frac{(x-u_m)}{\sigma_m}\right)\right] \quad (12.4.1)$$

where σ_m is the variance of the shadowing of the mth diversity branch, and the u_m are the mean values of the shadowing. The outage probability is then computed by first computing (analytically) the outage probability for each average signal level x (i.e., with the equations of Chapter 9), and then integrating over (12.4.1). Figure 12.1 shows an example of the resulting outage probability.

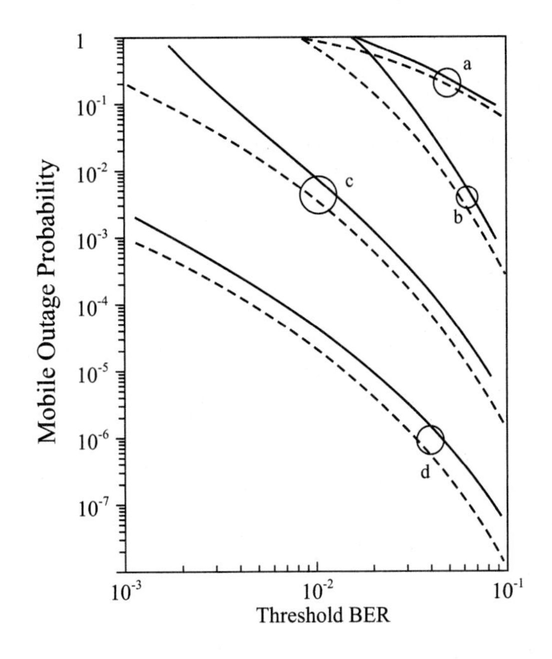

Figure 12.1. Outage probability of L_P-port macrodiversity with BPSK (solid) and QPSK (dashed) with L-branch EGC. Two-delay channel with $S = 0.1$T. Lognormal shadowing with variance 6 dB and $E_b/N_0 = 15$ dB. Curves a: $L = 1$, $L_p = 1$; curves b: $L = 1$, $L_p = 3$; curves c: $L = 2$, $L_p = 3$; curves d: $L = 4$, $L_p = 3$. Adapted from [33], copyright IEEE.

12.5 RESULTS FOR DIVERSITY WITH FIXED SAMPLING

Diversity leads to an error floor $K(S/T)^{2L}$ in a two-delay Rayleigh-fading channel [11]. The proportionality constant K depends on the type of diversity and the modulation format. Similar reductions of the error rate can also be achieved for errors due to the noise and the random FM (although the latter quantity is of little importance in high-data-rate systems). Figure 12.2 shows the error floor for selection diversity and maximum ratio combining for MSK. These tendencies were also confirmed experimentally [34].

Assuming differentially detected MSK, two-branch diversity, and fixed sampling, $K = 3(\pi/4)^4$ for RSSI-driven selection diversity and $K = (\pi/4)^4$ for BER-driven selection diversity. $\pi/4-$DQPSK is treated in [13]; [19] showed that for this modulation format, SC is only about a factor of 1.5 worse than equal-gain combining for reducing the error floor and also for combatting noise effects. 16-STAR-QAM is treated in [35], [36].

The efficiency of diversity is reduced if the two diversity branches are correlated or if the powers in the two diversity branches are unequal. The equations for $\pi/4$–DQPSK are given in [14]; the effect can be seen in Figure 12.3.

The results for FSK show very similar tendencies. A further choice lies here in the method of detection: either frequency-discriminator detection or differential detection. Frequency-discriminator detection performs better in purely time-dispersive environments (Figure 12.4) but (usually) worse in AWGN environments (flat fading). Figure 12.4 also demonstrates that the type of diversity combining has a rather small influence on the actual error probability.

12.6 DIVERSITY WITH ADAPTIVE SAMPLING

There are various ways to implement selection diversity with adaptive sampling. The most straightforward way is RSSI-driven diversity, where the signal with the largest envelope is selected. For BER-driven diversity, we need two complete receivers and a training sequence that must be repeated in intervals that are shorter than the time over which the channel changes considerably. The error floor is also proportional to $(S/T)^4$; the proportionality constant depends on the filtering. The difference between RSSI-driven and BER-driven diversity becomes more pronounced as the filtering becomes softer. These results were detailed in [32], which also interpreted the experiments of [37].

Safavi and Lopes [38], [39] proposed a slightly different receiver structure for DECT that is also based on the evaluation of the preamble. It is based on the squared envelope of the preamble signal. However, these references take not only the average value (this would be the usual RSSI-driven diversity), but compute the correlation coefficients

$$w_0 = \int |r(t)|^4 \mathrm{d}t \qquad w_1 = \int |r(t)|^2 |r(t-T)|^2 \mathrm{d}t. \tag{12.6.1}$$

Then, for both antenna signals, they check whether w_1/w_0 is larger or smaller than 0.5. If one is larger and one is smaller than 0.5, then the antenna whose signal gives the larger w_1/w_0 is selected. If both are larger or both are smaller, then the signal with the larger RSSI is selected. With this structure, an error floor of 10^{-3} for a normalized delay spread of about 0.4 for a DECT system is obtained in simulations.

The combination of equalization with diversity is discussed in [24], [25], and [40], which also includes synchronization aspects.

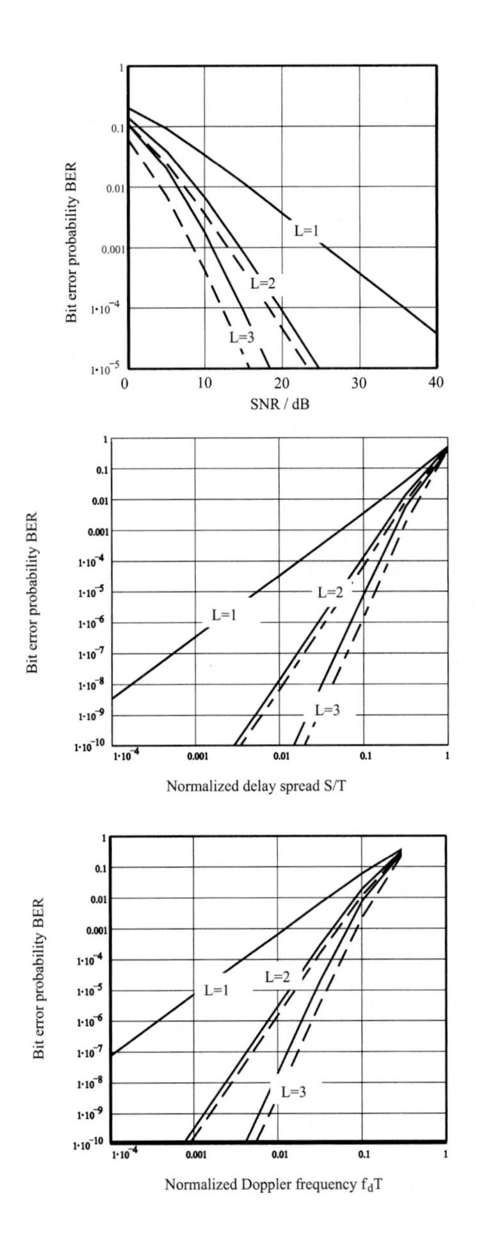

Figure 12.2. Error probability for selection combining (solid) and maximum ratio combining (dashed) for various number of diversity antennas L. Upper figure: only noise; middle figure: only time dispersion; lower figure: only random FM.

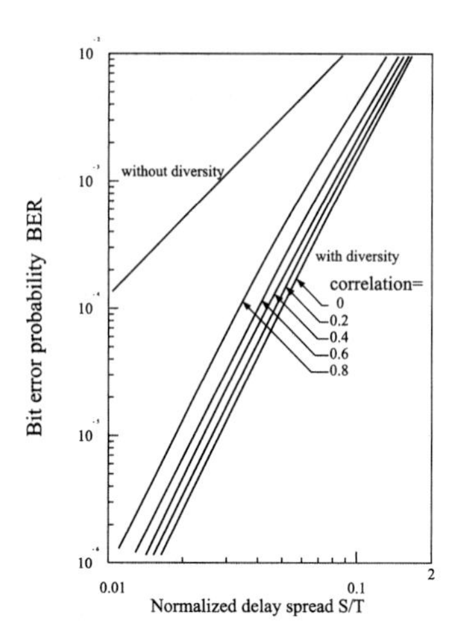

Figure 12.3. Influence of fading correlation on the error floor. Modulation format: $\pi/4$-DQPSK with raised-cosine filtering, $\alpha = 0.5$. Two-branch selection diversity, equal average power in diversity branches. Two-delay Rayleigh-fading channel. Adapted from [14], copyright IEEE.

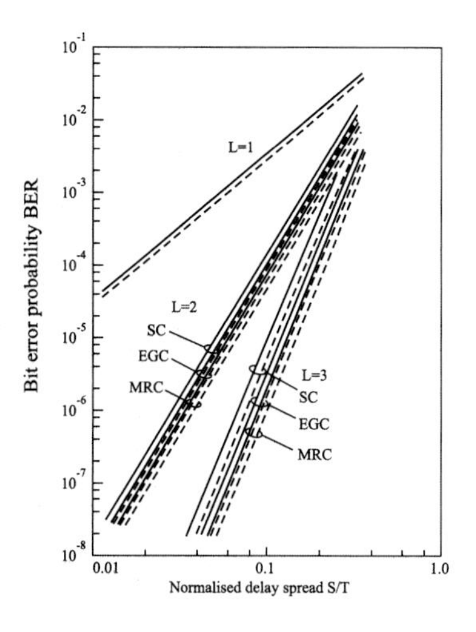

Figure 12.4. Error floor of MSK with optimum receiver filter BT product in two-delay, Rayleigh-fading channel. Differential phase detection (solid) and frequency-discriminator detection (dashed). Adapted from [11], copyright IEEE.

Bibliography

[1] P. H. Lehne and M. Pettersen, "Space diversity measurements for DECT in indoor and outdoor microcells," in *Proc. PIMRC'96*, pp. 728–732, 1996.

[2] J. H. Winters, "Optimum combining in digital mobile radio with co-channel interference," *IEEE J. Selected Areas Comm.*, vol. 2, pp. 528–539, 1984.

[3] M. Clark, L. Greenstein, W. Kennedy, and M. Shafi, "MMSE diversity combining for wide-band digital cellular radio," *IEEE Trans. Comm.*, vol. 40, pp. 1128–1135, 1992.

[4] M. V. Clark, L. J. Greenstein, W. K. Kennedy, and M. Shafi, "Matched filter performance bounds for diversity combining receivers in digital mobile radio," *IEEE Trans. Vehicular Techn.*, vol. 41, pp. 356–362, 1992.

[5] M. V. Clark, L. J. Greenstein, W. K. Kennedy, and M. Shafi, "Optimum linear diversity receivers for mobile communications," *IEEE Trans. Vehicular Techn.*, vol. 43, pp. 47–56, 1994.

[6] J. Wigard, P. E. Mogensen, F. Fredericksen, and O. Norklit, "Evaluation of optimum combining diversity in DECT," in *Proc. PIMRC'95*, pp. 507–511, 1995.

[7] F. Adachi, "Postdetection optimal diversity combiner for DPSK differential detection," *IEEE Trans. Vehicular Techn.*, vol. 42, pp. 326–337, 1993.

[8] M. Ikura and F. Adachi, "Postdetection phase combining diversity," *IEEE Trans. Vehicular Techn.*, vol. 43, pp. 298–303, 1994.

[9] F. Adachi and M. Ikura, "Postdetection diversity using differential phase detection of M-ary DPSK," *Electronics Letters*, vol. 30, pp. 1745–1746, 1994.

[10] F. Adachi, "Phase-combining diversity using adaptive decision-aided branch-weight estimation for reception of faded M-ary DPSK signals," *IEEE Trans. Vehicular Techn.*, vol. 46, pp. 786–790, 1997.

[11] F. Adachi and J. Parsons, "Unified analysis of postdetection diversity for binary digital FM mobile radio," *IEEE Trans. Vehicular Techn.*, vol. 37, pp. 189–198, 1988.

[12] F. Adachi and J. Parsons, "Error rate performance of digital FM mobile radio with postdetection diversity," *IEEE Trans. Comm.*, vol. 37, pp. 200–210, 1989.

[13] F. Adachi and K. Ohno, "BER performance of QDPSK with postdetection diversity reception in mobile radio channels," *IEEE Trans. Vehicular Techn.*, vol. 40, pp. 237–249, 1991.

[14] F. Adachi, K. Ohno, and M. Ikura, "Postdetection selection diversity reception with correlated, unequal average power Rayleigh fading signals for pi/4-shift QDPSK mobile radio," *IEEE Trans. Vehicular Techn.*, vol. 41, pp. 199–210, 1992.

[15] K. Ohno and F. Adachi, "QDPSK signal transmission performance with postdetection selection diversity reception in land mobile radio," *IEEE Trans. Vehicular Techn.*, vol. 40, pp. 798–804, 1991.

[16] J. Proakis, "On the probability of error for multichannel reception of binary signals," *IEEE Trans. Comm.*, vol. 16, pp. 68–71, 1968.

[17] I. Korn, "M-ary CPFSK-DPD with L-diversity maximum ratio combining in Rician fast-fading channels," *IEEE Trans. Vehicular Techn.*, vol. 45, pp. 613–621, 1996.

[18] D. Noneaker and M. Pursley, "Error probability bounds for M-PSK and M-DPSK and selective fading diversity channels," *IEEE Trans. Vehicular Techn.*, vol. 43, pp. 997–1005, 1994.

[19] N. Benvenuto and L. Tomba, "Performance comparison of space diversity and equalization techniques for indoor radio systems," *IEEE Trans. Vehicular Techn.*, vol. 46, pp. 358–368, 1997.

[20] S. Chennakeshu and J. B. Anderson, "Error rates for Rayleigh fading multichannel reception of MPSK signals," *IEEE Trans. Comm.*, vol. 43, pp. 338–346, 1995.

[21] A. Abu-Dayya and N. Beaulieu, "Comparison of diversity with simple block coding on correlated frequency-selective fading channels," *IEEE Trans. Comm.*, vol. 43, pp. 2704–2713, 1995.

[22] A. Abu-Dayya and N. Beaulieu, "MDPSK on frequency-selective Ricean channels with diversity and cochannel interference," *European Trans. Telecomm.*, vol. 8, pp. 221–230, 1997.

[23] K. A. M. Soliman, "BER analysis of GMSK with one-bit differential detection and offset receiver diversity in frequency-selective Rayleigh channels with CCI," in *URSI 15th National Radio Science Conf., Egypt*, pp. C33; 1–8, 1998.

[24] P. Balaban and J. Salz, "Optimum diversity combining and equalization in digital transmission with applications to cellular mobile radio—Part II: Theoretical considerations," *IEEE Trans. Comm.*, vol. 40, pp. 885–894, 1992.

[25] P. Balaban and J. Salz, "Optimum diversity combining and equalization in digital data transmission with applications to cellular mobile radio—Part II: Numerical results," *IEEE Trans. Comm.*, vol. 40, pp. 895–907, 1992.

[26] J. Ventura-Traveset, G. Caire, E. Biglieri, and G. Taricco, "Impact of diversity reception on fading channels with coded modulation—Part I: Coherent detection," *IEEE Trans. Comm.*, vol. 45, pp. 563–572, 1997.

[27] J. Ventura-Traveset, G. Caire, E. Biglieri, and G. Taricco, "Impact of diversity reception on fading channels with coded modulation—Part II: Differential block detection," *IEEE Trans. Comm.*, vol. 45, pp. 676–686, 1997.

[28] J. Ventura-Traveset, G. Caire, E. Biglieri, and G. Taricco, "Impact of diversity reception on fading channels with coded modulation—Part III: Co-channel interference," *IEEE Trans. Comm.*, vol. 45, pp. 809–819, 1997.

[29] A. R. Nix and J. P. McGeehan, "Modelling and simulation of frequency selective fading using switched antenna diversity," *Electronics Letters*, vol. 26, pp. 1868–1869, 1990.

[30] G. Foschini and J. Salz, "Digital communications over fading radio channels," *Bell Syst. Techn. Jour.*, vol. 62, pp. 429–456, 1983.

[31] M. K. Simon and M. S. Alouini, *Digital Communications over Generalized Fading Channels: A Unified Approach to Performance Analysis.* New York: Wiley, 2000.

[32] A. F. Molisch, H. Novak, J. Fuhl, and E. Bonek, "Reduction of the error floor of MSK by selection diversity," *IEEE Trans. Vehicular Techn.*, vol. 47, pp. 1281–1291, 1998.

[33] A. A. Abu-Dayya and N. Beaulieu, "Micro- and macrodiversity MDPSK on shadowed frequency-selective channels," *IEEE Trans. Vehicular Trans.*, vol. 43, pp. 2334–2343, 1995.

[34] A. Afrashteh, J. Chuang, and D. Chukurov, "Measured performance of 4-QAM with 2-branch selection diversity in frequency-selective fading using only one receiver," in *Proc. VTC'89*, pp. 458–463, 1989.

[35] X. Dong, T. T. Tjhung, F. Adachi, and C. C. Ko, "Diversity reception of 16STAR-QAM in frequency-selective Rician fading," *Electronics Letters*, vol. 31, pp. 1317–1319, 1995.

[36] X. Dong, T. T. Tjhung, and F. Adachi, "Error probability analysis for 16 STAR QAM in frequency-selective Rician fading with diversity reception," *IEEE Trans. Vehicular Techn.*, vol. 47, pp. 924–935, 1998.

[37] G. Schultes, J. Radlbauer, H. Egger, D. Walter, K. Pillekamp, and B. Pauli, "Performance of the Siemens DECT-prototype gigaset 95x in a dispersive indoor environment," in *Proc. IEEE VTC'94*, pp. 1079–1084, 1994.

[38] S. Safavi and L. Lopes, "Novel techniques for performance optimisation of DECT receivers with non-linear front-end," in *Proc. VTC'96*, vol. 5, pp. 824–828, 1996.

[39] S. Safavi and L. B. Lopes, "Predetection quality diversity scheme for DECT outdoor applications," *Electronics Letters*, vol. 32, pp. 966–968, 1996.

[40] F. Gulgiemi, C. Luschi, and A. Spalvieri, "Joint clock recovery and baseband combining for the diversity radio channel," *IEEE Trans. Comm.*, vol. 44, pp. 114–116, 1996.

Chapter 13

SUMMARY AND CONCLUSIONS

In this part of the book, we have given an overview of the computation methods and the results for the error probability of unequalized wireless systems in time-dispersive environments.

Concerning the computation methods, we arrive at the following conclusions:

- There exist a variety of different computation methods. Each has certain advantages and disadvantages for various practical situations, and some are designed only for very specific situations. There is no master method that is optimum for all situations of interest. An overview of the various methods is given in Table 13.1.

- The most commonly used method is the QFGV method, which can cover most situations where fixed sampling is used. The only restrictions in that case are (i) the channel must be Gaussian, and (ii) the decision variable at the receiver must be represented as a sum of quadratic forms of complex Gaussian variables. The latter condition implies that M-ary angle modulation (with $M > 4$) can be treated only in an approximate way.

- The ABGV method is also suitable for differentially detected signals. It too requires that the channel is Gaussian, and inclusion of diversity does not seem straightforward. M-ary angle modulation can be computed exactly, but higher-order QAM cannot be treated.

- Another alternative is the TPEM method. As with the QFGV and ABGV methods, it requires that the channel is Gaussian; however, it *is* able to deal with M-ary modulation and higher-order QAM, as long as we perform differential detection or frequency-discriminator detection. Also, processing of the received signal (e.g., quantization) can be included. It is not possible, however, to include diversity.

- The EVCM method is suitable for coherent reception of signals. Inclusion of diversity does not seem straightforward.

None of the above methods is suitable for the inclusion of adaptive sampling. The following methods are either approximate or use a rather restrictive channel model; however, they *can* deal with adaptive sampling.

- The error region method is suitable only for a two-delay channel model. It can include the effects of diversity and can deal with most modulation formats and postprocessing of the received signal. Inclusion of noise is possible in an approximate way by eye-pattern analysis.

- The group-delay method is suitable only when the delay spread is much smaller than the symbol width. Furthermore, it is only approximate. However, inclusion of noise is much easier (and CPU-time saving) than in the two-delay channel model.

The results that we quoted here are of importance to the cordless communications systems that are currently under deployment all over the world. The most important examples are the DECT, PHS, and PWT systems, i.e., the European, Japanese, and American standards. Typical signal-to-noise ratios are on the order of 30 dB or higher. For the DECT system, which uses GMSK as modulation format, the symbol duration is about 850 ns; for the PHS and PWT systems, which use $\pi/4$-DQPSK, it is 5.2 μs and 1.7 μs, respectively. In indoor environments, typical delay spreads are on the order of 100 ns; for outdoor environments that are relevant for WLL applications and low-mobility PCS applications, they are on the order of 500 ns. We thus have exactly those situations that we have analyzed in this part, and all results can be used almost without restrictions. For JDC, which is a cellular, not a cordless, system, the symbol duration is 48 μs, so that time dispersion has an effect in some outdoor environments, like hilly terrain or mountaineous regions.

We thus can make the following important conclusions:

- For many environments, both GMSK and $\pi/4-$DQPSK with fixed sampling suffer an error floor that is much higher than the 10^{-3} that is tolerable for (uncoded) speech communication even for "classical" cordless applications. Thus, some form of error reduction scheme is necessary [1].

- The modulation formats proposed for this goal, Section 10.3, are not possible except for future systems, because they would require changes in the standards.

- Training-sequence-based adaptive sampling is preferable to fixed sampling, and is actually implemented in some form in most commercial products. Special receiver structures, like fractional-bit detection, are also a good option for reducing the error floor. The receiver filter width plays an important role for the error floor; wide receiver filters help to keep the error floor low. Trade-offs with the noise and CCI performance are possible, suggesting adaptive digital receiver filters.

Method	Includes Noise?	N-path?	LOS Comp.?	M-ary Modems?	Exact?	Adaptive Sampling ?	Diversity?
QFGV	✓	✓	✓	(✓)	✓	-	✓
ABGV	✓	✓	✓	✓	✓	-	?
EVCM	✓	✓	?	✓	✓	-	?
Group delay	✓	✓	✓	✓	-	✓	?
Error region	(✓)	-	✓	✓	✓	✓	✓
TPEM	✓	✓	✓	✓	✓	-	-
Eye-pattern analysis	✓	-	?	✓	-	✓	-

Table 13.1: Summary of the various computation methods. ✓ signifies that a method is capable of dealing with a certain aspect; - signifies that it is not easily possible. ? signifies that it is likely that the method is capable of this generalization, but this capability has not been described explicitly in the literature.

- The most important scheme for BER reduction is diversity, which is often used for base stations, and is also desirable for the mobile [2]. With these schemes, a good speech quality can be achieved for most situations of practical interest. Diversity not only reduces the error floor but also improves performance in the presence of noise and CCI. Furthermore, it can be used in conjunction with other error-floor-reduction methods (like adaptive sampling). Some authors have also proposed equalizer structures for use in cordless systems (DECT): References [3], [4], [5] proposed a combination of a two-state Viterbi equalizer with RSSI-driven diversity; References [6] and [7] suggested the use of a decision-feedback equalizer.

- Further research and further application of existing results are required for next-generation systems, which will probably work at even higher data rates and should provide better quality at even lower costs.

Bibliography

[1] A. F. Molisch, H. Novak, and E. Bonek, "The DECT air interface," *Teletronikk*, vol. 94, pp. 45–53, 1998.

[2] L. Lopes, "An overview of DECT radio link research in COST 231," in *Proc. PIMRC'94*, pp. 99–104, 1994.

[3] S. Safavi, L. Lopes, P. Mogensen, and F. Frederiksen, "A hierarchy of receiver options for DECT system," in *Proc. PIMRC'95*, pp. 1351–1356, 1995.

[4] S. Safavi, L. Lopes, P. Mogensen, and F. Frederiksen, "An advanced base station receiver concept for DECT," in *Proc. VTC'95*, pp. 150–154, 1995.

[5] P. Mogensen, F. Frederiksen, P. Thomsen, S. Safavi, and L. Lopes, "Evaluation of an advanced receiver concept for DECT," in *Proc. VTC'95*, pp. 514–519, 1995.

[6] J. Fuhl, G. Schultes, and W. Kozek, "Adadptive equalization for DECT systems operating in low time-dispersive channels," in *Proc. VTC'94*, pp. 714–718, 1994.

[7] N. Benvenuto, A. Salloum, and L. Tomba, "Performance of non-coherent digital equalisers in DECT systems," *Electronics Lett.*, vol. 32, pp. 97–98, 1996.

Appendix A

EQUATIONS FOR BER COMPUTATION WITH FIXED SAMPLING

A.1 SOLUTIONS OF THE OFGV METHOD

As mentioned in Chapter 8, we consider the probability that $D < 0$, where D is a general quadratic form [1]

$$D = A|X|^2 + B|Y|^2 + CXY^* + C^*X^*Y \tag{A.1.1}$$

where A, B, and C are complex constants and X and Y are Gaussian statistical variables (they need not be zero-mean). Introducing the following variables

$$\begin{aligned}
w &= \frac{A\mu_{xx} + B\mu_{yy} + C\mu_{xy}^* + C^*\mu_{xy}}{4\left(\mu_{xx}\mu_{yy} - |\mu_{xy}|^2\right)\left(|C|^2 - AB\right)} \\
v_{1,2} &= \sqrt{w^2 + \frac{1}{4\left(\mu_{xx}\mu_{yy} - |\mu_{xy}|^2\right)\left(|C|^2 - AB\right)}} \mp w \\
\alpha_1 &= 2\left(|C|^2 - AB\right)\left(|\overline{X}|^2\mu_{yy} + |\overline{Y}|^2\mu_{xx} - \overline{X}^*\overline{Y}\mu_{xy} - \overline{X}\,\overline{Y}^*\mu_{xy}^*\right) \\
\alpha_2 &= A|\overline{X}|^2 + B|\overline{Y}|^2 + C\overline{X}^*\overline{Y} + C^*\overline{X}\,\overline{Y}^* \\
p_1 &= \frac{\sqrt{2v_1^2 v_2(\alpha_1 v_2 - \alpha_2)}}{v_1 + v_2} \\
p_2 &= \frac{\sqrt{2v_1 v_2^2(\alpha_1 v_1 + \alpha_2)}}{v_1 + v_2},
\end{aligned} \tag{A.1.2}$$

we get the following probability that D is smaller than 0:

$$P\{D<0\} = Q_m(p_1,p_2) - \frac{v_2/v_1}{1+v_2/v_1} I_0(p_1 p_2) \exp\left(-\frac{p_1^2+p_2^2}{2}\right) \tag{A.1.3}$$

where I_0 is the modified Bessel function zero-order, first kind, and $Q_m(p_1,p_2)$ is Marcum's Q-function

$$Q_m(p_1,p_2) = \int_{p_2}^{\infty} \exp\left[-\frac{p_1^2+x^2}{2}\right] I_0(p_1 x) x \mathrm{d}x.$$

In the case that we have Rayleigh fading, the above equations simplify considerably. α_1 and α_2 become 0, so that $P(D<0) = v_1/(v_1+v_2)$.

For independent antenna diversity, we also need the quadratic form

$$D = \sum_{l=1}^{L} A|X_l|^2 + B|Y_l|^2 + CX_lY_l^* + C^*X_l^*Y_l \tag{A.1.4}$$

where the X_l and Y_l are independent, identically distributed Gaussian random variables. The solution reads

$$\begin{aligned} P\{D<0\} &= Q_m(p_1,p_2) - I_0(p_1p_2)\exp\left(-\frac{p_1^2+p_2^2}{2}\right) \\ &+ \frac{I_0(p_1p_2)\exp\left(-\frac{p_1^2+p_2^2}{2}\right)}{(1+v_2/v_1)^{2L-1}} \sum_{l=0}^{L-1} \binom{2L-1}{l} \left(\frac{v_2}{v_1}\right)^l \\ &+ \frac{\exp\left(-\frac{p_1^2+p_2^2}{2}\right)}{(1+v_2/v_1)^{2L-1}} \sum_{n=1}^{L-1} I_n(p_1p_2) \\ &\left\{\sum_{l=0}^{L-1-n} \binom{2L-1}{l} \left[\left(\frac{p_2}{p_1}\right)^n \left(\frac{v_2}{v_1}\right)^l - \left(\frac{p_1}{p_2}\right)^n \left(\frac{v_2}{v_1}\right)^{2L-1-l}\right]\right\} \end{aligned} \tag{A.1.5}$$

for $L>1$.

A generalization of these equations, namely, the pdf and cdf of the quadratic form (not just the probability that it is larger than zero), is given in [2].

A.2 THE ABGV METHOD

Reference [3] considers the problem sketched in Figure A.1: two deterministic vectors A_1, A_2 are disturbed by complex Gaussian noise; we want to compute the pdf of the angle between the disturbed vectors.

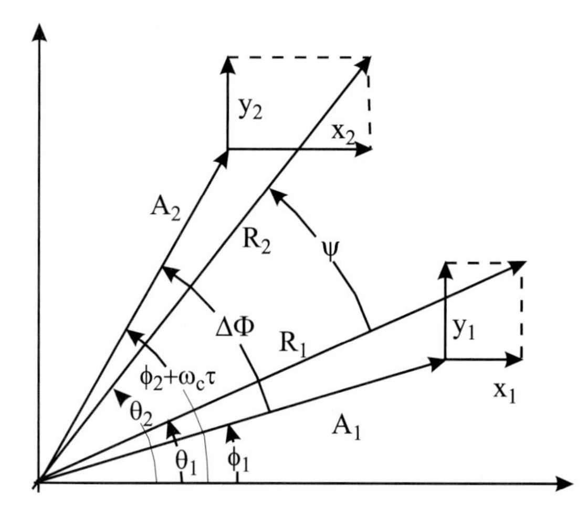

Figure A.1. Definitions for the problem of the angle between two vectors perturbed by Gaussian noise. Adapted from [3], copyright IEEE.

For the definition of the various quantities, consider Figure A.1; to allow easier access to the original source, we use in this subsection the nomenclature of [3], even though it is not completely consistent with the rest of Part II of this book. $\Delta\Phi$ denotes the phase difference between the two nominal (undisturbed) vectors, A_i the amplitudes of the undisturbed vectors, and $\Gamma_i = A_i^2/(2\sigma_i^2)$ the instantaneous SNRs for the two vectors, where σ_i is the variance of the noise and $i = 1, 2$. We furthermore have the parameters

$$\begin{aligned} U &= \frac{1}{2}(\Gamma_2 + \Gamma_1) \\ V &= \frac{1}{2}(\Gamma_2 - \Gamma_1) \\ W &= \sqrt{\Gamma_1\Gamma_2} = \sqrt{U^2 - V^2}. \end{aligned} \tag{A.2.1}$$

The correlation coefficients of the noise are

$$\begin{aligned} \rho_c &= \frac{E\{x_1x_2\}}{\sigma_1\sigma_2} = \frac{E\{y_1y_2\}}{\sigma_1\sigma_2} \\ \rho_s &= \frac{E\{x_1y_2\}}{\sigma_1\sigma_2} = -\frac{E\{y_1x_2\}}{\sigma_1\sigma_2}. \end{aligned} \tag{A.2.2}$$

We now want to compute the cumulative distribution function of the actual angle ψ, i.e., the probability that ψ lies between ψ_1 and ψ_2. It is given as

$$P(\psi_1 \leq \psi \leq \psi_2) = \begin{cases} F(\psi_2) - F(\psi_1) + 1 & \psi_1 < \Delta\Phi < \psi_2 \\ F(\psi_2) - F(\psi_1) & \text{otherwise} \end{cases} \tag{A.2.3}$$

where by definition $F(\psi)$ is periodic with 2π and has a jump at $\psi = \Delta\Phi$. Averaging over all symbols sequences gives the total error probability. The function F is given in the most general case as

$$F(\psi) = \frac{1}{4\pi} \int_{-\pi/2}^{\pi/2} \exp\left[-\frac{U - V\sin(t) - W\cos(\Delta\Phi - \psi)\cos(t)}{1 - (\rho_c\cos(\psi) + \rho_s\sin(\psi))\cos(t)}\right] \tag{A.2.4}$$
$$\left[\frac{W\sin(\Delta\Phi - \psi)}{U - V\sin(t) - W\cos(\Delta\Phi - \psi)\cos(t)}\right.$$
$$\left. + \frac{\rho_c\sin(\psi) - \rho_s\cos(\psi)}{1 - (\rho_c\cos(\psi) + \rho_s\sin(\psi))\cos(t)}\right] \mathrm{dt}.$$

Various simplifications are possible for special situations [3], [4]. In particular, the formulation where the deterministic components are zero, but including (correlated) noise, is given in [5], [6].

A.3 THE ERROR REGION METHOD

For a direct implementation of the error region method, we require the coordinates of the error circles. For MSK, they are given as [7]

$$x_0 = -\frac{1}{2} \frac{\alpha_{2,k+1}\alpha_{1,k}\sin(\Phi_{2,k+1} - \Phi_{1,k}) + \alpha_{1,k+1}\alpha_{2,k}\sin(\Phi_{1,k+1} - \Phi_{2,k})}{\alpha_{2,k+1}\alpha_{2,k}\sin(\Phi_{2,k+1} - \Phi_{2,k})} \tag{A.3.1}$$

$$y_0 = -\frac{1}{2} \frac{\alpha_{2,k+1}\alpha_{1,k}\cos(\Phi_{2,k+1} - \Phi_{1,k}) - \alpha_{1,k+1}\alpha_{2,k}\cos(\Phi_{1,k+1} - \Phi_{2,k})}{\alpha_{2,k+1}\alpha_{2,k}\sin(\Phi_{2,k+1} - \Phi_{2,k})} \tag{A.3.2}$$

$$R = \frac{1}{2}\left[\alpha_{2,k+1}^2\alpha_{1,k}^2 + \alpha_{1,k+1}^2\alpha_{2,k}^2\right. \tag{A.3.3}$$
$$\left. -2\alpha_{1,k}\alpha_{2,k}\alpha_{1,k+1}\alpha_{2,k+1}\cos(-\Phi_{1,k} + \Phi_{2,k} + \Phi_{1,k+1} - \Phi_{2,k+1})\right]^{1/2}$$
$$\left[\alpha_{2,k+1}\alpha_{2,k}|\sin(\Phi_{2,k+1} - \Phi_{2,k})|\right]^{-1}.$$

For M-ary PSK, they are given in [8]. The distribution of the amplitude ratios is for a two-path Rayleigh channel

$$pdf_{\tilde{r}}(\tilde{r}) = \frac{2\tilde{r}(\sigma_2/\sigma_1)^2}{[\tilde{r}^2 + (\sigma_2/\sigma_1)^2]^2}. \tag{A.3.4}$$

Bibliography

[1] J. G. Proakis, *Digital Communications.* New York, McGraw Hill, 3rd ed., 1995.

[2] K. H. Biyari and W. C. Lindsey, "Statistical distributions of Hermitian quadratic forms in complex Gaussian variables," *IEEE Trans. Comm.*, vol. 39, pp. 1076–1082, 1993.

[3] R. F. Pawula, S. O. Rice, and J. H. Roberts, "Distribution of the phase angle between two vectors perturbed by Gaussian noise," *IEEE Trans. Comm.*, vol. 30, pp. 1828–1841, 1982.

[4] M. K. Simon and C. C. Wang, "Differential detection of Gaussian MSK in a mobile radio environment," *IEEE Trans. Vehicular Techn.*, vol. 33, pp. 307–320, 1984.

[5] C. S. Ng, T. T. Tjhung, F. Adachi, and K. M. Lye, "On the error rates of differentially detected narrowband pi/4-DQPSK in Rayleigh fading and Gaussian noise," *IEEE Trans. Vehicular Techn.*, vol. 42, pp. 259–265, 1993.

[6] C. S. Ng, T. T. Tjhung, F. Adachi, and F. P. S. Chin, "Closed-form error probability formula for narrow-band FSK, with limiter-discriminator-integrator detection, in Rayleigh fading," *IEEE Trans. Comm.*, vol. 42, pp. 2795–2802, 1994.

[7] A. F. Molisch, L. Lopes, M. Paier, J. Fuhl, and E. Bonek, "Error floor of unequalized wireless personal communications systems with MSK modulation and training-sequence-based adaptive sampling," *IEEE Trans. Comm.*, vol. 45, pp. 554–562, 1997.

[8] A. F. Molisch and E. Bonek, "Reduction of error floor of differential PSK in mobile radio channels by adaptive sampling," *IEEE Trans. Vehicular Techn.*, vol. 47, pp. 1276–1280, 1998.

Appendix B

NOMENCLATURE FOR PART II

small Latin letters:

a_{LOS}	amplitude of LOS component
a_i	amplitude of the ith echo
a_A	magnitude of normalized phasor in diversity branch A
b_m	transmitted bits
c_m	transmitted complex coefficients
$\mathrm{d}(t)$	transmitted signal. Note that d is written upright to distinguish it from the differential dT.
d_e	Euclidean distance
f	frequency
f_D	maximum Doppler frequency
$g(t)$	basis pulse for PAM transmission
$\widetilde{g}(t)$	phase-shaping pulse
$h(t, \tau)$	channel impulse response
h_{mod}	modulation index
i	index counter for echoes
j	imaginary unit
k	index counter for sampling instants
l	index counter for diversity antennas
m	index counter for symbols
n	index counter for oversampling or auxiliary counter
$n(t)$	noise
p_i	auxiliary parameters for QFGV method
$p_x(x)$	pdf of x
q	rotated detector output
$\overrightarrow{q}$	eigenvector of array of complex Gaussian variables
$r(t)$	received signal
$\widetilde{r}$	a_2/a_1
s	Laplace transform variable

s_n	sequence of received signals
t	time
t_s	sampling time
u	auxiliary variable
u_n	mean value of shadowing
w	auxiliary variable or function
x, y	auxiliary variables

capital Latin letters

A, B	complex constant
B_n	noise bandwidth
B_G	bandwidth of Gaussian filter
B_r	receiver filter bandwidth
BER	average bit error probability
$BER(\tilde{r}, \varphi)$	instantaneous bit error probability
C	rotation of complex plane
D	decision variable
$E\{\}$	expected value
E_s	symbol energy
E_b	bit energy
$F()$	metric for training sequence
G	eye height
H	transfer function
H_r	transfer function of receiver filter
I_n	modified Bessel function of n-th order
I^{id}	in-phase component of ideal signal
I^{dev}	in-phase component of deviation from ideal signal
K	proportionality constant
K_{rice}	Rice factor
L	number of diversity antennas
L_P	number of macrodiversity ports
M	degree of modulation format
N	number of taps for channel model
N_{sym}	number of symbols that influence a decision
N_{samp}	oversampling factor
N_{int}	number of integration points
N_0	one-sided noise power spectral density
$N(f)$	noise spectrum
$P()$	probability

P_d	power attenuation of diffuse path
P_M	probability that signal falls into certain region of complex plane
Q^{id}	quadrature component of ideal signal
Q^{dev}	quadrature component of deviation
Q_m	Marcum Q function
R	$\|r(t_s)\|$
$\underline{R}$	channel correlation matrix
S	rms delay spread
$S(\nu,\tau)$	spreading function
T	symbol duration
T_g	group delay
$\underline{U}$	transformation matrix
$\underline{W}$	signal correlation matrix
X, Y	complex Gaussian variables

Greek letters

α	roll-off parameter
α_i	auxiliary parameter
$\alpha_{i,k}$	amplitude of transmitted signal in ith path at kth sampling time
β	phase shift for modulation
$\delta(t)$	Dirac function
ε	auxiliary variable
φ_ι	phase shift of the ith echo
φ	$\varphi_2 - \varphi_1$
ϕ_m	modulation angles for PSK
ϕ_{rot}	rotation angle
$\phi_{err-reg}$	angle for rotation of error region
λ	eigenvalues of $\underline{R}\,\underline{W}$
μ_{xy}	second central moment, $\frac{1}{2}E\{xy^*\}$
ν	Doppler frequency
ν_i	auxiliary parameters
ω	angular frequency
ρ	normalized correlation coefficient, $\rho = \rho_c + j\rho_s$
σ	variance of Gaussian variables
σ_n	variance of shadowing
τ	delay
τ_i	delay of the ith echo
ξ_h	delay cross-power spectral density
ξ_s	scattering function
ξ_n	noise correlation function
Γ	signal-to-noise ratio

Φ	phase of $d(t)$
Φ_r	phase of $r(t)$
Φ_c	phase of channel transfer function
$\underline{\Lambda}$	matrix of eigenvalues of $\underline{RW}$
$\Pi_D(s)$	Laplace transform of the pdf of D
θ_{DSK}	phase shift for DSK modulation
ϑ	phase shift by the channel
$\varkappa$	auxiliary constant

The symbol $\tilde{}$ denotes normalization. Δx denotes a change in x.

Part III

Equalization Techniques for Single Carrier, Unspread Digital Modulations

Giorgio Vitetta, Brian Hart, Aarne Mammela, Desmond Taylor

Chapter 14

PRELIMINARY TOPICS

As in the other Parts of this book, we are concerned with the design, interpretation, implementation and performance assessment of communication systems for wireless channels. Unlike Part II, this Part considers the case where the channel's delay spread or time variation are sufficiently large that equalization can significantly improve performance. Unlike Parts IV and V, Part III focuses on classical single-carrier, unspread digital modulation schemes even though many principles can be directly applied in other schemes. In addition, we mainly focus on *linear* digital modulations transmitted over *linear* time-varying channels.

14.1 INTRODUCTION

Equalization means estimating, implicitly or explicitly, the impulse response of the channel, then using the estimate to compensate for the channel's distortion with the objective of improving the transmission link's performance. One simple example is a time invariant, *frequency-selective* (FS) channel followed by a *zero-forcing linear equalizer* [1]. Such an equalizer is a transversal filter adapting its transfer function via its coefficients to compensate for both the amplitude and the phase distortion of the communication channel. In this way, a signal transmitted through the cascade of channel and equalizer is ideally undistorted, provided that the equalizer length is infinite and the channel noise can be neglected.

Equalization for frequency-selective channels has a rich history. It dates back to the early work of Lucky and others, summarized in [2]. Their work was aimed almost entirely at the *telephone channel*, which may be characterized as an essentially linear, time-invariant, frequency-selective channel. Its amplitude and phase distortions were compensated for by a linear equalizer producing a reduction of the *intersymbol interference* (ISI) at the decision device input. Later work examined the line-of-sight microwave channel [3], which may be considered as a very slowly time-varying channel. Thus, most of the earlier theory on channel equalization could be directly applied, albeit at the much higher transmission rates. This work is discussed in detail in [1] and its references. Finally, other work, e.g., [4], [5], [6], considered the *ionospheric* and the *tropospheric channels*, both of which are time-varying wireless channels.

The mobile digital wireless channel is in part the same as these channels, and in part different [7]. We may draw from the existing body of knowledge, but we must also extend it to address the new challenges and higher expectations of modern, mobile, wideband wireless communications [3].

As discussed in Part I, the line-of-sight path between the transmitter and receiver is often obstructed, so that the propagation is via reflection, diffraction, and scattering off interspaced hills, vegetation, buildings, and vehicles. Multiple paths exist, and their subtly different delays inevitably lead to amplitude fading and unpredictably modify the carrier phase. When the scatterer separation increases, the spread between the paths' delays increases. Moreover, the transmitter and receiver are mobile with respect to each other. According to the geometry of the mobile terminals and scatterers, each path is characteristically Doppler-shifted up to some maximum. The net effect is delay and Doppler spread, or frequency selectivity and time variation.

The degree to which frequency selectivity and time variation are important depends on the signalling rate $R_s \triangleq 1/T_s$ [5], where T_s is the symbol interval. To clarify this point, let us define the *normalized delay spread* d and the *normalized Doppler spread* b as

$$d \triangleq \tau_{ds} R_s \tag{14.1.1}$$

$$b \triangleq B_D / R_s \tag{14.1.2}$$

respectively, where τ_{ds} and B_D are the *channel delay spread* and the *fading Doppler bandwidth*, respectively (see Part I). In any equalization problem, a random communication channel can be considered as *slowly fading* over an observation interval if $b \ll 1$ and as *frequency flat* if $d \ll 1$. This allows us to partition a Cartesian plane having d as abscissa and b as ordinate into four rectangular regions corresponding to four different physical situations, as shown in Figure 14.1. For low enough transmission rates, d is small so that there is often significant time variability but little frequency selectivity. Under this assumption, the channel may be idealized as time varying and *frequency flat* (FF). At higher rates, the channel is typically frequency selective, but it undergoes slow variations with respect to the transmission rate. Then, it is idealized as slowly fading and *frequency selective* (FS). There is an extreme third case, where both time variation and frequency selectivity are nontrivial, and here the channel is sometimes known as *doubly selective* (DS). In all cases the channel is modelled as a linear system.

It is difficult to devise a rational method of presenting all the equalization algorithms for multipath fading channels. They may be organized according to the chronology of their discovery or categorized according to how much a priori information they assume or whether they are linear or nonlinear, optimal or suboptimal, for the frequency-flat, frequency-selective, or doubly selective channels, or some other criterion. Ultimately, they all require information about the time-varying channel impulse response (CIR) or some equivalent quantity: either by assuming it *known*,

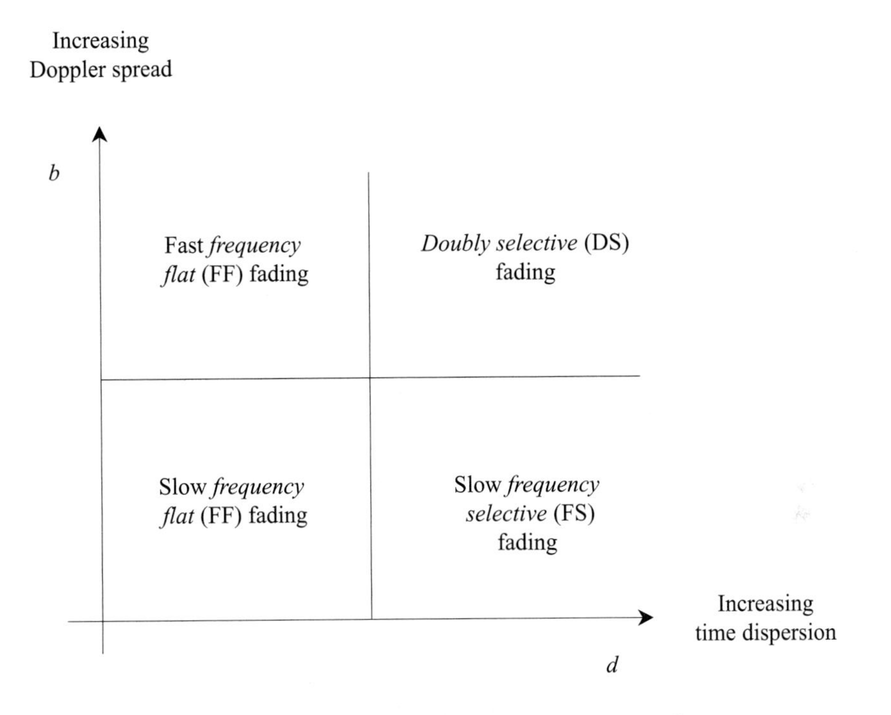

Figure 14.1. Four cases of wireless channels

by *estimating it*, or by *averaging over it* if the *statistical properties* of the CIR are known. These lead to fundamental differences in the nature of the equalizer structures, and so we choose to make this the primary division.

Any equalization algorithm processes the received signal, producing a set of real quantities, known as *metrics*, evaluated by the receiver to make decisions on the transmitted data. After some preliminary topics in this chapter, in Chapter 15 we derive, interpret and analyze the performance of the metrics computed by *optimal*[1] detectors under the assumption that the CIR (or some equivalent quantity) is known, estimated, or averaged over for the doubly selective wireless channel and its special cases: the frequency-flat and frequency-selective channels. Both the *maximum likelihood* (ML) and *maximum a posteriori* (MAP) methods are discussed as optimality criteria, and their application to bits, symbols and sequences of symbols is illustrated. In particular, MAP *bit detectors* (MAPBDs), MAP *symbol detectors* (MAPSDs), and ML *sequence detectors* (MLSDs) are considered. These ideas are summarized in Figure 14.2.

In Chapter 16, we describe various *equalization algorithms* corresponding to dif-

[1]In general, the decision rules provided in Section 15.3 are not rigorously optimal because the detection strategies developed under the assumption of estimated CIR do not often minimize symbol or sequence error probabilities.

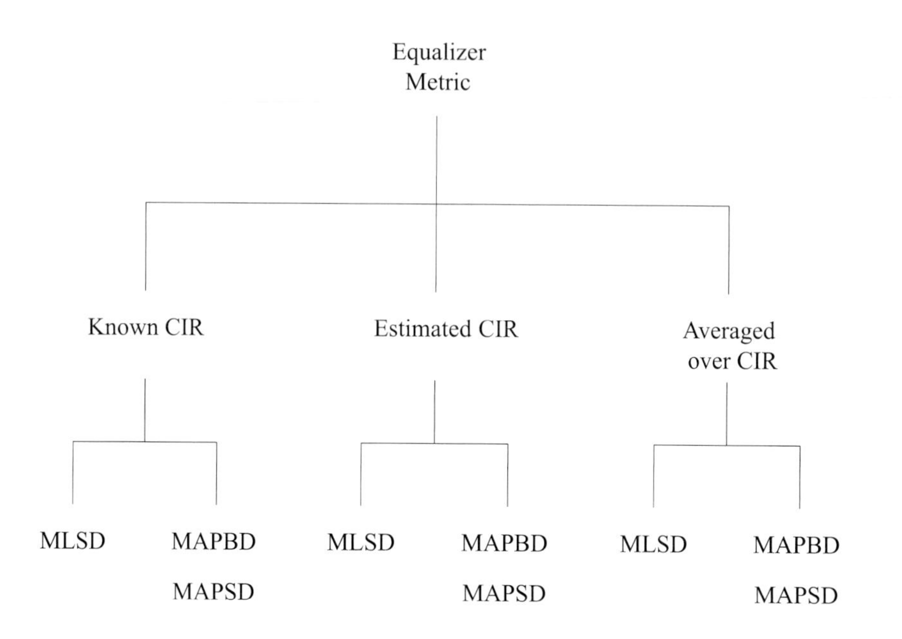

Figure 14.2. Organization of the chapter on equalizer metrics

ferent ways of implementing the computation of the derived metrics, ranging from optimal to highly suboptimal. Again, the primary division into sections is related to how the CIR is treated. Within each section, the structure, complexity, and performance measures of each type of equalizer are described. In particular, all the theoretically and practically important classes of equalizers for FS and DS channels, including maximum likelihood sequence detectors, maximum *a posteriori* symbol detectors, reduced complexity techniques, *decision-feedback equalizers* (DFEs), and *linear equalizers* (LEs) are considered. In addition one-shot, block, and sequence detectors for FF channels are described. Figure 14.3 presents this layout diagrammatically.

14.2 SIGNAL MODEL

14.2.1 Transmitted signals

We address two classes of transmitted signal, namely, *linear modulation* [4] and *Continuous Phase Modulation* (CPM) [8].

In the case of linear modulation, a sequence of information bits $\{b_i\}$ enters a constellation mapper,[2] which outputs a sequence of constellation points or *channel*

[2]Mapping may be memoryless or may entail differential encoding. With memoryless mapping, a known signal (such a training sequence, pilot symbols, or a pilot tone) must be transmitted to allow the receiver to solve the so-called *phase ambiguity problem*.

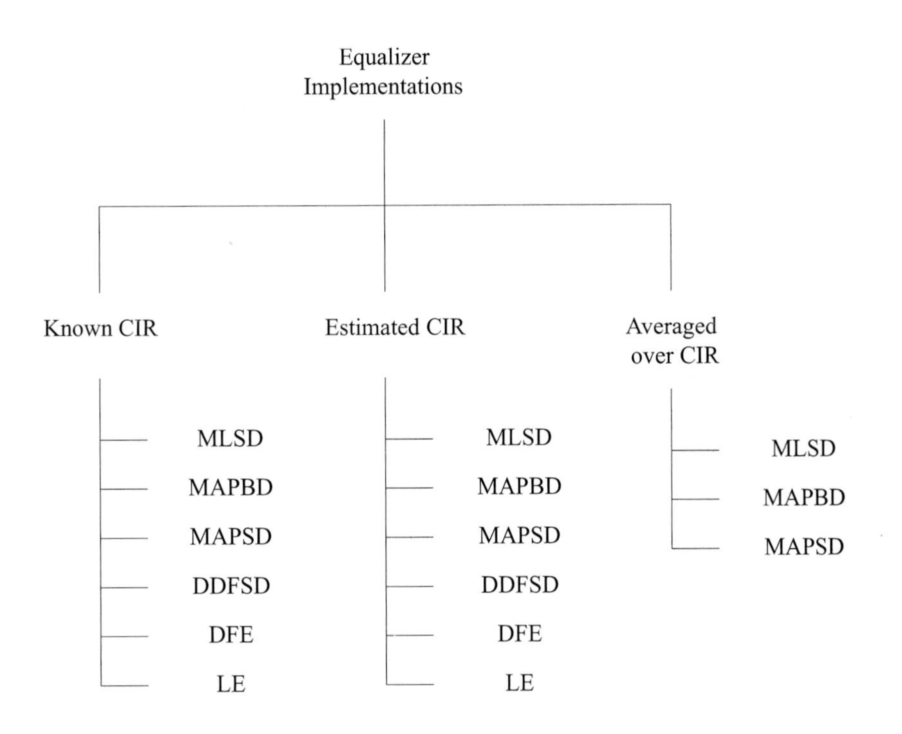

Figure 14.3. Organization of the chapter on the implementation of equalization techniques

symbols $\{c_m\}$. These are filtered by the *transmitter pulse shape* $g(t)$, so that in complex baseband the transmitted signal $d(t)$ equals

$$d(t) \triangleq \sum_{m=0}^{L_c-1} c_m \, g(t - mT) \tag{14.2.1}$$

where L_c is the number of symbols transmitted in this burst. The parameter L_c takes on a finite value and is typically under a thousand in *time division multiple access* (TDMA) systems [9]; for *frequency division multiple access* (FDMA) systems, L_c can be arbitrarily large.

The symbols c_m are taken from an M-ary constellation, where M is normally a power of two, so that $\log_2 M$ bits are needed for every symbol. The practical, important modulations are as follows: Amplitude Shift Keying (ASK), where c_m is taken from the set $\{0, 1, ..., M-1\}$; Pulse Amplitude Modulation (PAM), where c_m is taken from the set $\{\pm 1, \pm 3, ..., \pm(M-1)\}$; Phase Shift Keying (PSK), where c_m belongs to the set $\{\exp(j2\pi i/M), \, i = 0, \, 1, \, ..., \, M-1\}$; and Quadrature Amplitude Modulation (QAM) [10], [4]. For square QAM constellations, M is a power of 4 and the symbols have the form $c_m = a_m + jb_m$, with $a_m, \, b_m \in \{\pm 1, \pm 3, ..., \pm(\sqrt{M} - 1)\}$. In the following we denote the symbol sequence of eq. (14.2.1) as $\mathbf{c} \triangleq [c_0, c_1, ..., c_{L_c-1}]^T$.

The pulse shape $g(t)$ is normally designed according to criteria appropriate for the single-user AWGN channel (i.e., an ideal, time-invariant, frequency-flat channel). First, it is designed to have minimum bandwidth yet still allow the cascade of transmitter pulse shape, ideal channel, and receiver filter to satisfy Nyquist's First Criterion for zero ISI [11]. In this case, the overall filter is referred to as *Nyquist.* Second, it can be shown that the optimal distribution of the overall transfer function between transmitter and receiver is half-and-half, so that the transmitter and receiver filters are *root Nyquist.* A popular choice is the *root-raised cosine* (RRC) filter which, when cascaded with a matched filter at the receiver, produces a Nyquist overall impulse response [2]. Therefore, one pulse does not interfere with other pulses (zero ISI), as long as each is slid in time by multiples of the symbol period, and the matched filtered received signal is sampled only once per symbol period, at these regularly spaced locations.

Unfortunately, this zero ISI property is lost once significant time variation and/or frequency selectivity is encountered (which is inevitable in wideband wireless channels). Nonetheless, in a narrow sense, we find that most standards and systems keep the same pulse shape design. In a wider sense, both the *orthogonal frequency division multiplexing* (OFDM) and *direct sequence-code division multiple access* (DS-CDMA) modulations of Part IV and Part V are linear modulations, but with considerably narrower and wider bandwidth pulse shapes, respectively. OFDM requires multiple subcarriers and prior serial-to-parallel symbol conversion to maintain the data rate, whereas DS-CDMA requires multiple overlapping users to maintain spectral efficiency.

We also note that the pulse shapes having root-raised-cosine or raised-cosine spectrum are noncausal and have infinite extent. For implementation purposes, it is necessary to window them and to delay the windowed waveforms to make them causal.

CPM signals may be expressed in complex baseband as [8]

$$d(t) \triangleq A \exp\left(j \int_{-\infty}^{t} \phi(\tau, \mathbf{c})\, d\tau\right) \tag{14.2.2}$$

where A is the signal amplitude, $\mathbf{c} \triangleq [c_0, c_1, ..., c_{L_c-1}]^T$ is a vector of the transmitted symbols, c_m, taken from an M-ary alphabet $\{\pm1, \pm3, ..., \pm(M-1)\}$. The function $\phi(\tau, \mathbf{c})$ is the *information-bearing phase* and can be expressed as

$$\phi(\tau, \mathbf{c}) \triangleq 2\pi h_{mod} \sum_{m=0}^{L_c-1} c_m\, \tilde{g}(t - mT) \tag{14.2.3}$$

where h_{mod} is the *modulation index* and $\tilde{g}(t)$ is called the *phase-shaping pulse.*

Equations (14.2.2) and (14.2.3) show that the relationship between the symbols and the transmitted signal is nonlinear. This nonlinearity makes equalizing CPM signals transmitted over frequency-selective channels more complicated. Nonetheless, CPM signals may be represented as the superposition of a finite number of

PAM signals: this representation is known as the *Laurent expansion* [12]. Furthermore, a truncated expansion is still highly accurate, and this can be used as a mathematical basis for designing equalizers.

However, except where otherwise noted, the discussion within Part III is concerned with linearly modulated signals. Most of the concepts can be extended to CPM signals by the Laurent expansion or, simply, the model (14.2.2)—(14.2.3), but the topic is beyond the scope of this book.

14.2.2 Wireless channels

The properties of wireless channels are described fully in Part I. For convenience we repeat here that the received signal equals the transmitted signal, distorted by the time-varying, frequency-selective channel and then corrupted by additive stationary Gaussian noise, as (see Figure 14.4)

$$y\,(t) = s(t) + w(t) \tag{14.2.4}$$

where $w(t)$ is complex Additive White Gaussian Noise (AWGN) with two-sided noise spectral density N_0, and $s(t)$, the noiseless or data-bearing portion of the received signal, can be expressed as

$$s(t) = \int_{-\infty}^{+\infty} d(t-\tau)\, h(t,\tau)\, d\tau. \tag{14.2.5}$$

Here $h(t,\tau)$ is the channel *time-varying impulse response* (TVIR). The received signal $y\,(t)$ feeds a front-end filter (with impulse response $h_R\,(t)$), producing the filtered received signal

$$r\,(t) = z(t) + n(t) \tag{14.2.6}$$

where $z(t) \triangleq s(t) \otimes h_R\,(t)$ is the useful signal component (the symbol $\otimes$ denotes the convolution operator) and $n(t) \triangleq w(t) \otimes h_R\,(t)$ is the filtered noise.

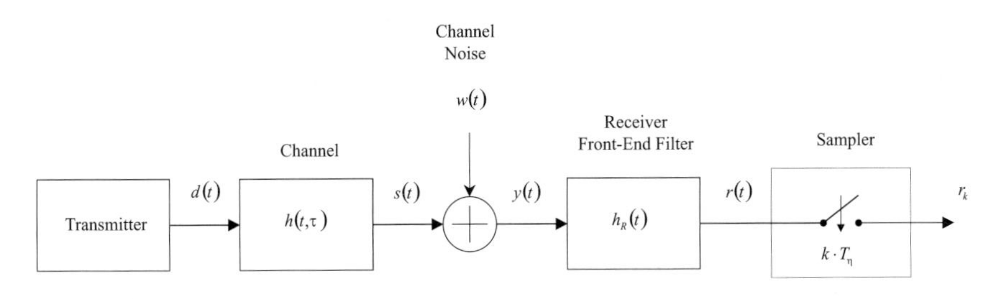

Figure 14.4. Sketch of transmitter, channel, and receiver front-end processing in a wireless communication system

14.2.3 Receiver filtering and sampling

An optimal receiver must not discard useful information present in the continuous-time received signal $y(t)$ (14.2.4); digital processing, however, is an inevitable requirement of any modern receiver. Therefore, some attention must be paid to filtering and discretizing the received signal [13], [14], [15].

There is some ambiguity here: we cannot identify what is useful until we know what we want to achieve. For communications, we seek to detect the transmitted information, but first we may have to estimate the channel impulse response or some other parameter. Therefore, any part of the data-bearing part of the received signal may be useful, and we should not discard it yet.

First, we consider an idealized case. We assume that the noiseless portion $z(t)$ of the filtered received signal is absolutely bandlimited with single-sided bandwidth B_z. This imposes stringent constraints on the transmitter pulse $g(t)$ in (14.2.4), and indicates also that the bandwidth of $h(t, \tau)$ in t (i.e, the *channel Doppler spread*) must be limited. Given that the noiseless received signal is absolutely bandlimited, it may be reconstructed in the mean-square sense from the infinite set of its samples $\{z_k = z(kT_\eta)\}$, as

$$\tilde{z}(t) = \sum_{k=-\infty}^{+\infty} z_k \operatorname{sinc}\left(\frac{t - kT_\eta}{T_\eta}\right) \tag{14.2.7}$$

where $T_\eta \triangleq T/\eta$ is the *sampling period* and η is the *number of samples per symbol.* If the sampling rate satisfies the Nyquist rate, i.e., $f_\eta \triangleq 1/T_\eta > 2B_z$, the interpolation formula (14.2.7) expresses an exact representation of $z(t)$ in the mean-square sense, that is,

$$E\left\{|z(t) - \tilde{z}(t)|^2\right\} = 0. \tag{14.2.8}$$

Equations (14.2.7)—(14.2.8) express the classic *sampling theorem* for stochastic processes [4]. Since the useful component $z(t)$ of the received signal may be reconstructed exactly by its samples, sampling does not entail any information loss.

The noise spectral density over the bandwidth B_z may be required for equalization, but outside it is irrelevant if the noise is Gaussian and stationary. Therefore, all the required information is captured without degradation if (a) the receiver front filter (see Figure 14.4) limits the frequency components of the received signal to the one-sided bandwidth B_r, where $B_r \geqslant B_z$; (b) the receiver filter's transfer function does not drop to zero within its bandwidth; and (c) the filtered signal is sampled quickly enough so that the noise beyond B_z does not alias into B_z. Thus, the sampling frequency is required to satisfy the inequality $f_\eta > B_r + B_z$.

Second, we study the realistic situation. Typically, part of the Radio Frequency (RF) spectrum is allocated to an FDMA system or a TDMA system. The spectrum is divided into channels (in this case, "channel" refers to a defined bandwidth), and, in the TDMA case, further divided into time slots. We assume that the channels are equispaced by f_{ch}. To avoid one user interfering with another user in an adjacent channel, the bandwidth, $2B_z$, of each user's signal after Doppler spreading is designed to be concentrated within f_{ch}.

If the transmitted signal is not absolutely bandlimited, weak sidelobes spill over into adjacent channels (and cause adjacent channel interference there). In this instance, a sampling rate equal to f_{ch} is information lossy. However, the desired signal's power spectral density at more distant frequencies is small and swamped by the large power spectral densities of the additive Gaussian noise and interference anyway. For this reason, the total power outside the nominal channel bandwidth is small and therefore does not provide much additional information. Accordingly, an engineering trade-off is assumed: the receiver filter strongly attenuates all signals outside the bandwidth $f_{ch}/2$, and the sampling rate satisfies the inequality $f_\eta > f_{ch}$. We cannot now claim that sampling is information lossless, but we do note that by increasing the sampling rate f_η, the information loss may be made arbitrarily small.

We still discuss bandwidths (and thus Nyquist sampling rates), but these must henceforth be interpreted as engineering definitions, such as the 99.99% power bandwidth or, more appropriately, the range of frequencies where the data-bearing signal's power spectral density exceeds some fraction of the power spectral density of the Gaussian noise plus interference.

In this case, since nonspread modulations are as spectrally efficient as possible by design, their (engineering) bandwidths are normally between $1.1/2T$ and $1.5/2T$, so we find that $\eta = 2$ is usually high enough to make information loss negligibly small.

Two popular receiver filters are (a) one matched to the transmitter filter and (b) one with a flat amplitude response and linear phase over the data-bearing signal's bandwidth, and an amplitude response that tapers to zero beyond that. The former has too narrow a bandwidth if the Doppler spread is high, but it is sufficient for the unknown-time invariant channel as long as its output is sampled at the Nyquist rate [14]. For practical reasons, it is also appropriate when the adjacent channel signals are strong. However, in fast time-varying channels, the latter is more appropriate (and convenient, since it negligibly distorts the received data-bearing signal), e.g., [16], [17], [18]. Finally, it is worth remembering that if the receive filter has an RRC transfer function with bandwidth $B_r = (1+\alpha)/2T_\eta$ and the channel noise is white over the filter bandwidth, then the noise samples at the receiver filter output make up a white sequence too.

Minimal complexity rather than high performance may be the objective in receiver design. In this case, the receiver filter may alias additive noise into the data-bearing signal's bandwidth, or the filtered signal may be sampled below its Nyquist rate, typically at one sample per symbol, e.g., [19], [20]. In both cases there is a mild power penalty. When the sampling rate is below the Nyquist rate for the Doppler spread (an unlikely event except in extraordinarily fast fading), the receiver's error rate levels out to an irreducible error floor as the noise variance is decreased [21], [22]. In such circumstances, increasing the transmitted signal power does not improve detection performance.

14.2.4 Quantities and notations for the complete link

In this section we introduce notations for the various scalar and matrix quantities needed in the remainder of Part III.

As described in Section 14.2.3, it is possible to design a receiver filter such that the frequency components of the data-bearing signal in $y(t)$ (14.2.4) are not distorted,[3] yet the noise is filtered to some narrow bandwidth. Thus, sampling[4] the filtered received signal of (14.2.6) at the instants $t = kT_\eta$ produces

$$r_k \triangleq r(kT_\eta) = z_k + n_k \tag{14.2.9}$$

where $n_k \triangleq n(kT_\eta)$ is a stationary (possibly colored) Gaussian noise sequence and

$$z_k \triangleq z(kT_\eta) = \int_{-\infty}^{+\infty} d(kT_\eta - \tau)\, h(kT_\eta, \tau)\, d\tau. \tag{14.2.10}$$

Substituting (14.2.1) into (14.2.10) (to evidence the dependence of z_k on the transmitted data $\{c_m\}$) produces

$$z_k = \sum_{m=0}^{L_c-1} c_m\, q_{k,k-m\eta} \tag{14.2.11}$$

where

$$q_{k,l} \triangleq \int_{-\infty}^{+\infty} g(lT_\eta - \tau)\, h(kT_\eta, \tau)\, d\tau \tag{14.2.12}$$

The RHS of (14.2.11) can be interpreted as a "convolution" between the transmitted symbol sequence $\{c_m\}$ and a discrete-time filter with time-varying impulse response $\{q_{k,l}\}$. Since the sequence $\{q_{k,l\eta}\}$ is itself generated through the convolution of $g(t)$ and the CIR $h(t, \tau)$, we call it the *overall channel impulse response* (OCIR). Its second subscript parameter $(k - m\eta)$ arises by analogy with the transmitted signal $d(t)$ (14.2.1). In fact, sampling $d(t)$ at $t = kT_\eta$ produces

$$d_k \triangleq d(kT_\eta) = \sum_{m=0}^{L_c-1} c_m\, g_{k-m\eta} \tag{14.2.13}$$

where $g_k \triangleq g(kT_\eta)$. Therefore, we identify $(k - m\eta)$ (or, equivalently, l) with the usual delay index of the impulse response of a time-invariant, discrete-time filter. Given that the channel is actually time varying, then the OCIR must have an additional parameter for time, i.e., its first subscript parameter k.

[3] In the following the delay due to the front-end filter is neglected so that $z(t) = s(t)$.
[4] Symbol synchronization problems are ignored here.

In the following, the OCIR is assumed to have finite duration (i.e., the channel is modelled as a time-varying FIR) to facilitate the derivation of optimal receivers in Chapter 15. Accordingly, $q_{k,l}$ is assumed to be zero in its delay parameter outside L_q symbol periods,[5] that is,

$$q_{k,l} = 0 \tag{14.2.14}$$

if $l < 0$ or $l \geqslant L_q\,\eta$, and the expression (14.2.9) can be simplified as

$$z_k = \sum_{m=\lfloor k/\eta \rfloor - L_q + 1}^{\lfloor k/\eta \rfloor} c_m\, q_{k,k-m\eta} \tag{14.2.15}$$

where $\lfloor x \rfloor$ denotes the largest integer smaller than the real number x.

The FIR property requires that the transmitter pulse and channel be causal and that the durations of the transmitter pulse and channel's delay spread be both finite and upper bounded. On physical grounds, this is a reasonable approximation for the delay spread, as discussed in Part I. Unfortunately, the requirement that the transmitter pulse's duration be finite is inconsistent (a) with the first, theoretical part of Section 14.2.3, where we constrained the continuous-time received signal to be an absolutely bandlimited signal in order to justify sampling it, and (b) with the second, practical part of Section 14.2.3, where we noted that the digital-to-analog converter's reconstruction filter smears out the transmitter pulse indefinitely. For this reason, we must interpret L_q by some engineering definition, such as the received pulses' 99.99% power duration or, more appropriately, the duration where the squared magnitude of the received pulse exceeds some fraction (e.g., 10%) of the power of the sampled noise. This interpretation involves some loss of accuracy, but the error can be made arbitrarily small by increasing L_q.

Substituting (14.2.15) into (14.2.9) yields

$$r_k = \sum_{m=\lfloor k/\eta \rfloor - L_q + 1}^{\lfloor k/\eta \rfloor} c_m\, q_{k,k-m\eta} + n_k. \tag{14.2.16}$$

The signal model (14.2.16) is general in that it applies to any wireless channel. In the particular case of an FF channel, the time-varying channel impulse response specializes as $h(t,\tau) = \alpha(t)\,\delta(\tau)$ [23], where $\alpha(t)$ is a complex random process expressing the channel distortion. In this case, (14.2.16) simplifies as

$$r_k = \alpha_k\, d_k + n_k \tag{14.2.17}$$

where $\alpha_k \triangleq \alpha\,(kT_\eta)$. In particular, if the received signal undergoes baud-rate sampling (i.e., $\eta = 1$) and the overall pulse $g\,(t) \otimes h_R\,(t)$ is Nyquist, then (14.2.16)

[5] L_q is assumed integer throughout this Part.

becomes[6]

$$r_k = \alpha_k\, c_k + n_k \tag{14.2.18}$$

Although (14.2.17) is derived under the assumption of a linearly modulated $d(t)$, Nyquist channel, and baud-rate sampling, it does not really rely on the assumption. In fact, (14.2.17) also applies to any signal transmitted over an FF channel, including CPM.

The time-invariant, frequency-selective channel is another important special case. The physical channel is a linear time-invariant (LTI) filter like the transmitter pulse-shaping filter so that

$$h(t, \tau) = h(0, \tau) = h(\tau) \tag{14.2.19}$$

and (14.2.16) becomes

$$q_{k,l} \triangleq \int_{-\infty}^{+\infty} g(lT_\eta - \tau)\, h(0, \tau)\, d\tau = g(\tau) \otimes h(\tau)\big|_{\tau = lT_\eta} \tag{14.2.20}$$

Then, the subscript parameter k can also be discarded, as $q_{k,l} = q_l$, in both (14.2.16) and (14.2.12).

The definition of the convolution of transmitter pulse shaping and channel as $q_{k,l}$ in (14.2.12) is not unique. It is, however, most useful when we think in terms of estimating a stationary channel's evolution in time. For fixed values of l and τ, $q_{k,l}$ and $h(t, \tau)$ are both stationary processes in k and t, respectively, and therefore Kalman or Wiener filtering/prediction follow naturally. However, as we see in Section 15.2.1, the presence of time k in both subscripts of $q_{k,k-m\eta}$ is unhelpful. Therefore, we introduce another OCIR sequence $\{\bar{q}_{k,l}\}$, equivalent to the OCIR $\{q_{k,l}\}$ and defined as

$$\bar{q}_{k,l} \triangleq q_{k\eta+l,l} \tag{14.2.21}$$

or, equivalently,

$$\bar{q}_{(k-l)/\eta,l} \triangleq q_{k,l} \tag{14.2.22}$$

so that $q_{k,k-m\eta} = \bar{q}_{m,k-m\eta}$. Then, we can rewrite (14.2.16) as

$$r_k = \sum_{m=\lfloor k/\eta \rfloor - L_q + 1}^{\lfloor k/\eta \rfloor} c_m\, \bar{q}_{m,k-m\eta} + n_k \tag{14.2.23}$$

This signal model has a useful interpretation, as follows. The mth symbol c_m is associated with the transmitted pulse $g(t - mT)$ or, equivalently, with the sequence

[6] If the receiver filter is a matched filter, this equation holds only if the fading distortion is slow [24].

$\{g_{k-m\eta}\}$ of samples taken at the time epochs $\{t = kT_\eta\}$. After transmission through the time-varying, frequency-selective channel, receiver filtering, and sampling, the contribution in r_k (14.2.23) due to c_m is received as the sequence $\{c_m \bar{q}_{m,k-m\eta}\}$. The received signal comprises the superposition of received pulses, the mth being slid in time by mT (i.e., $m\,\eta\,T_\eta$) and scaled by the symbol c_m. Unlike the transmitted pulse, the received pulse has an extra parameter, i.e., the first subscript parameter m, because each transmitted pulse is distorted differently by the time-varying channel.

It is convenient to introduce matrix representations for the received signal to simplify the derivation of optimal MLSD or MAPSD detectors. To begin, we note that the L_c symbols are smeared out by the length L_q OCIR, so that the data-bearing sequence $\{z_k\}$ (14.2.11) is non-negligible for

$$L_z \triangleq L_c + L_q - 1 \tag{14.2.24}$$

symbol periods, or, equivalently,

$$N_z \triangleq (L_c + L_q - 1)\eta \tag{14.2.25}$$

samples. Within these samples there are

$$N_q \triangleq (L_c + L_q - 1)L_q\eta \tag{14.2.26}$$

OCIR samples. Thus, we can write (14.2.11), (14.2.16), and (14.2.23) compactly by stacking the samples $\{r_k\}$ and $\{n_k\}$ into the vectors $\mathbf{r}$ and $\mathbf{n}$, respectively. We also note that the terms involving the data-bearing signal are discrete convolutions between channel parameters and source parameters, as $q_{k,k-m\eta}$ and c_m (see (14.2.11) and (14.2.16)) or $\bar{q}_{m,k-m\eta}$ and c_m (see (14.2.23)). To simplify the representation of such terms, (a) the samples of the sequences related to the channel, that is $\{z_k\}$, $\{d_k\}$, $\{q_{k,k-m\eta}\}$, and $\{\bar{q}_{m,k-m\eta}\}$, are stacked into the vectors $\mathbf{z}$, $\mathbf{d}$, $\mathbf{q}$, and $\bar{\mathbf{q}}$, respectively; (b) channel matrices for the corresponding source parameters, namely, $\mathbf{F}_z$, $\mathbf{F}_d$, $\mathbf{F}_q$, and $\bar{\mathbf{F}}_q$, are defined. The matrices $\mathbf{F}_d$, $\mathbf{F}_q$, and $\bar{\mathbf{F}}_q$ are constructed assuming a particular symbol sequence $\mathbf{c}$, and, to stress this dependence, it is useful to write, for instance, $\mathbf{F}_d(\mathbf{c})$ in place of $\mathbf{F}_d$.

Formally, assuming that N_z (14.2.25) samples are taken at the receiver side, we define the observation vector as $\mathbf{r} \triangleq [r_0, r_1, ..., r_{N_z-1}]^T$, the noise sample vector as $\mathbf{n} \triangleq [n_0, n_1, ..., n_{N_z-1}]^T$, the noiseless observations as $\mathbf{z} \triangleq [z_0, z_1, ..., z_{N_z-1}]^T$, the OCIR samples as $\mathbf{q} \triangleq [\mathbf{q}_0^T, \mathbf{q}_1^T, ..., \mathbf{q}_{N_z-1}^T]^T$, where

$$\mathbf{q}_k \triangleq [q_{k,k-(\lfloor k/\eta \rfloor - L_q+1)\eta}, q_{k,k-(\lfloor k/\eta \rfloor - L_q+2)\eta}, ..., q_{k,k-\lfloor k/\eta \rfloor \eta}]^T \tag{14.2.27}$$

and the received pulse samples as $\bar{\mathbf{q}} \triangleq [\bar{\mathbf{q}}_0^T, \bar{\mathbf{q}}_1^T, ..., \bar{\mathbf{q}}_{L_c-1}^T]^T$, where

$$\bar{\mathbf{q}}_m \triangleq [\bar{q}_{m,0}, \bar{q}_{m,1}, ..., \bar{q}_{m,L_q\eta-1}]^T \tag{14.2.28}$$

In addition, we define the convolution matrix for the noiseless samples as

$$\mathbf{F}_z \triangleq \mathbf{I}_{N_z} \tag{14.2.29}$$

where $\mathbf{I}_N$ is the identity matrix of order N, the convolution matrix for the OCIR samples as

$$\mathbf{F}_q \triangleq \text{diag}(\mathbf{c}_0, ..., \mathbf{c}_{N_z-1}) \tag{14.2.30}$$

where

$$\mathbf{c}_m \triangleq \left[c_{\lfloor k/\eta \rfloor - L_q+1}, c_{\lfloor k/\eta \rfloor - L_q+2}, ..., c_{\lfloor k/\eta \rfloor}\right]^T \tag{14.2.31}$$

with the convention that $c_m \equiv 0$ if $m < 0$ or $m \geqslant L_c$, and the convolution matrix for the received pulse samples as

$$\bar{\mathbf{F}}_q \triangleq [\bar{\mathbf{F}}_{q,0}, \bar{\mathbf{F}}_{q,1}, ..., \bar{\mathbf{F}}_{q,L_c-1}] \tag{14.2.32}$$

where

$$\bar{\mathbf{F}}_{q,m} \triangleq c_m \begin{bmatrix} \mathbf{0}_{(m-1)\eta, L_q\eta} \\ \mathbf{I}_{L_q\eta} \\ \mathbf{0}_{(L_c-L_q-m+1)\eta, L_q\eta} \end{bmatrix}. \tag{14.2.33}$$

Then, the received sample vector $\mathbf{r}$ can be expressed as

$$\mathbf{r} = \mathbf{F}_z \, \mathbf{z} + \mathbf{n} = \mathbf{F}_q \, \mathbf{q} + \mathbf{n} = \bar{\mathbf{F}}_q \, \bar{\mathbf{q}} + \mathbf{n}. \tag{14.2.34}$$

All these representations are formally equivalent. It is also worth noting that with frequency flat channels the CIR samples and the convolution matrix for the CIR samples can be defined as $\boldsymbol{\alpha} \triangleq [\alpha_0, ..., \alpha_{N_z-1}]^T$ and $\mathbf{F}_\alpha \triangleq \text{diag}(d_0, ..., d_{N_z-1})$, respectively. Then, (14.2.34) becomes

$$\mathbf{r} = \mathbf{F}_\alpha . \boldsymbol{\alpha} + \mathbf{n} \tag{14.2.35}$$

The relations (14.2.34) and (14.2.35) are presented diagrammatically in Figure 14.5 (the receiver filter or *Noise-Limiting Filter* is denoted by the acronym NLF).

In Chapter 15, the CIR (or some equivalent quantity) is alternately known, estimated, and averaged over. In the case of known CIR, or equivalently, known received pulses, we are interested in the multivariate ***probability density function*** (pdf, briefly) of the received signal vector $\mathbf{r}$ conditioned on the transmitted data and the received pulses. In this case, the conditional received signal's mean vector $\boldsymbol{\eta}_r(\mathbf{c}, \mathbf{q})$ equals

$$\boldsymbol{\eta}_r(\mathbf{c}, \mathbf{q}) \triangleq E\{\mathbf{r}|\mathbf{c}, \mathbf{q}\} = \mathbf{z} = \mathbf{F}_q(\mathbf{c}) \, \mathbf{q} = \bar{\mathbf{F}}_q(\mathbf{c}) \, \bar{\mathbf{q}} \overset{FF}{=} \mathbf{F}_\alpha(\mathbf{c}) \, \boldsymbol{\alpha} \tag{14.2.36}$$

and so the remainder $[\mathbf{r} - E\{\mathbf{r}|\mathbf{c}, \mathbf{q}\}] = \mathbf{n}$ is a Gaussian noise vector. Therefore, we complete the description of the desired pdf by noting that the conditional received signal's autocovariance $\mathbf{R}_r(\mathbf{c}, \mathbf{q})$ equals

$$\mathbf{R}_r(\mathbf{c}, \mathbf{q}) \triangleq \frac{1}{2} E\left\{[\mathbf{r} - \boldsymbol{\eta}_r(\mathbf{c}, \mathbf{q})] \, [\mathbf{r} - \boldsymbol{\eta}_r(\mathbf{c}, \mathbf{q})\}]^H |\mathbf{c}, \mathbf{q}\right\} = \frac{1}{2} E\{\mathbf{n} \, \mathbf{n}^H\}. \tag{14.2.37}$$

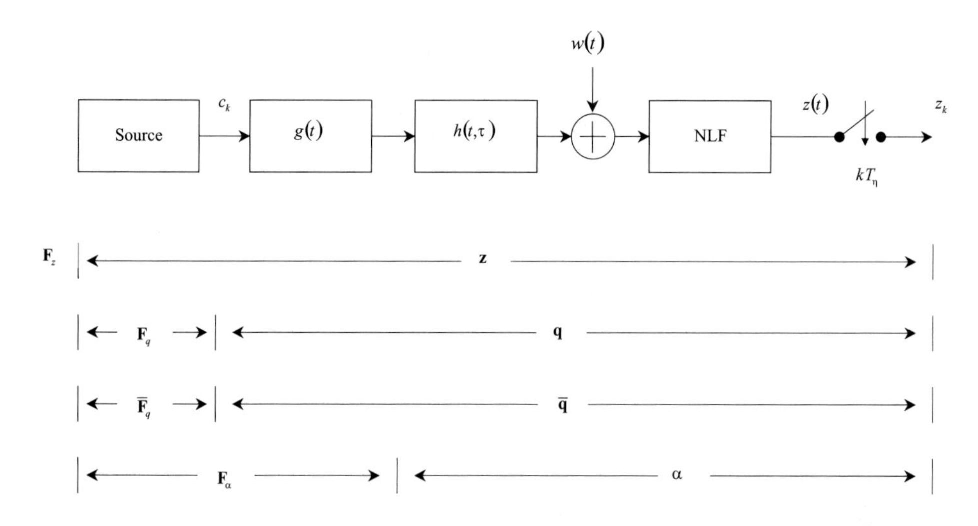

Figure 14.5. Notation for transmitter, channel, and receiver front end

The noise autocovariance $\mathbf{R}_n \triangleq (1/2)\, E\{\mathbf{n}\,\mathbf{n}^H\}$ depends both on the noise spectral density of the additive Gaussian noise $w\,(t)$ in (14.2.4) and on the receiver filter.

When the CIR is averaged over, the pdf of the received signal vector $\mathbf{r}$ depends on the pdf of all the OCIR taps for all sample instants, conditioned on the transmitted data. The pdfs illustrated in Part I, Chapter 3 refer to the amplitude of one channel tap only and so are insufficiently general. Instead, by modelling the channel as a large number of paths with independent complex attenuations, we can invoke the Central Limit Theorem to justify a complex Gaussian multivariate pdf for the OCIR taps. Luckily, this is mathematically tractable and a good model for fading channels. Under this assumption, the mean vector $\boldsymbol{\eta}_r\,(\mathbf{c})$ of $\mathbf{r}$, conditioned on the data vector $\mathbf{c}$ is

$$\boldsymbol{\eta}_r\,(\mathbf{c}) \triangleq E\{\mathbf{r}|\mathbf{c}\} = \mathbf{F}_z\,(\mathbf{c})\;\boldsymbol{\eta}_z = \mathbf{F}_q\,(\mathbf{c})\;\boldsymbol{\eta}_q = \bar{\mathbf{F}}_q\,(\mathbf{c})\;\boldsymbol{\eta}_{\bar{q}} \overset{FF}{=} \mathbf{F}_\alpha\,(\mathbf{c})\;\boldsymbol{\eta}_\alpha \qquad (14.2.38)$$

where $\boldsymbol{\eta}_x \triangleq E\{\mathbf{x}|\mathbf{c}\}$ and x is a placeholder for z, q or α. It can be immediately checked that $\boldsymbol{\eta}_r\,(\mathbf{c})$ is equal to zero in Rayleigh fading. In Rician fading, the elements in the vector $E\{\mathbf{q}\}$ are given by

$$E\{q_{k,k-m\eta}\} = \int_{-\infty}^{+\infty} g\left((k-m\eta)T_\eta - \tau\right)\, h_d(kT_\eta,\tau)\, d\tau \qquad (14.2.39)$$

where $h_d(t,\tau) \triangleq E\left\{h(t,\tau)\right\}$ is the *deterministic component* of the channel TVIR. Similarly, under the assumption that the channel TVIR and the channel noise are independent, the conditional received signal's autocovariance matrix $\mathbf{R}_r\,(\mathbf{c})$ is given by

$$\begin{aligned}\mathbf{R}_r(\mathbf{c}) &\triangleq \frac{1}{2}E\left\{[\mathbf{r}-\boldsymbol{\eta}_r(\mathbf{c})\}]\,[\mathbf{r}-\boldsymbol{\eta}_r(\mathbf{c})]^H\,|\mathbf{c}\right\}\\ &= \mathbf{F}_z(\mathbf{c})\,\mathbf{R}_z\,\mathbf{F}_z^H(\mathbf{c})+\mathbf{R}_n = \mathbf{F}_q(\mathbf{c})\,\mathbf{R}_q\,\mathbf{F}_q^H(\mathbf{c})+\mathbf{R}_n\\ &\overset{FF}{=} \mathbf{F}_\alpha(\mathbf{c})\,\mathbf{R}_\alpha\,\mathbf{F}_\alpha^H(\mathbf{c})+\mathbf{R}_n \end{aligned} \tag{14.2.40}$$

where $\mathbf{R}_x(\mathbf{c}) \triangleq (1/2)\,E\{(\mathbf{x}-E\{\mathbf{x}|\mathbf{c}\})\,(\mathbf{x}-E\{\mathbf{x}|\mathbf{c}\})^H\,|\mathbf{x}\}$ and x is a placeholder for z, q or α. The $(k_1L_q+\lfloor k_1/\eta\rfloor-m_1, k_2L_q+\lfloor k_2/\eta\rfloor-m_2)$th entry in the OCIR autocovariance matrix $\mathbf{R}_q$ is given by

$$\begin{aligned}&[\mathbf{R}_q]_{k_1L_q+\lfloor k_1/\eta\rfloor-m_1,k_2L_q+\lfloor k_2/\eta\rfloor-m_2}\\ &\quad= \frac{1}{2}E\{(q_{k_1,k_1-m_1\eta}-E\{q_{k_1,k_1-m_1\eta}\})(q_{k_2,k_2-m_2\eta}-E\{q_{k_2,k_2-m_2\eta}\})^*\}\\ &= \iint_{-\infty}^{+\infty} g((k_1-m_1\eta)T_\eta-\tau_1)\,g^*((k_2-m_2\eta)T_\eta-\tau_2)\,R_h(k_1T_\eta,k_2T_\eta;\tau_1,\tau_2)\,d\tau_1\,d\tau_2\end{aligned}$$

where

$$R_h(t_1,t_2;\tau_1,\tau_2) \triangleq \frac{1}{2}E\left\{[h(t_1,\tau_1)-h_d(t_1,\tau_1)]\,[h^*(t_2,\tau_2)-h_d^*(t_2,\tau_2)]\right\} \tag{14.2.41}$$

is the *autocovariance function* of the TVIR [23]. The meaning of $R_h(t_1,t_2;\tau_1,\tau_2)$ is discussed in detail in Part I.

Finally, it is worth noting that the average ratio of bit energy-to-noise spectral density E_b/N_0 at the receiver input equals

$$\frac{E_b}{N_0} = \frac{1}{N_0\log_2 M}E\left\{\int_{-\infty}^{+\infty}\left|\int_{-\infty}^{+\infty} c_m\,g(t-\tau)\,h(t,\tau)\,d\tau\right|^2 dt\right\}. \tag{14.2.42}$$

14.2.5 Channel parameterization

In Chapter 15 we investigate detectors that estimate the CIR (or some equivalent quantity). Unfortunately, the representations of the received signal illustrated in the previous section may not be suitable for channel estimation. As stated previously, the received signal should be sampled at its Nyquist rate in order to prevent loss of information. Then, the discretized channel is represented by the OCIR samples $\{q_{k,k-m\eta}\}$, i.e., by a *tapped delay line model*. However, in a sense, the T_η spacing is unnecessarily short with respect to the slow variations of the OCIR sequence in its first subscript parameter k, since they are only due to the channel Doppler spread. For this reason, $\{q_{k,k-m\eta}\}$ may be an inefficient parameterization of the OCIR, and it is reasonable to seek more parsimonious representations or parameterizations, that is, *reduced complexity channel models* [23], [25], [26], [27], [28], [29], [30], [31], [32], [33]. Such models can reveal which channel parameters are important and which are not. As we shall see, this distinction greatly assists with the channel estimation problem.

The Karhunen-Lóeve expansion

The optimal parameterization comes from the so-called *Karhunen-Lóeve* (KL) expansion [26], [27], [29]. The OCIR autocovariance matrix $\mathbf{R}_q$ is eigendecomposed as

$$\mathbf{R}_q = \mathbf{U}_{KL}\,\mathbf{D}\,\mathbf{U}_{KL}^H \tag{14.2.43}$$

where $\mathbf{U}_{KL}$ is a unitary matrix having the eigenvectors of $\mathbf{R}_q$ as its columns and $\mathbf{D}$ is a diagonal matrix of the eigenvalues of $\mathbf{R}_q$. Since $\mathbf{R}_q$ is an autocovariance matrix, it is positive semidefinite and its eigenvalues are real and non-negative. For convenience, the diagonal entries in $\mathbf{D}$, $\{d_k\}$, are sorted by magnitude, with the largest eigenvalue first, and the columns of $\mathbf{U}_{KL}$ are reordered accordingly. In Rayleigh fading (i.e., with a zero mean CIR), the Karhunen-Loeve expansion of the OCIR vector $\mathbf{q}$ is defined as

$$\mathbf{q} = \mathbf{U}_{KL}\,\mathbf{u} \tag{14.2.44}$$

where $\mathbf{u}$ is a zero-mean vector of Gaussian parameters, $\{u_k\}$, with autocovariance matrix

$$\mathbf{R}_{\mathbf{u}} \triangleq \frac{1}{2}E\{\mathbf{u}\,\mathbf{u}^H\} = \mathbf{U}_{KL}^H\frac{1}{2}E\{\mathbf{q}\,\mathbf{q}^H\}\mathbf{U}_{KL} = \mathbf{D}, \tag{14.2.45}$$

i.e., the components of the random vector $\mathbf{u}$ are uncorrelated. If some of the eigenvalues $\{d_k\}$ are zero, the variances of their matching parameters $\{u_k\}$ are zero and they can be ignored. Therefore, the number of channel parameters is $N_{KL} \triangleq \text{rank}(\mathbf{D})$. Formally, we can delete the zero eigenvalues and their matching parameters by defining the masking matrix

$$\mathbf{M}_{KL} \triangleq [\mathbf{I}_{N_{KL}}, \mathbf{0}_{N_{KL}, N_z - N_{KL}}] \tag{14.2.46}$$

and rewriting the representation (14.2.34) of the received vector $\mathbf{r}$ as

$$\mathbf{r} = \mathbf{F}_q\mathbf{q} + \mathbf{n} = \mathbf{F}_q\mathbf{U}_{KL}\mathbf{u} + \mathbf{n} = \mathbf{F}_q\mathbf{U}_{KL}\,\mathbf{M}_{KL}^H\,\mathbf{M}_{KL}\,\mathbf{u} + \mathbf{n}. \tag{14.2.47}$$

Then, if we define the matrix $\mathbf{V}_{KL} \triangleq \mathbf{U}_{KL}\mathbf{M}_{KL}^H$, the new convolution matrix

$$\mathbf{F}_{KL} \triangleq \mathbf{F}_q\,\mathbf{U}_{KL}\,\mathbf{M}_{KL}^H = \mathbf{F}_q\,\mathbf{V}_{KL}, \tag{14.2.48}$$

and the vector of the nonzero parameters $\mathbf{q}_{KL} \triangleq \mathbf{M}_{KL}\mathbf{u}$, eq.(14.2.47) can be rewritten as

$$\mathbf{r} = \mathbf{F}_{KL}\mathbf{q}_{KL} + \mathbf{n}, \tag{14.2.49}$$

evidencing the dependence of $\mathbf{r}$ on a reduced complexity representation of the OCIR.

Complex exponential parameterization

It is known from Part I that in a wireless link there are multiple paths between the transmitter and receiver due to surrounding obstacles. The multipaths experience different complex gains, delays, and Doppler shifts, depending on the geometry of the obstacles and the mobile terminal. This a priori knowledge of the channel leads to the CIR being parameterized as a finite number N_P of paths, that is, as [34]

$$h(kT_\eta, \tau) = \sum_{i=0}^{N_P-1} a_i \exp(j\phi_i) \exp(j2\pi k f_{D,i} T_\eta)\, \delta(\tau - \tau_i) \tag{14.2.50}$$

where $a_i \exp(j\phi_i)$, $f_{D,i}$, and τ_i are the complex gain, Doppler shift, and delay, respectively, of the ith path. From (14.2.12) and (14.2.50), the OCIR samples (14.2.12) are given by

$$q_{k,k-m\eta} = \sum_{i=0}^{N_P-1} a_i \exp(j\phi_i)\, g((k-m\eta)T_\eta - \tau_i)\, \exp(j2\pi k f_{D,i} T_\eta). \tag{14.2.51}$$

This representation suggest that complex exponentials can be used as *basis functions* for the OCIR. It can be easily shown that the coefficients $\{a_i \exp(j\phi_i)\ g((k-m\eta)\,T_\eta\ -\tau_i)\}$ and complex exponential samples $\{\exp(j2\pi k f_{D,i} T_\eta)\}$ can be written as a length $N_{CE} = N_P L_q$ vector $\mathbf{q}_{CE}$ of coefficients (in descending order of magnitude) and a matrix $\mathbf{V}_{CE}$ of complex exponential basis vectors, respectively. Therefore, the OCIR vector $\mathbf{q}$ can be expressed as

$$\mathbf{q} = \mathbf{V}_{CE}\, \mathbf{q}_{CE} \tag{14.2.52}$$

so that the representation (14.2.34) of the received vector $\mathbf{r}$ is given by

$$\mathbf{r} = \mathbf{F}_q\, \mathbf{V}_{CE}\, \mathbf{q}_{CE} + \mathbf{n} = \mathbf{F}_{CE}\, \mathbf{q}_{CE} + \mathbf{n} \tag{14.2.53}$$

with $\mathbf{F}_{CE} \triangleq \mathbf{F}_q \mathbf{V}_{CE}$. There is no guarantee, however, that N_{CE} is actually smaller than N_q, so this parameterization may not be parsimonious at all. It does, however, identify which parameters are weak and places them at the bottom of the vector $\mathbf{q}_{CE}$.

Power series

Power series models have found wide application in the representation of channel's time or frequency selectivity without regard to the pulse shape [31].

A simple application of such models is the power series representation of an FF fading channel [23], [35], [36]. A time-selective fading process $\alpha(t)$ can be expanded in its Taylor series around the time epoch $t = t_0$ as

$$\alpha(t) = \sum_{n=0}^{+\infty} \alpha_n \left(\frac{t-t_0}{T_0}\right)^n \tag{14.2.54}$$

where T_0 is a parameter for time normalization and the nth coefficient α_n is given by

$$\alpha_n \triangleq \frac{T_0^m}{n!} \frac{d^n\, \alpha(t)}{dt^n}\bigg|_{t=t_0} \tag{14.2.55}$$

The stochastic parameters $\{\alpha_n, n = 0, 1, ...\}$ are correlated Gaussian random variables with zero mean (with Rayleigh fading) and correlation [35]

$$R_{n,m}\,(i) \triangleq E\,\{\alpha_n\, \alpha_m^*\} = w_i \, \frac{(-1)^m\, T_0^{n+m}}{n!\, m!} \, \frac{d^{n+m}\, R_D(\tau)}{d\tau^{n+m}}\bigg|_{\tau=0} \tag{14.2.56}$$

where $R_D(\tau)$ is the *Doppler autocovariance function* [23]. Truncating the power series of (14.2.54) to N_{TS} terms produces an N_{TS}th order reduced-complexity model $\alpha_{N_{TS}}(t)$,

$$\alpha_{N_{TS}}(t) = \sum_{n=0}^{N_{TS}-1} \alpha_n \left(\frac{t - t_0}{T_0}\right)^n \tag{14.2.57}$$

of the process $\alpha(t)$.

Polynomial models can be also employed to represent the *time-variant frequency response* (TVFR) $H\,(f,t)$ of a DS fading channel [23]. In [25], [33], for instance, $H\,(f,t)$ is approximated as

$$H_{N_{PS}}(t,f) \triangleq \sum_{i=0}^{N_{PS}-1} q_i\; (t)\; f^i \tag{14.2.58}$$

where

$$q_i\,(t) \triangleq \frac{1}{i!}\, \frac{\partial^i}{\partial f^i} H\,(t,f)\bigg|_{f=0},\; i = 0, 1, \ldots, N_{PS} - 1 \tag{14.2.59}$$

are correlated Gaussian processes having zero mean with Rayleigh fading.

Generally speaking, the CIR of an absolutely bandlimited channel can be represented by a Taylor series converging to the CIR itself in the mean-square sense [33]. This general result suggests that the OCIR samples $q_{k,l}$ (14.2.12) of a DS channel may be represented by a power series in their first subscript parameter k (expressing time evolution) over a window of L_w symbol periods [23], while keeping their second subscript parameter l (representing the tap number) fixed, that is [33], as

$$q_{k,l} \simeq \sum_{p=0}^{N_l-1} q_{PS}\,[p, l, k \bmod \eta] \left(\frac{\lfloor k/\eta \rfloor}{L_w}\right)^p \tag{14.2.60}$$

for $l = 0, 1, ..., L_q\eta - 1$, where $q_{PS}\,[p, l, k \bmod \eta]$ is the coefficient for the pth power term of the lth tap in the symbol interval $(k \bmod \eta)$. It is worth noting that the parameter L_w has been inserted in the denominator of all the fractions in (14.2.60) to minimize numerical problems. If $L_w = L_c$ is assumed, (14.2.60) allows us to approximate the received samples as

$$\mathbf{r} = \mathbf{F}_q\,\mathbf{q} + \mathbf{n} \simeq \mathbf{F}_q\,\mathbf{V}_{PS}\,\mathbf{q}_{PS} + \mathbf{n} = \mathbf{F}_{PS}\,\mathbf{q}_{PS} + \mathbf{n} \tag{14.2.61}$$

where $\mathbf{F}_{PS} \triangleq \mathbf{F}_q\,\mathbf{V}_{PS}$ is the convolution matrix and

$$\mathbf{q}_{PS} \triangleq [\mathbf{q}_{PS,0}^T, \mathbf{q}_{PS,1}^T, ..., \mathbf{q}_{PS,N_l-1}^T]^T \tag{14.2.62}$$

is a vector containing $N_{PS} \triangleq N_l\,L_q\,\eta$ channel parameters. Here

$$\mathbf{q}_{PS,p} \triangleq \left[\mathbf{q}_{PS}^T(p,0), \mathbf{q}_{PS}^T(p,0), ..., \mathbf{q}_{PS}^T(p,\eta-1)\right]^T \tag{14.2.63}$$

is made of the coefficients referring to the pth power in (14.2.60) and $\mathbf{q}_{PS}(p, k \bmod \eta)$ contains, in turn, the coefficients having the ith sample position within the $(k \bmod \eta)$-th symbol period, as

$$\begin{aligned}
&\mathbf{q}_{PS}(p, k \bmod \eta) \\
&\triangleq [q_{PS}[p,l,k-(\lfloor k/\eta \rfloor - L_q + 1)\eta], q_{PS}[p,l,k-(\lfloor k/\eta \rfloor - L_q + 1)\eta + 1], \\
&..., q_{PS}[p,l,k-\lfloor k/\eta \rfloor\,\eta]]^T
\end{aligned} \tag{14.2.64}$$

The vector $\mathbf{V}_{PS}$ is a matrix of basis vectors, defined by a Kronecker matrix product as

$$\mathbf{V}_{PS} \triangleq \bar{\mathbf{V}}_{PS} \odot \mathbf{I}_{L_q\eta} = \begin{bmatrix} \bar{V}_{PS,0,0}\mathbf{I}_{L_q\eta} & \cdots & \bar{V}_{PS,0,N_l-1}\mathbf{I}_{L_q\eta} \\ \vdots & & \vdots \\ \bar{V}_{PS,L_c-1,0}\mathbf{I}_{L_q\eta} & \cdots & \bar{V}_{PS,L_c-1,N_l-1}\mathbf{I}_{L_q\eta} \end{bmatrix} \tag{14.2.65}$$

where

$$\bar{\mathbf{V}}_{PS} \triangleq \begin{bmatrix} 1 & \frac{0}{L_c} & \cdots & \left(\frac{0}{L_c}\right)^{N_l-1} \\ 1 & \frac{1}{L_c} & \cdots & \left(\frac{1}{L_c}\right)^{N_l-1} \\ \vdots & \vdots & & \vdots \\ 1 & \frac{L_c-1}{L_c} & \cdots & \left(\frac{L_c-1}{L_c}\right)^{N_l-1} \end{bmatrix}. \tag{14.2.66}$$

As $N_l \longrightarrow \infty$, the power series model of the vector $\mathbf{r}$ converges on the true $\mathbf{r}$, so $N_{PS} \longrightarrow \infty$ and the power series model is not parsimonious at all. However, the later entries are due to higher and higher powers of k, so their mean-square power must diminish as the sequence $\{q_{k,l}\}$ is bandlimited in the parameter k.

Reduced rank parameterizations

There is no guarantee that any of the parameterizations presented so far are actually more parsimonious than the tapped delay line model $\{q_{k,k-m\eta}\}$. In statistical estimation, this problem is commonly circumvented by *rank reduction.* In other words, the parameters contributing least to $\mathbf{z}$ are set to zero and not further considered. However, rank reduction introduces a bias, which is traded off against reduced variance in the estimation error [37].

In all three models—the Karhunen-Lóeve expansion (KL), the complex exponential parameterization (CE), or the power series (PS)—the weak parameters are relegated to the end of the parameter vector $\mathbf{q}_x$ (x is used as a placeholder for KL, CE, or PS). Therefore, in a procedure closely following (14.2.47), we delete weak parameters by defining the masking matrix (see (14.2.46))

$$\mathbf{M}_{x,RR} \triangleq [\mathbf{I}_{N_{x/RR}}, \mathbf{0}_{N_{x/RR},N_x-N_{x/RR}}] \tag{14.2.67}$$

and rewriting (14.2.34) as

$$\mathbf{r} = \mathbf{F}_x(\mathbf{c})\,\mathbf{q}_x + \mathbf{n} \simeq \mathbf{F}_x(\mathbf{c})\,\mathbf{M}^H_{x/RR}\mathbf{M}_{x/RR}\mathbf{q}_x + \mathbf{n} = \mathbf{F}_{x/RR}(\mathbf{c})\,\mathbf{q}_{x/RR} + \mathbf{n} \tag{14.2.68}$$

where $\mathbf{F}_{x/RR}(\mathbf{c}) \triangleq \mathbf{F}_x(\mathbf{c})\,\mathbf{M}^H_{x/RR}$, $\mathbf{q}_{x/RR} \triangleq \mathbf{M}_{x/RR}\mathbf{q}_x$, and $N_{x/RR}$ is the number of parameters in the reduced rank model. In this way, the least important parameters are always disregarded and it is possible to ensure that the number of received samples or observations N_z exceeds the number of parameters N_x.

Collecting the results

By employing x as a placeholder for z, q, α, the KL, CE, and PS, equations (14.2.34), (14.2.49), (14.2.53), and (14.2.61) can be collected as in (14.2.68). The received signal's autocovariance matrix $\mathbf{R}_r(\mathbf{c})$, conditioned on the symbol vector $\mathbf{c}$, can also be written compactly as

$$\mathbf{R}_r(\mathbf{c}) = \mathbf{F}_x(\mathbf{c})\,\mathbf{R}_{q,x}\mathbf{F}^H_x(\mathbf{c}) + \mathbf{R}_n \simeq \mathbf{F}_{x/RR}(\mathbf{c})\,\mathbf{R}_{q,x/RR}\mathbf{F}^H_{x/RR}(\mathbf{c}) + \mathbf{R}_n \tag{14.2.69}$$

where $\mathbf{R}_{q,x} \triangleq (1/2)\,E\{\mathbf{q}_x\mathbf{q}_x^H\}$ is the parameter autocovariance matrix.

Bibliography

[1] S. U. H. Qureshi, "Adaptive equalization," *Proc. IEEE*, vol. 73, pp. 1349–1387, Sep. 1985.

[2] R. W. Lucky, J. Salz, and E. J. Weldon, *Principles of Data Communication.* New York: McGraw Hill, 1968.

[3] W. C. Jakes, *Microwave Mobile Communications.* Piscataway, NJ: IEEE Press, 1993.

[4] J. G. Proakis, *Digital Communications.* New York: McGraw-Hill, 2nd ed., 1989.

[5] S. Stein, "Fading channel issues in system engineering," *IEEE J. Select. Areas Comm.*, vol. 5, pp. 68–89, Feb. 1987.

[6] M. Schwartz, W. R. Bennett, and S. Stein, *Communication Systems and Techniques.* New York: McGraw-Hill, 1966.

[7] E. Biglieri, G. Caire, G. Taricco, and G. Ventura-Traveset, "Computing error probabilities over fading channels: A unified approach," *European Trans. Telecomm.*, vol. 9, pp. 15–25, 1998.

[8] J. B. Anderson, T. Aulin, and C. E. Sundberg, *Digital Phase Modulation.* Plenum, 1986.

[9] T. S. Rappaport, *Wireless Communications Principles and Practice.* Piscataway, NJ: IEEE Press, 1996.

[10] E. Arthurs and H. Dym, "On the optimum detection of digital signals in the presence of white Gaussian noise—a geometric interpretation and a study of three basic data transmission systems," *IRE Trans. Commun. Syst.*, pp. 386–372, Dec. 1962.

[11] H. Nyquist, "Certain topics in telegraph transmission theory," *AIEE Trans.*, vol. 47, pp. 617–644, 1928.

[12] P. A. Laurent, "Exact and approximate construction of digital phase modulations by superposition of amplitude modulated pulses," *IEEE Trans. Comm.*, vol. 34, pp. 150–160, Feb. 1986.

[13] J. C. Hancock and P. A. Wintz, *Signal Detection Theory.* New York: McGraw-Hill, 1966.

[14] K. M. Chugg and A. Polydoros, "MLSE for an unknown channel—Part I: Optimality considerations," *IEEE Trans. Comm.*, vol. 44, pp. 836–846, July 1996.

[15] G. D. Forney, "Maximum-likelihood sequence estimation of digital sequences in the presence of intersymbol interference," *IEEE Trans. Inf. Theory*, vol. 18, pp. 363–378, May 1971.

[16] Q. Dai and E. Shwedyk, "Detection of bandlimited signals Over frequency-selective Rayleigh fading channels," *IEEE Trans. Comm.*, vol. 42, pp. 941–950, 1994.

[17] G. M. Vitetta and D. P. Taylor, "Maximum likelihood decoding of uncoded and coded PSK signal sequences transmitted over Rayleigh flat-fading channels," *IEEE Trans. Comm.*, vol. 43, pp. 2750–2758, Nov. 1995.

[18] X. Yu and S. Pasupathy, "Innovations-based MLSE for Rayleigh fading channels," *IEEE Trans. Comm*, vol. 43, pp. 1534–1544, Feb./Mar./Apr. 1995.

[19] F. R. Magee and J. G. Proakis, "Adaptive maximum-likelihood sequence estimation for digital signalling in the presence of intersymbol interference," *IEEE Trans. Inf. Theory*, vol. 19, pp. 120–124, Jan. 1973.

[20] F. Davarian, "Mobile digital communication via tone calibration," *IEEE Trans. Vehicular Techn.*, vol. 36, pp. 55–62, May 1987.

[21] M. L. Moher and J. H. Lodge, "TCMP—a modulation and coding strategy for Rician fading channels," *IEEE J. Select. Areas Comm.*, vol. 7, pp. 1347–1355, Dec. 1989.

[22] B. D. Hart and D. P. Taylor, "On the irreducible error floor in fast fading channels," *IEEE Trans. Vehicular Techn.*, Apr. 1997.

[23] P. Bello, "Characterization of Randomly Time-Variant Linear Channels," *IEEE Trans. Comm.*, vol. 11, pp. 360–393, 1963.

[24] J. K. Cavers, "On the validity of the slow and moderate fading models for matched filter detection of Rayleigh fading signals," *Can. J. of Elect. & Comp. Eng*, vol. 17, pp. 183–189, 1992.

[25] W. S. Leon and D. P. Taylor, "An adaptive receiver for the time- and frequency-selective fading channel," *IEEE Trans. Comm.*, vol. 45, pp. 1548–1559, Dec. 1997.

[26] K. W. Yip and T. S. Ng, "Efficient simulation of digital transmission over WSSUS channels," *IEEE Trans. Comm.*, vol. 43, pp. 2907–2912, Dec. 1995.

[27] K. W. Yip and T. S. Ng, "Karhunen-Loeve expansion of the WSSUS channel output and its application to efficient simulation," *IEEE J. Sel. Areas Comm.*, vol. 15, pp. 640–646, May 1997.

[28] P. M. Crespo and J. Jimenez, "Computer simulation of radio channels using an harmonic decomposition technique," *IEEE Trans. Vehicular Techn.*, vol. 44, pp. 414–419, Aug. 1995.

[29] M. Clark, L. Greenstein, W. Kennedy, and M. Shafi, "MMSE diversity combining for wide-band digital cellular radio," *IEEE Trans. Comm.*, vol. 40, pp. 1128–1135, 1992.

[30] A. M. Sayeed, E. N. Onggosanusi, and B. D. V. Veen, "A canonical space-time characterization of mobile wireless channels," *IEEE Comm. Lett.*, vol. 3, pp. 94–96, Apr. 1999.

[31] D. P. Taylor, G. M. Vitetta, B. D. Hart, and A. Mammela, "Wireless channel equalization," *European Trans. Telecomm.*, vol. 9, pp. 117–143, Mar./Apr. 1998.

[32] S. A. Fechtel, "A novel approach to modeling and efficient simulation of frequency-selective fading radio channels," *IEEE J. Sel. Areas Comm.*, vol. 11, pp. 422–431, Apr. 1993.

[33] D. K. Borah and B. D. Hart, "Frequency-selective fading channel estimation with a polynomial time-varying channel model," *IEEE Trans. Comm.*, vol. 47, pp. 862–871, June 1999.

[34] G. B. Giannakis and C. Tepedelenlioglu, "Basis expansion models and diversity techniques for blind identification and equalization of time-varying channels," *IEEE Proc.*, vol. 86, pp. 1969–1985, Oct. 1998.

[35] G. M. Vitetta, D. P. Taylor, and U. Mengali, "Double-filtering receivers for PSK signals transmitted over Rayleigh frequency-flat fading channels," *IEEE Trans. Comm.*, vol. 44, pp. 686–695, June 1996.

[36] G. M. Vitetta, U. Mengali, and D. P. Taylor, "Optimal noncoherent detection of FSK signals transmitted Over linearly time-selective Rayleigh fading channels," *IEEE Trans. Comm.*, vol. 45, pp. 1417–1425, Nov. 1997.

[37] L. L. Scharf, *Statistical Signal Processing: Detection, Estimation and Time Series Analysis.* Reading, MA: Addison-Wesley, 1991.

Chapter 15

OPTIMAL DECISION RULES

In this chapter, optimal decision rules for the signal model of Section 14.2.4 are illustrated and intuitive interpretations of these rules are discussed. In addition, we present the analytical tools for characterizing how well the rules perform. We shall see that such optimal rules may require an unreasonable amount of a priori information, or their implementation complexity may be infeasibly high. It is useful to study their properties nonetheless, since (a) being optimal, they define a fundamental lower bound on the performance that other, more practical, detectors may achieve, (b) their performance may be easy to determine analytically, and (c) they provide insight into what an implementable equalizer should attempt.

15.1 OPTIMALITY

The first question that arises is "what is the criterion of optimality?" Statistical detection theory and the Bayes criterion lead to the *maximum likelihood* (ML) and *maximum a posteriori* (MAP) criteria [1], [2]. In communications, the goal is to detect the digital transmitted data, i.e., the bits into the constellation mapper or, equivalently, the symbols generated by it. Moreover, these symbols can be treated individually or as a complete sequence (there is a one-to-one mapping between the sequence of symbols and sequence of bits, so it is unnecessary to consider the sequence of bits separately). By combining all these different options, we have six different criteria of optimality.

Without regard to the CIR, we formally write the decision rules as

$$\hat{b}_i = \arg\max_{\tilde{b}_i} P(\tilde{b}_i|\mathbf{r}) \qquad \text{MAP Bit Detector (MAPBD)} \tag{15.1.1}$$

$$\hat{c}_m = \arg\max_{\tilde{c}_m} P(\tilde{c}_m|\mathbf{r}) \qquad \text{MAP Symbol Detector (MAPSD)} \tag{15.1.2}$$

$$\hat{\mathbf{c}} = \arg\max_{\tilde{\mathbf{c}}} P(\tilde{\mathbf{c}}|\mathbf{r}) \qquad \text{MAP Sequence Detector} \tag{15.1.3}$$

$$\hat{b}_i = \arg\max_{\tilde{b}_i} p(\mathbf{r}|\tilde{b}_i) \qquad \text{ML Bit Detector} \tag{15.1.4}$$

$$\hat{c}_m = \arg\max_{\tilde{c}_m} p(\mathbf{r}|\tilde{c}_m) \qquad \text{ML Symbol Detector} \tag{15.1.5}$$

$$\hat{\mathbf{c}} = \arg\max_{\tilde{\mathbf{c}}} p(\mathbf{r}|\tilde{\mathbf{c}}) \qquad \text{ML Sequence Detector (MLSD)} \tag{15.1.6}$$

where $\tilde{x}$ ($\tilde{\mathbf{x}}$) denotes a trial parameter and $\hat{x}$ ($\hat{\mathbf{x}}$) denotes its detected value, whereas $p(\cdot)$ and $P(\cdot)$ represent a joint pdf and a probability, respectively. The argument of the arg max operator is referred to as the bit/symbol/symbol sequence *metric* and the arg max operator signifies "choose the argument maximizing the metric." Therefore, as an example, the MAP bit detector of (15.1.1) searches over both possible values of the ith bit $\tilde{b}_i$ and chooses as $\hat{b}_i$ the one that is most probable given the received sample vector $\mathbf{r}$ (i.e., maximizes $P(\tilde{b}_i|\mathbf{r})$). The ML decision rules (15.1.4)—(15.1.6) are a little more subtle: the first parameter in each likelihood expression is treated as constant, and the decision rule searches over the *conditioned* parameter to find which value is most likely to lead to the received samples $\mathbf{r}$.

This is an unduly large list of criteria of optimality, so what is the best criterion of optimality? First, as discussed below, the MAP criterion is more appropriate than the ML criterion if our interest is in the reliability of information data. Nonetheless, it is straightforward to show that the MAP and ML criteria are equivalent under the assumption of a priori equiprobable bits/symbols/symbol sequences [1]. For example, the MAP sequence detection criterion may be rewritten as the ML sequence detection criterion, as

$$\hat{\mathbf{c}} = \arg\max_{\tilde{\mathbf{c}}} P(\tilde{\mathbf{c}}|\mathbf{r}) = \arg\max_{\tilde{\mathbf{c}}} \frac{p(\mathbf{r}|\tilde{\mathbf{c}})\, P(\tilde{\mathbf{c}})}{p(\mathbf{r})} = \arg\max_{\tilde{\mathbf{c}}} p(\mathbf{r}|\tilde{\mathbf{c}}) \tag{15.1.7}$$

since scale factors independent of the parameter being maximized do not affect the location of the maximum and hence may be discarded.

However, *training sequences* and *pilot symbols* are commonly multiplexed with the data symbols [3], [4], [5], [6]. Both are groups of known symbols (with established a priori values), but training sequences are generally contiguous, whereas pilot symbols are regularly but sparsely inserted. In this case, the ML criterion is inappropriate *in theory.* In practice, it is easy to derive the ML decision rule assuming a priori equiprobable symbols and then amend it for the known symbols by forcing these to take their a priori values.

The field of *turbo decoding* or iterative MAP decoding [7], [8] creates a more significant difference, as follows. We note from Figure 15.1 that the modulator/equalizer pair we have studied in isolation so far is commonly augmented by a

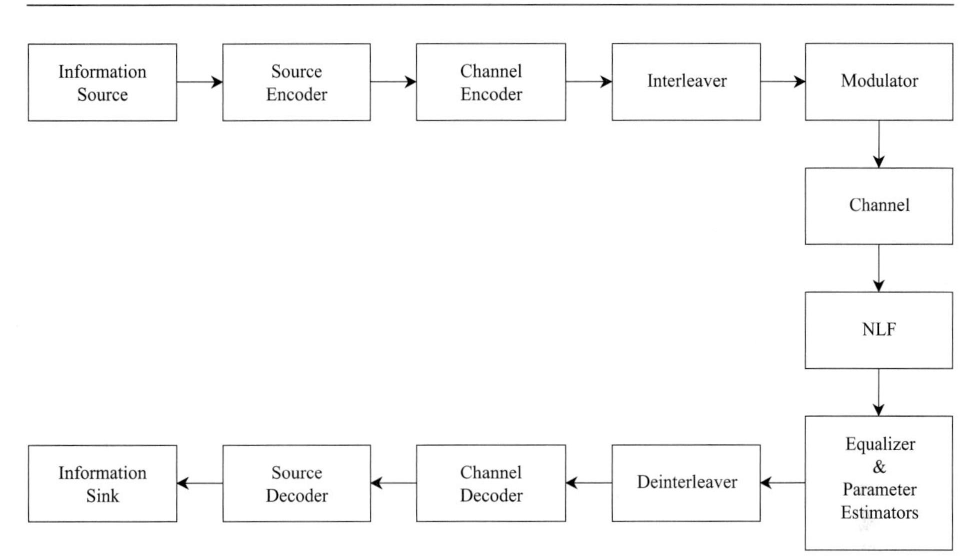

Figure 15.1. Block diagram of a typical communications link

channel encoder/decoder pair [9]. The channel decoder can correct isolated errors, yet in wireless channels, the channel fades give rise to bursts of errors by the equalizer. Therefore, an interleaver/deinterleaver pair is normally present also to break up bursts of equalizer errors [10], [11], [12], [13], [14], [15]. For complexity reasons, equalization is traditionally undertaken without regard to the channel code. However, in turbo decoding, the improved decisions from the channel decoder are fed back to the equalizer as modified (non-equiprobable) a priori bit (or symbol) probabilities to improve the equalizer's performance [16]. This description omits important details because we want to make only one point: turbo decoding needs an equalizer that can exploit non-equiprobable a priori bit probabilities, but the ML criterion does not lead to an equalizer that can.

Second, the bit and symbol criteria are preferred above the symbol sequence criterion. The channel decoder following the equalizer must correct any decision errors made by the equalizer, and this correction is made harder if it can work only with the equalizer's decisions of (15.1.1)—(15.1.6) (these are called *hard decisions.*) If the equalizer can provide reliability information to the channel decoder also, then the channel decoder can be redesigned with improved performance. In the AWGN channel, $10 \log_{10} \pi/2 \approx 2$dB of gain in signal-to-noise ratio is achieved [17].

Since the equalizer and channel decoder are separated by a deinterleaver in fading channels, the reliability information must be tagged to these bits/symbols so it can be deinterleaved also. Thus, the best information the equalizer can deliver is the raw bit or symbol probabilities: that is, the equalizer should calculate the bit/symbol metrics in (15.1.1), (15.1.2), (15.1.4), or (15.1.5) and send them to the decoder without actually making a decision. These probabilities are often called *soft decisions*. By comparison, the sequence metrics of the MAP or ML sequence detectors are not useful since they do not pinpoint which bits or symbols are likely to be in error.

There is little to choose from between the MAP bit and symbol criteria of optimality. We present both, because MAP symbol detection is well known and widely discussed, whereas MAP bit detection is useful to derive a lower bound on a system's bit error rate (BER). By comparison, the MAPSD can be used to lower-bound the symbol error rate (SER).

Notwithstanding these comments, we shall study the ML sequence detector (MLSD) as well as the MAP bit/symbol detector (MAPBD/MAPSD). The reasons are that (1) an implementation of a maximum likelihood sequence decision rule is more attractive than a MAPSD/MAPBD implementation in terms of complexity, (2) there is an enormous body of research on MLSDs, and (3) an MLSD implementation can in fact be extended to provide approximate symbol probabilities [18], [19], [20]. The other decision rules are not studied further.

15.2 DETECTION METRICS WHEN THE CIR IS KNOWN

In many cases of practical interest, the channel impulse response can be estimated with very low mean-square error before detection commences. In this case, it is useful to idealize the CIR, OCIR, or received pulses as known ideally. (The transmitter pulse shape is always assumed known, so there is a one-to-one mapping among these three quantities.)

In practice, the OCIR samples must be estimated with a known (*reference*, *training*, or *pilot*) signal transmitted with the data-bearing signal [21] such that they are separable at the receiver. In DS channels, regularly spaced training sequences may be embedded.[1] By interpolating between them, the OCIR is accurately estimated everywhere [4], [5], [6], [22], [23], [24]. A comb of pilot tones with interpolation in frequency provides equivalent information [25], [23]. In frequency-flat channels, these two schemes reduce to regularly spaced pilot symbols (i.e., a training sequence of unity length) [3], [26], and to a single pilot tone. In time-invariant or slowly time-varying channels, a single training sequence suffices, as long as the data-bearing signal is not separated from the training sequence by more than the channel's coherence time [4], [27], [28], [29].

15.2.1 ML sequence detector

In this section we study the ML sequence detector (MLSD) [30], [31], [23]. Its decision rule is given by (15.1.6). Since the received pulses are assumed known (they are provided as side information, or the receiver is *genie-aided*), we can recognize

[1] The separation between consecutive pilot symbols should be spaced at the Nyquist rate for the maximum Doppler spread.

this explicitly by conditioning on the received pulses $\bar{\mathbf{q}}$, as

$$\hat{\mathbf{c}} = \arg\max_{\tilde{\mathbf{c}}} p(\mathbf{r}|\tilde{\mathbf{c}}, \bar{\mathbf{q}}). \tag{15.2.1}$$

From (14.2.34), the received signal equals

$$\mathbf{r} = \mathbf{z} + \mathbf{n} = \bar{\mathbf{F}}_q \bar{\mathbf{q}} + \mathbf{n}$$

where $\mathbf{n}$ is a zero-mean Gaussian vector with autocovariance matrix $\mathbf{R}_n$ and $\mathbf{z}$ is the mean of $\mathbf{r}$, conditioned on $\mathbf{c} = \tilde{\mathbf{c}}$, that is

$$\eta_r(\tilde{\mathbf{c}}, \bar{\mathbf{q}}) \triangleq E\{\mathbf{r}|\tilde{\mathbf{c}}, \bar{\mathbf{q}}\} = \mathbf{z} = \bar{\mathbf{F}}_q(\tilde{\mathbf{c}})\,\bar{\mathbf{q}}. \tag{15.2.2}$$

Therefore, the vector $\mathbf{r}$ conditioned on the data has a complex, multivariate, Gaussian likelihood. Assuming that the real and imaginary parts of the noise are identically distributed, then the likelihood equals

$$p(\mathbf{r}|\tilde{\mathbf{c}}, \bar{\mathbf{q}}) = \frac{1}{\det(2\pi \mathbf{R}_n)} \exp\left[-\frac{1}{2}\left(\mathbf{r} - \bar{\mathbf{F}}_q(\tilde{\mathbf{c}})\,\bar{\mathbf{q}}\right)^H \mathbf{R}_n^{-1}\left(\mathbf{r} - \bar{\mathbf{F}}_q(\tilde{\mathbf{c}})\,\bar{\mathbf{q}}\right)\right]. \tag{15.2.3}$$

The decision rule of (15.2.1) can be simplified. The location of the maximum is unchanged if the sequence metric is transformed by a monotonically increasing function, such as the logarithm function. For this reason, a sequence metric equivalent to (15.2.3) is the log-likelihood

$$\log p(\mathbf{r}|\tilde{\mathbf{c}}, \bar{\mathbf{q}}) = -\log\det(2\pi\mathbf{R}_n) - \frac{1}{2}\left(\mathbf{r} - \bar{\mathbf{F}}_q(\tilde{\mathbf{c}})\,\bar{\mathbf{q}}\right)^H \mathbf{R}_n^{-1}\left(\mathbf{r} - \bar{\mathbf{F}}_q(\tilde{\mathbf{c}})\,\bar{\mathbf{q}}\right) \tag{15.2.4}$$

so that a functionally identical, but computationally more attractive decision rule, is

$$\hat{\mathbf{c}} = \arg\min_{\tilde{\mathbf{c}}} \left(\mathbf{r} - \bar{\mathbf{F}}_q(\tilde{\mathbf{c}})\,\bar{\mathbf{q}}\right)^H \mathbf{R}_n^{-1}\left(\mathbf{r} - \bar{\mathbf{F}}_q(\tilde{\mathbf{c}})\,\bar{\mathbf{q}}\right) \tag{15.2.5}$$

since data-independent terms and positive scalars may also be discarded under the arg max operator, and a negative scalar affects the decision rule only insofar as the arg max operator is replaced by the arg min operator.

Equation (15.2.5) is part-way to a feasible decision rule. The inverse noise autocovariance matrix $\mathbf{R}_n^{-1}$ is assumed to be known also and precomputed. Therefore, the sequence metric involves only the relatively simple operations of addition/subtraction and multiplication. In Chapter 16, we show how to simplify the brute force comparison of the M^{L_c} different sequences' metrics in (15.2.5).

In white sampled noise, $\mathbf{R}_n^{-1}$ is a scaled identity matrix, so a simplified decision rule is

$$\hat{\mathbf{c}} = \arg\min_{\tilde{\mathbf{c}}} \left(\mathbf{r} - \bar{\mathbf{F}}_q(\tilde{\mathbf{c}})\,\bar{\mathbf{q}}\right)^H \left(\mathbf{r} - \bar{\mathbf{F}}_q(\tilde{\mathbf{c}})\,\bar{\mathbf{q}}\right) = \arg\min_{\tilde{\mathbf{c}}} \left\|\mathbf{r} - \bar{\mathbf{F}}_q(\tilde{\mathbf{c}})\,\bar{\mathbf{q}}\right\|^2, \tag{15.2.6}$$

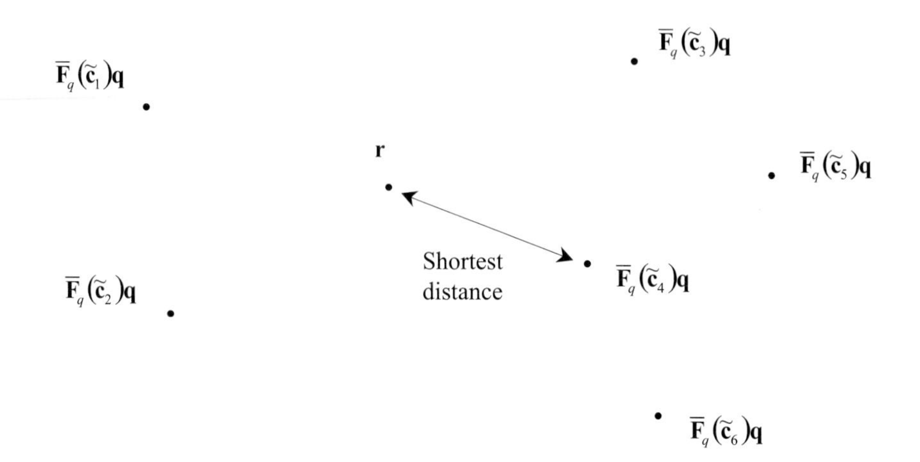

Figure 15.2. Euclidean distance interpretation of MLSD decision rule

which has a convenient geometrical interpretation illustrated in Figure 15.2. Each trial sequence $\tilde{\mathbf{c}}$ can be represented by the point $\bar{\mathbf{F}}_q(\tilde{\mathbf{c}})\,\bar{\mathbf{q}}$ in an N_z-dimensional signal space. The received samples $\mathbf{r}$ define another point. The detected sequence is the one whose point is closest in the *Euclidean* sense to the received samples' point [30].

By comparison, the sequence metric $\left(\mathbf{r} - \bar{\mathbf{F}}_q(\tilde{\mathbf{c}})\,\bar{\mathbf{q}}\right)^H \mathbf{R}_n^{-1}\left(\mathbf{r} - \bar{\mathbf{F}}_q(\tilde{\mathbf{c}})\,\bar{\mathbf{q}}\right)$ in (15.2.5) is the less familiar *Mahalanobis distance* [32]. The additive noise is normally colored by the receiver filter as described in Section 14.2.3, so at first glance (15.2.6) is interesting but less relevant. However, a more intuitive interpretation of the decision rule (15.2.5) can be provided as follows. The inverse noise autocovariance matrix $\mathbf{R}_n^{-1}$ can be factored according to the Cholesky decomposition [32] as

$$\mathbf{R}_n^{-1} = \mathbf{L}^H \mathbf{L} \tag{15.2.7}$$

where $\mathbf{L}$ is a lower triangular matrix with real entries on its main diagonal. It can be easily shown that the vector

$$\mathbf{v} = \mathbf{L}\,\mathbf{n} \tag{15.2.8}$$

generated by a linear and causal[2] transformation of the noise vector $\mathbf{n}$ is made of uncorrelated jointly Gaussian (and, consequently, independent) random variables. Therefore, the linear transformation of (15.2.8) decorrelates noise samples and is denoted as *noise whitening.*

Substituting (15.2.7) into (15.2.5) produces

$$\hat{\mathbf{c}} = \arg\max_{\tilde{\mathbf{c}}} \left\|\mathbf{L}\,\mathbf{r} - \mathbf{L}\,\bar{\mathbf{F}}_q(\tilde{\mathbf{c}})\,\bar{\mathbf{q}}\right\|^2. \tag{15.2.9}$$

[2]The transformation is causal because $\mathbf{L}$ is lower triangular.

This decision rule is still to choose the symbol sequence whose point in multidimensional space is closest in the *Euclidean* sense to some reference point, but now each symbol sequence's point $\mathbf{L}\,\bar{\mathbf{F}}_q\,(\tilde{\mathbf{c}})\;\bar{\mathbf{q}}$ and the reference point $\mathbf{L}\,\mathbf{r}$ are transformed (rotated and scaled) versions of the points in white noise.

Actually, as we shall see, noise whitening is not needed where the signal's spectral density equals zero, since the sequence metric implicitly filters out these signals anyway. Noise whitening is needed only within the signal bandwidth.

Equation (15.2.5) has not been favored historically because it processes the received samples at the faster Nyquist rate (*fractionally sampled*) instead of the slower symbol rate (*T-spaced* or *symbol-spaced*). In many modern applications, this is a minor consideration, however.

Historically, (15.2.5) is expanded [31] and the data-independent $\mathbf{r}^H\mathbf{R}_n^{-1}\mathbf{r}$ term is discarded. This yields the equivalent decision rule

$$\hat{\mathbf{c}} = \arg\max_{\tilde{\mathbf{c}}} 2\Re\left\{\left(\bar{\mathbf{F}}_q\,(\tilde{\mathbf{c}})\;\bar{\mathbf{q}}\right)^H\;\mathbf{R}_n^{-1}\,\mathbf{r}\right\} - \left(\bar{\mathbf{F}}_q\,(\tilde{\mathbf{c}})\;\bar{\mathbf{q}}\right)^H\;\mathbf{R}_n^{-1}\;\left(\bar{\mathbf{F}}_q\,(\tilde{\mathbf{c}})\;\bar{\mathbf{q}}\right) \quad (15.2.10)$$

where $\Re\{x\}$ denotes the real part of the complex number x. Given that the sampled noise is white across the signal bandwidth, then a functionally equivalent decision rule is

$$\begin{aligned} \hat{\mathbf{c}} &= \arg\max_{\tilde{\mathbf{c}}} 2\Re\left\{\bar{\mathbf{q}}^H\,\bar{\mathbf{F}}_q\,(\tilde{\mathbf{c}})^H\;\mathbf{r}\right\} - \left\|\bar{\mathbf{F}}_q\,(\tilde{\mathbf{c}})\;\bar{\mathbf{q}}\right\|^2 \\ &= \arg\max_{\tilde{\mathbf{c}}} 2\Re\left\{\mathbf{z}\,(\tilde{\mathbf{c}})^H\;\mathbf{r}\right\} - \left\|\mathbf{z}\,(\tilde{\mathbf{c}})\right\|^2 \end{aligned} \quad (15.2.11)$$

where

$$\mathbf{z}\,(\tilde{\mathbf{c}}) \triangleq \bar{\mathbf{F}}_q\,(\tilde{\mathbf{c}})\;\bar{\mathbf{q}}. \quad (15.2.12)$$

The first term may be expanded as (see (14.2.11), (14.2.21) and (14.2.34))

$$\begin{aligned} 2\Re\left\{\bar{\mathbf{q}}^H\,\bar{\mathbf{F}}_q\,(\tilde{\mathbf{c}})^H\;\mathbf{r}\right\} &= 2\Re\left\{\sum_{k=0}^{N_z-1}\sum_{m=0}^{L_c-1}\tilde{c}_m^*\,\bar{q}_{m,k-m\eta}^*\,r_k\right\} \\ &= 2\Re\left\{\sum_{m=0}^{L_c-1}\tilde{c}_m^*\sum_{k=0}^{N_z-1}r_k\,\bar{q}_{m,k-m\eta}^*\right\} \\ &= 2\Re\left\{\sum_{m=0}^{L_c-1}\tilde{c}_m^*\,\left[r_k\otimes\bar{q}_{m,-k}^*\right]_{k=m\eta}\right\} \\ &= 2\Re\left\{\sum_{m=0}^{L_c-1}\tilde{c}_m^*\,v_m\right\} \end{aligned} \quad (15.2.13)$$

where

$$v_m = \left[r_k\otimes\bar{q}_{m,-k}^*\right]_{k=m\eta}. \quad (15.2.14)$$

In (15.2.13) the received samples are first convolved with the time-reversed, complex-conjugated received pulses $\{\bar{q}^*_{m,-k}\}$, then sampled at the symbol rate, i.e., at the discrete epochs $k = m\eta$. In time-varying channels, the received pulses are changing, so a different matched filter is required for every transmitted symbol.

We make three observations. First, from (15.2.11) and (15.2.13), the received samples $\{r_k\}$ are used only to calculate the symbol-spaced matched filter outputs $\{v_m\}$ (15.2.14), and therefore such outputs are a set of *sufficient statistics* for ML sequence detection when the OCIR is a priori known.

Second, the matched filters conceal from the sequence metric the behavior of the received signal outside their bandwidth B_z. Therefore, the out-of-band noise may be white or not: the sequence metric is unchanged. This observation confirms that noise whitening is not needed wherever the data-bearing signal has no spectral content.

Third, the matched filter output involves intersymbol interference (ISI). In fact, inserting (14.2.23) into the matched filter expression (15.2.14) yields

$$\begin{aligned} v_m &= \sum_{k=0}^{N_z-1} \left[\sum_{l=0}^{L_c-1} c_l \, q_{k,k-l\eta} + n_k \right] \bar{q}^*_{m,k-m\eta} \\ &= \sum_{k=0}^{N_z-1} \sum_{l=0}^{L_c-1} c_l \, q_{k,k-l\eta} \, \bar{q}^*_{m,k-m\eta} + \sum_{k=0}^{N_z-1} n_k \, \bar{q}^*_{m,k-m\eta} \qquad (15.2.15) \\ &= \sum_{l=0}^{L_c-1} c_l \sum_{k=0}^{N_z-1} q_{k,k-l\eta} \, \bar{q}^*_{m,k-m\eta} + \sum_{k=0}^{N_z-1} n_k \, \bar{q}^*_{m,k-m\eta} \\ &= c_m \sum_{k=0}^{N_z-1} q_{k,k-m\eta} \, \bar{q}^*_{m,k-m\eta} + \sum_{\substack{l=0 \\ l\neq m}}^{L_c-1} c_l \sum_{k=0}^{N_z-1} q_{k,k-l\eta} \, \bar{q}^*_{m,k-m\eta} \\ &\quad + \sum_{k=0}^{N_z-1} n_k \, \bar{q}^*_{m,k-m\eta}. \qquad (15.2.16) \end{aligned}$$

The second term in the RHS of the last equation expresses ISI at the matched filter output sample in the mth symbol interval. When $\mathbf{c} = \tilde{\mathbf{c}}$, this ISI is partially removed by the second term in (15.2.11). In fact, substituting (15.2.15) into (15.2.13) produces the following expansion for the first term in (15.2.11):

$$\begin{aligned} &2\Re\left\{ \bar{\mathbf{q}}^H \, \bar{\mathbf{F}}_q \, (\tilde{\mathbf{c}})^H \, \mathbf{r} \right\} \\ &= 2\Re\left\{ \sum_{m=0}^{L_c-1} \sum_{l=0}^{L_c-1} c_l \, \tilde{c}^*_m \sum_{k=0}^{N_z-1} q_{k,k-l\eta} \, \bar{q}^*_{m,k-m\eta} + \sum_{m=0}^{L_c-1} \tilde{c}^*_m \sum_{k=0}^{N_z-1} n_k \, \bar{q}^*_{m,k-m\eta} \right\}. \end{aligned}$$

The non-noise (ISI) term in the last expression is twice the second term in (15.2.11), as

$$\left\|\bar{\mathbf{F}}_q(\tilde{\mathbf{c}})\,\bar{\mathbf{q}}\right\|^2 = \|\mathbf{z}(\tilde{\mathbf{c}})\|^2 = \sum_{m_1=0}^{L_c-1}\sum_{m_2=0}^{L_c-1} c_{m_1} c^*_{m_2} \sum_{k=0}^{N_z-1} q_{k,k-m_1\eta}\, q^*_{k,k-m_2\eta}. \tag{15.2.17}$$

The discrepancy arises because the data-independent term $\mathbf{r}^H\,\mathbf{R}_n^{-1}\,\mathbf{r}$ was neglected in (15.2.10).

15.2.2 Union bound

The *bit error probability* (BER) of the MLSD can be upper-bounded by a standard technique called the *union bound* [30], and developed further in [31], [33], [34]. In this subsection the union bound is derived and its application to MLSD with known CIR is illustrated. Our initial development closely follows the work of [35].

To begin let us define the difference vector

$$\boldsymbol{\varepsilon} = \mathbf{c} - \hat{\mathbf{c}} \tag{15.2.18}$$

between the transmitted sequence $\mathbf{c}$ and detected sequence $\hat{\mathbf{c}}$, each having length L_c. Then, the exact expression of the MLSD BER is given by

$$\text{BER} = \frac{1}{L_c \log_2 M} \sum_{\mathbf{c}} \sum_{\boldsymbol{\varepsilon}\in\mathcal{F}} w(\boldsymbol{\varepsilon})\; P(\hat{\mathbf{c}} = \mathbf{c} - \boldsymbol{\varepsilon}|\mathbf{c})\; P(\mathbf{c}) \tag{15.2.19}$$

where $\mathcal{F}$ is the set $\{\boldsymbol{\varepsilon}\}$ of all possible error vectors and $w(\boldsymbol{\varepsilon})$ is the number of bit errors associated with the error event $\boldsymbol{\varepsilon}$. It is worth noting that not all the error vectors are possible with any $\mathbf{c}$, and, in these instances, the conditional probability $P(\hat{\mathbf{c}} = \mathbf{c} - \boldsymbol{\varepsilon}|\mathbf{c})$ is equal to zero.

The RHS of (15.2.19) can be also expressed in a different form as follows. Let us introduce $\mathcal{G}$ as the set of error vectors that do not contain more than $(L_q - 1)$ consecutive zero entries (L_q is the channel memory in symbol intervals) amid non-zero entries. The entries including and within the outermost non-zero entries are regarded as an *error event*, so $\mathcal{G}$ is the set of error event vectors. Then, (15.2.19) can be written as

$$\text{BER} = \frac{1}{L_c \log_2 M} \sum_{\mathbf{c}} \left[\sum_{\boldsymbol{\varepsilon}\in\mathcal{G}} w(\boldsymbol{\varepsilon})\; P(\hat{\mathbf{c}} = \mathbf{c} - \boldsymbol{\varepsilon}|\mathbf{c}) + \sum_{\boldsymbol{\varepsilon}\in\bar{\mathcal{G}}} w(\boldsymbol{\varepsilon})\; P(\hat{\mathbf{c}} = \mathbf{c} - \boldsymbol{\varepsilon}|\mathbf{c})\right] P(\mathbf{c}) \tag{15.2.20}$$

where $\bar{\mathcal{G}}$ is the set of members of $\mathcal{F}$ not already in $\mathcal{G}$ (i.e., the set of error vectors that are made up of several error events).

The RHS of (15.2.19) or (15.2.20) cannot be easily evaluated because the probability $P(\hat{\mathbf{c}} = \mathbf{c} - \boldsymbol{\varepsilon}|\mathbf{c})$ is difficult to compute. Then, an upper bound is derived. The union bound on the BER expression (15.2.20) is given by

$$\text{BER} \leqslant \frac{1}{L_c \log_2 M} \sum_{\mathbf{c}} \sum_{\boldsymbol{\varepsilon}\in\mathcal{G}} w(\boldsymbol{\varepsilon})\; P(\Lambda(\mathbf{c} - \boldsymbol{\varepsilon}) \geqslant \Lambda(\mathbf{c})|\mathbf{c})\; P(\mathbf{c}) \tag{15.2.21}$$

where $\Lambda(\tilde{\mathbf{c}})$ denotes the sequence metric assigned to $\tilde{\mathbf{c}}$. In the last equation it is assumed that the decision strategy selects the sequence with the largest metric; when the optimal metric is the smallest, then the event $\{\Lambda(\mathbf{c}-\varepsilon) \geqslant \Lambda(\mathbf{c})\}$ is replaced by the complementary one, $\{\Lambda(\mathbf{c}-\varepsilon) \leqslant \Lambda(\mathbf{c})\}$.

Equation (15.2.21) can be derived from (15.2.20) as follows. First, we note that any error vector $\varepsilon_{\bar{\mathcal{G}}} \in \bar{\mathcal{G}}$ can be written as the union of disjoint error events as

$$\varepsilon_{\bar{\mathcal{G}}} = \bigcup_{i=0}^{I-1} \varepsilon_{\mathcal{G}}^{(i)} \tag{15.2.22}$$

for some integer I, where $\{\varepsilon_{\mathcal{G}}^{(i)}\}$ are error vectors in $\mathcal{G}$ corresponding to non-overlapping error events. Second, we consider an example error vector $\varepsilon_{\bar{\mathcal{G}}} = \varepsilon_{\mathcal{G}}^{(0)} + \varepsilon_{\mathcal{G}}^{(1)}$ $(I=2)$. The probability $P(\Lambda(\mathbf{c}-\varepsilon_{\mathcal{G}}^{(i)}) \geqslant \Lambda(\mathbf{c})|\mathbf{c})$ can be evaluated as

$$P\left(\Lambda(\mathbf{c}-\varepsilon_{\mathcal{G}}^{(i)}) \geqslant \Lambda(\mathbf{c})|\mathbf{c}\right) = P\left(\hat{\mathbf{c}} = \mathbf{c}-\varepsilon_{\mathcal{G}}^{(i)}|\mathbf{c}\right) + P\left(\hat{\mathbf{c}} = \mathbf{c}-\varepsilon_{\mathcal{G}}^{(0)}-\varepsilon_{\mathcal{G}}^{(1)}|\mathbf{c}\right) \tag{15.2.23}$$

for $i = 0,1$ since $\{\hat{\mathbf{c}} = \mathbf{c}-\varepsilon_{\mathcal{G}}^{(i)}|\mathbf{c}\}$ and $\{\hat{\mathbf{c}} = \mathbf{c}-\varepsilon_{\mathcal{G}}^{(0)}-\varepsilon_{\mathcal{G}}^{(1)}|\mathbf{c}\}$ are disjoint events; if either is true, then the event $\{\Lambda(\mathbf{c}-\varepsilon_{\mathcal{G}}^{(i)}) \geqslant \Lambda(\mathbf{c})|\mathbf{c}\}$ occurs. Accordingly, we can write

$$\begin{aligned} w(\varepsilon_{\mathcal{G}}^{(1)})\, P(\hat{\mathbf{c}} = \mathbf{c}-\varepsilon_{\mathcal{G}}^{(1)}|\mathbf{c}) + w(\varepsilon_{\mathcal{G}}^{(2)})\, P(\hat{\mathbf{c}} = \mathbf{c}-\varepsilon_{\mathcal{G}}^{(2)}|\mathbf{c}) + w(\varepsilon_{\bar{\mathcal{G}}})\, P(\hat{\mathbf{c}} = \mathbf{c}-\varepsilon_{\bar{\mathcal{G}}}|\mathbf{c}) \\ \leqslant w(\varepsilon_{\mathcal{G}}^{(1)})\, P\left(\Lambda(\mathbf{c}-\varepsilon_{\mathcal{G}}^{(1)}) \geqslant \Lambda(\mathbf{c})|\mathbf{c}\right) + w(\varepsilon_{\mathcal{G}}^{(2)})\, P\left(\Lambda(\mathbf{c}-\varepsilon_{\mathcal{G}}^{(2)}) \geqslant \Lambda(\mathbf{c})|\mathbf{c}\right) \end{aligned}$$

Finally, we obtain (15.2.21) by applying a similar argument to all other $\varepsilon_{\bar{\mathcal{G}}} \in \bar{\mathcal{G}}$ (15.2.22) in (15.2.20).

It is important to note that the derivation of (15.2.21) does not require any special properties about the decision metric $\Lambda(\tilde{\mathbf{c}})$: it may be optimal or suboptimal; it may rely on the CIR being known, estimated, or averaged over. In fact, (15.2.21) applies to any detector that assigns each sequence a metric $\Lambda(\tilde{\mathbf{c}})$, then exhaustively searches over all sequences to find the one with largest metric. For this reason, different union bounds are obtained in different situations.

The evaluation of the bound (15.2.21) requires the computation of the factor $P(\Lambda(\mathbf{c}-\varepsilon) \geqslant \Lambda(\mathbf{c})|\mathbf{c})$. This term represents the probability that the erroneous sequence $(\mathbf{c}-\varepsilon)$ has a larger metric than the transmitted sequence $\mathbf{c}$; it is known as the *pairwise error probability* (PEP).

Let us apply now the union bound to MLSD with known CIR. The sequence metrics $\Lambda(\mathbf{c}-\varepsilon)$ and $\Lambda(\mathbf{c})$ can be expressed as (see (15.2.11))

$$\Lambda(\mathbf{c}-\varepsilon) = \mathbf{z}^H(\tilde{\mathbf{c}})\,\mathbf{r} + \mathbf{r}^H\,\mathbf{z}(\tilde{\mathbf{c}}) - \|\mathbf{z}(\tilde{\mathbf{c}})\|^2 \tag{15.2.24}$$

$$\Lambda(\mathbf{c}) = \mathbf{z}^H\,\mathbf{r} + \mathbf{r}^H\,\mathbf{z} - \|\mathbf{z}\|^2 \tag{15.2.25}$$

respectively, where $\tilde{\mathbf{c}} \triangleq \mathbf{c} - \boldsymbol{\varepsilon}$ and $\mathbf{z}(\tilde{\mathbf{c}}) \triangleq \bar{\mathbf{F}}_q(\tilde{\mathbf{c}})\,\bar{\mathbf{q}}$. If the error vector

$$\begin{aligned}\mathbf{e}(\boldsymbol{\varepsilon}) &\triangleq \mathbf{z} - \tilde{\mathbf{z}}(\tilde{\mathbf{c}}) = \left(\bar{\mathbf{F}}_q(\mathbf{c}) - \bar{\mathbf{F}}_q(\tilde{\mathbf{c}})\right)\bar{\mathbf{q}} = \left(\mathbf{F}_q(\mathbf{c}) - \mathbf{F}_q(\tilde{\mathbf{c}})\right)\mathbf{q} \\ &= \bar{\mathbf{F}}_q(\boldsymbol{\varepsilon})\,\bar{\mathbf{q}} = \mathbf{F}_q(\boldsymbol{\varepsilon})\,\mathbf{q}\end{aligned} \tag{15.2.26}$$

is defined and it is noted that $\mathbf{r} = \mathbf{z} + \mathbf{n}$, the PEP can be evaluated as

$$\begin{aligned}&P(\Lambda(\mathbf{c}-\boldsymbol{\varepsilon}) \geqslant \Lambda(\mathbf{c})|\mathbf{c}) \\ &= P\left(\mathbf{z}^H(\tilde{\mathbf{c}})\,\mathbf{r} + \mathbf{r}^H\,\mathbf{z}(\tilde{\mathbf{c}}) - \|\mathbf{z}(\tilde{\mathbf{c}})\|^2 \geqslant \mathbf{z}^H\,\mathbf{r} + \mathbf{r}^H\,\mathbf{z} - \|\mathbf{z}\|^2\,|\mathbf{c}\right) \\ &= P(\kappa(\boldsymbol{\varepsilon}) \leq 0)\end{aligned} \tag{15.2.27}$$

where the random variable $\kappa(\boldsymbol{\varepsilon})$ is defined as

$$\kappa(\boldsymbol{\varepsilon}) \triangleq \mathbf{n}^H\,\mathbf{e}(\boldsymbol{\varepsilon}) + \mathbf{e}(\boldsymbol{\varepsilon})^H\,\mathbf{n} + \|\mathbf{e}(\boldsymbol{\varepsilon})\|^2 . \tag{15.2.28}$$

If the CIR is known and fixed, the error vector $\mathbf{e}(\boldsymbol{\varepsilon})$ is fixed for a given error event $\boldsymbol{\varepsilon}$ and the random variable $\kappa(\boldsymbol{\varepsilon})$ is a real Gaussian with mean and variance[3] (see (15.2.28))

$$\mu_\kappa(\boldsymbol{\varepsilon}) = \|\mathbf{e}(\boldsymbol{\varepsilon})\|^2 \tag{15.2.29}$$

$$\sigma_\kappa^2(\boldsymbol{\varepsilon}) = 2\,\frac{N_0}{T_\eta}\,\|\mathbf{e}(\boldsymbol{\varepsilon})\|^2 , \tag{15.2.30}$$

respectively. Then, the PEP is given by

$$P(\Lambda(\mathbf{c}-\boldsymbol{\varepsilon}) \geqslant \Lambda(\mathbf{c})|\mathbf{c}) = \frac{1}{\sqrt{2}}\,\mathrm{erfc}\left(\frac{\mu_\kappa(\boldsymbol{\varepsilon})}{\sqrt{2}\,\sigma_\kappa(\boldsymbol{\varepsilon})}\right) \tag{15.2.31}$$

where $\mathrm{erfc}(x)$ is the *complementary error function.* By a combination of (15.2.29), (15.2.30), (15.2.31), the PEP equals

$$\mathrm{PEP}(\boldsymbol{\varepsilon}) \triangleq P(\Lambda(\mathbf{c}-\boldsymbol{\varepsilon}) \geqslant \Lambda(\mathbf{c})|\mathbf{c}) = \frac{1}{\sqrt{2}}\,\mathrm{erfc}\left(\sqrt{\frac{T_\eta}{4\,N_0}\,\|\mathbf{e}(\boldsymbol{\varepsilon})\|^2}\right). \tag{15.2.32}$$

The last expression evidences that the PEP depends on $\mathbf{e}$ and, in particular, on the symbol difference $\boldsymbol{\varepsilon}$. Then, it is independent of the transmitted sequence $\mathbf{c}$. Accordingly, (15.2.21) can be simplified as

$$\mathrm{BER} \leqslant \frac{1}{L_c \log_2 M}\sum_{\boldsymbol{\varepsilon}\in\mathcal{G}} \frac{N(\mathbf{c}|\boldsymbol{\varepsilon})}{N(\mathbf{c})}\,w(\boldsymbol{\varepsilon})\,P(\Lambda(\mathbf{c}-\boldsymbol{\varepsilon}) \geqslant \Lambda(\mathbf{c})) \tag{15.2.33}$$

[3]The NLF is an ideal low pass filter with bandwidth $B_r = 1/2T_\eta$.

involving an arbitrary transmitted sequence $\mathbf{c}$. Here, $N(\mathbf{c}|\varepsilon)$ denotes the number of symbol sequences that can have ε as an error vector, whereas $N(\mathbf{c}) = M^{L_c}$ is the total number of possible transmitted sequences. Furthermore, the PEP of an error event beginning at one instant is the same as if it had occurred at the next, so evaluating (15.2.33) still involves redundant calculations. If we define $\mathcal{G}_0$ as the set of error vectors with a non-zero first entry (i.e., their error events begin at the start of the burst) and note that an error event of length L_ε symbol periods may occur at $L_c - L_\varepsilon + 1$ different positions within a burst, we can eliminate this redundancy by writing, in place of (15.2.33), the bound

$$\text{BER} \leqslant \frac{L_c - L_\varepsilon + 1}{L_c \log_2 M} \sum_{\varepsilon \in \mathcal{G}_0} \frac{N(\mathbf{c}|\varepsilon)}{N(\mathbf{c})} w(\varepsilon)\, P\left(\Lambda(\mathbf{c}-\varepsilon) \geqslant \Lambda(\mathbf{c})\right). \tag{15.2.34}$$

The number of terms to compute in (15.2.34) is still large. It has been shown that the summation in (15.2.34) is dominated by the error event with the smallest value of $\|\mathbf{e}\|^2$ at high SNRs (i.e., the shortest error events), and error events with larger values contribute little to the value of the upper bound [30]. If the summation in (15.2.34) is truncated, a lower bound on the upper bound is obtained (a somewhat meaningless result). The truncated bound may appear to be tight, but this is illusory: in the limit that all but the shortest error event are discarded, the truncated bound converges to a tight *lower* bound, known as the *matched filter bound* (MFB). The MFB is illustrated in Section 15.2.4.

As proved above, if the CIR is known and given, we can bound the BER of MLSDs by resorting to union bound (15.2.34) with the PEP expression (15.2.32) [30]. However, the same line of reasoning can be also applied when the CIR is known and *random* in order to derive an upper bound on the *average* BER. With random channels the evaluation of the PEP is known to be a tractable problem in two cases:

1. The channel is affected by multipath Rayleigh fading.

2. The channel is time invariant and frequency flat (i.e., it is characterized by a single complex gain).

In the first case, the error vector $\mathbf{e}$ (15.2.26) is zero-mean complex Gaussian and the PEP is calculated as the ensemble average of (15.2.32) over all possible instances of $\mathbf{e}$ [23]. In the second case, the BER is computed as a function of the channel's fading amplitude, then it is averaged over the pdf of the fading amplitude [36]. For Rayleigh fading, this case is a simplified version of the first case, but the method applies to other fading pdfs also, such as those mentioned in Part I, Chapter 3. This second case is little relevant to wideband communications, so we do not discuss it further.

Let us now examine the first case. When the channel is Rayleigh fading, the random variable $\kappa(\varepsilon)$ (15.2.28) is a quadratic form in the zero-mean Gaussian random variables $\mathbf{e}$ and $\mathbf{n}$ and can be expressed as

$$\kappa(\varepsilon) = \mathbf{k}(\varepsilon)^H \mathbf{A}\, \mathbf{k}(\varepsilon) \tag{15.2.35}$$

where $\mathbf{k}(\varepsilon) \triangleq [\mathbf{e}(\varepsilon)^T, \mathbf{n}^T]^T$ and

$$\mathbf{A} \triangleq \begin{bmatrix} \mathbf{I}_{N_z} & \mathbf{I}_{N_z} \\ \mathbf{I}_{N_z} & \mathbf{0}_{N_z,N_z} \end{bmatrix}. \tag{15.2.36}$$

The vector $\mathbf{k}$ is zero-mean Gaussian with autocovariance matrix (the channel and white noise are assumed independent)

$$\mathbf{R}_k(\varepsilon) \triangleq \frac{1}{2} E\{\mathbf{k}(\varepsilon)\, \mathbf{k}(\varepsilon)^H\} = \begin{bmatrix} \mathbf{R}_e(\varepsilon) & \mathbf{0}_{N_z,N_z} \\ \mathbf{0}_{N_z,N_z} & \mathbf{R}_n \end{bmatrix} \tag{15.2.37}$$

where $\mathbf{R}_e(\varepsilon) \triangleq (1/2)\, E\{\mathbf{e}(\varepsilon)\, \mathbf{e}(\varepsilon)^H\}$ is the autocovariance matrix of $\mathbf{e}(\varepsilon)$. From the right-hand equality of (15.2.26), the error vector autocovariance equals

$$\mathbf{R}_e(\varepsilon) = [\mathbf{F}_q(\mathbf{c}) - \mathbf{F}_q(\tilde{\mathbf{c}})]\, \mathbf{R}_q\, [\mathbf{F}_q(\mathbf{c}) - \mathbf{F}_q(\tilde{\mathbf{c}})]^H \tag{15.2.38}$$

where $\mathbf{R}_q \triangleq (1/2)\, E\{\mathbf{q}\mathbf{q}^H\}$ is defined in Section 14.2.4 where its entries are given.

In this case, the PEP is still expressed by (15.2.27) and can be evaluated with the methods illustrated in [37], [38]. Some numerical results are available in [23]. Unluckily, it is found that the BER union bound for the Rayleigh fading channels is unhelpfully loose. In fact, at any average SNR, a proportion of sample channels have very low SNRs and, when taking the ensemble average over good and bad channels, these cases are non-negligible. Therefore, the union bound in (15.2.34) may not be usefully tight. Further insight is provided by the observation that the probability of longer error events does not diminish as fast as with a fixed channel. Best results are obtained for high SNRs and fast fading, so that the instantaneous SNR is generally high and fades are brief. When mean fade durations are a matter of symbol periods, long error events are relatively unlikely.

15.2.3 MAP bit and symbol detectors

In this section, we study the MAP bit/symbol detectors (MAPBD/MAPSD) [39], [40], [41], [42]. Their decision rules are given by (15.1.1) and (15.1.2) but are modified to account for the known received pulses $\bar{\mathbf{q}}$, as

$$\hat{b}_i = \arg\max_{\tilde{b}_i} P(\tilde{b}_i|\mathbf{r}, \bar{\mathbf{q}}) \tag{15.2.39}$$

$$\hat{c}_m = \arg\max_{\tilde{c}_m} P(\tilde{c}_m|\mathbf{r}, \bar{\mathbf{q}}) \tag{15.2.40}$$

respectively. Recall that the soft bit/symbol metrics $P(\tilde{b}_i|\mathbf{r}, \bar{\mathbf{q}})$ and $P(\tilde{c}_m|\mathbf{r}, \bar{\mathbf{q}})$ are at least as important as the decisions when a channel code is present. Using total

$b_0b_1b_2$	φ *(rad)*
000	0
001	$5\pi/4$
010	$\pi/2$
011	$3\pi/4$
100	π
101	$\pi/4$
110	$3\pi/2$
111	$7\pi/4$

Table 15.1. Unconventional 8-PSK mapping for MAP bit detection example

probability, we can rewrite the desired metrics as

$$P(\tilde{b}_i|\mathbf{r},\bar{\mathbf{q}}) = \sum_{\tilde{\mathbf{c}}\to\tilde{b}_i} P(\tilde{\mathbf{c}}|\mathbf{r},\bar{\mathbf{q}}) = \sum_{\tilde{\mathbf{c}}\to\tilde{b}_i} \frac{p(\mathbf{r}|\tilde{\mathbf{c}},\bar{\mathbf{q}})P(\tilde{\mathbf{c}})}{p(\mathbf{r}|\bar{\mathbf{q}})} \tag{15.2.41}$$

$$P(\tilde{c}_m|\mathbf{r},\bar{\mathbf{q}}) = \sum_{\tilde{\mathbf{c}}\to\tilde{c}_m} P(\tilde{\mathbf{c}}|\mathbf{r},\bar{\mathbf{q}}) = \sum_{\tilde{\mathbf{c}}\to\tilde{c}_m} \frac{p(\mathbf{r}|\tilde{\mathbf{c}},\bar{\mathbf{q}})P(\tilde{\mathbf{c}})}{p(\mathbf{r}|\bar{\mathbf{q}})} \tag{15.2.42}$$

respectively, assuming that $\tilde{\mathbf{c}}$ is independent of $\bar{\mathbf{q}}$. The summations are over all sequences, $\tilde{\mathbf{c}}$, which are consistent with $\{b_i = \tilde{b}_i\}$ or $\{c_m = \tilde{c}_m\}$, respectively. The data bits/symbols are a priori equiprobable; and non-equiprobable in the second and subsequent iterations of a turbo decoder [7], [8]. The conditional pdf $p(\mathbf{r}|\bar{\mathbf{q}})$ can be expressed as

$$p(\mathbf{r}|\bar{\mathbf{q}}) = \sum_{\tilde{\mathbf{c}}} P(\tilde{\mathbf{c}})\, p(\mathbf{r}|\tilde{\mathbf{c}},\bar{\mathbf{q}}) \tag{15.2.43}$$

and can be interpreted as a scale factor, ensuring that the probabilities assigned to the possible values of each bit/symbol sum to unity. In many cases, this normalization can be omitted, and even if normalization is required, the unnormalized bit/symbol probabilities are available so normalization can proceed by their direct use. Thus, $p(\mathbf{r}|\bar{\mathbf{q}})$ and any other data-independent scale factor can be disregarded without penalty.

The common probability factor $p(\mathbf{r}|\tilde{\mathbf{c}},\bar{\mathbf{q}})$ in (15.2.41) and (15.2.42) is given by (15.2.3). It is repeated here for convenience as

$$p(\mathbf{r}|\tilde{\mathbf{c}},\bar{\mathbf{q}}) = \frac{1}{\det(2\pi\mathbf{R}_n)} \exp\left[-\frac{1}{2}\left(\mathbf{r}-\bar{\mathbf{F}}_q(\tilde{\mathbf{c}})\,\bar{\mathbf{q}}\right)^H \mathbf{R}_n^{-1}\left(\mathbf{r}-\bar{\mathbf{F}}_q(\tilde{\mathbf{c}})\,\bar{\mathbf{q}}\right)\right] \tag{15.2.44}$$

so we see that the metrics in (15.2.41) and (15.2.42) constitute the sum of scaled exponentials of scaled Mahalanobis distances. The metric is not as intuitive as the metrics used in ML sequence detection.

Finally, the concept of MAP bit detection can be exemplified as follows. Let us consider a communication system in which a bit triad $(b_0\, b_1\, b_2)$ is unconventionally mapped to a single complex 8-PSK symbol $c_0 = \exp(j\varphi)$ according to Table 15.1. The symbol is transmitted through a time-invariant, frequency-flat channel that is corrupted by complex additive Gaussian noise with zero mean and variance $\sigma_n^2 = 0.5$. The receiver performs MAP bit detection for each of the three information bits (assumed to be equiprobable). Because each bit can take on two values, the MAPBD computes six metrics. Figure 15.3 plots these metrics for all possible values of the complex received sample r_0. Each sequence (in this case, a single symbol) contributes a Gaussian-shaped peak, and it is the point-wise sum of these peaks that is plotted. For a given bit and received sample, one of the bit values will have a higher metric and be detected. Therefore, decision regions can be constructed, as Figure 15.3 also shows. The decision boundaries are curved in general, and they change with SNR. When more than one received sample is available (as is typically the case), then the decision boundaries are highly complicated, multidimensional surfaces.

15.2.4 Matched filter bound

The MAPSD is optimal in that it minimizes the symbol error probability. However, its *symbol error rate* (SER) performance is difficult to compute, because ISI makes detecting the transmitted symbols more difficult and degrades the MAPSD's SER. Therefore, an improved symbol error performance would be achieved if the interfering received pulses were magically eliminated from the received samples. At the same time, the probability of symbol error would be easier to calculate. This idea is the *matched filter bound* (MFB) concept: an isolated pulse is transmitted by a linear modulator and is detected optimally assuming a known OCIR. The SER of the optimal (i.e., matched-filter) detector for this situation is dubbed MFB and provides a lower bound on the MAPSD's SER.[4] In this context the MAPSD turns into a "one shot" detector.

Since the MAPSD is the optimal equalizer in terms of SER, the MFB also lower-bounds the SER of all other equalizers, including the MLSD. This presumes the same modulation method and channel model, so there are different MFBs for different circumstances [43], [44], [45], [46], [47], [48], [49], [50], [51], [52]. In particular, MFBs for FS and DS Rayleigh channels are derived in [43], [46], [47], [48],[49], [50], [51], [52]. Closed-form expressions of such bounds are available only for discrete FS channels [43], [47], [48], [52], whereas approximate error formulas have been derived for continuous *power delay profiles* (PDPs) [49], [50], [51], [52].

The MFB is complementary and often superior to the union bound. In Rayleigh fading channels, it is tighter than the union bound, especially at low SNRs, and it requires fewer computations. Being a lower bound, it is also more relevant. With large delay spreads, as shown in Section 16.1.6, the most computationally

[4]The MFB may be invoked for any modulation method that transmits digital data as a (possibly overlapping) sequence of waveforms taken from a finite alphabet.

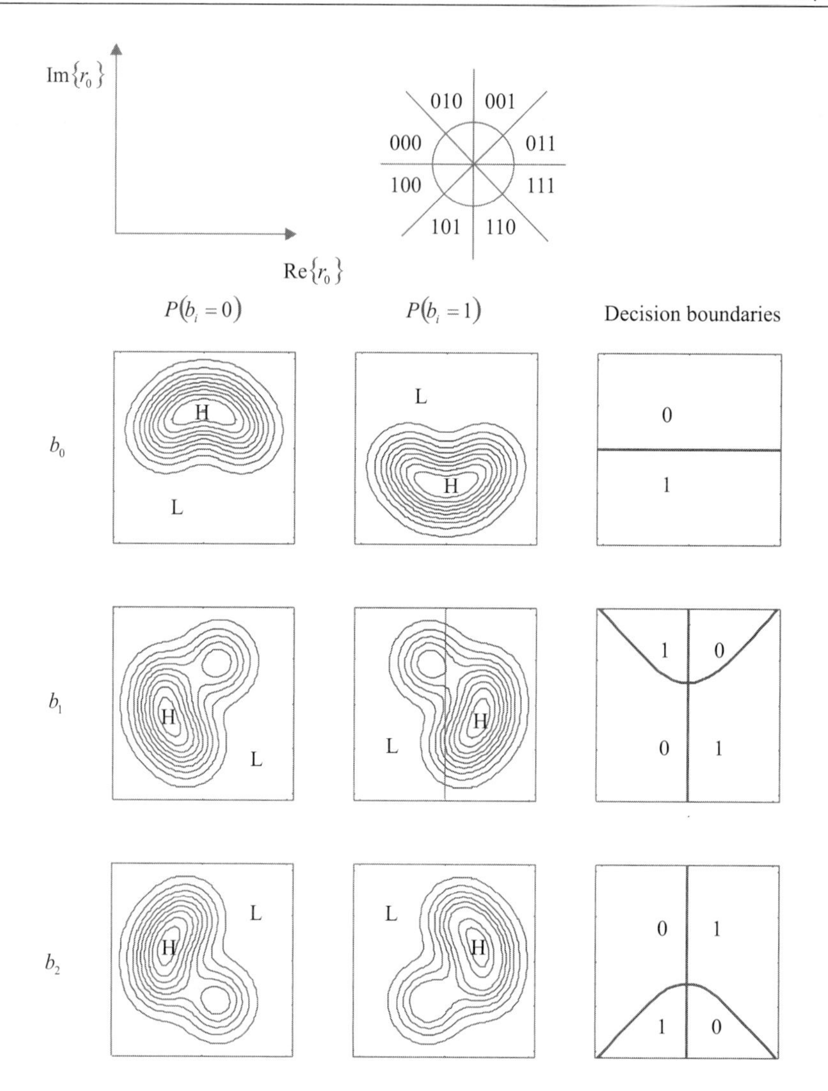

Figure 15.3. Metrics and decision boundaries for a MAPBD, assuming one 8-ary transmitted symbol and one complex received sample. "H" and "L" refer to the pdf peak and skirts, respectively.

efficient MLSD, MAPBD or MAPSD are still infeasible and will always remain so. Therefore, the MAPBD's BER and the MAPSD's SER really only establish the limits of what is achievable—they provide a *benchmark lower bound* for the more practical, important equalizers, such as the linear and decision-feedback equalizers. The MFB is meaningful in that it lower-bounds such a benchmark performance, whereas the union bound is only an upper bound on it.

In this section, we present an extension of the MFB, derived as a lower bound on the MAPBD error performance. Like the initial development of the union bound, its applicability extends beyond the case of known CIR, and in fact we shall again apply the following ideas in Section 15.4.4. To facilitate this and because our emphasis is on BER (and not SER), our presentation differs somewhat from that illustrated in the previous references.

Let us now derive the BER lower bound. To begin, we assume the transmission of a bit burst $\mathbf{b} \triangleq [b_0, b_1, ..., b_{L_c \log_2 M-1}]^T$, and we concentrate on detection of its ith bit b_i. The average BER of the MAPBD (see (15.2.39)) equals the average probability that the transmitted bit has a worse metric than the other possible bits $\{\bar{b}_i \neq b_i\}$, that is,

$$\mathrm{BER}_{MAPBD} = \sum_{b_i} \sum_{\mathbf{b}_o} P\left(P(\bar{b}_i|\mathbf{r}, \bar{\mathbf{q}}) \geqslant P(b_i|\mathbf{r}, \bar{\mathbf{q}}) \middle| b_i, \mathbf{b}_o\right) P(b_i, \mathbf{b}_o) \qquad (15.2.45)$$

where $\mathbf{b}_o \triangleq [b_0, b_1, .., b_{i-1}, b_{i+1}, ..., b_{L_c \log_2 M-1}]^T$ contains all the transmitted bits in the burst $\mathbf{b}$, b_i excluded. In (15.2.45), the averaging over the transmitted data is made explicit, but there is also an *implicit* averaging over the noise and possibly the channel too. It is not known how to compute (15.2.45), since the metrics, $P(b_i|\mathbf{r}, \bar{\mathbf{q}})$ and $P(\bar{b}_i|\mathbf{r}, \bar{\mathbf{q}})$, are complicated functions of correlated Gaussian random variables, as (15.2.44) shows. To circumvent this problem, we introduce a useful result. Given that minimizing the average BER is the criterion of optimality, an optimal detector, with a certain amount of information, cannot outperform, in terms of average BER, another optimal detector endowed with a superset of that information. The reason is that the first detector's decision rule is a candidate decision rule for the second detector.

We apply this result to the problem at hand by introducing a second, *genie-aided* detector. When detecting a bit, say b_i, such a detector is given the same a priori information as the actual MAPBD, *as well as* the values of the other transmitted bits, $\mathbf{b}_o$. Accordingly, the genie-aided receiver's average BER (denoted as BER_{ga}) is a lower bound[5] on the average BER of the actual MAPBDs, that is,

$$\mathrm{BER}_{MAPBD} \geqslant \mathrm{BER}_{ga} = \sum_{b_i} \sum_{\mathbf{b}_o} P\left(P(\bar{b}_i|\mathbf{r}, \bar{\mathbf{q}}, \mathbf{b}_o) \geqslant P(b_i|\mathbf{r}, \bar{\mathbf{q}}, \mathbf{b}_o) \middle| \mathbf{b}\right) P(b_i, \mathbf{b}_o). \qquad (15.2.46)$$

We introduce now $\bar{\mathbf{b}} \triangleq [b_0, b_1, .., b_{i-1}, \bar{b}_i, b_{i+1}, ..., b_{L_c \log_2 M-1}]^T$ as the vector of bits transmitted in the burst but with $\bar{b}_i$ in place of b_i (i.e., $\mathbf{b}$ and $\bar{\mathbf{b}}$ constitute $\mathbf{b}_o$ together with b_i and $\bar{b}_i$, respectively) and note that there is a one-to-one mapping between $\mathbf{b}$ and $\bar{\mathbf{b}}$ and a symbol sequence, denoted by $\mathbf{c}$ and $(\mathbf{c} - \boldsymbol{\varepsilon})$, respectively. Then, the first probability factor in (15.2.46) can be interpreted as a *pairwise error probability* (PEP) in the choice between $\mathbf{b}$ and $\bar{\mathbf{b}}$ (or, equivalently, between $\mathbf{c}$ and

[5]Such a bound can be interpreted as an extension of the MFB.

$\mathbf{c}-\boldsymbol{\varepsilon}$), as

$$\begin{aligned}
&P\left(P(\bar{b}_i|\mathbf{r},\mathbf{b}_o,\bar{\mathbf{q}}) \geqslant P(b_i|\mathbf{r},\mathbf{b}_o,\bar{\mathbf{q}})\middle|\, \mathbf{b}\right) \\
&= P\left(\frac{p(\mathbf{r}|\bar{b}_i,\mathbf{b}_o,\bar{\mathbf{q}})\,P(\bar{b}_i,\mathbf{b}_o)}{p(\mathbf{r},\mathbf{b}_o|\bar{\mathbf{q}})} \geqslant \frac{p(\mathbf{r}|b_i,\mathbf{b}_o,\bar{\mathbf{q}})\,P(b_i,\mathbf{b}_o)}{p(\mathbf{r},\mathbf{b}_o|\bar{\mathbf{q}})}\,|\mathbf{b}\right) \\
&= P\left(p(\mathbf{r}|\mathbf{c}-\boldsymbol{\varepsilon},\bar{\mathbf{q}}) \geqslant p(\mathbf{r}|\mathbf{c},\bar{\mathbf{q}})|\mathbf{c}\right). \qquad (15.2.47)
\end{aligned}$$

The quantity in the RHS of the last equation is just the probability that $(\mathbf{c}-\boldsymbol{\varepsilon})$ has a larger likelihood than the transmitted sequence $\mathbf{c}$. This means that the PEP needed for the MAPBD and its MFB is exactly the same as the PEP needed for the MLSD (denoted as PEP $(\boldsymbol{\varepsilon})$, see (15.2.32)) and its union bound. Therefore, the results of Section 15.2.2 can be employed directly in solving this problem. Following such an approach, the likelihoods are converted to log-likelihoods, irrelevant terms and scale factors are discarded, and the final result matches (15.2.27).

Like the MLSD case, the PEP calculation can be carried out with (a) Rayleigh fading channels (see (15.2.36)—(15.2.38)), (b) a given channel impulse response (see (15.2.29)—(15.2.32)) [30], and (c) slow, frequency-flat, non-Gaussian channels [36].

Equation (15.2.46) can be simplified for single-bit errors and the known CIR case as follows. Given that the constellation mapper is memoryless, the bit vector $\mathbf{b}_m \triangleq [b_{m\log_2 M}, b_{m\log_2 M+1}, \ldots, b^T_{(m+1)\log_2 M-1}]^T$ uniquely identifies the mth channel symbol c_m, and vice versa. Therefore, a single-bit error creates only a single symbol error, so the corresponding error vector $\boldsymbol{\varepsilon}$ has one non-zero entry, at its mth location. Since, from (15.2.27), the PEP depends on the error vector $\boldsymbol{\varepsilon}$ and not the transmitted sequence $\mathbf{c}$, $P\left(P(\bar{b}_i|\mathbf{r},\bar{\mathbf{q}},\mathbf{b}_o) \geqslant P(b_i|\mathbf{r},\bar{\mathbf{q}},\mathbf{b}_o)\middle|\,\mathbf{b}\right)$ in (15.2.46) cannot depend on $\mathbf{b}_o$. Similarly, for random data $P(b_i,\mathbf{b}_o) = P(b_i)\,P(\mathbf{b}_o)$, so (15.2.46) can be rewritten as

$$\begin{aligned}
\mathrm{BER}_{MAPBD} &\geqslant \sum_{b_i} P(b_i)\,P\left(P(\bar{b}_i|\mathbf{r},\bar{\mathbf{q}},\mathbf{b}_o) \geqslant P(b_i|\mathbf{r},\bar{\mathbf{q}},\mathbf{b}_o)\middle|\,\mathbf{b}\right)\sum_{\mathbf{b}_o} P(\mathbf{b}_o) \\
&= \sum_{b_i} P(b_i)\,P\left(P(\bar{b}_i|\mathbf{r},\bar{\mathbf{q}},\mathbf{b}_o) \geqslant P(b_i|\mathbf{r},\bar{\mathbf{q}},\mathbf{b}_o)\middle|\,\mathbf{b}\right). \qquad (15.2.48)
\end{aligned}$$

15.2.5 Refinement to FF channel

All the results illustrated in this section apply to doubly selective channels and, consequently, to frequency-flat fading channels. With time-selective fading the optimal decision rule (15.2.5) turns into

$$\hat{\mathbf{c}} = \arg\min_{\tilde{\mathbf{c}}} \left(\mathbf{r} - \mathbf{F}_\alpha(\tilde{\mathbf{c}})\,\boldsymbol{\alpha}\right)^H \mathbf{R}_n^{-1} \left(\mathbf{r} - \mathbf{F}_\alpha(\tilde{\mathbf{c}})\,\boldsymbol{\alpha}\right). \qquad (15.2.49)$$

In particular, with white sampled noise, the optimal rule (see (15.2.6)) simplifies as

$$\hat{\mathbf{c}} = \arg\min_{\tilde{\mathbf{c}}} \|\mathbf{r} - \mathbf{F}_\alpha(\tilde{\mathbf{c}})\,\boldsymbol{\alpha}\|^2 \qquad (15.2.50)$$

or (see (14.2.17))

$$\hat{\mathbf{c}} = \arg\min_{\tilde{\mathbf{c}}} \sum_{k=0}^{N_z} |r_k - \alpha_k\, d_k(\tilde{\mathbf{c}})|^2 \tag{15.2.51}$$

where $d_k(\tilde{\mathbf{c}})$ represents the value of d_k (14.2.13) for $\mathbf{c} = \tilde{\mathbf{c}}$. With baud rate sampling ($\eta = 1$) and Nyquist pulse shaping, (15.2.51) leads to (see (14.2.18))

$$\hat{\mathbf{c}} = \arg\min_{\tilde{\mathbf{c}}} \sum_{k=0}^{L_c-1} |r_k - \alpha_k\, \tilde{c}_k|^2 \tag{15.2.52}$$

so that the MLSD rule turns into a symbol-by-symbol decision strategy, namely,

$$\hat{c}_k = \arg\min_{\tilde{c}_k} |r_k - \alpha_k\, \tilde{c}_k|^2 \tag{15.2.53}$$

for $k = 0, 1, ..., L_c - 1$. Other details on optimal detection with FF fading are given in Section 16.1.5, where diversity reception techniques are illustrated.

15.3 DETECTION METRICS WHEN THE CIR IS ESTIMATED

Although it is convenient to idealize the CIR as being known exactly by invocation of a friendly genie, in practice the CIR (or some equivalent quantity) must be estimated, as well as any other unknown parameters, such as the noise variance. The estimate is then used as if it were exact. In channels with significant time variation or delay spread, estimation becomes more difficult. We recall that, in general, wireless channels can be classified by their *spread factor* (SF), defined as [17] (see also Part I)

$$SF \triangleq B_D\, \tau_{ds}, \tag{15.3.1}$$

which plays an essential role in defining the measurability of a channel [53], [54], [55], [56]. When $SF < 1$, the channel is *underspread* [57] and its impulse response can be estimated (e.g., from pilot tones or pilot symbols), although the difficulty increases as SF nears unity. When $SF > 1$, the channel is overspread and cannot be estimated.[6]

We must account for the estimation error when seeking a realistic assessment of performance [58], [59], [60], [61], [62], [63]. Accordingly, in this section we are interested in effective methods to estimate the channel from the received samples and the BER degradation due to imperfect estimation. Finally, it is worth noting that in this case, the derived metrics are not rigorously optimal as in known CIR and averaged-over CIR cases. In fact the detection strategies derived under the estimated CIR assumption do not really minimize bit/symbol error probability or sequence error probability.

[6]This criterion was introduced by Kailath [53]. A more accurate criterion replaces the spread factor with the area under the Doppler delay spread function [55].

15.3.1 ML sequence detector

We modify the MLSD's decision rule from (15.2.1) by replacing known quantities with estimated quantities, as

$$\hat{\mathbf{c}} = \arg\max_{\tilde{\mathbf{c}}} p(\mathbf{r}|\tilde{\mathbf{c}}, \hat{\mathbf{q}}, \hat{\mathbf{R}}_n) = \arg\min_{\tilde{\mathbf{c}}} (\mathbf{r} - \mathbf{F}_q(\tilde{\mathbf{c}})\hat{\mathbf{q}})^H \hat{\mathbf{R}}_n^{-1} (\mathbf{r} - \mathbf{F}_q(\tilde{\mathbf{c}})\hat{\mathbf{q}}) \quad (15.3.2)$$

where $\hat{\mathbf{q}}$ and $\hat{\mathbf{R}}_n$ are estimates of the OCIR and the noise autocovariance matrix, respectively. Assuming white analog noise, the filtered, sampled noise's power spectral density can be deduced a priori up to a scale factor because the transfer function of the receiver filter is known. By examining (15.3.2), we see that this scale factor does not affect the MLSD's decisions. Therefore, unless there is additional colored interference, $\hat{\mathbf{R}}_n$ is effectively known.

In (15.3.2), $\hat{\mathbf{q}}$ is only an intermediate step to the noiseless received samples

$$\hat{\mathbf{z}}(\tilde{\mathbf{c}}) = \mathbf{F}_q(\tilde{\mathbf{c}})\,\hat{\mathbf{q}}. \quad (15.3.3)$$

Accordingly, we are free to select whichever channel parameterization leads to easiest or best estimation, such as one from Section 14.2.5, that is,

$$\mathbf{r} = \mathbf{F}_q(\tilde{\mathbf{c}})\,\mathbf{q} + \mathbf{n} = \mathbf{F}_{KL}(\tilde{\mathbf{c}})\,\mathbf{q}_{KL} + \mathbf{n} = \mathbf{F}_{CE}(\tilde{\mathbf{c}})\,\mathbf{q}_{CE} + \mathbf{n} = \mathbf{F}_{PS}(\tilde{\mathbf{c}})\,\mathbf{q}_{PS} + \mathbf{n} \quad (15.3.4)$$

from (14.2.34) and (14.2.68).

Since the $\{\mathbf{F}_x(\tilde{\mathbf{c}})\}$ matrices (where x stands for q, KL, CE or PS) are constructed with a trial sequence $\tilde{\mathbf{c}}$, any channel estimate must implicitly depend on $\tilde{\mathbf{c}}$. Given that the channel estimate is used in (15.3.2) to form the metric for $\tilde{\mathbf{c}}$, it is appropriate for the channel estimate to be calculated assuming that $\tilde{\mathbf{c}}$ was indeed transmitted so that *every possible transmitted sequence needs a different channel estimate* $\hat{\mathbf{q}}_x(\tilde{\mathbf{c}})$ [64], [65].

Estimating $\hat{\mathbf{q}}_x(\tilde{\mathbf{c}})$, given the linear convolution/additive Gaussian noise model of (15.3.4), is a standard problem. The usual estimators are the *maximum likelihood estimator* (MLE), the *reduced rank maximum likelihood estimator* (RRE), and the *Bayesian estimator* (BE) [32], [66]. The *autocovariance-preserving estimator* (APE) [67] is insightful also. Without describing them in detail, we note that they all produce $\hat{\mathbf{z}}_x$ as

$$\hat{\mathbf{z}}_x(\tilde{\mathbf{c}}) = \mathbf{F}_x(\tilde{\mathbf{c}})\,\hat{\mathbf{q}}_x(\tilde{\mathbf{c}}) \quad (15.3.5)$$

through the common framework of

$$\hat{\mathbf{q}}_x(\tilde{\mathbf{c}}) = \arg\min_{\tilde{\mathbf{q}}_x} \left[(\mathbf{r} - \mathbf{F}_x(\tilde{\mathbf{c}})\,\tilde{\mathbf{q}}_x)^H \mathbf{R}_n^{-1} (\mathbf{r} - \mathbf{F}_x(\tilde{\mathbf{c}})\,\tilde{\mathbf{q}}_x) + \tilde{\mathbf{q}}_x^H \mathbf{B}_x(\tilde{\mathbf{c}})\,\tilde{\mathbf{q}}_x\right] \quad (15.3.6)$$

where $\mathbf{B}_x(\tilde{\mathbf{c}})$ is a positive semidefinite matrix characterizing the different estimators and $\tilde{\mathbf{q}}_x$ is a trial OCIR. For notational simplicity, we omit any dependence on $\tilde{\mathbf{c}}$ in the remainder of the section.

In (15.3.6) the arg min argument or *cost function* consists of two terms, working in competition. The first term $(\mathbf{r}-\mathbf{F}_x\tilde{\mathbf{q}}_x)^H\mathbf{R}_n^{-1}(\mathbf{r}-\mathbf{F}_x\tilde{\mathbf{q}}_x)$ is a Mahalanobis distance and is minimized by the $\hat{\mathbf{q}}_x$, such that $\mathbf{F}_x\hat{\mathbf{q}}_x$ is closest in the Mahalanobis sense (Euclidean sense in white noise) to $\mathbf{r}$. The second, quadratic, term $\tilde{\mathbf{q}}_x^H\mathbf{B}_x\tilde{\mathbf{q}}_x$ is an estimator-dependent measure of the energy in the parameters $\tilde{\mathbf{q}}_x$, so it is uniquely minimized by $\tilde{\mathbf{q}}_x = \mathbf{0}_{N_z,1}$ if $\mathbf{B}_x\,(\tilde{\mathbf{c}})$ is a positive definite. The final estimate from (15.3.6) balances the competing objectives of these two terms.

Differentiating the cost function in (15.3.6) with respect to $\tilde{\mathbf{q}}_x$, then setting the result to zero, yields

$$\mathbf{M}_{B,x}\,\hat{\mathbf{q}}_x = \mathbf{F}_x^H\mathbf{R}_n^{-1}\mathbf{r} \tag{15.3.7}$$

where

$$\mathbf{M}_{B,x} \triangleq \mathbf{F}_x^H\mathbf{R}_n^{-1}\mathbf{F}_x + \mathbf{B}_x. \tag{15.3.8}$$

If the matrix $\mathbf{M}_{B,x}$ is full rank, it is inferred from (15.3.4) that

$$\hat{\mathbf{q}}_x = \mathbf{M}_{B,x}^{-1}\,\mathbf{F}_x^H\,\mathbf{R}_n^{-1}\mathbf{r} \tag{15.3.9}$$

and

$$\hat{\mathbf{z}}_x = \mathbf{P}_{\mathbf{B},x}\,\mathbf{r} \tag{15.3.10}$$

where

$$\mathbf{P}_{\mathbf{B},x} \triangleq \mathbf{F}_x\,\mathbf{M}_{B,x}^{-1}\,\mathbf{F}_x^H\,\mathbf{R}_n^{-1} \tag{15.3.11}$$

is the *estimator matrix*.

The quality of this estimate is characterized in several ways. The estimate $\hat{\mathbf{z}}$ of a random vector $\mathbf{z}$ is itself a random variable, so it can be characterized by its mean $\boldsymbol{\eta}_{\hat{\mathbf{z}}} \triangleq E\{\hat{\mathbf{z}}\}$ and the autocovariance matrix

$$\mathbf{R}_{\mathbf{e}_{\hat{\mathbf{z}}}} \triangleq \frac{1}{2}E\{\mathbf{e}_{\hat{\mathbf{z}}}\,\mathbf{e}_{\hat{\mathbf{z}}}^H\} \tag{15.3.12}$$

of the corresponding estimation error

$$\mathbf{e}_{\hat{\mathbf{z}}} = \hat{\mathbf{z}} - E\{\hat{\mathbf{z}}\}. \tag{15.3.13}$$

The estimator mean variance $\boldsymbol{\sigma}_{\hat{\mathbf{z}}}^2$ can be derived from $\mathbf{R}_{\mathbf{e}_{\hat{\mathbf{z}}}}$ as

$$\boldsymbol{\sigma}_{\hat{\mathbf{z}}}^2 \triangleq \frac{1}{2}E\{\mathbf{e}_{\hat{\mathbf{z}}}^H\mathbf{e}_{\hat{\mathbf{z}}}\} = \frac{1}{2}\,\mathrm{tr}\left(E\{\mathbf{e}_{\hat{\mathbf{z}}}\mathbf{e}_{\hat{\mathbf{z}}}^H\}\right) = \frac{1}{2}\,\mathrm{tr}\left(\mathbf{R}_{\mathbf{e}_{\hat{\mathbf{z}}}}\right) \tag{15.3.14}$$

where $\mathrm{tr}\,(\mathbf{A})$ denotes the trace of matrix $\mathbf{A}$. The difference between the estimator mean $\boldsymbol{\eta}_{\hat{\mathbf{z}}}$ and its true value is called the estimation *bias* $\mathbf{b}_{\hat{\mathbf{z}}}$, defined as

$$\mathbf{b}_{\hat{\mathbf{z}}} \triangleq \mathbf{z} - E\{\hat{\mathbf{z}}\}. \tag{15.3.15}$$

Estimator	$\mathbf{B}_x$
MLE	$\mathbf{0}_{N_x,N_x}$
RRE	$\text{diag}\left(\left[\underbrace{0,...0}_{N_{x/RR}}, \underbrace{\infty,...\infty}_{N_x-N_{x/RR}}\right]\right)$
BE	Λ^{-1}
APE	$\Lambda^{-1} + \mathbf{J}^{-\frac{1}{2}}\left(\mathbf{J}^{\frac{1}{2}}\Lambda^{-1}\mathbf{J}^{\frac{1}{2}}\right)^{\frac{1}{2}}\mathbf{J}^{-\frac{1}{2}}$

Table 15.2. Characteristic matrix $\mathbf{B}_x$

Estimator	$\mathbf{P}_{\mathbf{B},x}$
MLE	$\mathbf{F}_x\left(\mathbf{F}_x^H\,\mathbf{F}_x\right)^{-1}\mathbf{F}_x^H$
RRE	$\mathbf{F}_{x/RR}\left(\mathbf{F}_{x/RR}^H\,\mathbf{F}_{x/RR}\right)^{-1}\mathbf{F}_{x/RR}^H$
BE	$\mathbf{F}_x(\mathbf{F}_x^H\mathbf{R}_n^{-1}\mathbf{F}_x + \mathbf{R}_{q,x}^{-1})^{-1}\mathbf{F}_x^H\mathbf{R}_n^{-1} = \mathbf{R}_z\,\mathbf{R}_r^{-1} = \mathbf{I}_{N_z} - \frac{N_0}{T_r}\mathbf{R}_r^{-1}$
APE	$\mathbf{F}_{KL}(\mathbf{F}_{KL}^H\,\mathbf{R}_n^{-1}\mathbf{F}_{KL} + \mathbf{B}_{KL})^{-1}\mathbf{F}_{KL}^H\,\mathbf{R}_n^{-1} = \mathbf{I}_{N_z} - \left(\frac{N_0}{T_r}\right)^{\frac{1}{2}}\mathbf{R}_r^{-\frac{1}{2}}$

Table 15.3. Estimator matrix $\mathbf{P}_{\mathbf{B},x}$

The combined contribution of the bias and estimator variance is given by the *mean squared error* $MSE\left(\hat{\mathbf{z}}\right)$, defined as

$$MSE\left(\hat{\mathbf{z}}\right) \triangleq \frac{1}{2}E\{|\mathbf{b}_{\hat{\mathbf{z}}}|^2\} = \frac{1}{2}\,\text{tr}\left(\mathbf{R}_{\mathbf{e}_{\hat{\mathbf{z}}}}\right) + \boldsymbol{\sigma}_{\hat{\mathbf{z}}}^2. \tag{15.3.16}$$

These definitions focus on $\mathbf{z}_x$ rather than $\mathbf{q}$, $\mathbf{q}_{KL}$, $\mathbf{q}_{CE}$, or $\mathbf{q}_{PS}$ since that is the desired quantity in the detection problem.

Tables 15.2—15.5 summarize the properties of the different estimators in *white noise*. In Table 15.2, the matrix Λ is equal to the matrix $\mathbf{D}$ of the KL eigendecomposition in (14.2.43) and the matrix $\mathbf{J}$ of the APE is defined as

$$\mathbf{J} \triangleq \left[\left(\frac{N_0}{T_r}\right)^{-1}\mathbf{F}_{KL}^H\,\mathbf{F}_{KL} + \Lambda^{-1}\right]^{-1} \tag{15.3.17}$$

The following comments are in order.

- **MLE:** The maximum likelihood estimator is unbiased and has minimum estimator variance for this bias [66]. It is most appropriate for time-invariant channels, where the number of unknown parameters is fixed at $N_x = L_q\,\eta$ and the number of observations N_z increases with transmission duration. Unfortunately, in general DS channels there are more unknown parameters than observations, i.e., $N_x > N_z$, whatever the parameterization, and the matrix $\left(\mathbf{F}_x^H\mathbf{F}_x\right)^{-1}$ is not well defined. Returning to (15.3.7)—(15.3.8), we see that

Estimator	$\mathbf{b}_{\hat{\mathbf{z}}}$
MLE	$\mathbf{0}_{N_z,1}$
RRE	$(\mathbf{I} - \mathbf{P}_{\mathbf{0},x/RR})\,\mathbf{z}$
BE	$\frac{N_0}{T_\eta}\mathbf{R}_r^{-1}\mathbf{z}$
APE	$\left(\frac{N_0}{T_r}\right)^{\frac{1}{2}}\mathbf{R}_r^{-1}\,\mathbf{z}$

Table 15.4. Bias $\mathbf{b}_{\hat{\mathbf{z}}}$.

Estimator	$MSE\,(\hat{\mathbf{z}_x})$
MLE	$\frac{N_0}{T_\eta}N_z$
RRE	$\text{tr}\left(\left(\mathbf{I}_{N_z} - \mathbf{P}_{\mathbf{0},x/RR}\right)\mathbf{R}_z\left(\mathbf{I}_{N_z} - \mathbf{P}_{\mathbf{0},x/RR}\right)^H\right) + \frac{N_0}{T_\eta}N_{x/RR}$
BE	$\left(\frac{N_0}{T_\eta}\right)^2\text{tr}\left(\mathbf{R}_r^{-1}\,\mathbf{R}_z\,\mathbf{R}_r^{-1}\right) + \frac{N_0}{T_\eta}\text{tr}\left(\mathbf{R}_z\,\mathbf{R}_r^{-2}\,\mathbf{R}_z\right)$
APE	$2\,\frac{N_0}{T_r}N_z - 2\,\text{tr}\left(\left(\frac{N_0}{T_r}\right)^{\frac{3}{2}}\mathbf{R}_r^{-\frac{1}{2}}\right)$

Table 15.5. Mean-square error $MSE\,(\hat{\mathbf{z}_x})$

its solution is nonunique and we assert that detection in such uncertainty is unlikely to be successful. For example, consider the typical case of white noise and an $N_z\times\,N_x$ convolution matrix $\tilde{\mathbf{F}}_x$ whose columns are able to span the N_z-dimensional observation space. In this case, (15.3.7) can be rewritten as

$$\mathbf{F}_x^H\left(\mathbf{r} - \mathbf{F}_x\hat{\mathbf{q}}_x\right) = \mathbf{0}_{N_z,1}. \tag{15.3.18}$$

There are N_z, equations, but $N_x \geqslant N_z$ unknowns. Since $\mathbf{F}_x$ can span the observation space, there always exists a $\hat{\mathbf{q}}_x$, such that $\mathbf{r} = \mathbf{F}_x\hat{\mathbf{q}}_x$ is a solution to (15.3.18). When substituted into (15.3.2), all trial sequences share the same metric and sequence detection is impossible.

- **RRE:** Reduced rank ML estimation is applied when most parameters are weak and can be discarded without much penalty. Then the reduced number of parameters $N_{x/RR}$ to be estimated is chosen so that $\mathbf{F}_{x/RR}^H\mathbf{F}_{x/RR}$ is invertible. At a minimum, this requires $N_{x/RR} < N_z$. The RRE's $\mathbf{B}_x$ matrix makes the metric of any trial $\tilde{\mathbf{q}}_{x/RR}$ infinitely large whenever any of its last $\left(N_x - N_{x/RR}\right)$ parameters is non-zero. In this way, the final estimated vector $\hat{\mathbf{q}}_{x/RR}$ is forced to have only $N_{x/RR}$ non-zero parameters. The number of estimator parameters $N_{x/RR}$ is usually optimized to maximize the quality of the estimate. Further, we note the following:

    1. The tapped delay line representation of (14.2.34) does not identify weak parameters; it is for this reason that alternative parameterizations are studied.

2. The Karhunen-Loeve expansion optimally identifies weak parameters, but it implicitly requires a priori the OCIR autocovariance matrix and its eigendecomposition. Therefore, an auxiliary procedure is needed to estimate these quantities: either jointly with the data and OCIR or from previous bursts. Both procedures are computationally expensive. Estimation with previous bursts assumes they exist, that the OCIR autocovariance changes little over the bursts, and that the data symbols have been reliably detected, even though the KL expansion is not known yet (so it may be necessary for early bursts to contain training symbols only). Joint data detection, OCIR estimation, and OCIR autocovariance matrix estimation are highly complicated and it is not at all clear how to accomplish them.

3. The complex exponential parameterization has one strength, but two major weaknesses. First, if there are few paths with distinct delays and Doppler shifts, then the parameterization may be a highly parsimonious representation. Second, this saving cannot be guaranteed, especially in environments dominated by scattering and diffraction, so rank reduction is needed. Third, the representation implicitly assumes the Doppler shifts $\{f_{D,i}\}$ to be known a priori. In practice, they must be estimated beforehand. At the same time, the parameters weak enough to be deleted must also be identified.

4. The power series model cannot isolate the weak parameters as efficiently as the KL expansion, but it needs considerably less a priori information: only the duration of the received pulse L_q and the polynomial order N_{PS}. Accordingly, it is the most pragmatically attractive parameterization proposed so far.

- **BE:** The Bayesian estimator does not delete weak parameters, but it discourages their squared magnitudes from exceeding their mean squared values. We show this as follows. First, from Table 15.2, whatever channel parameters $\hat{\mathbf{q}}_x$ are estimated, the same $\hat{\mathbf{z}}$ is obtained. Therefore, the channel parameterization is arbitrary and we can choose the KL expansion. The penalty term (i.e., the second term in the RHS of (15.3.6)) equals

$$\tilde{\mathbf{q}}_{KL}^H \Lambda^{-1} \tilde{\mathbf{q}}_{KL} = \sum_{k=0}^{N_{KL}-1} \frac{|q_{KL,k}|^2}{d_k}, \tag{15.3.19}$$

which guides the search in (15.3.6) away from solutions for which $|q_{KL,k}|^2 \gg d_k = (1/2)\, E\{|q_{KL,k}|^2\}$.

- **APE:** The autocovariance preserving estimator is a poor estimator on the classical measures of bias or MSE. Nonetheless, its channel estimate is good for data detection, and it is useful for interpretation of the data detection metric obtained in Section 15.4. In particular, it is designed so that the estimate $\hat{\mathbf{z}}_{KL}$ satisfies

$$\mathbf{r}^H \mathbf{R}_r^{-1} \mathbf{r} = (\mathbf{r} - \hat{\mathbf{z}}_{KL})^H \mathbf{R}_n^{-1} (\mathbf{r} - \hat{\mathbf{z}}_{KL}). \tag{15.3.20}$$

The $\mathbf{B}_{KL}$ expression involves matrix powers. The ath power (a being an integer) of a positive semidefinite matrix $\mathbf{R}$ with eigendecomposition $\mathbf{R} = \mathbf{UDU}^H$ is defined as $\mathbf{R}^a = \mathbf{UD}^a\mathbf{U}^H$. Matrix powers are neither distributive nor commutative unless the matrices involved share the same eigenvectors (the columns of $\mathbf{U}$). Diagonal matrices are a special case, as $\mathbf{U} = \mathbf{I}$.

- *All estimators:* Equation (15.3.20) raises an important point. The metrics used by detectors estimating the CIR are supposed to be close to the Mahalanobis distance in (15.2.5), yet, in fact, they are always quadratic forms in $\mathbf{r}$. We can generalize (15.3.20) to all the estimators discussed so far by applying the common estimator formulation of (15.3.10) in (15.3.2) as

$$\begin{aligned}\hat{\mathbf{c}} &= \arg\min_{\tilde{\mathbf{c}}} (\mathbf{r} - \mathbf{P}_{\mathbf{B},x}(\tilde{\mathbf{c}})\,\mathbf{r})^H\mathbf{R}_n^{-1}(\mathbf{r} - \mathbf{P}_{\mathbf{B},x}(\tilde{\mathbf{c}})\,\mathbf{r}) \\ &= \arg\min_{\tilde{\mathbf{c}}} \mathbf{r}^H (\mathbf{I}_{N_z} - \mathbf{P}_{\mathbf{B},x}(\tilde{\mathbf{c}}))^H \mathbf{R}_n^{-1} (\mathbf{I}_{N_z} - \mathbf{P}_{\mathbf{B},x}(\tilde{\mathbf{c}}))\,\mathbf{r}. \end{aligned} \tag{15.3.21}$$

15.3.2 Union bound

The union bound of (15.2.21) applies to the detector at hand, where the sequence metrics are interpreted as the metrics in (15.3.21). Therefore, the pairwise error probability equals

$$\text{PEP}(\mathbf{c},\boldsymbol{\varepsilon}) = P(\Lambda(\mathbf{c} - \boldsymbol{\varepsilon}) \leqslant \Lambda(\mathbf{c})\,|\mathbf{c}) = P\left(\mathbf{r}^H\mathbf{K}(\mathbf{c},\boldsymbol{\varepsilon})\,\mathbf{r} \geqslant 0|\mathbf{c}\right) \tag{15.3.22}$$

where

$$\begin{aligned}\mathbf{K}(\mathbf{c},\boldsymbol{\varepsilon}) \triangleq{}& (\mathbf{I}_{N_z} - \mathbf{P}_{\mathbf{B},x}(\mathbf{c}))^H \mathbf{R}_n^{-1} (\mathbf{I}_{N_z} - \mathbf{P}_{\mathbf{B},x}(\mathbf{c})) \\ &- (\mathbf{I}_{N_z} - \mathbf{P}_{\mathbf{B},x}(\mathbf{c} - \boldsymbol{\varepsilon}))^H \mathbf{R}_n^{-1} (\mathbf{I}_{N_z} - \mathbf{P}_{\mathbf{B},x}(\mathbf{c} - \boldsymbol{\varepsilon})). \end{aligned} \tag{15.3.23}$$

Defining the random variable as

$$\kappa(\mathbf{c},\boldsymbol{\varepsilon}) \triangleq \mathbf{r}^H \mathbf{K}(\mathbf{c},\boldsymbol{\varepsilon})\,\mathbf{r} \tag{15.3.24}$$

the PEP (15.3.22) equals

$$\text{PEP}(\mathbf{c},\boldsymbol{\varepsilon}) = P(\kappa(\mathbf{c},\boldsymbol{\varepsilon}) \geqslant 0|\mathbf{c}). \tag{15.3.25}$$

The methods outlined in Section 15.2.2 for the evaluation of the PEP can be applied in the following two cases: the PEP is averaged across the CIR or computed for a particular CIR.

In the first case, κ is a quadratic form in the zero-mean Gaussian random variables $\mathbf{r}$, as in (15.2.35). In the second case, only the channel noise $\mathbf{n}$ is regarded as random. Therefore, the PEP expression (15.3.22) is rearranged as (the dependence of $\mathbf{K}$ on $(\mathbf{c},\varepsilon)$ is dropped for simplicity)

$$\text{PEP} = P\left(\mathbf{z}^H\mathbf{K}\mathbf{z} + \mathbf{n}^H\mathbf{K}\mathbf{z} + \mathbf{z}^H\mathbf{K}\mathbf{n} + \mathbf{n}^H\mathbf{K}\mathbf{n} \geqslant 0|\mathbf{c}\right) \tag{15.3.26}$$

since $\mathbf{r} = \mathbf{z} + \mathbf{n}$. The left side in the inequality is problematic since its terms are constant, linear, and quadratic in $\mathbf{n}$. However, at high SNRs, the $\mathbf{n}^H\mathbf{K}\mathbf{n}$ term is negligible compared to the other noise terms. Then, we can approximate (15.3.26) by

$$\text{PEP} \approx P\left(\xi \geqslant 0|\mathbf{c}\right) \tag{15.3.27}$$

where

$$\xi \triangleq \mathbf{z}^H\,\mathbf{K}\,\mathbf{z} + \mathbf{n}^H\,\mathbf{K}\,\mathbf{z} + \mathbf{z}^H\,\mathbf{K}\,\mathbf{n} \tag{15.3.28}$$

is a real Gaussian random variable with mean and variance

$$\mu_\xi = \mathbf{z}^H\mathbf{K}\mathbf{z} \tag{15.3.29}$$

$$\sigma_\xi^2 = 2\mathbf{z}^H\,\mathbf{K}\,\mathbf{R}_n\,\mathbf{K}\,\mathbf{z} \tag{15.3.30}$$

respectively. Therefore, at high SNRs, the PEP can be approximated as

$$\text{PEP}\left(\mathbf{c},\varepsilon\right) \approx \frac{1}{\sqrt{2}}\,\text{erfc}\left(-\frac{\mathbf{z}^H\,\mathbf{K}\left(\mathbf{c},\varepsilon\right)\mathbf{z}}{2\sqrt{\mathbf{z}^H\,\mathbf{K}\left(\mathbf{c},\varepsilon\right)\mathbf{R}_n\,\mathbf{K}\left(\mathbf{c},\varepsilon\right)\,\mathbf{z}}}\right). \tag{15.3.31}$$

Other performance evaluations are presented in [68], [69].

15.3.3 MAP bit and symbol detectors

The MAP bit/symbol detectors are modified from (15.2.39) and (15.2.40) to account for the estimated channel information, as

$$\hat{b}_i = \arg\max_{\tilde{b}_i} P(\tilde{b}_i|\mathbf{r},\hat{\mathbf{q}}_x,\hat{\mathbf{R}}_n) \qquad \text{MAPBD} \tag{15.3.32}$$

$$\hat{c}_m = \arg\max_{\tilde{c}_m} P(\tilde{c}_m|\mathbf{r},\hat{\mathbf{q}}_x,\hat{\mathbf{R}}_n) \qquad \text{MAPSD} \tag{15.3.33}$$

where $\hat{\mathbf{R}}_n$ is the estimated noise autocovariance. The notation $\hat{\mathbf{q}}_x$ denotes any one of $\mathbf{q}$, $\mathbf{q}_{KL}$, $\mathbf{q}_{CE}$, or $\mathbf{q}_{PS}$.

In parallel with the MLSD case, $\hat{\mathbf{q}}_x$ should be estimated conditioned on $\tilde{b}_i$ and $\tilde{c}_m$, respectively. Unlike the MLSD case, there are other bits/symbols, and these are nuisance parameters, like the CIR is in the data detection problem. Therefore, the other bits/symbols may be assumed known, their detected values may be used as if correct, or they may be averaged over. The training sequences and pilot symbols are certainly known a priori. The detected value of a data bit/symbol can only be used once the bit/symbol has actually been detected. Averaging over a bit/symbol is always possible, where the a priori bit/symbol probabilities are $\frac{1}{2}$ and $\frac{1}{M}$, respectively. There is a fourth scheme, a hybrid of the last two, in which the data bit/symbols are averaged over using their *a posteriori* bit/symbol probabilities.

The presence of these nuisance parameters makes optimal channel estimation considerably more difficult. Indeed, it is difficult to even think of an estimator structure. Fortunately, it is always clear how to obtain an ML estimator: simply maximize the received samples' likelihood over the unknown parameters. Bearing in mind the problems with the MLE in the MLSD case, we substitute the true observation equation for a reduced rank version. Therefore, the channel parameters for bit and symbol detectors can be estimated by

$$\begin{aligned}\left(\hat{\mathbf{q}}_{x/RR}\left(\tilde{b}_i\right), \hat{\mathbf{R}}_n\left(\tilde{b}_i\right)\right) &= \arg \max_{\tilde{\mathbf{q}}_{x/RR}, \tilde{\mathbf{R}}_n} p\left(\mathbf{r}, \tilde{b}_i | \tilde{\mathbf{q}}_{x/RR}, \tilde{\mathbf{R}}_n\right) \\ &= \arg \max_{\tilde{\mathbf{q}}_{x/RR}, \tilde{\mathbf{R}}_n} \sum_{\tilde{\mathbf{c}} \to \tilde{b}_i} p(\mathbf{r}|\tilde{\mathbf{c}}, \tilde{\mathbf{q}}_{x/RR}, \tilde{\mathbf{R}}_n) P(\tilde{\mathbf{c}})\end{aligned} \tag{15.3.34}$$

and

$$\begin{aligned}\left(\hat{\mathbf{q}}_{x/RR}\left(\tilde{c}_m\right), \hat{\mathbf{R}}_n\left(\tilde{c}_m\right)\right) &= \arg \max_{\tilde{\mathbf{q}}_{x/RR}, \tilde{\mathbf{R}}_n} p\left(\mathbf{r}, \tilde{c}_m | \tilde{\mathbf{q}}_{x/RR}, \tilde{\mathbf{R}}_n\right) \\ &= \arg \max_{\tilde{\mathbf{q}}_{x/RR}, \tilde{\mathbf{R}}_n} \sum_{\tilde{\mathbf{c}} \to \tilde{c}_m} p(\mathbf{r}|\tilde{\mathbf{c}}, \tilde{\mathbf{q}}_{x/RR}, \tilde{\mathbf{R}}_n) P(\tilde{\mathbf{c}})\end{aligned} \tag{15.3.35}$$

respectively. The cost function is the summation of Gaussian pdfs. Little is known about how to optimize it directly, and so an iterative algorithm, such as the *expectation-maximization* (EM) algorithm [70], [71], is probably required.

15.3.4 Matched filter bound

Since the matched filter bound is a lower bound, it still applies when the channel is estimated rather than known exactly. A tighter lower bound that takes into account the channel estimation error has not been devised, for the following reasons.

The BER lower bound derived with the MAPBD in Section 15.2.4 relies on the principle that an optimal detector given a certain amount of information cannot outperform, in terms of average BER, an optimal detector given a superset of that information. A uniform definition of optimality is assumed. However, a detector that *estimates* the CIR or some related quantity cannot claim optimality: the choice of channel estimation and estimator is somewhat arbitrary. We have seen many different channel parameterizations and estimators, and it is unclear which, if any, can claim optimality.

The usual matched filter bound using the MAPSD and only one transmitted symbol does not lead to a lower bound either. Certainly intersymbol interference is removed, but the shortened transmission interval means that there are fewer observations involving the channel. Channel estimation is unlikely to be as good, so overall the detector may perform better or worse when only one symbol is transmitted.

In summary, we do not know how to tighten the lower bound of Section 15.2.4, and it must be used "as is" to characterize detectors that estimate the CIR. Finally, it is worth noting that a lower bound for a specific symbol-by-symbol receiver can be derived under the assumptions that (a) all the symbols in the sequence but the one to be estimated are known to the receiver; (b) all the symbols (including the unknown data symbol) are known to its channel estimator. The real receiver cannot outperform such a bound [72].

15.4 DETECTION METRICS WHEN THE CIR IS AVERAGED-OVER

Instead of estimating the CIR or some parameterization of it, another approach is to average over it. Therefore, the pdf of the CIR must be known and it must be tractable. For all practical purposes, this class of detectors is limited to complex Gaussian channels (i.e., Rayleigh or Rician fading), and we shall see that it requires a priori the received signal autocovariance matrices $\mathbf{R}_r$ for all symbol sequences, or, equivalently, the channel and noise autocovariance matrices. They must be estimated, a harder task than CIR estimation. This requirement makes the implementation of this class of equalizers extremely complicated in many circumstances.

15.4.1 ML sequence detector

Data detection

The MLSD's decision rule from (15.2.1) is rewritten to account for averaging over the channel parameters as

$$\begin{aligned}\hat{\mathbf{c}} &= \arg\max_{\tilde{\mathbf{c}}} \int_{\mathbf{q}} p(\mathbf{r}|\tilde{\mathbf{c}},\mathbf{q})\, p(\mathbf{q})\; d\mathbf{q} = \arg\max_{\tilde{\mathbf{c}}} \int_{\mathbf{q}} \frac{p(\mathbf{r},\mathbf{q}|\tilde{\mathbf{c}})}{p(\mathbf{q}|\tilde{\mathbf{c}})}\, p(\mathbf{q})\; d\mathbf{q} \\ &= \arg\max_{\tilde{\mathbf{c}}} \;\; p(\mathbf{r}|\tilde{\mathbf{c}}) = \arg\min_{\tilde{\mathbf{c}}} \left[-\log p(\mathbf{r}|\tilde{\mathbf{c}})\right] = \arg\min_{\tilde{\mathbf{c}}} \Lambda(\tilde{\mathbf{c}}) \end{aligned} \tag{15.4.1}$$

where

$$\Lambda(\tilde{\mathbf{c}}) \triangleq -\log p(\mathbf{r}|\tilde{\mathbf{c}}) = \frac{1}{2}\left(\mathbf{r} - E\left\{\mathbf{r}|\tilde{\mathbf{c}}\right\}\right)^{H} \mathbf{R}_r^{-1}\left(\tilde{\mathbf{c}}\right)\left(\mathbf{r} - E\left\{\mathbf{r}|\tilde{\mathbf{c}}\right\}\right) + \log\det(2\pi\mathbf{R}_r\left(\tilde{\mathbf{c}}\right)). \tag{15.4.2}$$

Then, in Rician- and Rayleigh-fading channels, the metric is expressed as

$$\Lambda(\tilde{\mathbf{c}}) = \frac{1}{2}\left(\mathbf{r} - E\{\mathbf{r}|\tilde{\mathbf{c}}\}\right)^H \mathbf{R}_r^{-1}(\tilde{\mathbf{c}})\left(\mathbf{r} - E\{\mathbf{r}|\tilde{\mathbf{c}}\}\right) + \log\det(2\pi\mathbf{R}_r(\tilde{\mathbf{c}})) \qquad (15.4.3)$$

and

$$\Lambda(\tilde{\mathbf{c}}) = \frac{1}{2}\mathbf{r}^H\mathbf{R}_r^{-1}(\tilde{\mathbf{c}})\,\mathbf{r} + \log\det(2\pi\mathbf{R}_r(\tilde{\mathbf{c}})) \qquad (15.4.4)$$

respectively. In the remainder of this section, we restrict our attention to the Rayleigh fading case. This problem has been thoroughly studied in [73], [74], [75], [76], [77], [78], [79], [80], [81], [82], [83], [84], [85], [86], [87], [88], [89], although the extension to Rician fading is straightforward [90] and arbitrary fading pdfs may be handled also [91], [92], [93]. In (15.4.3), the first term of the metric is a quadratic form in the received vector $\mathbf{r}$, and its second term is referred to as the *bias* term [94]. As written, the meaning of the metric is not intuitively obvious, nor is it amenable to implementation. Therefore, it is appropriate to study various reformulations of it.

Estimation-correlation

The earliest interpretation of the metric (15.4.3) is due to Kailath [79], [95]. Given the various parameterizations in (15.3.4), the autocovariance of the received signal equals

$$\mathbf{R}_r(\tilde{\mathbf{c}}) = \mathbf{F}_x(\tilde{\mathbf{c}})\,\mathbf{R}_{q,x}\,\mathbf{F}_x^H(\tilde{\mathbf{c}}) + \mathbf{R}_n \qquad (15.4.5)$$

where x is a placeholder for z, q, KL, CE or PS and $\mathbf{R}_{q,x}$ is the autocovariance matrix of the unknown parameters. From the Matrix Inversion Lemma [96], the inverse of $\mathbf{R}_r(\tilde{\mathbf{c}})$ is rewritten as

$$\mathbf{R}_r^{-1}(\tilde{\mathbf{c}}) = \mathbf{R}_n^{-1} - \mathbf{R}_n^{-1}\mathbf{F}_x(\tilde{\mathbf{c}})\left[\mathbf{F}_x^H(\tilde{\mathbf{c}})\,\mathbf{R}_n^{-1}\mathbf{F}_x(\tilde{\mathbf{c}}) + \mathbf{R}_{q,x}^{-1}\right]^{-1}\mathbf{F}_x^H(\tilde{\mathbf{c}})\,\mathbf{R}_n^{-1} \qquad (15.4.6)$$

so that (15.4.1) can be expressed as

$$\begin{aligned}\hat{\mathbf{c}} = \arg\max_{\tilde{\mathbf{c}}}\Big\{&\frac{1}{2}\left[\mathbf{r}^H\,\mathbf{R}_n^{-1}\,\mathbf{F}_x(\tilde{\mathbf{c}})\left(\mathbf{F}_x^H(\tilde{\mathbf{c}})\,\mathbf{R}_n^{-1}\mathbf{F}_x(\tilde{\mathbf{c}}) + \mathbf{R}_{q,x}^{-1}\right)^{-1}\mathbf{F}_x^H(\tilde{\mathbf{c}})\,\mathbf{R}_n^{-1}\,\mathbf{r}\right] \\ &- \log\det(2\pi\,\mathbf{R}_r(\tilde{\mathbf{c}}))\Big\}. \end{aligned} \qquad (15.4.7)$$

Now, it was shown in Table 15.3 that

$$\hat{\mathbf{z}}(\tilde{\mathbf{c}}) = \mathbf{F}_x(\tilde{\mathbf{c}})\,\hat{\mathbf{q}}_x = \mathbf{F}_x(\tilde{\mathbf{c}})\left(\mathbf{F}_x^H(\tilde{\mathbf{c}})\,\mathbf{R}_n^{-1}\,\mathbf{F}_x(\tilde{\mathbf{c}}) + \mathbf{R}_{q,x}^{-1}\right)^{-1}\mathbf{F}_x^H(\tilde{\mathbf{c}})\,\mathbf{R}_n^{-1}\,\mathbf{r} \qquad (15.4.8)$$

is the estimate of the noiseless received vector $\mathbf{z}$, we obtained by computing the Bayesian estimate of the channel parameters $\mathbf{q}_x$, under the assumtion that $\mathbf{c} = \tilde{\mathbf{c}}$. Therefore, (15.4.1) can be written as

$$\hat{\mathbf{c}} = \arg\max_{\tilde{\mathbf{c}}}\left[\frac{1}{2}\mathbf{r}^H\,\mathbf{R}_n^{-1}\,\hat{\mathbf{z}}(\tilde{\mathbf{c}}) - \log\det(2\pi\,\mathbf{R}_r)\right]. \qquad (15.4.9)$$

Then, if the channel noise is white and the bias term $\log\det(2\pi\mathbf{R}_r)$ is neglected, the optimal metric is expressed by the correlation $\mathbf{r}^H\hat{\mathbf{z}}(\tilde{\mathbf{c}})$ between the received sample vector and an estimate of its noiseless value conditioned on the assumption that the vector $\tilde{\mathbf{c}}$ has been transmitted. For this reason, the operation of the ML detector, when the CIR is averaged over, can be summarized as *estimation-correlation.*

The Mahalanobis distance

It has already been shown (see (15.3.20)) that

$$\mathbf{r}^H\,\mathbf{R}_r^{-1}(\tilde{\mathbf{c}})\,\mathbf{r} = (\mathbf{r}-\hat{\mathbf{z}}(\tilde{\mathbf{c}}))^H\,\mathbf{R}_n^{-1}(\mathbf{r}-\hat{\mathbf{z}}(\tilde{\mathbf{c}})) \tag{15.4.10}$$

where $\hat{\mathbf{z}}(\tilde{\mathbf{c}})$ is an estimate of the noiseless received vector $\mathbf{z}$ computed with the APE. Therefore, the first term of the metric $\Lambda(\tilde{\mathbf{c}})$ in (15.4.4) can be interpreted as a Mahalanobis distance between the received vector and $\hat{\mathbf{z}}(\tilde{\mathbf{c}})$, or a squared Euclidean distance in white noise. The APE's structure is generally similar to other estimators, such as the MLE, RE, and BE, in that it can be obtained from the same general framework of (15.3.6). However, at high SNRs at least, the APE has somewhat greater bias and its MSE is twice that of the BE.

The innovations process and the Cholesky decomposition

The most useful interpretation comes from the so-called *innovations process*, which, as shown below, is closely related to recursive conditional probabilities, the Cholesky decomposition, and Wiener (linear, minimum mean-square error) filtering [96].

To derive the new interpretation, let us factor the positive definite matrix $\mathbf{R}_r(\tilde{\mathbf{c}})$ in (15.4.5) by means of the Cholesky decomposition [96] as

$$\mathbf{R}_r(\tilde{\mathbf{c}}) = \mathbf{U}_C^H(\tilde{\mathbf{c}})\,\mathbf{U}_C(\tilde{\mathbf{c}}) \tag{15.4.11}$$

where $\mathbf{U}_C(\tilde{\mathbf{c}})$ is an upper triangular matrix. The upper triangular matrix $\mathbf{U}_C^{-1}(\tilde{\mathbf{c}})$ can be factored as

$$\mathbf{U}_C^{-1}(\tilde{\mathbf{c}}) = \mathbf{W}^H(\tilde{\mathbf{c}})\,\mathbf{S}^{-\frac{1}{2}}(\tilde{\mathbf{c}}) \tag{15.4.12}$$

where $\mathbf{W}(\tilde{\mathbf{c}})$ is a lower triangular matrix with ones on its main diagonal and $\mathbf{S}^{-\frac{1}{2}}(\tilde{\mathbf{c}}) = \mathrm{diag}[\sqrt{s_k(\tilde{\mathbf{c}})}]$ is a diagonal matrix and is made of the diagonal entries of $\mathbf{U}_C^{-1}(\tilde{\mathbf{c}})$. Then, the inverse of the matrix $\mathbf{R}_r(\tilde{\mathbf{c}})$ can be expressed as

$$\mathbf{R}_r^{-1}(\tilde{\mathbf{c}}) = \mathbf{U}_C^{-1}(\tilde{\mathbf{c}})\,\mathbf{U}_C^{-H}(\tilde{\mathbf{c}}) = \mathbf{W}^H(\tilde{\mathbf{c}})\,\mathbf{S}^{-\frac{1}{2}}(\tilde{\mathbf{c}})\,\mathbf{S}^{-\frac{1}{2}}(\tilde{\mathbf{c}})\,\mathbf{W}(\tilde{\mathbf{c}})\,. \tag{15.4.13}$$

Substituting (15.4.13) into (15.4.4) yields the following expression for the optimal metric:

$$\Lambda(\tilde{\mathbf{c}}) = \frac{1}{2}\left[\mathbf{S}^{-\frac{1}{2}}(\tilde{\mathbf{c}})\,\mathbf{W}(\tilde{\mathbf{c}})\,\mathbf{r}\right]^H\left[\mathbf{S}^{-\frac{1}{2}}(\tilde{\mathbf{c}})\,\mathbf{W}(\tilde{\mathbf{c}})\,\mathbf{r}\right] + \log\det(\mathbf{S}(\tilde{\mathbf{c}})). \tag{15.4.14}$$

Let us now provide an interpretation of the last expression. First, we write the entries of $\mathbf{W}(\tilde{\mathbf{c}})$ as

$$\mathbf{W}(\tilde{\mathbf{c}}) = \begin{bmatrix} 1 & 0 & 0 & \cdots & 0 \\ -w_{1,1}(\tilde{\mathbf{c}}) & 1 & 0 & & 0 \\ -w_{2,2}(\tilde{\mathbf{c}}) & -w_{2,1}(\tilde{\mathbf{c}}) & 1 & \ddots & \vdots \\ \vdots & & \ddots & \ddots & 0 \\ -w_{N_z-1,N_z-1}(\tilde{\mathbf{c}}) & -w_{N_z-1,N_z-2}(\tilde{\mathbf{c}}) & \cdots & -w_{N_z-1,1}(\tilde{\mathbf{c}}) & 1 \end{bmatrix} \quad (15.4.15)$$

so that the product of the kth row of $\mathbf{W}(\tilde{\mathbf{c}})$ with $\mathbf{r}$ in (15.4.14) is given as

$$e_k(\tilde{\mathbf{c}}) \triangleq r_k - \sum_{m=1}^{k} w_{k,m}(\tilde{\mathbf{c}})\, r_{k-m}. \quad (15.4.16)$$

Accordingly, (15.4.14) can be expressed as

$$\Lambda(\tilde{\mathbf{c}}) = \sum_{k=0}^{N_z-1} \left| \frac{r_k - \sum_{m=1}^{k} w_{k,m}(\tilde{\mathbf{c}})\, r_{k-m}}{\sqrt{2\, s_k(\tilde{\mathbf{c}})}} \right|^2 + \log\left(2\pi\, s_k(\tilde{\mathbf{c}})\right). \quad (15.4.17)$$

Second, we revert to the probability expression of (15.4.1) and use the conditional probability to isolate the first received sample, then the second, and so forth, as

$$\begin{aligned} \hat{\mathbf{c}} &= \arg\max_{\tilde{\mathbf{c}}} p(\mathbf{r}|\tilde{\mathbf{c}}) = \arg\max_{\tilde{\mathbf{c}}} \prod_{k=0}^{N_z-1} p(r_k|\mathcal{R}_{k-1}, \tilde{\mathbf{c}}) \\ &= \arg\min_{\tilde{\mathbf{c}}} \Lambda(\tilde{\mathbf{c}}) \end{aligned} \quad (15.4.18)$$

where

$$\Lambda(\tilde{\mathbf{c}}) = -\log \prod_{k=0}^{N_z-1} p(r_k|\mathcal{R}_{k-1}, \tilde{\mathbf{c}}) = -\sum_{k=0}^{N_z-1} \log p(r_k|\mathcal{R}_{k-1}, \tilde{\mathbf{c}}) \quad (15.4.19)$$

and $\mathcal{R}_k \triangleq [r_0, r_1, ..., r_k]^T$. Now, we note that the random variable r_k $(k = 0, 1, ..., N_z - 1)$ conditioned on the transmitted sequence $\mathbf{c}$ and on $\mathcal{R}_{k-1}$ is a Gaussian random variable with mean $\eta_{k|\mathcal{R},\tilde{\mathbf{c}}} \triangleq E\{r_k|\mathcal{R}_{k-1}, \tilde{\mathbf{c}}\}$ and variance $\sigma^2_{k|\mathcal{R},\tilde{\mathbf{c}}} \triangleq (1/2)\, E\{|r_k - \eta_{k|\mathcal{R},\tilde{\mathbf{c}}}|^2\}$ and, consequently,

$$p(r_k|\mathcal{R}_{k-1}, \tilde{\mathbf{c}}) = \frac{1}{2\pi\sigma^2_{k|\mathcal{R},\tilde{\mathbf{c}}}} \exp\left[-\frac{\left| r_k - \eta_{k|\mathcal{R},\tilde{\mathbf{c}}} \right|^2}{2\,\sigma^2_{k|\mathcal{R},\tilde{\mathbf{c}}}} \right]. \quad (15.4.20)$$

Substituting (15.4.20) into (15.4.19) produces

$$\Lambda(\tilde{\mathbf{c}}) = \sum_{k=0}^{N_z-1} \left| \frac{r_k - E\{r_k|\,\mathcal{R}_{k-1}, \tilde{\mathbf{c}}\}}{\sqrt{2\,\sigma^2_{k|\mathcal{R},\tilde{\mathbf{c}}}}} \right|^2 + \log\left(2\pi\,\sigma^2_{k|\mathcal{R},\tilde{\mathbf{c}}}\right). \tag{15.4.21}$$

Finally, by comparing (15.4.17) with (15.4.21), we match up $\sigma^2_{k|\mathcal{R},\tilde{\mathbf{c}}}$ with $s_k(\tilde{\mathbf{c}})$ and $\eta_{k|\mathcal{R},\tilde{\mathbf{c}}}$ with $\sum_{m=1}^{k} w_{k,m}(\tilde{\mathbf{c}})\, r_{k-m}$. Therefore, $\eta_{k|\mathcal{R},\tilde{\mathbf{c}}}$ is a linear, one-step prediction of r_k given all past samples $\mathcal{R}_{k-1}$, assuming a particular transmitted data sequence $\tilde{\mathbf{c}}$ and having a prediction variance $2\sigma^2_{k|\mathcal{R},\tilde{\mathbf{c}}}$.

The quantity $[\mathbf{S}^{-\frac{1}{2}}(\mathbf{c})\,\mathbf{W}(\mathbf{c})\,\mathbf{r}]$ in (15.4.14) is referred to as the *innovations vector* [74]. Its entries are the scaled prediction errors or *innovations*, namely,

$$i_k \triangleq \frac{r_k - \eta_{k|\mathcal{R},\mathbf{c}}}{\sqrt{2\,\sigma^2_{k|\mathcal{R},\mathbf{c}}}} = \frac{\left[-\mathbf{w}_k^T(\mathbf{c}), 1\right]\mathcal{R}_k}{\sqrt{2\,s_k(\mathbf{c})}} \tag{15.4.22}$$

where $\mathbf{w}_k(\mathbf{c}) \triangleq [w_{k,k}(\mathbf{c}), w_{k,k-1}(\mathbf{c}), ..., w_{k,1}(\mathbf{c})]^T$. Since the numerator of (15.4.22) comprises r_k minus its expected value $E\{r_k|\,\mathcal{R}_{k-1}, \tilde{\mathbf{c}}\}$, conditioned on the past received samples, it can be shown that the entries in the sequence

$$r_0 - E\{r_0|\,\mathbf{c}\}, r_1 - E\{r_1|\,\mathcal{R}_0, \mathbf{c}\}, r_2 - E\{r_2|\,\mathcal{R}_1, \mathbf{c}\}, ... \tag{15.4.23}$$

are uncorrelated with one another. Due to the Gaussian nature of the entries, the normalized sequence of innovations given by (15.4.22) contains independent complex Gaussian random variables with unit variance. It is easily verified that the innovations are uncorrelated with each other and past received samples, as

$$\frac{1}{2}E\{i_k\, i_l^*\} = \delta_{kl} \tag{15.4.24}$$

$$\frac{1}{2}E\{r_k\, i_l^*\} = 0, \qquad k < l. \tag{15.4.25}$$

Moreover,

$$\frac{1}{2}E\{r_k\, i_k^*\} = \sqrt{2\,s_k(\mathbf{c})} \tag{15.4.26}$$

since r_k can be written as

$$\begin{aligned} r_k &= \sqrt{2s_k(\mathbf{c})}\left(\frac{r_k - E\{r_k|\,\mathcal{R}_{k-1}, \mathbf{c}\}}{\sqrt{2\,s_k(\mathbf{c})}} + \frac{E\{r_k|\,\mathcal{R}_{k-1}, \mathbf{c}\}}{\sqrt{2\,s_k(\mathbf{c})}}\right) \\ &= \sqrt{2s_k(\mathbf{c})}\left(i_k + \frac{E\{r_k|\,\mathcal{R}_{k-1}, \mathbf{c}\}}{\sqrt{2\,s_k(\mathbf{c})}}\right) \end{aligned} \tag{15.4.27}$$

and $E\{r_k | \mathcal{R}_{k-1}, \mathbf{c}\}$ is computed from $\mathcal{R}_{k-1}$ only. Therefore, from (15.4.24) and (15.4.26), it follows that

$$\begin{aligned}\sqrt{2\, s_k(\mathbf{c})}\frac{1}{2}E\{\mathcal{R}_k i_k^*\} &= \frac{1}{2}E\{\mathcal{R}_k\mathcal{R}_k^H\}\begin{bmatrix}-\mathbf{w}_k(\mathbf{c})\\ 1\end{bmatrix}\\ &= \mathbf{R}_{r,k}(\mathbf{c})\begin{bmatrix}-\mathbf{w}_k(\mathbf{c})\\ 1\end{bmatrix} = \begin{bmatrix}\mathbf{0}_{N_z,1}\\ 2\, s_k(\mathbf{c})\end{bmatrix}\end{aligned} \tag{15.4.28}$$

where $\mathbf{R}_{r,k} \triangleq (1/2)\, E\{\mathcal{R}_k\mathcal{R}_k^H\}$. By partitioning $\mathbf{R}_{r,k}$ as

$$\mathbf{R}_{r,k} = \begin{bmatrix}\mathbf{R}_{r,k-1} & \mathbf{R}_{r,k-1,k}\\ \mathbf{R}_{r,k-1,k}^H & \frac{1}{2}E\left\{|r_k|^2\right\}\end{bmatrix} \tag{15.4.29}$$

where the column vector $\mathbf{R}_{r,k-1,k}$ is defined as $\mathbf{R}_{r,k-1,k} \triangleq (1/2)\, E\{\mathcal{R}_{k-1}r_k^*\}$, we can rewrite (15.4.28) as

$$\mathbf{R}_{r,k-1}\mathbf{w}_k(\mathbf{c}) = \mathbf{R}_{r,k-1,k} \tag{15.4.30}$$

$$2\, s_k(\mathbf{c}) = \frac{1}{2}E\left\{|r_k|^2\right\} - \mathbf{R}_{r,k-1,k}^H\mathbf{w}_k(\mathbf{c})\,. \tag{15.4.31}$$

Equations (15.4.30) and (15.4.31) are used to compute $\mathbf{w}_k(\mathbf{c})$ and $2s_k(\mathbf{c})$, respectively, for all $\mathbf{c}$. These equations can be compared with the Wiener or minimum-MSE (MMSE) linear prediction coefficients derived as

$$\begin{aligned}\mathbf{w}_{MMSE,k}(\mathbf{c}) &= \arg\min_{\tilde{\mathbf{w}}_k}\frac{1}{2}E\left\{\left\|r_k - \tilde{\mathbf{w}}_k^H(\mathbf{c})\,\mathcal{R}_{k-1}\right\|^2\right\}\\ &= \frac{1}{2}E\{\mathcal{R}_{k-1}\mathcal{R}_{k-1}^H\}^{-1}\frac{1}{2}E\{\mathcal{R}_{k-1}r_k^*\}\\ &= \mathbf{R}_{r,k-1}^{-1}\,\mathbf{R}_{r,k-1,k}.\end{aligned} \tag{15.4.32}$$

We see that these equations are identical to those obtained in (15.4.30). In summary, the first term in the metric (15.4.4) can be interpreted as the sum of squared prediction errors, each normalized by the expected squared prediction error. The predictions are computed as the linear sum of past samples, and they have minimum mean-squared prediction error.

15.4.2 Union bound

The union bound of (15.2.21) applies to the detector at hand [90], [89]. The sequence metrics are given in (15.4.4), so the pairwise error probability in (15.2.21) equals

$$\begin{aligned}\mathrm{PEP}(\mathbf{c},\varepsilon) &\triangleq P(\Lambda(\mathbf{c}-\varepsilon) \leqslant \Lambda(\mathbf{c})\,|\mathbf{c})\\ &= P(-\log p(\mathbf{r}|\mathbf{c}-\varepsilon) \leqslant -\log p(\mathbf{r}|\mathbf{c})|\mathbf{c})\\ &= P(\kappa(\mathbf{c},\varepsilon) \geqslant \kappa_{\min}(\mathbf{c},\varepsilon)\,|\mathbf{c})\end{aligned} \tag{15.4.33}$$

where

$$\kappa_{\min}(\mathbf{c},\varepsilon) = \log\det(2\pi\mathbf{R}_r(\mathbf{c}-\varepsilon)) - \log\det(2\pi\mathbf{R}_r(\mathbf{c})) \\ = \log\det\left(\mathbf{R}_r(\mathbf{c}-\varepsilon)\,\mathbf{R}_r^{-1}(\mathbf{c})\right) \tag{15.4.34}$$

$$\kappa(\mathbf{c},\varepsilon) \triangleq \mathbf{r}^H\,\mathbf{K}(\mathbf{c},\varepsilon)\,\mathbf{r} \tag{15.4.35}$$

with

$$\mathbf{K}(\mathbf{c},\varepsilon) \triangleq \frac{1}{2}\mathbf{R}_r^{-1}(\mathbf{c}) - \frac{1}{2}\mathbf{R}_r^{-1}(\mathbf{c}-\varepsilon). \tag{15.4.36}$$

This problem is closely related to the one studied in Section 15.3.2, up to the different matrix $\mathbf{K}$ and the non-zero threshold $\kappa_{\min}$. We look at two tractable cases: the PEP is (a) averaged across the CIR, or (b) computed for a particular CIR.

In the first case, κ is a quadratic form in the zero-mean Gaussian random variables $\mathbf{r}$, exactly as in Section 15.3.2. In the second case, we observe that only $\mathbf{n}$ is regarded as random. Following the same approach as (15.3.26), we can rewrite the PEP expression as (the dependence on $(\mathbf{c},\varepsilon)$ is dropped for simplicity)

$$\text{PEP} = P\left(\mathbf{z}^H\,\mathbf{K}\,\mathbf{z} + \mathbf{n}^H\,\mathbf{K}\mathbf{z} + \mathbf{z}^H\,\mathbf{K}\,\mathbf{n} + \mathbf{n}^H\,\mathbf{K}\mathbf{n} \geqslant \kappa_{\min}|\mathbf{c}\right). \tag{15.4.37}$$

At high SNRs, the $\mathbf{n}^H\,\mathbf{K}\mathbf{n}$ term is negligible compared to the other noise terms, and so we can approximate (15.4.37) by

$$\text{PEP} = P(\xi \geqslant \kappa_{\min}|\mathbf{c}) \tag{15.4.38}$$

where (see (15.3.28))

$$\xi \triangleq \mathbf{z}^H\,\mathbf{K}\,\mathbf{z} + \mathbf{n}^H\,\mathbf{K}\mathbf{z} + \mathbf{z}^H\,\mathbf{K}\,\mathbf{n} \tag{15.4.39}$$

is a real Gaussian random variable with mean and variance given by (15.3.29) and (15.3.30), respectively. Then, from (15.4.38) we evaluate the PEP as

$$\text{PEP}(\mathbf{c},\varepsilon) \approx \frac{1}{\sqrt{2}}\,\text{erfc}\left(\frac{\kappa_{\min}(\mathbf{c},\varepsilon) - \mathbf{z}^H\,\mathbf{K}(\mathbf{c},\varepsilon)\,\mathbf{z}}{2\sqrt{\mathbf{z}^H\,\mathbf{K}(\mathbf{c},\varepsilon)\,\mathbf{R}_n\,\mathbf{K}(\mathbf{c},\varepsilon)\,\mathbf{z}}}\right). \tag{15.4.40}$$

15.4.3 MAP bit and symbol detectors

A natural MAP bit detection rule, when averaging over the CIR, is

$$\hat{b}_i = \arg\max_{\tilde{b}_i} \int_{\mathbf{q}} p(\tilde{b}_i, |\mathbf{r},\mathbf{q})\,p(\mathbf{q})\,d\mathbf{q}, \tag{15.4.41}$$

but this is incorrect. Averaging over the CIR is strongly related to estimating it, so the metric must incorporate $\mathbf{r}$ and $\mathbf{q}$ as a likelihood, $p(..., \mathbf{r}|\mathbf{q}, ...)$, as in (15.3.34) and (15.3.35). Therefore, the correct MAPBD detection rule is given by [8], [97]

$$\begin{aligned}\hat{b}_i &= \arg\max_{\tilde{b}_i} \int_{\mathbf{q}} p(\tilde{b}_i, \mathbf{r}|\mathbf{q})\ p(\mathbf{q})\ d\mathbf{q} \\ &= \arg\max_{\tilde{b}_i} p\left(\tilde{b}_i, \mathbf{r}\right) \qquad (15.4.42) \\ &= \arg\max_{\tilde{b}_i} p\left(\tilde{b}_i \middle| \mathbf{r}\right) \qquad (15.4.43)\end{aligned}$$

since $p(\mathbf{r})$ is data-independent. As we naively sought in (15.4.41), the metric conditions on $\mathbf{r}$. Furthermore, (15.4.42) can be refined as

$$\hat{b}_i = \arg\max_{\tilde{b}_i} \sum_{\tilde{\mathbf{c}} \rightarrow \tilde{b}_i} p(\tilde{\mathbf{c}}, \mathbf{r}) = \arg\max_{\tilde{b}_i} \sum_{\tilde{\mathbf{c}} \rightarrow \tilde{b}_i} p(\mathbf{r}|\,\tilde{\mathbf{c}}) \qquad (15.4.44)$$

where

$$p(\mathbf{r}|\tilde{\mathbf{c}}) = \frac{1}{\det(2\pi \mathbf{R}_r)} \exp\left(-\frac{1}{2}\mathbf{r}^H\, \mathbf{R}_r^{-1}\, \mathbf{r}\right) \qquad (15.4.45)$$

in Rayleigh fading. Similarly for MAPSDs,

$$\hat{c}_m = \arg\max_{\tilde{c}_m} \int_{\mathbf{q}} p(\tilde{c}_m, \mathbf{r}|\mathbf{q})\, p(\mathbf{q})\ d\mathbf{q} = \arg\max_{\tilde{c}_m} \sum_{\tilde{\mathbf{c}} \rightarrow \tilde{c}_m} p(\mathbf{r}|\,\tilde{\mathbf{c}}) = \arg\max_{\tilde{c}_m} \sum_{\tilde{\mathbf{c}} \rightarrow \tilde{c}_m} p(\mathbf{r}|\,\tilde{\mathbf{c}}). \qquad (15.4.46)$$

Both metrics are the sum of scaled exponentials of quadratic forms. The metric is difficult to interpret beyond the discussion in Section 15.4.1.

15.4.4 Matched filter bound

From (15.4.43) and (15.4.46), averaging over the nuisance CIR still leads to optimal MAPBD/MAPSD detection rules, and therefore the matched filter bound of (15.2.21) can be extended to these detectors. It is evident from the Mahalanobis distance or estimator/correlator interpretations of the metric that simply transmitting one symbol only does not lead to a lower bound, since the noiseless received samples are implicitly estimated by making use of all received samples.

In brief, a genie-aided detector is invoked that is given the same a priori information as the actual MAPBDs (i.e., the channel and noise autocovariances), as well as the values of the other transmitted bits, $\mathbf{b}_o$, excluding the ith bit, b_i, to be estimated. Then, we denote by $\mathbf{b}$ and $\bar{\mathbf{b}}$ the transmitted bits that map to the symbol sequence $\mathbf{c}$, and the vector that equals $\mathbf{b}$ but with $\bar{b}_i$ in place of b_i, respectively.

We also assume that $\bar{\mathbf{b}}$ maps to $(\mathbf{c}-\boldsymbol{\varepsilon})$. Like (15.2.46), the genie-aided receiver's average BER is used to bound the desired BER, as

$$\mathrm{BER}_{MAPBD} \geqslant \mathrm{BER}_{ga} = \sum_{b_i,\mathbf{b}_o} P\left(P(\bar{b}_i|\mathbf{r},\mathbf{b}_o) \geqslant P(b_i|\mathbf{r},\mathbf{b}_o)\big|\,\mathbf{b}\right) P(b_i,\mathbf{b}_o). \tag{15.4.47}$$

The first probability factor in (15.4.47) is the PEP between $\mathbf{b}$ and $\bar{\mathbf{b}}$ (or, equivalently, between $\mathbf{c}$ and $(\mathbf{c}-\boldsymbol{\varepsilon})$) and can be evaluated as

$$\begin{aligned}
&P\left(P(\bar{b}_i|\mathbf{r},\mathbf{b}_o) \geqslant P(b_i|\mathbf{r},\mathbf{b}_o)\big|\,\mathbf{b}\right)\\
&= P\left(\frac{p(\mathbf{r}|\bar{b}_i,\mathbf{b}_o)P(\bar{b}_i,\mathbf{b}_o)}{p(\mathbf{r},\mathbf{b}_o)} \geqslant \frac{p(\mathbf{r}|b_i,\mathbf{b}_o)P(b_i,\mathbf{b}_o)}{p(\mathbf{r},\mathbf{b}_o)}\right)\\
&= P\left(p(\mathbf{r}|\mathbf{c}-\boldsymbol{\varepsilon}) \geqslant p(\mathbf{r}|\mathbf{c})|\mathbf{c}\right)\\
&= P\left(-\log p(\mathbf{r}|\mathbf{c}-\boldsymbol{\varepsilon}) \leqslant -\log p(\mathbf{r}|\mathbf{c})|\mathbf{c}\right).
\end{aligned} \tag{15.4.48}$$

In Section 15.4.2 we showed how to calculate this quantity. Unlike the known CIR case, the exhaustive summation over all the sequences $\{\mathbf{b}=(b_i,\mathbf{b}_o)\}$ in (15.4.47) cannot be avoided.

15.4.5 Refinement to FF channel

All the considerations illustrated in this section apply to DS fading channels. In this subsection we concentrate on the interpretation of some results when the channel is frequency flat and time varying. In this case, the received signal samples are expressed by (see (14.2.17))

$$r_k = \alpha_k\, d_k + n_k \tag{15.4.49}$$

where α_k and d_k are the kth samples of the fading distortion and the transmitted data signal, respectively. In this case, the mean $\eta_{\,k|\mathcal{R},\tilde{\mathbf{c}}}$ of (15.4.20) can be expressed as

$$\eta_{\,k|\mathcal{R},\tilde{\mathbf{c}}} = \tilde{d}_k\,\tilde{\alpha}\,\{k|\,\mathcal{R}_{k-1},\tilde{\mathbf{c}}\} \tag{15.4.50}$$

where $\tilde{\alpha}\,\{k|\,\mathcal{R}_{k-1},\tilde{\mathbf{c}}\}$ is the MMSE one-step prediction of the fading sample α_k, based on $\mathcal{R}_{k-1}$ and assuming that the sequence $\tilde{\mathbf{c}}$ has been transmitted; $\tilde{d}_k$ is the kth useful signal sample corresponding to the transmission of the trial sequence $\tilde{\mathbf{c}}$. In addition, the variance $\sigma^2_{k|\mathcal{R},\tilde{\mathbf{c}}}$ of (15.4.20) is given by

$$\sigma^2_{k|\mathcal{R},\tilde{\mathbf{c}}} \triangleq \frac{1}{2}E\left\{\left|r_k - \tilde{d}_k\,\tilde{\alpha}\,\{k|\,\mathcal{R}_{k-1},\tilde{\mathbf{c}}\}\right|^2\middle|\,\mathcal{R}_{k-1},\tilde{\mathbf{c}}\right\}. \tag{15.4.51}$$

With CPM signals and with baud rate sampled ($\eta=1$) PSK signals (and no ISI) we have $|\tilde{d}_k|\equiv 1$. Then, we can rewrite (15.4.51) as

$$\sigma^2_{k|\mathcal{R},\tilde{\mathbf{c}}} = \sigma^2_{k|\mathcal{R},\tilde{\mathbf{c}}_m} \triangleq \frac{1}{2}E\left\{\left|r_k\,\tilde{d}_k^* - \tilde{\alpha}\,\{k|\,\mathcal{R}_{k-1},\tilde{\mathbf{c}}_m\}\right|^2\middle|\,\mathcal{R}_{k-1},\tilde{\mathbf{c}}_m\right\}. \tag{15.4.52}$$

with $\tilde{\mathbf{c}}_m \triangleq [c_0, c_1, ..., c_m]^T$, under the assumption that the kth sample belongs to the mth symbol interval. Similarly, under the same assumptions, $\eta_{k|\mathcal{R},\tilde{\mathbf{c}}}$ (15.4.50) can be evaluated as

$$\eta_{k|\mathcal{R},\tilde{\mathbf{c}}} = \eta_{k|\mathcal{R},\tilde{\mathbf{c}}_m} = \tilde{d}_k \, \tilde{\alpha} \{ k | \mathcal{R}_{k-1}, \tilde{\mathbf{c}}_m \} . \quad (15.4.53)$$

In FF channels the evaluation of $\tilde{\alpha} \{ k | \mathcal{R}_{k-1}, \tilde{\mathbf{c}}_m \}$ and of the variance $\sigma^2_{k|\mathcal{R},\tilde{\mathbf{c}}_m}$ can be accomplished by a time-varying Wiener filter [98], [96]. However, if the process $\{\alpha_k\}$ can be characterized by a Gauss-Markov model [98], [99], both quantities can be computed recursively by means of a Kalman predictor [96] for a given sequence $\tilde{\mathbf{c}}_m$. These problems are addressed in Section 16.3.

Bibliography

[1] H. L. V. Trees, *Detection, Estimation and Modulation Theory, Part III.* New York: John Wiley & Sons, 1971.

[2] J. C. Hancock and P. A. Wintz, *Signal Detection Theory.* New York: McGraw-Hill, 1966.

[3] J. K. Cavers, "An analysis of pilot symbol assisted modulation for Rayleigh fading channels," *IEEE Trans. Vehicular Techn.*, vol. 40, pp. 686–693, Nov. 1991.

[4] N. W. K. Lo, D. D. Falconer, and U. H. Sheikh, "Adaptive equalization and diversity combining for mobile radio using interpolated channel estimates," *IEEE Trans. Vehicular Techn.*, vol. 40, pp. 636–645, Aug. 1991.

[5] S. A. Fechtel and H. Meyr, "Optimal parametric feedforward estimation of frequency-selective fading radio channels," *IEEE Trans. Comm.*, vol. 42, pp. 1639–1650, Feb./Mar./Apr. 1994.

[6] T. Eyceoz and A. Duel-Hallen, "Simplified block adaptive diversity equalizer for cellular mobile radio," *IEEE Comm. Letters*, vol. 1, pp. 15–19, Jan. 1987.

[7] C. Berrou and A. Glavieux, "Near optimum error correcting coding and decoding: Turbo-codes," *IEEE Trans. Comm.*, vol. 44, pp. 1261–1271, Oct. 1996.

[8] M. J. Gertsman and J. H. Lodge, "Symbol-by-symbol MAP demodulation of CPM and PSK signals on Rayleigh flat-fading channels," *IEEE Trans. Comm.*, vol. 45, pp. 788–799, July 1997.

[9] S. H. Jamali and T. Le-Ngoc, *Coded Modulation Techniques for Fading Channels.* Boston, MA: Kluwer Academic Publishers, 1994.

[10] E. Biglieri, D. Divsalar, P. J. McLane, and M. Simon, *Introduction to Trellis-Coded Modulation with Applications.* New York: MacMillan, 1991.

[11] C. Schlegel and D. J. Costello, "Bandwidth efficient coding for fading channels: Code construction and performance analysis," *IEEE J. Sel. Areas Comm.*, vol. 7, pp. 1356–1368, Dec. 1989.

[12] P. J. McLane, P. H. Whittke, P. K. Ho, and C. Loo, "PSK and DPSK trellis codes for fast fading, shadowed mobile satellite communication channels," *IEEE Trans. Comm.*, vol. 36, pp. 1242–1246, Nov. 1988.

[13] P. Ho and D. Fung, "Error performance of interleaved trellis-coded PSK modulations in correlated Rayleigh fading channels," *IEEE Trans. Comm.*, vol. 40, pp. 1800–1809, Dec. 1992.

[14] J. K. Cavers and P. Ho, "Analysis of the error performance of trellis-coded modulations in Rayleigh fading channels," *IEEE Trans. Comm.*, vol. 40, pp. 74–80, Jan. 1992.

[15] K. Chan and A. Bateman, "The performance of reference-based M-ary PSK with trellis coded modulation in Rayleigh fading," *IEEE Trans. Vehicular Techn.*, vol. 41, pp. 190–198, May 1992.

[16] C. Douillard, M. Jezequel, C. Berrou, A. Picart, P. Didier, and A. Glavieux, "Iterative correction of intersymbol interference: Turbo equalization," *Eur. Trans. Telecomm.*, vol. 6, pp. 507–511, Sept.-Oct. 1995.

[17] J. G. Proakis, *Digital Communications.* New York: McGraw-Hill, 2nd ed., 1989.

[18] J. Hagenauer and P. Hoeher, "A Viterbi algorithm with soft-decision outputs and Its applications," in *Proc. of GLOBECOM '89, IEEE Global Telecommun. Conf.*, (Dallas, TX, USA), pp. 1680–1686, 1989.

[19] P. Hoeher, "Advances in soft-output decoding," in *Proc. of GLOBECOM '93, IEEE Global Telecommun. Conf.*, (Houston, TX, USA), pp. 793–797, 1993.

[20] M. P. C. Fossorier, F. Burkert, L. Shu, and J. Hagenauer, "On the Equivalence Between SOVA and Max-log-MAP Decodings," *IEEE Comm. Letters*, vol. 2, pp. 137–139, May 1998.

[21] G. D. Hingorani and J. C. Hancock, "A transmitted-reference system for communication in random or unknown channels," *IEEE Trans. Comm. Systems*, vol. 13, pp. 293–301, Sep. 1965.

[22] R. A. Ziegler and J. M. Cioffi, "Estimation of time-varying digital radio channels," *IEEE Trans. Vehicular Techn.*, vol. 41, pp. 134–151, May 1992.

[23] B. D. Hart and D. P. Taylor, "Extended MLSE diversity receiver for the time- and frequency-selective channel," *IEEE Trans. Comm.*, vol. 45, pp. 322–333, March 1997.

[24] D. K. Borah and B. D. Hart, "Frequency-selective fading channel estimation with a polynomial time-varying channel model," *IEEE Trans. Comm.*, vol. 47, pp. 862–871, June 1999.

[25] E. H. Lin and A. J. Giger, "Radio channel characterization by three tones," *IEEE J. Sel. Areas Commun.*, vol. 5, pp. 402–415, Apr. 1987.

[26] M. L. Moher and J. H. Lodge, "TCMP—a modulation and coding strategy for Rician fading channels," *IEEE J. Select. Areas Comm.*, vol. 7, pp. 1347–1355, Dec. 1989.

[27] S. N. Crozier, D. D. Falconer, and S. A. Mahmoud, "Least Sum of Squared Errors (LSSE) Channel Estimation," *IEE Proc.-F*, vol. 138, pp. 371–378, Aug. 1991.

[28] C. Tellambura, Y. J. Guo, and S. K. Barton, "Channel estimation using aperiodic binary sequences," *IEEE Comm. Lett.*, vol. 2, pp. 140–142, May 1998.

[29] J. C. L. Ng, K. B. Letaief, and R. D. Murch, "Complex optimal sequences with constant magnitude for fast channel estimation initialization," *IEEE Trans. Comm.*, vol. 46, pp. 305–308, March 1998.

[30] G. D. Forney, "Maximum-likelihood sequence estimation of digital sequences in the presence of intersymbol interference," *IEEE Trans. Inf. Theory*, vol. 18, pp. 363–378, May 1971.

[31] G. Ungerboeck, "Adaptive maximum-likelihood receiver for carrier-modulated data-transmission systems," *IEEE Trans. Comm.*, vol. 22, pp. 624–636, May 1974.

[32] L. L. Scharf, *Statistical Signal Processing: Detection, Estimation and Time Series Analysis.* Reading, MA: Addison-Wesley, 1991.

[33] G. J. Foschini, "Equalising without altering or detecting data," *Bell Syst. Tech. J.*, vol. 64, pp. 1885–1912, Oct. 1985.

[34] S. Verdu, "Maximum likelihood sequence detection for intersymbol interference channels: A new upper bound on error probability," *IEEE Trans. Inf. Theory*, vol. 33, pp. 62–68, Jan. 1987.

[35] W.-H. Sheen, C.-C. Tseng, and C. S. Wang, "On the diversity, bandwidth and performance of digital transmission over frequency-selective slow fading channels," *IEEE Trans. Vehicular Techn.*, vol. 49, pp. 835–843, May 2000.

[36] P. J. Crepeau, "Asymptotic performance of M-ary orthogonal modulation in generalised fading channels," *IEEE Trans. Comm.*, vol. 36, pp. 1246–1248, Nov. 1988.

[37] M. Schwartz, W. R. Bennett, and S. Stein, *Communication Systems and Techniques.* New York: McGraw-Hill, 1966.

[38] M. J. Barrett, "Error probability for optimal and suboptimal quadratic receivers in rapid Rayleigh fading channels," *IEEE J. Sel. Areas Comm.*, vol. 5, pp. 302–304, Feb. 1987.

[39] K. Abend and B. D. Fritchman, "Statistical detection for communication channels with intersymbol interference," *Proc. IEEE*, vol. 58, pp. 779–785, May 1970.

[40] K. Abend, T. J. Hartley, B. D. Fritchman, and C. Gumacos, "On optimum receivers for channels having memory," *IEEE Trans. Inf. Theory*, vol. 14, pp. 152–157, Nov. 1968.

[41] G. D. Forney, "The Viterbi algorithm," *Proc. IEEE*, vol. 61, pp. 268–278, Mar. 1973.

[42] Y. Li, B. Vucetic, and Y. Sato, "Optimum soft-output detection for channels with intersymbol interference," *IEEE Trans. Inf. Theory*, vol. 41, pp. 704–713, May 1995.

[43] J. E. Mazo, "Exact matched filter bound for two-beam Rayleigh fading," *IEEE Trans. Comm.*, vol. 39, pp. 1027–1030, July 1991.

[44] C. Wong and L. Greenstein, "Multipath fading models and adaptive equalizers in microwave digital radio," *IEEE Trans. Comm.*, vol. 32, pp. 928–934, Aug. 1984.

[45] R. A. Valenzuela, "Performance of adaptive equalisation for indoor radio communications," *IEEE Trans. Comm.*, vol. 37, pp. 291–293, Mar. 1989.

[46] D. Dzung and W. Braun, "Performance of coherent data transmission in frequency-selective Rayleigh fading channels," *IEEE Trans. Comm.*, vol. 41, pp. 1335–1341, 1993.

[47] F. Ling, "Matched filter bound for time-discrete multipath Rayleigh fading channels," *IEEE Trans. Comm.*, vol. 43, pp. 710–713, Feb./Mar./Apr 1995.

[48] V. P. Kaasila and A. Mammela, "Bit error probability of a matched filter in a Rayleigh fading multipath channel," *IEEE Trans. Comm.*, vol. 42, pp. 826–828, Feb./Mar./Apr. 1994.

[49] M. Clark, L. Greenstein, W. Kennedy, and M. Shafi, "MMSE diversity combining for wide-band digital cellular radio," *IEEE Trans. Comm.*, vol. 40, pp. 1128–1135, 1992.

[50] P. Monsen, "Digital transmission performance on fading dispersive diversity channels," *IEEE Trans. Comm.*, vol. 21, pp. 33–39, Jan. 1973.

[51] W. Burchill and C. Leung, "Matched filter bound for OFDM on Rayleigh fading channels," *IEE Electron. Lett*, vol. 31, pp. 1716–1717, Sept. 1995.

[52] W. Yip and T. S. Ng, "Matched filter bound for multipath Rician fading channels," *IEEE Trans. Comm.*, vol. 31, pp. 441–445, Apr. 1998.

[53] T. Kailath, "Measurements on time-variant communication channels," *IRE Trans. Inf. Theory*, vol. 8, pp. 229–236, Sept. 1962.

[54] T. Kailath, "Time variant communication channels," *IRE Trans. Inf. Theory*, vol. 9, pp. 233–237, October 1963.

[55] P. A. Bello, "Measurement of random time-variant channels," *IEEE Trans. Inf. Theory*, vol. 15, pp. 469–475, July 1969.

[56] P. A. Bello and R. Esposito, "Measurement techniques for time-varying dispersive channels," *Alta Frequenza*, vol. XXXIX, pp. 980–996, Nov. 1970.

[57] R. S. Kennedy, *Fading Dispersive Communication Channels.* New York: Wiley-Interscience, 1969.

[58] M.-C. Chiu and C.-C. Chao, "Analysis of LMS-adaptive MLSE equalization on multipath fading channels," *IEEE Trans. Comm.*, vol. 44, pp. 1684–1692, Dec. 1996.

[59] P. Y. Kam and H. M. Ching, "Sequence estimation over the slow nonselective Rayleigh fading channel with diversity reception and Its application to Viterbi decoding," *IEEE J. Sel. Areas Comm.*, vol. 10, pp. 562–570, April 1992.

[60] K. Pahlavan and J. W. Matthews, "Performance of adaptive matched filter receivers over fading multipath channels," *IEEE Trans. Comm.*, vol. 38, pp. 2106–2113, Dec. 1990.

[61] P. K. Varshney and A. H. Haddad, "A receiver with memory for fading channels," *IEEE Trans. Comm.*, vol. 26, pp. 278–283, Feb. 1978.

[62] A. Wautier, J.-C. Dany, C. Mourot, and V. Kumar, "A new method for predicting the channel estimate influence on performance of TDMA mobile radio systems," *IEEE Trans. Vehicular Techn.*, vol. 44, pp. 594–602, Aug. 1995.

[63] K. M. Chugg and A. Polydoros, "MLSE for an unknown channel—Part II: Tracking performance," *IEEE Trans. Comm.*, vol. 44, pp. 949–958, Aug. 1996.

[64] R. Raheli, G. Marino, and P. Castoldi, "Per-survivor processing and tentative decisions: What Is in between?," *IEEE Trans. Comm.*, vol. 44, pp. 127–129, Feb. 1996.

[65] K. M. Chugg and A. Polydoros, "MLSE for an unknown channel—Part I: Optimality considerations," *IEEE Trans. Comm.*, vol. 44, pp. 836–846, July 1996.

[66] S. M. Kay, *Fundamentals of Statistical Signal Processing: Estimation Theory.* Englewood Cliffs, NJ: Prentice Hall, 1993.

[67] B. D. Hart, D. K. Borah, and S. Pasupathy, "Interpretation of the MLSD metric for time varying, frequency selective Rayleigh fading channels," in *VTC'99 - Fall*, pp. 1321–1325, Sept. 1999.

[68] D.-L. Liu and K. Feher, "Pilot-symbol aided coherent M-ary PSK in frequency-selective fast Rayleigh fading channel," *IEEE Trans. Comm.*, vol. 42, pp. 54–62, Jan. 1994.

[69] J. K. Cavers, "Pilot symbol assisted modulation and differential detection in fading and delay spread," *IEEE Trans. Comm.*, vol. 43, pp. 2206–2212, July 1995.

[70] J. C. Han and C. N. Georghiades, "Sequence estimation in the presence of random parameters via the EM algorithm," *IEEE Trans. Comm.*, vol. 45, pp. 300–308, Mar. 1997.

[71] T. K. Moon, "The expectation-maximization algorithm," *IEEE Sig. Proc. Mag.*, vol. 13, pp. 47–63, Nov. 1996.

[72] P. Y. Kam, "Optimal detection of digital data over the nonselective Rayleigh fading channel with diversity reception," *IEEE Trans. Comm.*, vol. 39, pp. 214–219, Feb. 1991.

[73] G. M. Vitetta and D. P. Taylor, "Multi-sampling receivers for uncoded and coded PSK signal sequences transmitted over Rayleigh frequency-flat fading channels," *IEEE Trans. Comm.*, vol. 44, pp. 130–133, Feb. 1996.

[74] X. Yu and S. Pasupathy, "Innovations-based MLSE for Rayleigh fading channels," *IEEE Trans. Comm*, vol. 43, pp. 1534–1544, Feb./Mar./Apr. 1995.

[75] W. S. Leon and D. P. Taylor, "An adaptive receiver for the time- and frequency-selective fading channel," *IEEE Trans. Comm.*, vol. 45, pp. 1548–1559, Dec. 1997.

[76] R. Price, "The detection of signals perturbed by scatter and noise," *IRE Trans. Inf. Theory*, vol. 4, pp. 163–170, Sept. 1954.

[77] R. Price, "Optimum detection of random signals in noise with application to scatter-multipath communications—I," *IRE Trans. Inf. Theory*, vol. 6, pp. 125–135, Dec. 1956.

[78] D. Middleton, "On the detection of stochastic signals in additive normal noise—Part 1," *IRE Trans. Inf. Theory*, vol. 3, pp. 86–121, June 1957.

[79] T. Kailath, "Correlation detection of signals perturbed by a random channel," *IRE Trans. Inf. Theory*, vol. 6, pp. 361–366, June 1960.

[80] F. C. Schweppe, "Evaluation of likelihood functions for Gaussian signals," *IRE Trans. Inf. Theory*, vol. 11, pp. 61–70, Jan. 1965.

[81] T. Kailath, "Likelihood ratios for Gaussian processes," *IEEE Trans. Inf. Theory*, vol. 16, pp. 276–287, May 1970.

[82] P. Y. Kam and C. H. Teh, "Reception of PSK signals over fading channels via quadrature amplitude estimation," *IEEE Trans. Comm.*, vol. 31, pp. 1024–1027, August 1983.

[83] P. Y. Kam and C. H. Teh, "An adaptive receiver with memory for slowly fading channels," *IEEE Trans. Comm.*, vol. 32, pp. 654–659, June 1984.

[84] J. H. Lodge and M. J. Moher, "Maximum likelihood sequence estimation of CPM signals transmitted over Rayleigh flat-fading channels," *IEEE Trans. Comm.*, vol. 38, pp. 787–794, June 1990.

[85] Q. Dai and E. Shwedyk, "Detection of bandlimited signals Over frequency-selective Rayleigh fading channels," *IEEE Trans. Comm.*, vol. 42, pp. 941–950, 1994.

[86] W. C. Dam and D. P. Taylor, "An adaptive maximum likelihood receiver for correlated Rayleigh-fading channels," *IEEE Trans. Comm.*, vol. 42, pp. 2684–2692, Sep. 1994.

[87] G. M. Vitetta and D. P. Taylor, "Maximum likelihood decoding of uncoded and coded PSK signal sequences transmitted over Rayleigh flat-fading channels," *IEEE Trans. Comm.*, vol. 43, pp. 2750–2758, Nov. 1995.

[88] G. M. Vitetta and D. P. Taylor, "Maximum likelihood sequence estimation of differentially encoded PSK signals transmitted over Rayleigh frequency-flat fading channels," *Int. J. of Wireless Inf. Networks*, vol. 2, pp. 71–81, April 1995.

[89] X. Yu and S. Pasupathy, "Error performance of innovations-based MLSE for Rayleigh fading channels," *IEEE Trans. Vehicular Techn.*, vol. 45, pp. 631–642, Nov. 1996.

[90] B. D. Hart and D. P. Taylor, "Maximum-likelihood synchronization, equalization, and sequence estimation for unknown time-varying frequency-selective Rician channels," *IEEE Trans. Comm.*, vol. 46, pp. 211–221, Feb. 1998.

[91] T. E. Duncan, "Evaluation of likelihood functions," *Information & Control*, vol. 13, pp. 62–74, July 1968.

[92] T. Kailath, "A general likelihood-ratio formula for random signals in Gaussian noise," *IEEE Trans. Inf. Theory*, vol. 15, pp. 350–361, May 1969.

[93] T. Kailath, "A further note on a general likelihood formula for random signals in Gaussian noise," *IEEE Trans. Inf. Theory*, vol. 16, pp. 393–396, July 1970.

[94] A. Mammela and D. P. Taylor, "Bias terms in the optimal quadratic receiver," *IEEE Comm. Lett.*, vol. 2, pp. 57–58, Feb 1998.

[95] T. Kailath, "Optimum receivers for randomly varying channels," in *Proc. of the Fourth London Symposium on Information Theory*, (London), pp. 109–122, Butterworth Scientific Press, 1961.

[96] S. Haykin, *Adaptive Filter Theory.* Englewood Cliffs, NJ: Prentice-Hall, 1986.

[97] J. P. Seymour and M. P. Fitz, "Near optimal symbol-by-symbol detection schemes for flat Rayleigh fading," *IEEE Trans. Comm.*, vol. 43, pp. 1525–1533, Feb./Mar./Apr. 1995.

[98] C. W. Therrien, *Discrete Random Signals and Statistical Signal Processing.* Englewood Cliffs: Prentice Hall, 1992.

[99] H. S. Wang and P.-C. Chang, "On verifying the first order Markovian assumption for a Rayleigh fading channel model," *IEEE Trans. Vehicular Techn.*, vol. 45, pp. 353–357, May 1996.

Chapter 16

EQUALIZATION ALGORITHMS

The previous chapter provides a thorough description of the optimal decision rules, but we are ultimately interested in implementable equalizer structures. That is the topic of this chapter.

16.1 EQUALIZATION WHEN THE CIR IS KNOWN

In this section we describe five important classes of equalizers: MLSDs, reduced complexity sequence detectors, decision feedback equalizers (DFEs), linear equalizers (LE), and MAPSD and MAPBDs. We begin by assuming that the CIR is known exactly *a priori*.

16.1.1 MLSD based on the Viterbi algorithm

For the ML sequence detection problem with known CIR, a white noise sequence at the sampler output (following the noise limiting filter) is assumed. Then, the metric in the decision rule of (15.2.6) is recognized as a summation of squared Euclidean distances, that is

$$\Lambda\left(\tilde{\mathbf{c}}\right) \triangleq \left\|\mathbf{r} - \bar{\mathbf{F}}_q\left(\tilde{\mathbf{c}}\right)\bar{\mathbf{q}}\right\|^2 = \sum_{m=0}^{L_z-1}\sum_{k=m\eta}^{(m+1)\eta-1}\left|r_k - z_k\left(\tilde{\mathbf{c}}\right)\right|^2 \tag{16.1.1}$$

where the quantity (see (15.2.12))

$$z_k(\tilde{\mathbf{c}}) = z_k\left(\tilde{\mathbf{c}}_{m-L_q+1}^{m}\right) = \sum_{i=m-L_q+1}^{m}\tilde{c}_i\,\bar{q}_{k,k-i\eta} \tag{16.1.2}$$

depends on the data sequence through the vector $\tilde{\mathbf{c}}_{m-L_q+1}^{m} \triangleq [\tilde{c}_{m-L_q+1}, \tilde{c}_{m-L_q+2}, \ldots, \tilde{c}_m]^T$ of L_q consecutive data symbols only, with the convention that $\tilde{c}_i = 0$ for $i < 0$

or $i \geq L_c$. The metric $\Lambda(\tilde{\mathbf{c}})$ (16.1.1) can be also rewritten as

$$\Lambda(\tilde{\mathbf{c}}) = \sum_{m=0}^{L_z-1} \lambda(x_m, x_{m+1}) \tag{16.1.3}$$

where

$$\lambda(x_m, x_{m+1}) = \sum_{k=m\eta}^{(m+1)\eta-1} \left| r_k - \sum_{i=m-L_q+1}^{m} \tilde{c}_i \, \bar{q}_{k,k-i\eta} \right|^2 \tag{16.1.4}$$

and x_m is the integer representation of the symbol vector $(\tilde{c}_{m-L_q+1}, \tilde{c}_{m-L_q+2}, \ldots, \tilde{c}_{m-1})$, consisting of $(L_c - 1)$ consecutive data symbols. Equations (16.1.3)—(16.1.4) express the optimal metric as a summation of partial metrics $\{\lambda(x_m, x_{m+1})\}$. The mth of these terms, $\lambda(x_m, x_{m+1})$, depends on x_m and x_{m+1}, i.e., on the vectors of consecutive trial symbols $(\tilde{c}_{m-L_q+1}, \tilde{c}_{m-L_q+2}, \ldots, \tilde{c}_{m-1})$ and $(\tilde{c}_{m-L_q+2}, \tilde{c}_{m-L_q+3}, \ldots, \tilde{c}_m)$, respectively. These considerations suggest a recursive formula for the evaluation of $\Lambda(\tilde{\mathbf{c}})$. In fact, if we define the recursive relation

$$\Lambda(\tilde{\mathbf{c}}_m) = \Lambda(\tilde{\mathbf{c}}_{m-1}) + \lambda(x_m, x_{m+1}) \tag{16.1.5}$$

with $\tilde{\mathbf{c}}_i \triangleq [\tilde{c}_0, \tilde{c}_1, \ldots, \tilde{c}_i]$ $(\tilde{\mathbf{c}}_{L_c-1} = \tilde{\mathbf{c}})$ and $\Lambda(\tilde{\mathbf{c}}_0) = 0$, then we have

$$\Lambda(\tilde{\mathbf{c}}) = \Lambda(\tilde{\mathbf{c}}_{L_z-1}) \tag{16.1.6}$$

after L_z iterations.

A geometrical representation to the problem of search of the optimal metric can be given as follows. A trellis diagram with $N_s = M^{L_q-1}$ states is drawn, as illustrated in Figure 16.1 for $M = 4$ and $L_q = 2$. In the mth interval, each trellis state represents one of the N_s possible values that x_m can take and is connected by M branches to the next states x_{m+1}. The branch connecting x_m and x_{m+1} is labelled by the trial symbol $\tilde{c}_m$ and by the quantity $\lambda(x_m, x_{m+1})$, dubbed the *branch metric*. In this context, each trial sequence $\tilde{\mathbf{c}}$ is in a one-to-one correspondence with a sequence of states in the trellis diagram, i.e., with a distinct path in the trellis. Moreover, looking for the optimal sequence decision is equivalent to searching the minimum distance path in the trellis. Such a search does not require an exhaustive analysis of all the possible trial sequences but can be carried out recursively, employing the so-called *Viterbi algorithm* (VA).

The VA is a recursive algorithm for determining the optimal state sequence of a discrete-time, Markov process observed in memoryless noise [1] and is an efficient tool for many communications problems, including ML sequence detection [2].

We now apply the VA algorithm to implement the MLSD with known CIR. The VA operates on the state trellis defined above. In the mth symbol interval, each branch, labelled by the couple of states (x_m, x_{m+1}), is assigned the corresponding branch metric $\lambda(x_m, x_{m+1})$. The VA's task is to find the sequence of branches through the trellis with the smallest metric (i.e., the shortest path). The VA accomplishes this by the following actions (see Figure 16.2)

1. maintaining one *survivor* path per state x_m in the mth symbol interval;
2. extending these paths one step along all the M branches (labelled by $\tilde{c}_m$) emanating from it;
3. pruning these by retaining only the path with the smallest[1] metric $\Lambda(\tilde{\mathbf{c}}_m)$ into each state x_{m+1}.

In the mth symbol interval, then, the Viterbi algorithm keeps track only of the one path (the so-called *survivor*) leading to each state x_m. Such a path, denoted as $\hat{x}(x_m)$, is the sequence of consecutive states belonging to the path and is characterized by an *accumulated metric* $\Lambda(x_m)$.

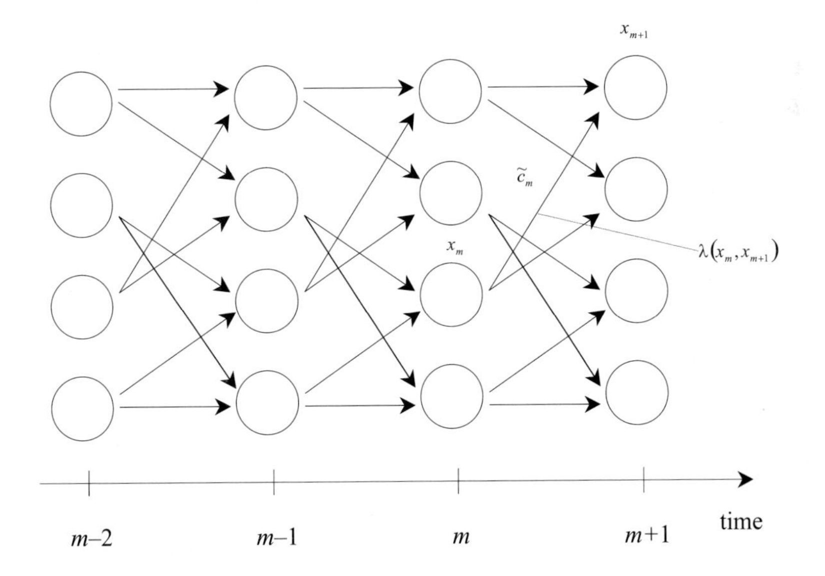

Figure 16.1. Four-state trellis ($M = 4$ and $L_q = 2$ are assumed)

The VA procedure can be summarized in the following steps (m denotes the time variable).

1. Set

$$m = 0, \hat{x}(x_0) = (x_0), \Lambda(x_0) = 0 \tag{16.1.7}$$

   to initialize the algorithm.

2. Repeat steps 3 through 7 until $m = L_z$.

[1]When a metric has to be maximized, the VA should instead be supplied with the negative metric.

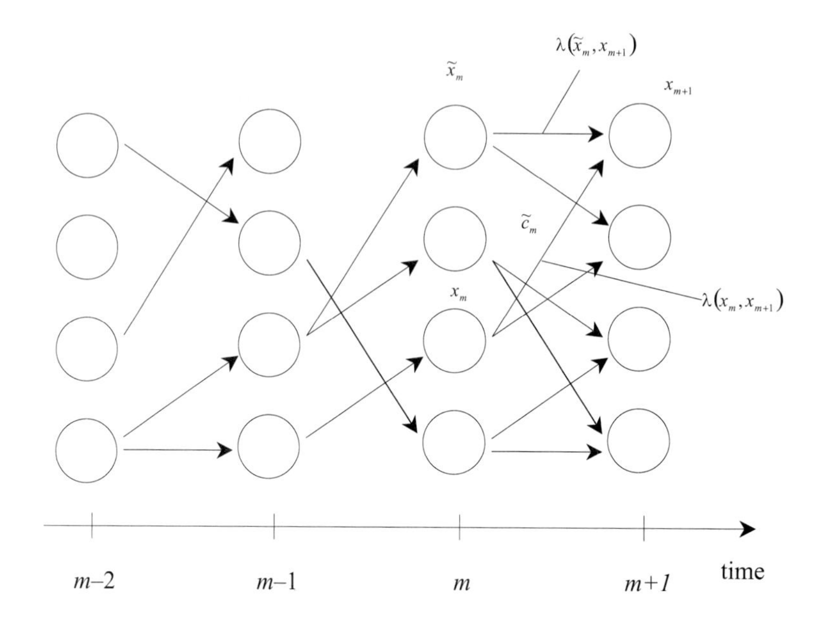

Figure 16.2. Time evolution of the VA

3. Extend path metrics according to (16.1.6), that is

$$\Lambda\left(x_{m+1}\right)=\Lambda\left(x_m\right)+\lambda\left(x_m, x_{m+1}\right) \tag{16.1.8}$$

for all the allowed state transitions $x_m \rightarrow x_{m+1}$.

4. For each destination state x_{m+1}, find the best (minimum metric) incoming path over all the previous state

$$\bar{x}_m=\arg \min_{x_m} \Lambda\left(x_{m+1}\right). \tag{16.1.9}$$

5. Update and store survivor paths as

$$\hat{x}\left(x_{m+1}\right)=\left(\hat{x}\left(\bar{x}_m\right), x_{m+1}\right). \tag{16.1.10}$$

6. Store the new survivor metrics as

$$\Lambda\left(x_{m+1}\right)=\Lambda\left(\bar{x}_m\right)+\lambda\left(\bar{x}_m, x_{m+1}\right). \tag{16.1.11}$$

7. Set $m=m+1$ (increment time counter).

8. Detect the ML decision for the symbol sequence as that associated with the survivor path $\hat{x}\left(x_{L_z}\right)$ with minimum metric $\Lambda\left(x_{L_z}\right)$ (termination).

It is worth noting that (a) branch metrics evaluated for state transitions inconsistent with known (i.e., training or pilot) symbols are set to a large value (virtually infinite) and (b) decisions are not available until time $k = K$. In practice, however, there is little degradation in making decisions after a decision delay of some 5—7 times L_q by tracing back from the survivor with the instantaneously best metric.

Other implementations using different front-ends and branch metrics are described in [2], [3], [4], [5]. To provide further insight into this problem, let us consider the optimal MLSD metric (15.2.11), that is,

$$\Gamma(\tilde{\mathbf{c}}) \triangleq 2\Re\left\{\bar{\mathbf{q}}^H \bar{\mathbf{F}}_q(\tilde{\mathbf{c}})^H \mathbf{r}\right\} - \left\|\bar{\mathbf{F}}_q(\tilde{\mathbf{c}})\,\bar{\mathbf{q}}\right\|^2. \tag{16.1.12}$$

Substituting (15.2.13) and (15.2.17) into (16.1.12) produces

$$\begin{aligned}\Gamma(\tilde{\mathbf{c}}) = {} & 2\Re\left\{\sum_{m=0}^{L_c-1} \tilde{c}_m^* v_m\right\} - \sum_{m_1=0}^{L_c-1}\sum_{m_2=0}^{L_c-1} c_{m_1} c_{m_2}^* \sum_{k=0}^{N_z-1} q_{k,k-m_1\eta}\, q_{k,k-m_2\eta}^* \\ & - \sum_{m=0}^{L_c-1} |c_m|^2 \sum_{k=0}^{N_z-1} q_{k,k-m\eta}\, q_{k,k-m\eta}^* \\ & -2\Re\left\{\sum_{m_1=0}^{L_c-1}\sum_{m_2=0}^{m_1-1} c_{m_1} c_{m_2}^* \sum_{k=0}^{N_z-1} q_{k,k-m_1\eta}\, q_{k,k-m_2\eta}^*\right\}\end{aligned} \tag{16.1.13}$$

where (see (15.2.13))

$$v_m \triangleq \left[r_k \otimes \bar{q}_{m,-k}^*\right]_{k=m\eta} \tag{16.1.14}$$

is the symbol rate output sample of a *time-varying matched filter* (TVMF) (see Section 15.2.1). After some manipulations, the metric can be expressed in a form similar to $\Lambda(\tilde{\mathbf{c}})$ (16.1.3), that is,

$$\Gamma(\tilde{\mathbf{c}}) = \sum_{m=0}^{L_z-1} \gamma(x_m, x_{m+1}) \tag{16.1.15}$$

where the branch metric $\gamma(x_m, x_{m+1})$ is given by

$$\begin{aligned}\gamma(x_m, x_{m+1}) \triangleq \Re\Bigg\{\tilde{c}_m^* \Bigg[& 2v_m - c_m \sum_{k=0}^{N_z-1} q_{k,k-m\eta}\, q_{k,k-m\eta}^* \\ & -2 \sum_{m_1=m-L_c+1}^{m-1} c_{m_1} \sum_{k=0}^{N_z-1} q_{k,k-m_1\eta}\, q_{k,k-m\eta}^*\Bigg]\Bigg\}.\end{aligned} \tag{16.1.16}$$

The MLSD implementation corresponding to (16.1.15) and (16.1.16) was first proposed by Ungerboeck [3] for FS channels and has been extended to time-varying

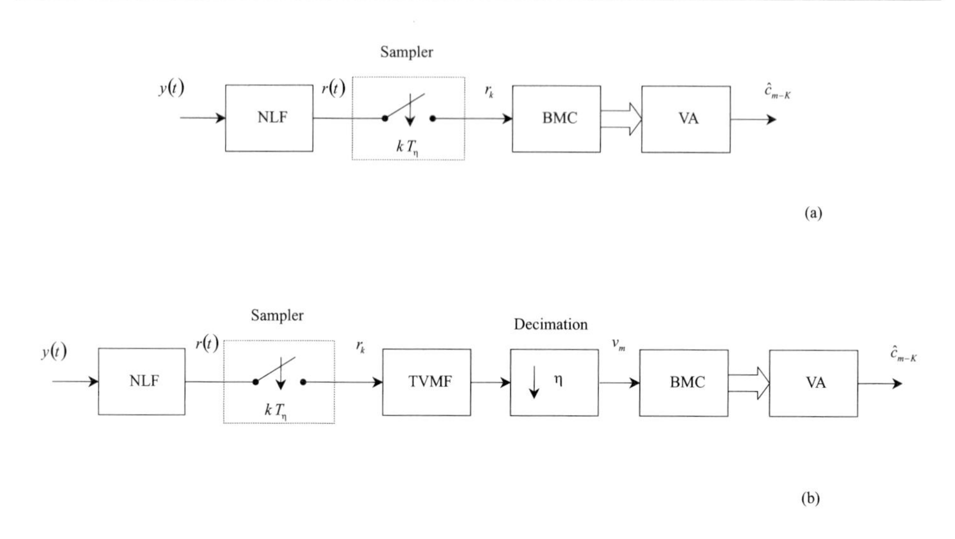

Figure 16.3. Two different implementations of the MLSD strategy: (a) structure employing the Euclidean metrics (16.1.4) and (b) Ungerboeck's receiver

channels in [6], [7], [8], [9]. The block diagrams of both the implementations described above are illustrated in Figure 16.3, where the subsystem evaluating the branch metrics for the VA is called *branch metric computer* (BMC).

Another well-known implementation for MLSD on FS channels, proposed by Forney [2], makes use of the *whitened matched filter* (WMF). Forney's receiver can be derived from Ungerboeck's one (or, vice versa), as shown in [9], where the relations between these historical implementations are thoroughly investigated. A brief derivation of Forney's approach is sketched in Section 16.1.2.

Finally, it is worth noting that MLSD detectors based on the VA have been extended to provide approximate symbol probabilities [10], [11], [12].

16.1.2 Reduced complexity sequence detection

The complexity of the VA (or, as shown in the following sections, of the forward-backward algorithm) is governed by the total number of branches, which is exponential in the length of the received pulses, L_q. Therefore, for typical wideband channels, the VA is unreasonably complicated [13]. Various schemes to reduce the VA's complexity have been proposed, and they usually involve searching only part of the trellis or simplifying the trellis.

Searching part of the trellis

Partway through a trellis processed by the VA, it can be shown that survivor sequences with metrics much worse than the best current survivor sequence are unlikely to end up forming part of the ML sequence. This observation leads us to the idea of searching the trellis more intelligently. In many cases, we find that an exhaustive search of the trellis with the VA is wasteful: other algorithms attain excellent performance at reduced complexity [14]. Their complete description is outside the scope of this book, but it worthwhile describing the *M-algorithm.*

Instead of preserving M^{L_q-1} survivors at the end of every symbol period, as in the VA, the M-algorithm keeps only a fixed number M_s, with $0 < M_s \leq M^{L_q-1}$. Next symbol period, each of the M_s survivors is extended along the M branches radiating from the ending state of each survivor. If more than one extended survivor enters the same next state, all but the best one are pruned, as in the VA. The survivors, numbering up to $M \cdot M_s$, are sorted by metric and further pruned so that only the best M_s are retained. The algorithm repeats this process each symbol period [15].

The M-algorithm is an example of *breadth-first* searching [16]: the algorithm views at once all the branches that it will ever view at that depth. Other search strategies are labelled *depth-first* (a single path is pursued continually until its metric exceeds a threshold), *metric-first* (the path with the instantaneously best metric is always pursued), or some hybrid of these. The most authoritative comparison of the single stack, Fano, 2-cycle, stack, merge, bucket and M- algorithms is [14]. Equalizer performance with the M-algorithm [17], [18], [19] and the Fano algorithm [20] has also been studied.

Simplifying the trellis

The received pulse has a peak around its middle, referred to as the *cursor*, energy tailing away beforehand, the *precursor*, and energy tailing away afterwards, the *postcursor*. The following complexity reduction schemes exploit this behavior.

For instance, in many cases the energy in the extreme tails of the precursor or postcursor is small and can be neglected without significant penalty. This neglect leads to *truncated sequence detection* (TSD) [21], [22], [23]. Mathematically, TSD defines the *truncated received pulses* as

$$\bar{q}^{tr}_{m,n} = \begin{cases} \bar{q}_{m,n} & 0 \leq n_s \leq n < n_f \\ 0 & n < n_s, n \geq n_f \end{cases} \tag{16.1.17}$$

where the parameters $n_s \geq 0$ and $n_f \leq L_q\eta$ control the truncation at the start and end of the pulses, as shown in Figure 16.4. The detector is then derived as if the received signal equaled (see (14.2.23))

$$r_k = \sum_{m=\lfloor (k-n_f+1)/\eta \rfloor}^{\lfloor (k-n_s)/\eta \rfloor} c_m \, \bar{q}^{tr}_{m,k-m\eta} + n_k \tag{16.1.18}$$

so that a part of the ISI contribution is neglected. Labelling the duration of $\bar{q}^{tr}_{m,n}$ in symbol periods as L^{tr}_q, with

$$L^{tr}_q = \left\lfloor \frac{n_f - n_s}{\eta} \right\rfloor + 1, \tag{16.1.19}$$

we have that $L_q^{tr} \le L_q$. For $n_s \gg 0$ and $n_f \ll L_q\eta$, $L_q^{tr} \ll L_q$ and the resulting trellis is considerably simpler.

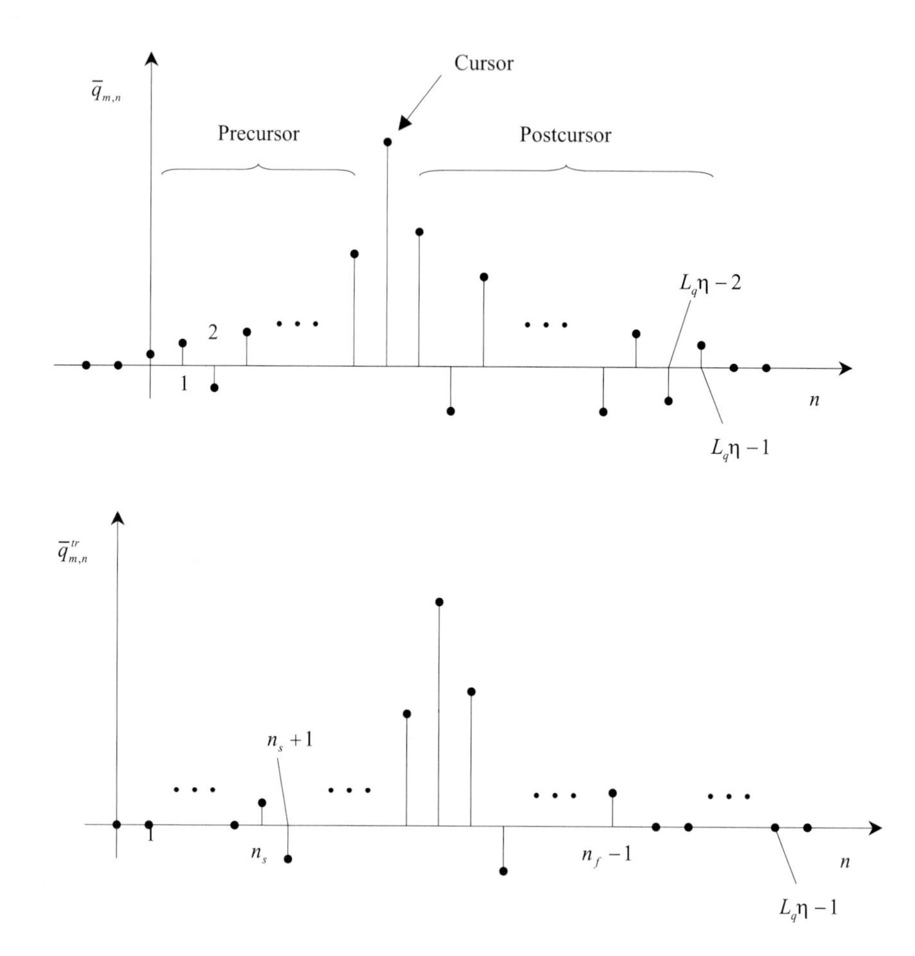

Figure 16.4. Time-varying OCIR and its truncated version

We also recall that the reliability of decisions within the survivor paths $\{\hat{x}(x_m)\}$ in the mth symbol interval increases with delay, and it was stated that beyond about 5—7 times L_q, there is no significant increase in reliability. Moreover, it is fair to say that *reliable* decisions are available much sooner: certainly after 2—3 times L_q, and even at delays of less than L_q as long as most of the received pulse's energy has been accounted for in the branch metrics. In fact, the path metrics $\{\Lambda(x_{m+1})\}$, evaluated in the mth symbol period, involve the mth trial symbol $\tilde{c}_m$ through the term $\tilde{c}_m\, \bar{q}_{m,k-m\eta}$ only for the received samples $\{r_k, k = m\eta, ..., (m+1)\eta - 1\}$ (see (16.1.4) and (16.1.8)), but the $(m-\Delta)$th symbol through $\tilde{c}_{m-\Delta}\, \bar{q}_{m-\Delta,k-(m-\Delta)\eta}$ over more samples, that is, $\{r_k, k = (m-\Delta)\eta, ..., (m+1)\eta - 1\}$. As long as the energy accounted for is much greater than the rest of the energy, namely,

$$\sum_{k=(m-\Delta)\eta}^{(m+1)\eta-1} \left|\bar{q}_{m-\Delta,k-(m-\Delta)\eta}\right|^2 \gg \sum_{k=(m+1)\eta}^{(L_q+m-1)\eta-1} \left|\bar{q}_{m-\Delta,k-(m-\Delta)\eta}\right|^2, \tag{16.1.20}$$

then reliable decisions on $\tilde{c}_{m-\Delta}$ are expected at the mth step.

With this result, we can describe *delayed decision-feedback sequence detection* (DDFSD) [24] and *reduced-state sequence detection* (RSSD) [25], [26].

In DDFSD, a reduced-state trellis is constructed. States in the VA's trellis, $\{x_m = (\tilde{c}_{m-L_q+1}, ..., \tilde{c}_{m-1})\}$, are collapsed together if they share the same "older" symbols. In other words, the DDFSD states are defined by the state vector

$$x_m^{RS} = (\tilde{c}_{m-L_{RS}+1}, ..., \tilde{c}_{m-1}) \tag{16.1.21}$$

where $L_{RS} \leq L_q$. The state definition leads to a trellis again, with branches and branch metrics associated with them. Trellis processing closely follows the VA, with survivor paths and survivor path metrics. The crucial difference of DDFSD over TSD is that in DDFSD the full branch metric of (16.1.4) (or (16.1.16)) is retained. However, for each state x_m and state transition labelled by the channel symbol $\tilde{c}_m$, the required symbols are obtained partly from the DDFSD state x_m^{RS} and partly (those unspecified by the state x_m^{RS}) from the corresponding survivor's path $\hat{x}\left(x_m^{RS}\right)$. This method represents an application of the *decision-feedback* concept [24], [27], [28], [22], [13]. As a first approximation, L_{RS} is designed to cover the precursor and cursor, and decision feedback is used for the postcursor.

RSSD is an elegant but minor extension of DDFSD. Instead of a full trellis for the precursor and a single decision history per survivor for the postcursor, set-partitioning principles [29] are applied to steadily reduce the number of hypothesized symbols as more of the received pulse arrives. In [30], still finer control of the number of states is achieved. One analysis of RSSD is presented in [25], and tighter performance bounds are provided in [31]. The general theme is gracefully degrading performance with decreasing L_{RS}.

Another way of working around a large L_q is to adaptively prefilter the signal to obtain shorter duration of the overall impulse response [32], by using a linear equalizer [33], [34] or a decision-feedback equalizer [35], [36], [37]. An MLSD based on the VA is then applied to this prefiltered signal. However, the prefilter colors the additive noise, thereby reducing performance if noise correlation is not taken into account [4], and the DFE prefilter exhibits error propagation. A hybrid structure that delivers the MLSD's soft outputs to the DFE is superior [38], [36], [27], but, in general, the performance attained with these prefiltering strategies has not been compelling.

Whitened matched filter

At this point, it is worthwhile introducing the *whitened matched filter* (WMF).[2] It can be viewed as the cascade of a matched filter, a symbol-rate sampler, and a symbol-spaced whitening filter [2]. These operations may replace or follow low pass filtering and fractional sampling. The resulting signal samples at the WMF output represent a set of sufficient statistics and are affected by additive white noise samples.

In addition, an important effect of WMF is redistributing the energy in the overall received pulse toward its front, although its duration remains unchanged. In short, the signal taken at the WMF output is far more amenable to DDFSD, since a relatively small value for L_{RS} still captures most of the received pulse energy. Unfortunately, the WMF is best suited to time-invariant channels and to slowly time-varying channels only by extension.

For compactness, the received pulse $\{\bar{q}_{m,k-m\eta}\}$, being time invariant, can be written simply as $\{\bar{q}_{k-m\eta}\}$. Its symbol-spaced autocorrelation $R_i \triangleq \bar{q}_{i\eta} \otimes \bar{q}^*_{-i\eta}$ and the autocorrelation's z-transform are given by

$$R_{i-m} = \sum_{k=\eta\,\max(i,m)}^{\eta\,\min(i,m)+L_q\eta-1} \bar{q}_{k-m\eta}\, \bar{q}^*_{k-i\eta} \tag{16.1.22}$$

and

$$R(z^{-1}) = \sum_{m=-L_q+1}^{L_q-1} R_m\, z^{-m}, \tag{16.1.23}$$

respectively. The autocorrelation sequence $\{R_i\}$ is Hermitian symmetric so that its z-transform satisfies the equality

$$R(z^{-1}) = R^*(z). \tag{16.1.24}$$

This entails that any root z_m of $R(z^{-1})$ is paired with the root z_m^{-1}, with the former inside the unit circle and the latter outside. Therefore, $R(z^{-1})$ admits the spectral factorization

$$R(z^{-1}) = f(z^{-1})\, f(z)^* \tag{16.1.25}$$

where $f(z^{-1})$ is an $(L_q - 1)$th order polynomial in z^{-1}containing only and all the roots that lie inside the unit circle. Then, it defines a *minimum phase* FIR filter. Inverting the z-transform (16.1.25) produces the following factorization of the pulse autocorrelation:

$$R_{i-m} = \sum_{p=\max(i,m)}^{\min(i,m)+L_q-1} f_{p-m}\, f^*_{p-i}. \tag{16.1.26}$$

[2]In most textbooks, the WMF is covered at the beginning of the study on channel equalization because of its profound significance in time-invariant channels. However, our focus is on wireless channels with appreciable time variation.

Let us now apply these results to derive another implementation of the MLSD strategy. In a time-invariant channel, the MLSD metric (16.1.3)-(16.1.4) can be rewritten as

$$\Lambda\left(\tilde{\mathbf{c}}\right)=\sum_{k=0}^{N_z-1}\left|r_k-\sum_{m=\lfloor k/\eta\rfloor-L_q+1}^{\lfloor k/\eta\rfloor}\tilde{c}_m\,\bar{q}_{k-m\eta}\right|^2. \tag{16.1.27}$$

Expanding (16.1.27), neglecting the term $\sum_{k=0}^{N_z-1}|r_k|^2$ (since it is independent of the trial sequence $\tilde{\mathbf{c}}$), and substituting (14.2.23) produces the equivalent MLSD metric

$$\Gamma\left(\tilde{\mathbf{c}}\right)=-2\Re\left(\sum_{m=0}^{L_c-1}\sum_{i=\max(0,m-L_q+1)}^{\min(L_c-1,m+L_q-1)}c_i^*\,\tilde{c}_m\,R_{i-m}+\sum_{m=0}^{L_c-1}\tilde{c}_m^*\sum_{k=m\eta}^{(m+L_q)\eta-1}n_k\,\bar{q}_{k-m\eta}^*\right)$$
$$+\sum_{m=0}^{L_c-1}\sum_{i=\max(0,m-L_q+1)}^{\min(L_c-1,m+L_q-1)}\tilde{c}_m\,\tilde{c}_i^*\,R_{i-m} \tag{16.1.28}$$

which is simply a reformulation of (16.1.13). The noise term

$$n_m'=\sum_{k=m\eta}^{(m+L_q)\eta-1}n_k\,\bar{q}_{k-m\eta}^* \tag{16.1.29}$$

in (16.1.28) is recognized as belonging to a colored noise sequence. We generate such a sequence by feeding a filter matched to the OCIR $\{\bar{q}_k\}$ with $\{n_k\}$ and sampling the filter output at symbol rate. The autocorrelation function of $\{n_m'\}$ is given by

$$R_{n'}\left[m-i\right]\triangleq\frac{1}{2}E\{n_m'\,n_i'^*\}=\frac{N_0}{T_\eta}R_{i-m} \tag{16.1.30}$$

assuming that the noise limiting filter is an ideal low pass filter with bandwidth $B_\eta=1/2T_\eta$. Due to the factorization (16.1.26) of $\{R_{i-m}\}$, the noise sequence $\{n_m'\}$ can be generated as

$$n_m'=\sum_{i=m}^{m+L_q-1}\nu_i\,f_{i-m}^*=\nu_m\otimes f_{-m}^* \tag{16.1.31}$$

where $\{\nu_m\}$ is a symbol spaced sequence of white noise samples, each with variance $\sigma_\nu^2=N_0/T_\eta$. This means that we can generate $\{n_m'\}$ by passing a white noise sequence $\{\nu_m\}$ through a filter whose z-transformed impulse response is $f'(z)^*$ (see (16.1.25)). Let us now define the inverse of such a filter as the *whitening filter* (WF). Then, the WF transfer function is

$$f_{WF}\left(z\right)\triangleq\frac{1}{f(z)^*} \tag{16.1.32}$$

and has all its poles inside the unit circle. Equation (16.1.32) entails that

$$f_m^{WF} \otimes f_{-m}^* = \delta_{m,0} \tag{16.1.33}$$

so that (see (16.1.31))

$$\nu_m = n'_m \otimes f_m^{WF}. \tag{16.1.34}$$

This means that the WF generates a white noise sequence $\{\nu_m\}$ when fed by the colored one $\{n'_m\}$.

Let us now define the *whitened matched filter* (WMF) as the cascade of a matched filter, a symbol-rate sampler, and a WF. The WMF output sample w_m in the mth symbol interval is given by

$$w_m = \left[r_k \otimes \bar{q}^*_{L_q\eta-1-k}\right]_{k=L_q\eta+m\eta} \otimes f_m^{WF} \sum_{i=\max\{0,m-L_q+1\}}^{m} c_i f_{m-i} + \nu_m. \tag{16.1.35}$$

It can be proved [2], [9] that

$$w_m = \sum_{i=\max\{0,m-L_q+1\}}^{m} c_i\, f_{m-i} + \nu_m \tag{16.1.36}$$

and that, if we define

$$\tilde{w}_m\left(x_m, x_{m+1}\right) = \sum_{i=\max\{0,m-L_q+1\}}^{m} \tilde{c}_i\, f_{m-i} \tag{16.1.37}$$

with $x_m = \left(\tilde{c}_{m-L_q+1}, ..., \tilde{c}_{m-1}\right)$, the metric $\Gamma\left(\tilde{\mathbf{c}}\right)$ (16.1.28) can be rewritten as

$$\Lambda\left(\tilde{\mathbf{c}}\right) = \sum_{m=0}^{L_c-1} \delta\left(x_m, x_{m+1}\right) \tag{16.1.38}$$

up to the term $\sum_{m=0}^{L_c-1} |w_m|^2$ (irrelevant, being independent of $\tilde{\mathbf{c}}$), with

$$\delta\left(x_m, x_{m+1}\right) \triangleq |w_m - \tilde{w}_m\left(x_m, x_{m+1}\right)|^2. \tag{16.1.39}$$

Equations (16.1.38) and (16.1.39) suggest an equivalent, but different, realization of the MLSD (for FS, time-invariant channels) to the ones obtained in Section 15.2.1 [2], [1]. This implementation is illustrated in Figure 16.5.

Finally, it is worth noting that (16.1.38) is verified by substituting into it (16.1.26), (16.1.31), (16.1.35) and (16.1.37) to obtain (16.1.28).

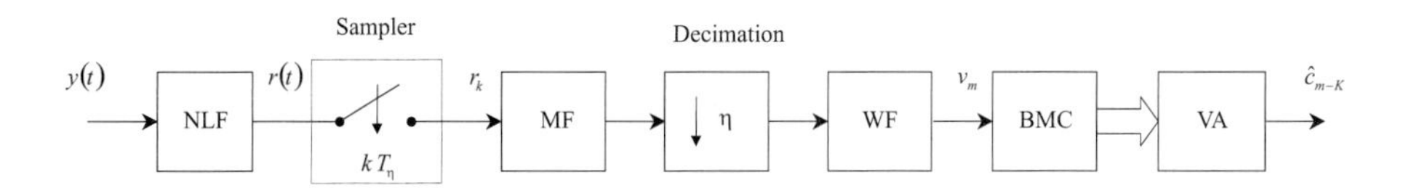

Figure 16.5. Forney's MLSD

The main point of these manipulations is that $\{w_m\}$, like $\{r_k\}$, also represents a set of sufficient statistics for ML sequence detection. Moreover, because the polynomial (in the variable z^{-1})

$$f(z^{-1}) = A \prod_{l=1}^{L_q - 1} \left(1 - \rho_l z^{-1}\right) \tag{16.1.40}$$

is a minimum phase function by design, we have $|\rho_m| < 1$ for any m. In this way, $f(z^{-1})$ tends to have larger coefficients associated with its lower powers of z^{-1} than with its higher powers. Thus, the energy in the sequence $\{f_m\}$ is concentrated at smaller values of m, making $\{w_m\}$ more amenable to DDFSD processing than is $\{r_k\}$.

16.1.3 Decision feedback equalizer

The decision feedback equalizer (DFE) is a special case of the prefilter plus DDFSD, where $L_{RS} = 1$ so that x_m^{RS} is an empty state vector. The trellis comprises one state with M branches, each returning to the same state. However, the DFE has a preeminent role in communications, thanks to its nice balance of complexity and performance. Therefore, we study it as a separate structure [39], [40], [41], [42].

In the DFE terminology, the prefilter is the *feedforward filter* (FFF), and the decision feedback of DDFSD is implemented by a *feedback filter* (FBF), as shown in Figure 16.6. In the time-varying channel, the impulse responses of FFF and FBF are time varying. Since a DFE's trellis has one state, a memoryless quantizer is sufficient as the trellis processor. As a simplification, the FFF in an $(L_f \cdot \eta)$-tap T_η-spaced FIR filter approximates a WMF to compensate for the channel's phase distortion, whereas the FBF an L_d-tap T-spaced FIR filter, adjusting the channel's amplitude distortion; in practice, both filters play a role in tackling both problems. Another viewpoint is illustrated in Figure 16.7 for a time-invariant channel and $\eta = 1$. The FFF shapes the final impulse response to be near zero until a large peak, the cursor; the FBF then eliminates the postcursor over the FBF's duration; and, finally, any postcursor beyond the end of the FBF is controlled via the FFF.

The impulse responses of the FFF and the FBF at the mth symbol period are given by

$$\mathbf{f}_m \triangleq \left[f_{m,0}, f_{m,1}, \ldots, f_{m,L_f\eta-1}\right]^T \tag{16.1.41}$$

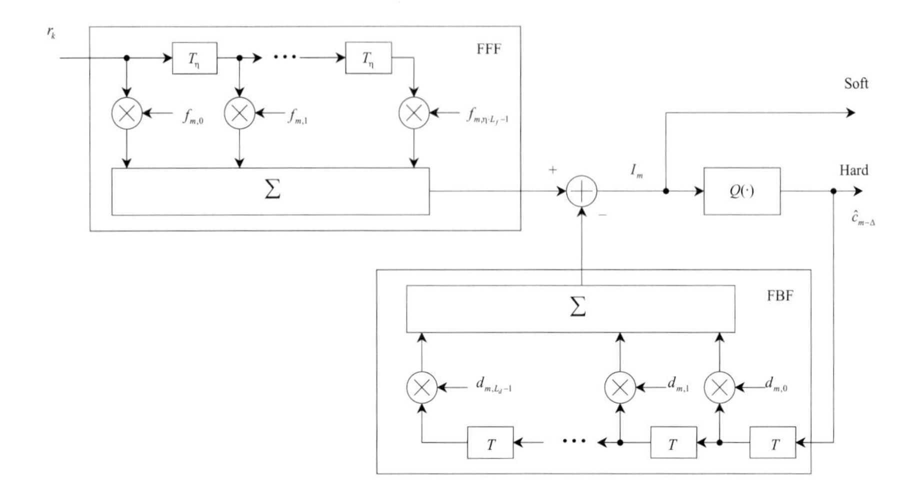

Figure 16.6. The decision-feedback equalizer. The function $Q(.)$ denotes the quantizer.

$$\mathbf{d}_m \triangleq \left[d_{m,0}, d_{m,1}, ..., d_{m,L_d-1}\right]^T, \tag{16.1.42}$$

respectively, where the feedforward filter is fractionally spaced [43], [44] and the feedback filter is symbol spaced. In the mth symbol interval, the fractionally spaced received samples within the span of the FFF are expressed by

$$\mathbf{r}_m = \mathbf{z}_m + \mathbf{n}_m = \mathbf{F}_{c,m}\,\mathbf{c}_m + \mathbf{n}_m \tag{16.1.43}$$

where

$$\mathbf{r}_m \triangleq \left[r_{(m-L_f+1)\eta}, r_{(m-L_f+1)\eta+1}, ..., r_{(m+1)\eta-1}\right]^T \tag{16.1.44}$$

$$\mathbf{z}_m \triangleq \left[z_{(m-L_f+1)\eta}, z_{(m-L_f+1)\eta+1}, ..., z_{(m+1)\eta-1}\right]^T \tag{16.1.45}$$

$$\mathbf{c}_m \triangleq \left[c_{m-L_q-L_f+2}, c_{m-L_q-L_f+3}, ..., c_m\right]^T \tag{16.1.46}$$

$$\mathbf{n}_m \triangleq \left[n_{(m-L_f+1)\eta}, ..., n_{(m+1)\eta-1}\right]^T \tag{16.1.47}$$

and $\mathbf{F}_{c,m}$ is defined as

$$\mathbf{F}_{c,m} \triangleq \left[\mathbf{q}^T_{(m-L_f+1)\eta}, \mathbf{q}^T_{(m-L_f+1)\eta+1}, ..., \mathbf{q}^T_{(m+1)\eta-1}\right]^T, \tag{16.1.48}$$

with

$$\mathbf{q}_k \triangleq \left[q_{k,k-(\lfloor k/\eta\rfloor-L_f-L_q+2)\eta}, q_{k,k+1-(\lfloor k/\eta\rfloor-L_f-L_q+2)\eta}, ..., q_{k,k-\lfloor k/\eta\rfloor\eta}\right]^T. \tag{16.1.49}$$

In the last equations, it is implicitly assumed that $q_{k,k-m\eta} \triangleq 0$ if $k - m\eta < 0$ or $k - m\eta \geqslant L_q\eta$ and that $c_m = 0$ for $m < 0$ or $m \geq L_c$. Equation (16.1.43) is similar

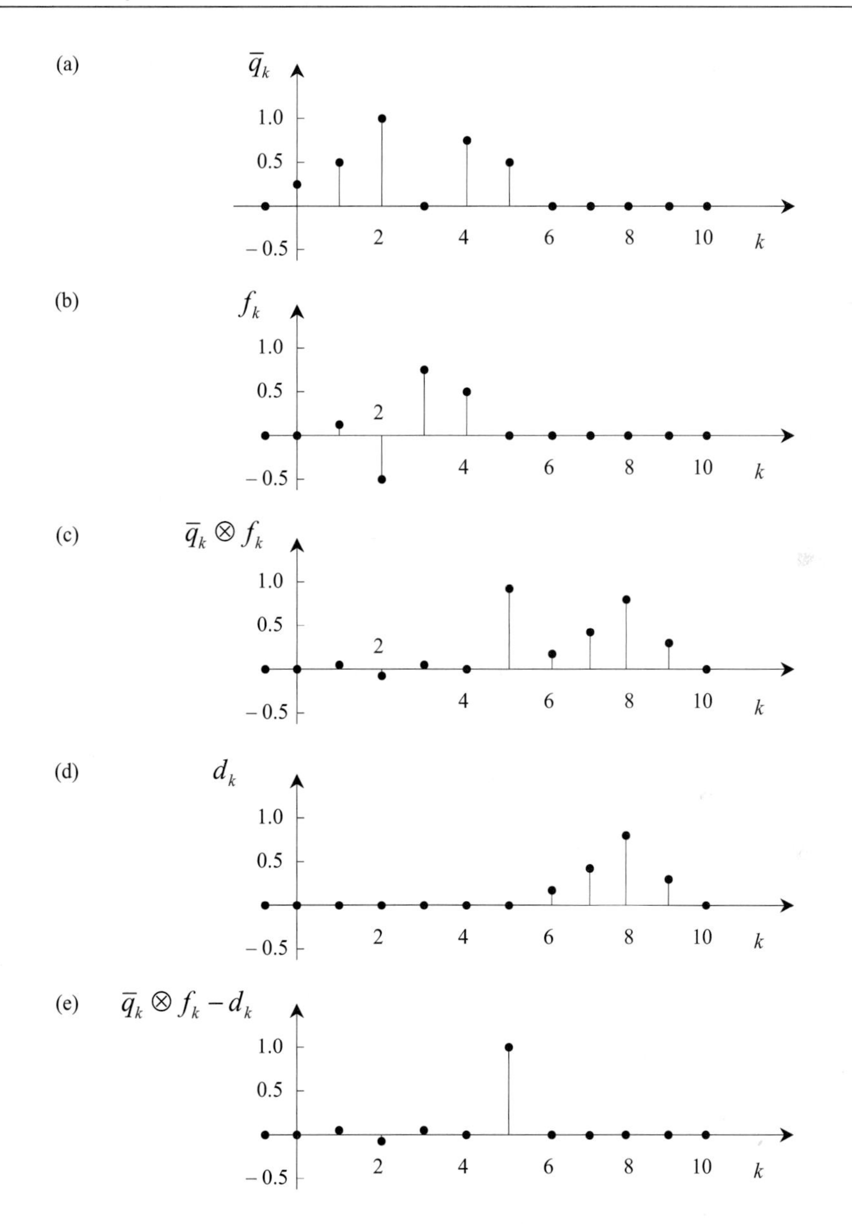

Figure 16.7. Impulse responses involved in a DFE: (a) channel; (b) feedforward filter; (c) cascade of CIR and FFF; (d) feedback filter; (e) channel and DFE

to (14.2.34), except now it is the channel parameters that make up the convolution matrix and the data symbols make up the vector of unknowns.

In any DFE there is always a delay Δ (in symbol periods) between the first received sample containing energy from a symbol and that symbol being detected. Such a parameter is the *decision delay* or *lag*. Therefore, the past decision vector $\hat{\mathbf{c}}_m$ employed in the FBF is given by

$$\hat{\mathbf{c}}_m = \left[\hat{c}_{m-\Delta-L_d}, \hat{c}_{m+1-\Delta-L_d}, ..., \hat{c}_{m-\Delta-1}\right]^T \tag{16.1.50}$$

and, if the decisions are correct, it can be put in the form

$$\hat{\mathbf{c}}_m = \mathbf{M}\,\mathbf{c}_m \tag{16.1.51}$$

when $L_d \leq L_q + L_f - \Delta - 2$ (i.e., the FBF is not pointlessly long), where

$$\mathbf{M} \triangleq \left[\mathbf{0}_{L_d, L_q+L_f-L_d-\Delta-2}, \mathbf{I}_{L_d}, \mathbf{0}_{L_d,\Delta+1}\right]. \tag{16.1.52}$$

From Figure 16.6, the input to the DFE's quantizer, $Q(.)$, is

$$I_m = \mathbf{f}_m^H\,\mathbf{r}_m - \mathbf{d}_m^H\,\hat{\mathbf{c}}_m, \tag{16.1.53}$$

and, ideally, we would like to have

$$Q\left(\mathbf{f}_m^H\,\mathbf{r}_m - \mathbf{d}_m^H\,\hat{\mathbf{c}}_m\right) = c_{m-\Delta} \tag{16.1.54}$$

where Δ must satisfy $0 \leq \Delta < L_q + L_f - 2$ from (16.1.43) and (16.1.46).

We optimize the DFE's performance by appropriately choosing L_f, L_d, Δ, $Q(.)$, $\mathbf{f}_m$, and $\mathbf{d}_m$. In known channels, L_f and L_d should be chosen to be as large as possible (i.e., infinite, in which case Δ becomes arbitrary). It is common to use a hard, nearest-neighbor quantizer (slicer) for $Q(.)$ and then design the DFE's filters so as to minimize the *mean-square error* (MSE) between the quantizer's input and output [41], [45], [46] under the assumption of correct past decisions, $\hat{\mathbf{c}}_m = \mathbf{c}_m$. The corresponding equalizer is called *minimum mean-square error - decision feedback* (MMSE-DFE).

The MMSE choice for the parameter vector $(\mathbf{f}_m, \mathbf{d}_m, \Delta)$ is the solution of the following problem:

$$(\mathbf{f}_m, \mathbf{d}_m, \Delta) = \arg \min_{\tilde{\mathbf{f}}_m, \tilde{\mathbf{d}}_m, \tilde{\Delta}} \frac{1}{2} E\left\{\left|c_{m-\tilde{\Delta}} - \tilde{\mathbf{f}}_m^H \mathbf{r}_m + \tilde{\mathbf{d}}_m^H \hat{\mathbf{c}}_m\right|^2 |\mathbf{F}_{c,m}\right\} \tag{16.1.55}$$

where the expectation is over the random data and channel noise but conditioned on the overall channel, $\mathbf{F}_{c,m}$. Substituting (16.1.43) and (16.1.51) into the cost function, expanding it, applying the expectation, differentiating the result with respect to $\tilde{\mathbf{f}}_m$ and $\tilde{\mathbf{d}}_m$, then setting it to zero yields

$$\mathbf{f}_m\left(\tilde{\Delta}\right) = \left[\mathbf{F}_{c,m}\left(\mathbf{I}_{L_q+L_f-1} - \mathbf{M}\left(\tilde{\Delta}\right)^H \mathbf{M}\left(\tilde{\Delta}\right)\right)\mathbf{F}_{c,m}^H + \frac{N_0}{T_\eta}\mathbf{I}_{L_f}\right]^{-1}$$
$$\cdot \mathbf{F}_{c,m}\,\mathbf{R}_{c,m}\left(\tilde{\Delta}\right) \tag{16.1.56}$$

$$\mathbf{d}_m\left(\tilde{\Delta}\right) = \mathbf{M}\left(\tilde{\Delta}\right) \mathbf{F}_{c,m}^{H} \mathbf{f}_m\left(\tilde{\Delta}\right) \tag{16.1.57}$$

$$\Delta = \arg\min_{\tilde{\Delta}} \frac{1}{2} E\left\{\left|c_{m-\tilde{\Delta}} - \mathbf{f}_m^H\left(\tilde{\Delta}\right) \mathbf{r}_m + \mathbf{d}_m^H\left(\tilde{\Delta}\right) \hat{\mathbf{c}}_m\right|^2\right\} \tag{16.1.58}$$

$$\mathbf{f}_m = \mathbf{f}_m\left(\Delta\right), \ \mathbf{d}_m = \mathbf{d}_m\left(\Delta\right) \tag{16.1.59}$$

where $(1/2)\, E\{|c_m|^2\} = 1$ by assumption,

$$\mathbf{R}_{c,m}\left(\tilde{\Delta}\right) \triangleq \frac{1}{2} E\left\{\mathbf{c}_m\, c_{m-\tilde{\Delta}}^*\right\} = \left[\mathbf{0}_{1,L_q+L_f-\tilde{\Delta}-2}, 1, \mathbf{0}_{1,\tilde{\Delta}}\right]^T \tag{16.1.60}$$

and

$$\mathbf{I}_{L_q+L_f-1} - \mathbf{M}\left(\tilde{\Delta}\right)^H \mathbf{M}\left(\tilde{\Delta}\right) = \operatorname{diag}\left(\left[\mathbf{1}_{1,L_q+L_f-L_d-\tilde{\Delta}-2}, \mathbf{0}_{1,L_d}, \mathbf{1}_{1,L_d,\tilde{\Delta}+1}\right]\right). \tag{16.1.61}$$

This last result shows that the FFF is designed without regard to the ISI nominally cancelled by the FBF. Since optimizing Δ involves calculating $\mathbf{f}_m(\tilde{\Delta})$ and $\mathbf{d}_m(\tilde{\Delta})$ for all allowed decision delays, a reasonable strategy is simply to preselect Δ near the middle of the FFF, as

$$\Delta \approx \frac{L_q + L_f - 1}{2}. \tag{16.1.62}$$

There are no closed form solutions or easily computed, tight bounds on the BER of the MMSE-DFEs, due to the non-Gaussianity of the residual ISI and the possibility of erroneous decisions being fed back through the FBF. In fact, it is for this reason that the FFF, FBF, and decision delay are designed with a minimum mean-squared error criterion in (16.1.55). Certain analytic techniques are available in the literature that account for error propagation [40], [47], [48], [49], [50], [51] or ignore it [52], [53], [54]. Monte Carlo simulation should always be considered to be a complementary technique. The simplest performance measure is the DFE's mean-squared error, assuming correct past decisions. It is calculated by substituting (16.1.56) and (16.1.57) into the cost function in (16.1.55) [41]. Furthermore, by assuming the error (see (16.1.53))

$$E_m \triangleq c_{m-\tilde{\Delta}} - I_m = c_{m-\tilde{\Delta}} - \mathbf{f}_m^H \mathbf{r}_m + \mathbf{d}_m^H \hat{\mathbf{c}}_m \tag{16.1.63}$$

to be Gaussian, we can calculate an optimistic BER estimate.

The "correct past decision" assumption has been repeatedly invoked, but has to be recognized as a strong assumption. In low SNRs, noise may cause a primary error, which is fed back into the FBF. Instead of cancelling the postcursor, the FBF can actually enhance it, which in turn increases the likelihood of subsequent, secondary errors. This error propagation phenomenon can be severe in wideband channels since the FBF may be hundreds of symbols long. More detailed studies are found in [55], [56], [57], [58], [59], [60].

A number of structural modifications to the classical fractionally spaced DFE have already been presented, including the use of a symbol-spaced FFF and/or a fractionally spaced FBF [39], [61], the predictor form [39], [62], a soft quantizer [63], [64], and a reduced complexity arrangement [13]. DFEs are also used in equalizing other signals such as CPM, where ISI is due in part to the memory in the modulation or demodulation scheme rather than to a dispersive channel [65], [66], [67]. Block equalizers, for short transmissions between training sequences, may also exploit past decisions [68], [69].

For wideband communications, reduced complexity DFEs for *sparse* channels are most interesting [70]. A sparse channel is one where there are a significant proportion of zero or near-zero OCIR taps relative to the total number of OCIR taps $(L_q \cdot \eta)$. In this case, the conventional DFE structure is unhelpful in that the FFF smears out the non-zero taps over $(L_f \cdot \eta)$ taps so that the convolution of OCIR and FFF tends not to be sparse. Figures 16.8 and 16.9 present two alternative feedback arrangements more suited to sparse channels. In the "partial" design (illustrated in Figure 16.8), as soon as a symbol is detected, its postcursor is subtracted from the received samples by FBF2, before it reaches the FFF. Some postcursor is already present in the FFF, which motivates the "complete" design, shown in Figure 16.9. In this structure, the postcursor is eliminated both before and from within the FFF as soon as decisions are made.

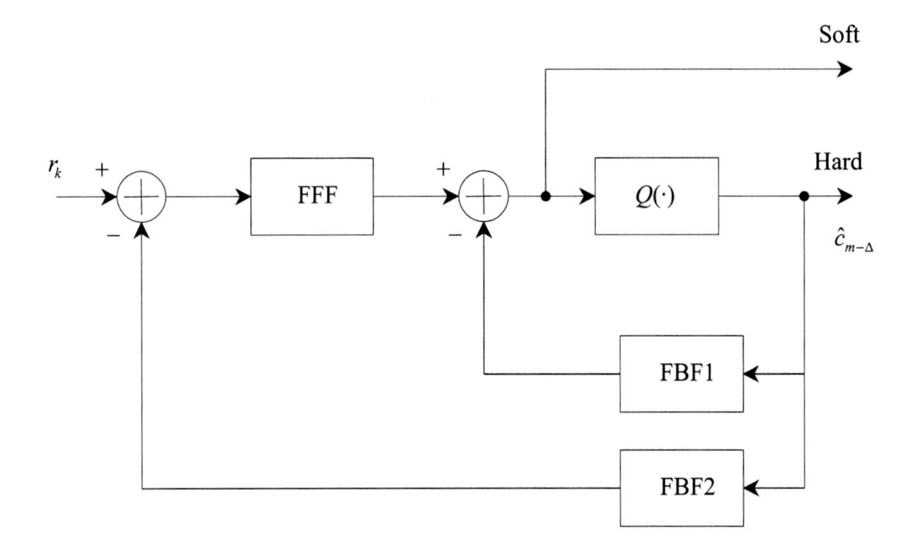

Figure 16.8. Structural modifications of the DFE for sparse channels: the *partial* DFE (PDFE)

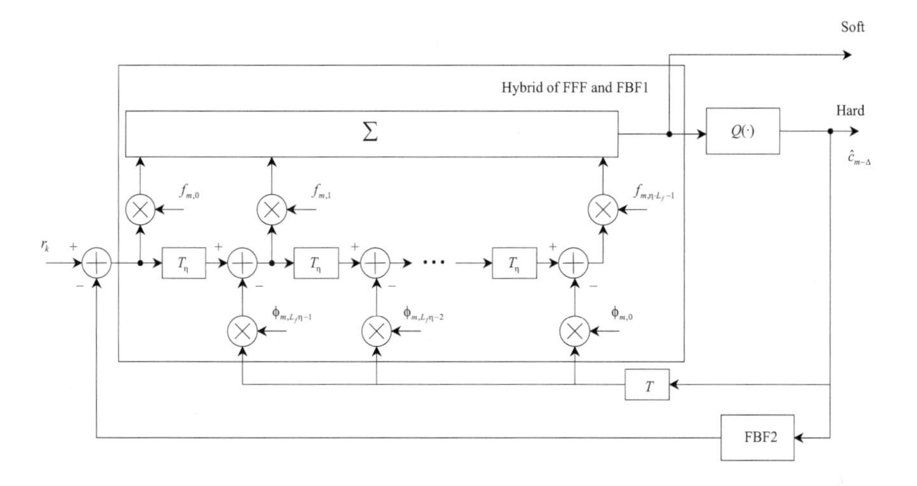

Figure 16.9. Structural modifications of the DFE for sparse channels: the *complete* DFE (CDFE)

16.1.4 Linear equalizer

The linear equalizer (LE) is a linear, normally transversal, filter followed by a nearest neighbor quantizer. The LE is the original equalizer and, arguably, the most intuitive. In the AWGN channel, matched filtering and symbol-rate sampling is optimal; when the channel is also frequency selective, then it is reasonable to incorporate an inverse or inverse-like filter. This is the LE. When it follows the matched filter and symbol-rate sampler, its taps are T-spaced. Alternatively, the LE may operate directly on the received samples, where it jointly performs matched filtering and channel equalization. In this case, its taps are fractionally spaced or T_η-spaced [39]. Both the symbol-spaced and the fractionally spaced structures are illustrated in Figure 16.10.

LEs are also distinguished according to their design criteria. In applications, two criteria are usually considered: the zero forcing[3] (ZF) and the *minimum mean-square error* MMSE one. The ZF method neglects the presence of the channel noise and minimizes the ISI contribution in the equalizer time span. This behavior is undesirable since wideband wireless channels are characterized by multiple, deep notches in frequency. In inverting these, a zero-forcing LE inevitably causes undue, considerable noise amplification and a degraded BER. For this reason, MMSE is a better design criterion, where the combination of noise and ISI is minimized. Both criteria are considered in the following.

Given the signal model of (16.1.43)—(16.1.48), the inputs to the quantizer of an

[3]The ZF criterion can be also applied in the design of the DFE equalizers. In this case, the noise contribution in (16.1.55) must be neglected.

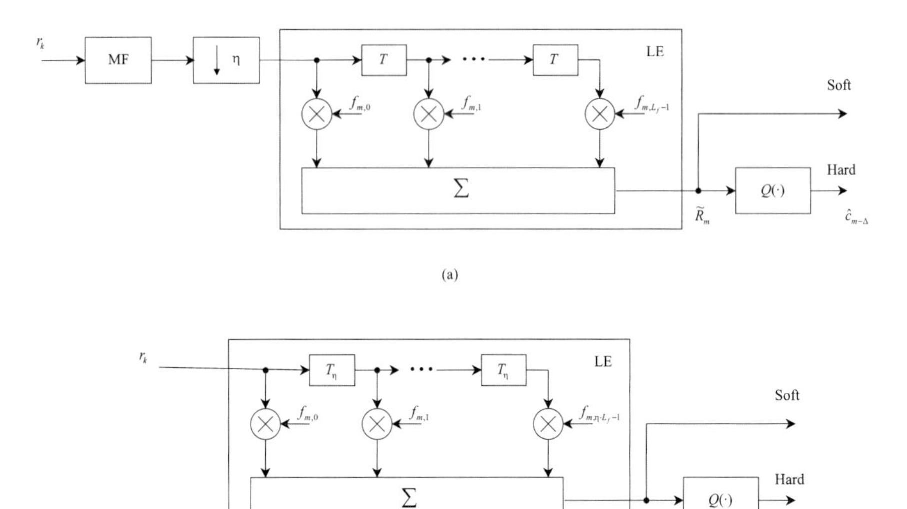

Figure 16.10. Structure of the linear equalizer: (a) symbol-spaced LE; (b) fractionally spaced LE

L_f-tap time-varying, fractionally spaced LE are given by

$$\tilde{R}_m \triangleq \mathbf{f}_m^H \, \mathbf{r}_m \tag{16.1.64}$$

where $\mathbf{f}_m$ is defined in (16.1.41). Then, a *zero-forcing* (ZF) is designed according to

$$(\mathbf{f}_m, \Delta) = \arg \min_{\tilde{\mathbf{f}}_m, \tilde{\Delta}} \frac{1}{2} E\left\{ \left| c_{m-\tilde{\Delta}} - \tilde{\mathbf{f}}_m^H \, \mathbf{z}_m \right|^2 \right\} \tag{16.1.65}$$

where $\tilde{\mathbf{f}}_m$ and $\tilde{\Delta}$ are trial values of $\mathbf{f}_m$ and Δ, respectively. Solving (16.1.65) leads to

$$\mathbf{f}_m\left(\tilde{\Delta}\right) = \left(\mathbf{F}_{c,m} \, \mathbf{F}_{c,m}^H\right)^{-1} \mathbf{F}_{c,m} \, \mathbf{R}_{c,m}\left(\tilde{\Delta}\right) \tag{16.1.66}$$

$$\Delta = \arg \min_{\tilde{\Delta}} \frac{1}{2} E\left\{ \left| c_{m-\tilde{\Delta}} - \mathbf{f}_m^H\left(\tilde{\Delta}\right) \mathbf{z}_m \right|^2 \right\} \tag{16.1.67}$$

$$\mathbf{f}_m = \mathbf{f}_m\left(\Delta\right) \tag{16.1.68}$$

for a full-rank matrix $\left(\mathbf{F}_{c,m} \, \mathbf{F}_{c,m}^H\right)$, where $\mathbf{R}_{c,m}\left(\tilde{\Delta}\right) \triangleq (1/2) \, E\{\mathbf{c}_m \, c_{m-\tilde{\Delta}}^*\}$ is expressed by (16.1.60). The channel convolution matrix $\mathbf{F}_{c,m}$ has dimension $(L_f \, \eta +$

$L_q - 1$), so a necessary condition is that there be fewer LE taps than symbols within the LE's span, that is,

$$L_f \eta \leq L_f + L_q - 1. \tag{16.1.69}$$

In a ZF equalizer, the minimum residual mean squared ISI equals zero when

$$\tilde{\mathbf{f}}_m^H \mathbf{F}_{c,m} \mathbf{c}_m = c_{m-\tilde{\Delta}} \tag{16.1.70}$$

or, equivalently,

$$\mathbf{F}_{c,m}^H \tilde{\mathbf{f}}_m = \left[\mathbf{0}_{1,L_q+L_f-\Delta-2}, 1, \mathbf{0}_{1,\Delta}\right]^T. \tag{16.1.71}$$

The channel convolution matrix $\mathbf{F}_{c,m}$ has dimension $(L_f \eta + L_q - 1)$, so a sufficient condition for (16.1.71) to hold is for $\mathbf{F}_{c,m}$ to satisfy

$$\text{rank}\left(\mathbf{F}_{c,m}\right) \geq L_q + L_f - 1 \tag{16.1.72}$$

which in turn requires that there be more LE taps than symbols within the LE's span, i.e., $(L_f \eta \geq L_f + L_q - 1)$. When the ZF LE can completely eliminate ISI [71], then the residual equalization error

$$\tilde{E}_m \triangleq c_{m-\tilde{\Delta}} - \tilde{R}_m = c_{m-\tilde{\Delta}} - \mathbf{f}_m^H \mathbf{r}_m \tag{16.1.73}$$

is purely Gaussian noise, i.e.,

$$\tilde{E}_m = \mathbf{f}_m^H \mathbf{n}_m \tag{16.1.74}$$

with variance $\sigma_{\tilde{E}}^2 = (N_0/T_\eta)\, \mathbf{f}_m^H \mathbf{f}_m$, and computing the symbol error rate is a straightforward calculation.

An MMSE equalizer is designed according to

$$(\mathbf{f}_m, \Delta) = \arg\min_{\tilde{\mathbf{f}}_m, \tilde{\Delta}} \frac{1}{2} E\left\{\left|c_{m-\tilde{\Delta}} - \tilde{\mathbf{f}}_m^H \mathbf{r}_m\right|^2\right\}. \tag{16.1.75}$$

This expression is identical to (16.1.55) with no FBF, i.e., $L_f = 0$. Therefore, the DFE design equations (16.1.56)—(16.1.59) are equally applicable to the fractionally spaced MMSE LE. Substituting (16.1.43) and (16.1.56) into (16.1.75) yields the MMSE as

$$\begin{aligned} \text{MMSE}\left(\tilde{\Delta}\right) &= \frac{1}{2} E\left\{\left| c_{m-\tilde{\Delta}} - \mathbf{R}_{c,m}\left(\tilde{\Delta}\right)^H \mathbf{F}_{c,m}^H \left[\mathbf{F}_{c,m} \mathbf{F}_{c,m}^H + \frac{N_0}{T_\eta} \mathbf{I}_{L_f}\right]^{-1} \right.\right. \\ &\qquad \left.\left. (\mathbf{F}_{c,m} \mathbf{c}_m + \mathbf{n}_m)\right|^2\right\} \\ &= 1 - \mathbf{R}_{c,m}\left(\tilde{\Delta}\right)^H \mathbf{F}_{c,m}^H \left[\mathbf{F}_{c,m} \mathbf{F}_{c,m}^H + \frac{N_0}{T_\eta} \mathbf{I}_{L_f}\right]^{-1} \mathbf{F}_{c,m} \mathbf{R}_{c,m}\left(\tilde{\Delta}\right). \end{aligned} \tag{16.1.76}$$

When the LE has an infinite number of T_η-spaced taps, it can synthesize any transfer function over the frequency interval $(-1/2T_\eta, 1/2T_\eta)$, and the LE coefficients $\mathbf{f}_m$ for an isolated pulse $\{\bar{q}_{m-\Delta,k-(m-\Delta)\eta}\}$ have a nice frequency domain interpretation. Let $\bar{F}_m(f)$ be the Fourier transform of $\mathbf{f}_m$ and $\bar{Q}_{m-\Delta}(f)$ be the Fourier transform of the sequence $\{\bar{q}_{m-\Delta,k-(m-\Delta)\eta}\}$ with respect to its second subscript parameter. Since both quantities are sampled every T_η seconds, their Fourier transforms are periodic with period $1/T_\eta$ Hz. The cascade of transmitter, channel, fractionally spaced LE, and symbol-rate sampler has the aliased spectrum $\bar{A}_m(f)$

$$\bar{A}_m(f) = \sum_{n=0}^{\eta-1} \bar{Q}_{m-\Delta}\left(f - \frac{n}{T}\right) \bar{F}_m\left(f - \frac{n}{T}\right). \tag{16.1.77}$$

It can be shown that the ZF and MMSE-LEs have the respective transfer functions [39]

$$\bar{F}_{ZF,m}(f) = \frac{\bar{Q}^*_{m-\Delta}(f)}{\sum_{i=0}^{\eta-1} \left|\bar{Q}_{m-\Delta}\left(f - \frac{i}{T}\right)\right|^2} \tag{16.1.78}$$

$$\bar{F}_{MMSE,m}(f) = \frac{\bar{Q}^*_{m-\Delta}(f)}{\sum_{i=0}^{\eta-1} \left|\bar{Q}_{m-\Delta}\left(f - \frac{i}{T}\right)\right|^2 + N_0} \tag{16.1.79}$$

so that the output spectra of the ZF-LE is unity, whereas that of the MMSE-LE is

$$\bar{O}_{MMSE,m}(f) = \frac{\sum_{m=0}^{\eta-1} \left|\bar{Q}_{m-\Delta}\left(f - \frac{m}{T}\right)\right|^2}{\left|\bar{Q}_{m-\Delta}\left(f - \frac{m}{T}\right)\right|^2 + N_0}. \tag{16.1.80}$$

From (16.1.78) and (16.1.79) it is easily inferred that the LE's transfer function performs matched filtering through the numerator $\bar{Q}^*_{m-\Delta}(f)$, then it compensates for the ISI present via the denominator. In the ZF case, it exactly inverts the channel's aliased transfer function

$$\bar{C}_{ZF,m}(f) = \sum_{i=0}^{\eta-1} \left|\bar{Q}_{m-\Delta}\left(f - \frac{i}{T}\right)\right|^2 \tag{16.1.81}$$

to satisfy Nyquist's (first) criterion for ISI-free transmission; in the MMSE case, the denominator contains N_0 to prevent the LE's gain from getting too large at the frequencies where the aliased transfer function nears zero.

Certain extensions to the basic LE structure are presented in [72], [73], [74].

16.1.5 Diversity reception over frequency-flat channels

Let us now consider reception of a linearly modulated signal (14.2.1) over a known frequency-flat fading channel. For generality, we assume Lth order diversity at the receiver,[4] i.e., that L faded replicas of transmitted signal are available at the receiver side [75], [76]. Then, if the overall transmitter and receiver filtering is Nyquist and the fading distortion is slow [77], the symbol-rate samples at the matched filter output of the l-th branch can be expressed as (see (14.2.18))

$$r_k(l) = \alpha_k(l)\, c_k + n_k(l),\ l = 1, 2, ..., L_c \tag{16.1.82}$$

where $\alpha_k(l)$ represents the fading distortion in the kth symbol interval, c_k is the kth transmitted symbol and $n_k(l)$ is AWGN with variance $\sigma_n^2(l)$. Under these assumptions, the ML rule (15.2.1), generalized to diversity reception, becomes

$$\hat{\mathbf{c}} = \arg\max_{\tilde{\mathbf{c}}} p(\mathbf{r}_1, \mathbf{r}_2, ..., \mathbf{r}_L | \tilde{\mathbf{c}}, \boldsymbol{\alpha}_1, \boldsymbol{\alpha}_2, ..., \boldsymbol{\alpha}_L) \tag{16.1.83}$$

where $\mathbf{r}_l \triangleq [r_0(l), r_1(l), ..., r_{L_c-1}(l)]$ and $\boldsymbol{\alpha}_l \triangleq [\alpha_0(l), \alpha_1(l), ..., \alpha_{L_c-1}(l)]$.

If the noise processes $\{n_k(l),\ l = 1, 2, ..., L\}$ are assumed to be mutually independent, the decision rule (16.1.83) becomes

$$\hat{\mathbf{c}} = \arg\max_{\tilde{\mathbf{c}}} \prod_{l=1}^{L} p(\mathbf{r}_l | \tilde{\mathbf{c}}, \boldsymbol{\alpha}_l) \tag{16.1.84}$$

or, equivalently, (see (2.2.5))

$$\hat{\mathbf{c}} = \arg\min_{\tilde{\mathbf{c}}} \sum_{l=1}^{L} (\mathbf{r}_l - \mathbf{F}_\alpha(\tilde{\mathbf{c}})\, \boldsymbol{\alpha}_l)^H\, \mathbf{R}_n^{-1}(l)\, (\mathbf{r}_l - \mathbf{F}_\alpha(\tilde{\mathbf{c}})\, \boldsymbol{\alpha}_l) \tag{16.1.85}$$

where $\mathbf{R}_n(l)$ is the noise covariance matrix for the lth branch and $\mathbf{F}_\alpha(\tilde{\mathbf{c}}) = diag(\tilde{c}_k)$. If the noise samples on each diversity branch are independent, i.e., $\mathbf{R}_n(l) = \sigma_n^2(l)\, \mathbf{I}_{L_c}$, the decision rule (16.1.85) becomes

$$\hat{\mathbf{c}} = \arg\min_{\tilde{\mathbf{c}}} \sum_{l=1}^{L} \frac{1}{\sigma_n^2(l)} \left\|\mathbf{r}_l - \mathbf{F}_\alpha(\tilde{\mathbf{c}})\, \boldsymbol{\alpha}_l\right\|^2 \tag{16.1.86}$$

or, equivalently,

$$\hat{c}_k = \arg\min_{\tilde{c}_k} \sum_{l=1}^{L} \frac{1}{\sigma_n^2(l)} \left|r_l - \alpha_k(l)\, c_k\right|^2,\ k = 1, 2, ..., L_c. \tag{16.1.87}$$

Dropping irrelevant terms in (16.1.87) produces

$$\hat{c}_k = \arg\min_{\tilde{c}_k} \sum_{l=1}^{L} \frac{1}{\sigma_n^2(l)} \left[|\alpha_k(l)|^2\, |c_k|^2 - 2\,\mathrm{Re}\left(r_l\, \alpha_k^*(l)\, c_k^*\right)\right],\ k = 1, 2, ..., L_c. \tag{16.1.88}$$

[4]Diversity reception on FF fading channels is a classic problem in communications history.

If the transmitted symbol belongs to a PSK constellation, the strategy (16.1.88) can be also rewritten as

$$\hat{c}_k = \arg \max_{\tilde{c}_k} \mathrm{Re}\left[Z_k\, c_k^*\right],\ k = 1, 2, ..., L_c \tag{16.1.89}$$

where

$$Z_k \triangleq \sum_{l=1}^{L} g_k\left(l\right) r_l,\ k = 1, 2, ..., L_c \tag{16.1.90}$$

and

$$g_k\left(l\right) \triangleq \frac{\alpha_k^*\left(l\right)}{\sigma_n^2\left(l\right)},\ k = 1, 2, ..., L_c,\ l = 1, 2, ..., L. \tag{16.1.91}$$

The algorithm (16.1.89) linearly combines (with coefficients (16.1.91)) all the matched filter outputs in the kth symbol interval to take an optimal decision on c_k, as illustrated in Figure 16.11. It is worth noting the following:

1. Multiplying the sample r_l by $\alpha_k^*\left(l\right)$ removes the channel phase distortion from the received signal. Therefore, the algorithm (16.1.89)—(16.1.91) can be interpreted as a form of coherent detection.

2. In the evaluation of the sufficient statistic Z_k (16.1.90), a larger weight is given to less noisy channels.

Concerning the last point, it can be proved [75] that the choice of the coefficients expressed by (16.1.90), among all the possible ones, maximizes the signal-to-noise ratio at the decision device input. For this reason, (16.1.89)—(16.1.91) express the *maximal ratio combining* (MRC) detection strategy [75], [78]. The use of the MRC requires knowledge of both the channel gains $\{\alpha_k\left(l\right)\}$ and the noise statistics $\{\sigma_n^2\left(l\right)\}$. If the last quantities are unknown, the choice

$$g_k\left(l\right) = \alpha_k^*\left(l\right),\ k = 1, 2, ..., L_c,\ l = 1, 2, ..., L \tag{16.1.92}$$

can be adopted. The strategy expressed by (16.1.89), (16.1.90), and (16.1.92) is known as *equal-gain combining* (ECG) and is suboptimal. The implementation of the MRC and EG strategies is illustrated in Figure 16.11.

Another well-known suboptimal strategy for diversity reception is the so-called *selection combining* (SC), in which the branch signal with the largest amplitude (or signal-to-noise ratio) is selected for demodulation. Clearly, SC and MRC (or ECG) represent the two extremes in diversity combining with respect to the number of signals used for detection. The SC strategy can be generalized to combine n of M branches with the largest amplitudes (multiple order SC).

A derivation of the pdfs of the signal-to-noise ratio at the output of maximal-ratio and equal-gain combiners is given in the textbook [75]. Error probability formulas for FSK and PSK signals with MRC and ECG are available in [79]. A comparison of diversity combining techniques (including multiple order SC) is given in [80]. A unified approach to evaluating the error performance of digital communication systems with diversity has been recently proposed by Simon and Alouini for both coherent and noncoherent detection [80].

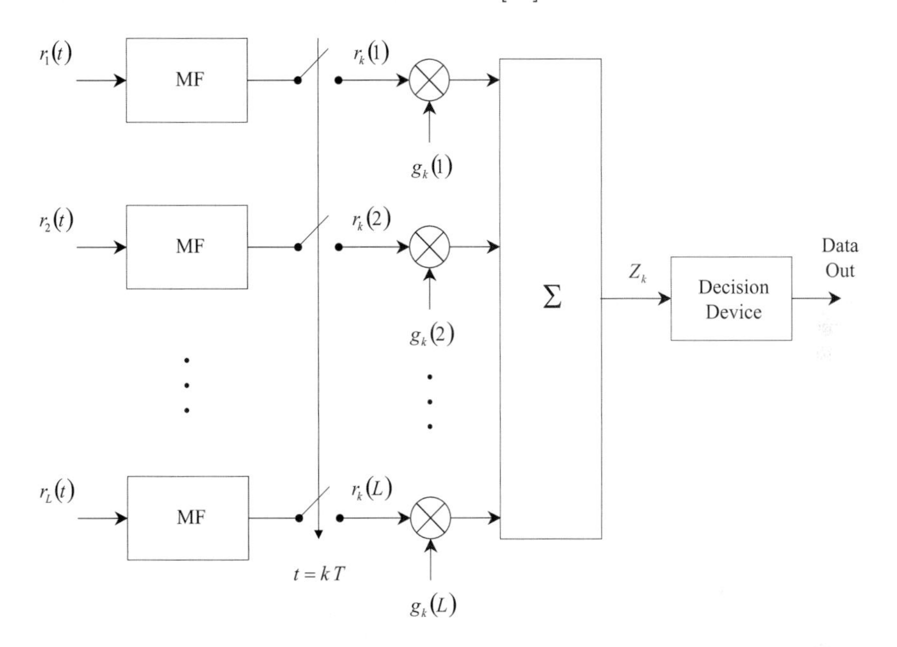

Figure 16.11. Block diagram of a combining system

All the strategies mentioned above exploit the *explicit* diversity provided by independently fading replicas of the same transmitted signal [75]. Such replicas can be provided by means of different mechanisms, including

- Space (spaced-antennas) diversity
- Frequency diversity
- Angle (of arrival) diversity
- Polarization diversity
- Time (signal repetition) diversity

A fading channel can itself be interpreted as a source of *implicit* diversity [81] to be related to both the phenomena of time dispersion [82],[83], [8], [84] and time variation of the channel [85]. In particular, exploiting the time diversity on FF fading channels can provide a substantial performance improvement [86], [87] (see also Section 16.3.2).

16.1.6 MAPBD/MAPSD: Forward-backward algorithm

When the MAP criterion replaces the ML criterion, the MAP *forward-backward algorithm* (MAP-FBA) replaces the VA [1]. The forward-backward operates on the same trellis as the VA, but it efficiently calculates the MAP bit or symbol probabilities, instead of just finding the ML sequence. The forward-backward algorithm requires all the received samples to be available before decisions can be made. If a large decision delay is to be avoided, a near-optimal, forward-only (fixed-lag) MAP strategy can be adopted. However, both MAP algorithms work with likelihoods rather than log-likelihoods, so their branch metrics are generally more expensive to compute and their operations are multiplication and addition instead of the addition and minimization characterizing the Viterbi algorithm. Much work on reduced complexity implementations has been undertaken, e.g., [11], [12], [88], [89], [90], [91], but this subject is outside the scope of this book.

Let us now describe the MAP algorithms mentioned above. A time-invariant, frequency-selective fading channel is assumed for simplicity.

The forward-backward algorithm

The FBA seeks the probabilities

$$P\left(c_m = \tilde{c}|\mathbf{r}, \bar{\mathbf{q}}\right), \; m = 0, 1, ..., L_c - 1 \tag{16.1.93}$$

for any possible trial symbol $\tilde{c}$ transmitted in the mth signalling interval. Here, it is assumed that the channel memory is not longer than $(L_q - 1)$ symbol periods, so any received signal sample depends on L_q consecutive channel symbols at most. Then, the FBA operates on the same M^{L_q-1}-state trellis as the VA and performs two basic recursions on the stream of received signal samples $\mathbf{r}$ to evaluate probabilities (16.1.93). In the following discsussion, keep in mind:

1. Each node in the trellis has M input branches and M output branches, with a branch corresponding to one of the M data symbols.

2. It is useful to associate the computed values of the FBA with nodes, states, and transitions in the trellis.

The evaluation of the probabilities (16.1.93) requires the computation of intermediate quantities, known as *state transition probabilities.* Given the trellis states x_m and x_{m+1} in the mth and in the $(m+1)$th symbol interval, respectively, the corresponding *state transition probability* is given by

$$P\left(x_m, \, x_{m+1}|\mathbf{r}, \bar{\mathbf{q}}\right) \tag{16.1.94}$$

and equals zero if the states are not connected. This quantity can be evaluated as

$$P\left(x_m, \, x_{m+1}|\mathbf{r}, \bar{\mathbf{q}}\right) = \frac{\sum_{\tilde{\mathbf{c}} \in C(x_m, x_{m+1})} P\left(\tilde{\mathbf{c}}|\mathbf{r}, \bar{\mathbf{q}}\right)}{\sum_{\tilde{\mathbf{c}} \in C} P\left(\tilde{\mathbf{c}}|\mathbf{r}, \bar{\mathbf{q}}\right)} \tag{16.1.95}$$

where $C(x_m, x_{m+1})$ is the subset (of the set $C \triangleq \{\tilde{\mathbf{c}}\}$) consisting of all the possible trial sequences that traverse the trellis branch connecting the states x_m and x_{m+1} (see Figure 16.12).

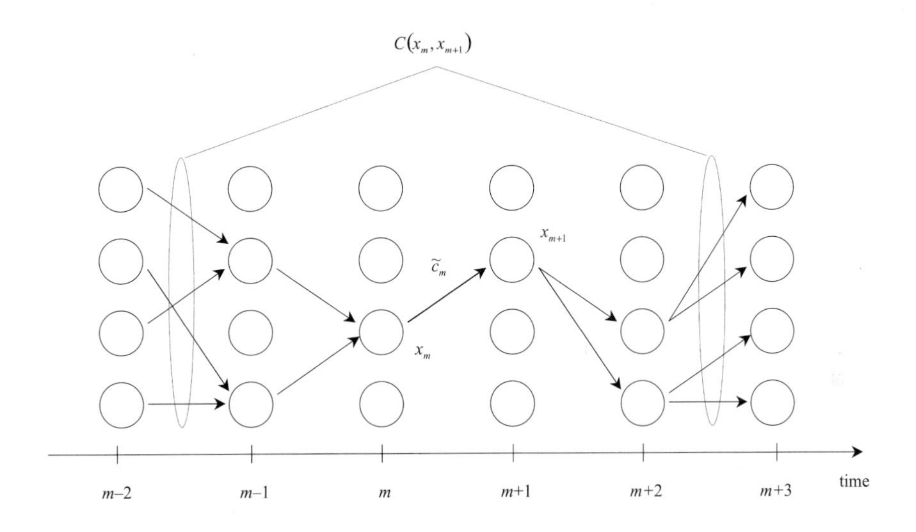

Figure 16.12. Subset $C_m(s, \tilde{s})$ of all the possible trial sequences traversing the trellis branch between the states x_m and x_{m+1}. $M = 4$ and $L_q = 2$ are assumed.

The probabilties $\{P(\tilde{\mathbf{c}}|\mathbf{r}, \bar{\mathbf{q}})\}$ are required in (16.1.95) and can be expanded by Bayes' theorem as

$$P(\tilde{\mathbf{c}}|\mathbf{r}, \bar{\mathbf{q}}) = \frac{p(\mathbf{r}|\tilde{\mathbf{c}}, \bar{\mathbf{q}})\, P(\tilde{\mathbf{c}})}{p(\mathbf{r}|\bar{\mathbf{q}})}. \tag{16.1.96}$$

If (16.1.96) is substituted into (16.1.95), all the terms have the same denominator $p(\mathbf{r}|\bar{\mathbf{q}})$, which, therefore, can be ignored. In evaluating the numerator of (16.1.96), we assume that all the channel symbols are independent, so that

$$P(\tilde{\mathbf{c}}) = \prod_{m=0}^{L_c-1} P(\tilde{c}_m). \tag{16.1.97}$$

Moreover, the first factor in the numerator of (16.1.96) can be evaluated as

$$p(\mathbf{r}|\tilde{\mathbf{c}}, \bar{\mathbf{q}}) = \prod_{m=0}^{L_c-1} \prod_{i=0}^{\eta-1} p(r_{m\,\eta+i}|\tilde{\mathbf{c}}, \bar{\mathbf{q}}) = \prod_{m=0}^{L_c-1} \prod_{i=0}^{\eta-1} p(r_{m\,\eta+i}|\tilde{c}_m, x_m, \bar{\mathbf{q}}) \tag{16.1.98}$$

because the received signal samples $\{r_{m\,\eta+i}, i = 0, 1, ..., \eta - 1\}$ in the mth symbol interval depend on the mth symbol $\tilde{c}_m$ and on the previous $(L_q - 1)$ symbols, i.e., on the trellis state $x_m \equiv (\tilde{c}_{m-L_q+1}, \tilde{c}_{m-L_q+2}, ..., \tilde{c}_{m-1})$. We also note that in (16.1.98) the product

$$\prod_{i=0}^{\eta-1} p\left(r_{m\,\eta+i}|\tilde{c}_m, x_m, \bar{\mathbf{q}}\right) \tag{16.1.99}$$

depends on the couple $(\tilde{c}_m, x_m)$, i.e., on the state transition $x_m \to x_{m+1}$. Then, we define

$$W_m\left(x_m, x_{m+1}\right) \triangleq \prod_{i=0}^{\eta-1} p\left(r_{m\,\eta+i}|\tilde{c}_m, x_m, \bar{\mathbf{q}}\right)$$
$$= \frac{1}{\left(\frac{N_0}{T_r}\right)^{\eta}} \exp\left[-\frac{1}{2\left(\frac{N_0}{T_r}\right)^{\eta}} \sum_{k=m\eta}^{(m+1)\eta-1} \left| r_k - \sum_{i=m-L_q+1}^{m} \tilde{c}_i\, \bar{q}_{i,k-i\eta}\right|^2\right] \tag{16.1.100}$$

so that the RHS of (16.1.98) can be rewritten as

$$p\left(\mathbf{r}|\tilde{\mathbf{c}}, \bar{\mathbf{q}}\right) = \prod_{m=0}^{L_c-1} W_m\left(x_m, x_{m+1}\right). \tag{16.1.101}$$

Here, the factors $\{W_m\left(x_m, x_{m+1}\right)\}$ are not dependent on the overall trial sequence $\tilde{\mathbf{c}}$ since the computation of each factor in (16.1.99) involves L_q symbols only.

Substituting (16.1.97) and (16.1.101) into the numerator of (16.1.96) yields

$$p\left(\mathbf{r}|\tilde{\mathbf{c}}, \bar{\mathbf{q}}\right)\, P\left(\tilde{\mathbf{c}}\right) = \prod_{m=0}^{L_c-1} \gamma_m\left(x_m, x_{m+1}\right) \tag{16.1.102}$$

where

$$\gamma_m\left(x_m, x_{m+1}\right) \triangleq P\left(\tilde{c}_m\right)\, W_m\left(x_m, x_{m+1}\right) \tag{16.1.103}$$

is a *weight function* depending on the trellis branch connecting the states x_m and x_{m+1}.

Finally, substituting (16.1.102) into (16.1.96) and (16.1.96) into (16.1.95) produces

$$P\left(x_m,\, x_{m+1}|\mathbf{r}, \bar{\mathbf{q}}\right) = \frac{\sum_{\tilde{\mathbf{c}} \in C(x_m, x_{m+1})} \prod_{m=0}^{L_c-1} \gamma_m\left(x_m, x_{m+1}\right)}{\sum_{\tilde{\mathbf{c}} \in C} \prod_{m=0}^{L_c-1} \gamma_m\left(x_m, x_{m+1}\right)}. \tag{16.1.104}$$

Then, the evaluation of the state transition probabilities requires the computation of (a) the sum of the products of the weights associated with all the paths containing the branch going out of x_m and entering x_{m+1} (see the numerator), and (b) the sum of the products of the weights associated with all the paths in the trellis (see the denominator). A computationally efficient method to solve this computational problem is the following [92]. Let us define the quantities $\{\alpha_m(x_m)\}$ and $\{\beta_m(x_m)\}$ through the recursive formula

$$\alpha_m(x_m) = \sum_{x_m} \alpha_{m-1}(x_{m-1}) \cdot \gamma_m(x_{m-1}, x_m) \tag{16.1.105}$$

with $m = 1, 2, ..., L_c - 1$, and

$$\beta_m(x_m) = \sum_{x_{m+1}} \beta_{m+1}(x_{m+1}) \cdot \gamma_m(x_m, x_{m+1}) \tag{16.1.106}$$

with $m = L_c - 2, L_c - 3, ..., 0$, respectively. Here it is assumed that $\{\alpha_0(x_0)\}$ and $\{\beta_{L_c-1}(x_{L_c-1})\}$ are known initial conditions.

The quantity $\alpha_m(x_m)$ expresses the sum of the products of the weights along all paths originating from all the possible past initial states $\{x_0\}$ and terminating in x_m in the mth signalling interval. Similarly, $\beta_m(x_m)$ represents the sum of the products of the weights along all paths ending in all the possible future terminal states $\{x_{L_c-1}\}$ and originating from x_m in the mth signalling interval. Then, the numerator of (16.1.95) can be evaluated as

$$\begin{aligned}\sigma_m(x_m, x_{m+1}) &\triangleq \sum_{\tilde{\mathbf{c}} \in C(x_m, x_{m+1})} \prod_{m=0}^{L_c-1} \gamma_m(x_m, x_{m+1}) \\ &= \alpha_m(x_m) \cdot \gamma_m(x_m, x_{m+1}) \cdot \beta_{m+1}(x_{m+1}),\end{aligned} \tag{16.1.107}$$

whereas its denominator is

$$\sum_{\tilde{\mathbf{c}} \in C} \prod_{m=0}^{L_c-1} \gamma_m(x_m, x_{m+1}) = \sum_{x_m, x_{m+1}} \sigma_m(x_m, x_{m+1}) \tag{16.1.108}$$

so that

$$P(x_m, x_{m+1} | \mathbf{r}, \bar{\mathbf{q}}) = \frac{\sigma_m(x_m, x_{m+1})}{\sum_{x_m, x_{m+1}} \sigma_m(x_m, x_{m+1})}. \tag{16.1.109}$$

The last result shows that all that is needed for the evaluation of the state transition probabilities are the quantities $\{\sigma_m(x_m, x_{m+1})\}$, computed as in (16.1.107). This computation, in turn, requires a forward (16.1.105) and a backward recursion (16.1.106) involving all the trellis states in each signalling interval, as illustrated in Figure 16.13. It is worth noting both recursions over all the trellis need to be performed only once.

For demodulation purposes, the quantity of interest is $P(\tilde{c}_m | \tilde{\mathbf{c}}, \bar{\mathbf{q}})$ for any possible value of the channel symbol $\tilde{c}_m$ (or bit $\tilde{b}_m$). This probability can be calculated by summing all the state transition probabilities (16.1.109) that correspond

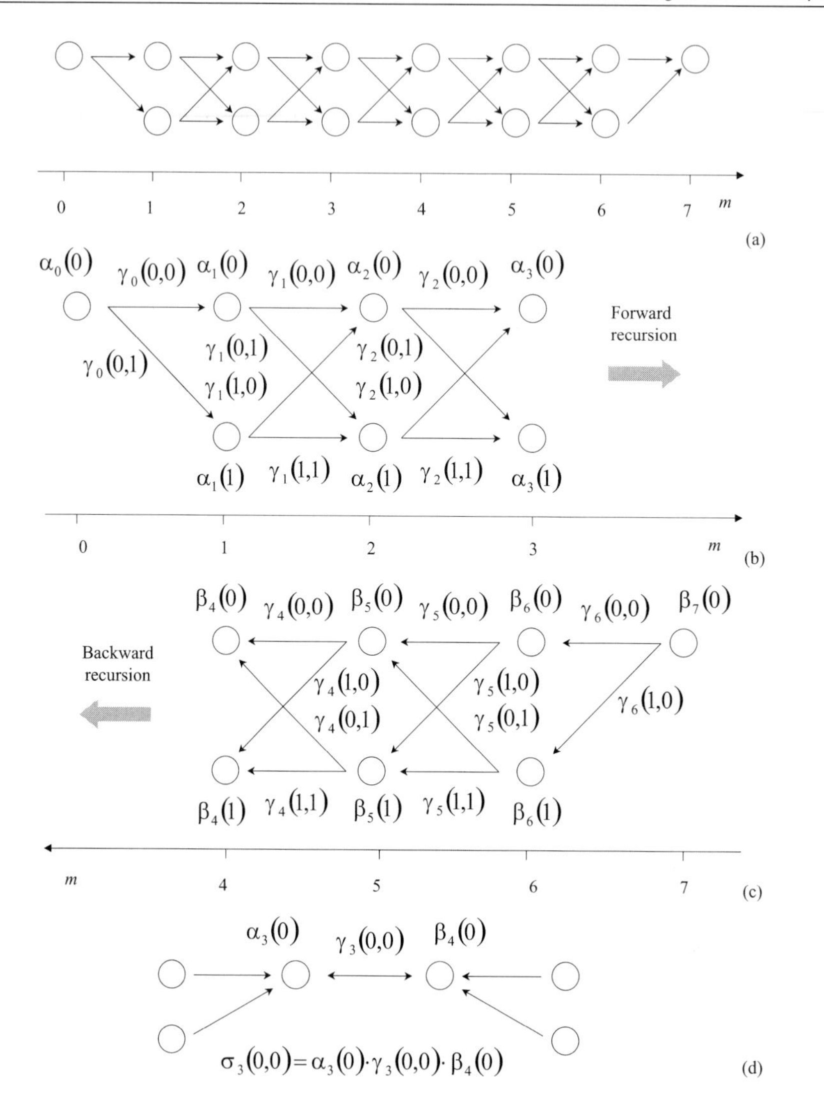

Figure 16.13. Application of the FBA to a 2-state trellis (a) ($M = 2$, $L_q = 2$). The quantities $\{\alpha_m(x_m)\}$ and $\{\beta_m(x_m)\}$ are computed recursively in the forward (b) and in the backward (c) recursion, respectively, using the branch probabilities $\{\gamma_m(x_m, x_{m+1})\}$. Finally, the state transition probabilities $\{\sigma_m(x_m, x_{m+1})\}$ are evaluated according to (16.1.107).

to branches associated with the symbol $\tilde{c}_m$ (or bit $\tilde{b}_m$). Then, if we define the set $S(\tilde{c}_m)$ of all state transitions (x_m, x_{m+1}) such that the channel symbol labelling the corresponding branch is $\tilde{c}_m$, $P(\tilde{c}_m|\tilde{\mathbf{c}}, \bar{\mathbf{q}})$ can be evaluated as

$$P\left(\tilde{c}_m|\bar{\mathbf{c}},\bar{\mathbf{q}}\right)=\sum_{(x_m,x_{m+1})\in S(\tilde{c}_m)} P\left(x_m,\,x_{m+1}|\mathbf{r},\bar{\mathbf{q}}\right)=\frac{\sum_{(x_m,x_{m+1})\in S(\tilde{c}_m)}\sigma_m\left(x_m,x_{m+1}\right)}{\sum_{x_m,x_{m+1}}\sigma_m\left(x_m,x_{m+1}\right)}. \tag{16.1.110}$$

Then, the FBA can be summarized in the following steps:

1. Evaluate the conditional probabilities $\{\gamma_m\left(x_m,x_{m+1}\right)\}$ for all the trellis branches, using (16.1.100) and (16.1.103).
2. Initialize the forward recursion setting
$$m=1,\ \alpha_0\left(x_0\right)=1. \tag{16.1.111}$$
3. Repeat steps 4 and 5 until $m=L_c$.
4. Compute and store the forward path probabilities $\{\alpha_m(x_m)\}$, using (16.1.105).
5. Set $m=m+1$.
6. Initialize the backward recursion setting
$$m=L_c-2,\ \beta_{L_c-1}\left(x_{L_c-1}\right)=1. \tag{16.1.112}$$
7. Repeat steps 8 and 9 until $m=0$.
8. Compute and store the backward path probabilities $\{\beta_m(x_m)\}$, using (16.1.106).
9. Set $m=m-1$.
10. For each trellis branch, compute the quantity $\sigma_m\left(x_m,x_{m+1}\right)$, using (16.1.107).
11. Evaluate the *a posteriori* symbol probabilities $\{P\left(\tilde{c}_m|\bar{\mathbf{c}},\bar{\mathbf{q}}\right)\}$ by means of (16.1.107).

16.1.7 Reduced complexity symbol detection

Fixed-lag MAPSD

The forward-backward algorithm is suited to short burst transmission because, otherwise, its delay and storage requirements are unsatisfactory. The MAP *fixed-lag algorithm* (MAP-FLA) detector [93] makes decisions at a fixed lag of L_{fl} symbols from the current received samples, as

$$\hat{c}_m=\arg\max_{\tilde{c}_m} P\left(\tilde{c}_m|\mathbf{r}_0^{(L_{fl}+m+1)\eta-1},\bar{\mathbf{q}}\right) \tag{16.1.113}$$

where the notation $\mathbf{x}_a^b$ denotes the vector $\left[x_a,x_{a+1},...,x_b\right]^T$. The parameter L_{fl} is akin to the decision delay in the VA, and so, for effectively optimal performance, it is expected to be as high as $5L_q$ or $7L_q$. The decision rule can be implemented with two good algorithms, both requiring a single forward pass only. The newer one, dubbed optimum *soft-output algorithm* (or OSA) [94], has a smaller computational complexity than the older one [95], [93] because the number of quantities to be stored and recursively updated increases linearly, rather than exponentially, with the decision delay. Nonetheless, we present only the older one because it is more suited to channel estimation, it is easier to grasp, and it has close parallels with the VA [96].

Let us now derive the MAP-FLA. The probability $P(\tilde{c}_m|\mathbf{r}_0^{(L_{fl}+m+1)\eta-1}, \bar{\mathbf{q}})$ in (16.1.113) is given by

$$P\left(\tilde{c}_m|\mathbf{r}_0^{(L_{fl}+m+1)\eta-1}, \bar{\mathbf{q}}\right) = \sum_{\tilde{\mathbf{c}}_0^{m-1}, \tilde{\mathbf{c}}_{m+1}^{m+1+L_f}} P\left(\tilde{\mathbf{c}}_0^{m+1+L_{fl}}|\mathbf{r}_0^{(L_{fl}+m+1)\eta-1}, \bar{\mathbf{q}}\right). \tag{16.1.114}$$

Using Bayes rule, we can express the term $P(\tilde{\mathbf{c}}_0^{m+1+L_{fl}}|\mathbf{r}_0^{(L_{fl}+m+1)\eta-1}, \bar{\mathbf{q}})$ in (16.1.114) as

$$P\left(\tilde{\mathbf{c}}_0^{m+1+L_{fl}}|\mathbf{r}_0^{(L_{fl}+m+1)\eta-1}, \bar{\mathbf{q}}\right) = \frac{p\left(\mathbf{r}_0^{(L_{fl}+m+1)\eta-1}|\tilde{\mathbf{c}}_0^{m+1+L_{fl}}, \bar{\mathbf{q}}\right) P\left(\tilde{\mathbf{c}}_0^{m+1+L_{fl}}\right)}{p\left(\mathbf{r}_0^{(L_{fl}+m+1)\eta-1}|\bar{\mathbf{q}}\right)} \tag{16.1.115}$$

where

$$P\left(\tilde{\mathbf{c}}_0^{m+1+L_{fl}}\right) = \prod_{m=0}^{m+1+L_{fl}} P\left(\tilde{c}_m\right) \tag{16.1.116}$$

is the a priori probability of $\tilde{\mathbf{c}}_0^{m+1+L_{fl}}$ and the conditional pdf $p\left(\mathbf{r}_0^{(L_{fl}+m+1)\eta-1}|\bar{\mathbf{q}}\right)$ can be evaluated as

$$p\left(\mathbf{r}_0^{(L_{fl}+m+1)\eta-1}|\bar{\mathbf{q}}\right) = \sum_{\tilde{\mathbf{c}}_0^{m+1+L_f}} p\left(\mathbf{r}_0^{(L_{fl}+m+1)\eta-1}|\tilde{\mathbf{c}}_0^{m+1+L_{fl}}, \bar{\mathbf{q}}\right) P\left(\tilde{\mathbf{c}}_0^{m+1+L_{fl}}\right). \tag{16.1.117}$$

Then, substituting (16.1.117) into (16.1.115) produces

$$P\left(\tilde{\mathbf{c}}_0^{m+1+L_{fl}}|\mathbf{r}_0^{(L_{fl}+m+1)\eta-1}, \bar{\mathbf{q}}\right) = \frac{p\left(\mathbf{r}_0^{(L_{fl}+m+1)\eta-1}|\tilde{\mathbf{c}}_0^{m+1+L_{fl}}, \bar{\mathbf{q}}\right) P\left(\tilde{\mathbf{c}}_0^{m+1+L_{fl}}\right)}{\sum_{\tilde{\mathbf{c}}_0^{m+1+L_f}} p\left(\mathbf{r}_0^{(L_{fl}+m+1)\eta-1}|\tilde{\mathbf{c}}_0^{m+1+L_{fl}}, \bar{\mathbf{q}}\right) P\left(\tilde{\mathbf{c}}_0^{m+1+L_{fl}}\right)} \tag{16.1.118}$$

so that (16.1.114) can be rewritten as

$$P\left(\tilde{c}_m|\mathbf{r}_0^{(L_{fl}+m+1)\eta-1},\bar{\mathbf{q}}\right)$$
$$=\frac{\sum_{\tilde{\mathbf{c}}_0^{m-1},\tilde{\mathbf{c}}_{m+1}^{m+1+L_{fl}}} p\left(\mathbf{r}_0^{(L_{fl}+m+1)\eta-1}|\tilde{\mathbf{c}}_0^{m+1+L_{fl}},\bar{\mathbf{q}}\right) P\left(\tilde{\mathbf{c}}_0^{m+1+L_{fl}}\right)}{\sum_{\tilde{\mathbf{c}}_0^{m+1+L_f}} p\left(\mathbf{r}_0^{(L_{fl}+m+1)\eta-1}|\tilde{\mathbf{c}}_0^{m+1+L_{fl}},\bar{\mathbf{q}}\right) P\left(\tilde{\mathbf{c}}_0^{m+1+L_{fl}}\right)}. \quad (16.1.119)$$

We note the following:

1. In the last equation, $P(\tilde{\mathbf{c}}_0^{m+1+L_{fl}})$ is not usually the same for all sequences because of the use of training or pilot symbols to solve the phase ambiguity of the signal constellation.

2. The quantities $\{p(\mathbf{r}_0^{(L_{fl}+m+1)\eta-1}|\tilde{\mathbf{c}}_0^{m+1+L_{fl}},\bar{\mathbf{q}})\,P(\tilde{\mathbf{c}}_0^{m+1+L_{fl}})\}$ represent a set of $M^{m+1+L_{fl}}$ sufficient statistics for making a MAP decision on c_m.

The evaluation of such statistics can be accomplished by means of a recursive algorithm [93]. In fact, assuming that the same quantities at the previous step, i.e., $\{p(\mathbf{r}_0^{(L_{fl}+m)\eta-1}|\tilde{\mathbf{c}}_0^{m+L_{fl}},\bar{\mathbf{q}})\,P(\tilde{\mathbf{c}}_0^{m+L_{fl}})\}$, are known and that the vector $\mathbf{r}_{(L_{fl}+m)\eta}^{(L_{fl}+m+1)\eta-1}$ has been observed, $p(\mathbf{r}_0^{(L_{fl}+m+1)\eta-1}|\tilde{\mathbf{c}}_0^{m+1+L_{fl}},\bar{\mathbf{q}})\,P(\tilde{\mathbf{c}}_0^{m+1+L_{fl}})$ can be written as

$$\begin{aligned}
& p\left(\mathbf{r}_0^{(L_{fl}+m+1)\eta-1}|\tilde{\mathbf{c}}_0^{m+1+L_f},\bar{\mathbf{q}}\right) P\left(\tilde{\mathbf{c}}_0^{m+1+L_{fl}}\right) \\
&= p\left(\mathbf{r}_0^{(L_{fl}+m)\eta-1},\mathbf{r}_{(L_{fl}+m)\eta}^{(L_{fl}+m+1)\eta-1}|\tilde{\mathbf{c}}_0^{m+L_{fl}},\tilde{c}_{L_f+m+1},\bar{\mathbf{q}}\right) P\left(\tilde{\mathbf{c}}_0^{m+L_{fl}},\tilde{c}_{L_f+m+1}\right) \\
&= P\left(\tilde{c}_{L_{fl}+m+1}\right) p\left(\mathbf{r}_{(L_{fl}+m)\eta}^{(L_{fl}+m+1)\eta-1}|\tilde{\mathbf{c}}_0^{m+L_f},\tilde{c}_{L_{fl}+m+1},\bar{\mathbf{q}}\right) \\
&\cdot\, p\left(\mathbf{r}_0^{(L_{fl}+m)\eta-1}|\tilde{\mathbf{c}}_0^{m+L_{fl}},\tilde{c}_{L_{fl}+m+1},\bar{\mathbf{q}}\right) P\left(\tilde{\mathbf{c}}_0^{m+L_{fl}}\right)
\end{aligned} \quad (16.1.120)$$

because the components of $\mathbf{r}_0^{(L_{fl}+m+1)\eta-1}$, conditioned on the transmitted data sequence, are independent random variables, provided that, as always assumed, all the noise samples are independent. Let us assume now that the algorithm lag is larger than the channel memory, i.e.,[5]

$$L_{fl} \geq L_q - 1 \quad (16.1.121)$$

Then, in (16.1.120) the conditional pdf $p(\mathbf{r}_{(L_{fl}+m)\eta}^{(L_{fl}+m+1)\eta-1}|\tilde{\mathbf{c}}_0^{m+L_f},\tilde{c}_{L_f+m+1},\bar{\mathbf{q}})$ sim-

[5] If the channel memory is so long that it exceeds the decision delay, this recursive procedure can be still applied with slight modifications. In this case, decision feedback can be required to reduce the computational complexity of the algorithm [93].

plifies as

$$p\left(\mathbf{r}_{(L_{fl}+m)\eta}^{(L_{fl}+m+1)\eta-1}|\tilde{\mathbf{c}}_0^{m+L_{fl}}, \tilde{c}_{L_{fl}+m+1}, \bar{\mathbf{q}}\right) = p\left(\mathbf{r}_{(L_{fl}+m)\eta}^{(L_{fl}+m+1)\eta-1}|\tilde{\mathbf{c}}_m^{m+L_{fl}}, \bar{\mathbf{q}}\right) \tag{16.1.122}$$

$$= \frac{1}{\left(\frac{N_0}{T_r}\right)^{\eta}} \exp\left[-\frac{1}{2\left(\frac{N_0}{T_r}\right)^{\eta}} \sum_{k=(L_{fl}+m)\eta}^{(L_{fl}+m+1)\eta-1} \left| r_k - \sum_{i=L_{fl}+m-L_q+1}^{L_{fl}+m} \tilde{c}_i\, \bar{q}_{i,k-i\eta} \right|^2\right] \tag{16.1.123}$$

since the received vector $\mathbf{r}_{(L_{fl}+m)\eta}^{(L_{fl}+m+1)\eta-1}$ depends on the symbol vector $\tilde{\mathbf{c}}_m^{m+L_{fl}}$ (i.e., on $L_{fl}+1$ symbols) at most. From (16.1.120) and (16.1.122) it is easily inferred that

$$\begin{aligned} & p\left(\mathbf{r}_0^{(L_{fl}+m+1)\eta-1}|\tilde{\mathbf{c}}_0^{m+1+L_{fl}}, \bar{\mathbf{q}}\right) P\left(\tilde{\mathbf{c}}_0^{m+1+L_{fl}}\right) \\ & = P\left(\tilde{\mathbf{c}}_0^{L_{fl}-1}\right) p\left(\mathbf{r}_0^{(L_{fl}-1)\eta-1}|\tilde{\mathbf{c}}_0^{L_{fl}}, \bar{\mathbf{q}}\right) \\ & \cdot \prod_{k=L_{fl}}^{m+L_{fl}} P\left(\tilde{c}_{k+1}\right) p\left(\mathbf{r}_{k\eta}^{(k+1)\eta-1}|\tilde{\mathbf{c}}_{k-L_{fl}}^{k}, \bar{\mathbf{q}}\right). \end{aligned} \tag{16.1.124}$$

The last result proves that the sufficient statistics $\{p(\mathbf{r}_0^{(L_{fl}+m+1)\eta-1}|\tilde{\mathbf{c}}_0^{m+1+L_{fl}}, \bar{\mathbf{q}})\, P(\tilde{\mathbf{c}}_0^{m+1+L_{fl}})\}$ in (16.1.119) can be evaluated by means of a forward recursion in an $M^{L_{fl}}$-state trellis. The trellis states in the mth symbol interval are defined as the vector $x_m = (\tilde{c}_{m-L_{fl}}, \tilde{c}_{m-L_{fl}+1}, ..., \tilde{c}_{m-1})$. The metric associated with the state transitions (x_m, x_{m+1}) is given by

$$\gamma_m\left(x_m, x_{m+1}\right) = P\left(\tilde{c}_{L_{fl}+m+1}\right) p\left(\mathbf{r}_{(L_{fl}+m)\eta}^{(L_{fl}+m+1)\eta-1}|\tilde{\mathbf{c}}_m^{L_{fl}+m}, \bar{\mathbf{q}}\right). \tag{16.1.125}$$

This forward recursion is the same as the forward recursion in the forward-backward algorithm, but now over the *enlarged* number of states when the inequality (16.1.121) holds.

Finally, let us compare some of the uses of the VA, the MAP-FLA, and the MAP-FBA [94]. Each of these algorithms can be employed for the detection of data sequences in the presence of ISI. Among these algorithms, however, the VA has the smallest complexity and this makes it the most reasonable choice in uncoded communication systems where soft output information is not required. The MAP algorithms have the following relevant disadvantages: (1) they require knowledge of the noise variance; (2) they have large memory and computation requirements. In particular, as already stated, MAP algorithms accomplish computations in the probability, instead of the logarithm, domain and, consequently, require a large number of multiplications and exponentiations.

The MAP-FBA detector performs two recursions and, consequently, it operates in block mode. It memorizes all the received signal samples of each block and

Features	**MAP-FBA**	**MAP-FLA**	**VA**
Minimization	symbol error	symbol error	sequence error
Decision type	soft	soft	hard
A priori information	noise variance	noise variance	-
Recursion requirement	forward and backward	forward	forward
Memory requirement	$\propto L_c \cdot M^{L_q}$	$\propto M^{L_{fl}}$	$\propto K \cdot M^{L_q}$
Computation requirement	$\propto L_c \cdot M^{L_q}$	$\propto L_c \cdot M^{L_{fl}}$	$\propto L_c \cdot M^{L_q}$

Table 16.1. Main features of the VA, the MAP-FBA, and the MAP- FB

afterwards processes the block. Its memory requirement grows linearly with the sequence length so that short data sequences should be processed. In addition, if the data sequence is long, the CIR cannot be assumed constant over the block duration and so would require channel tracking. This task seemingly cannot be incorporated in the detection algorithm because reliable data decisions are available at the end of both the recursions.

The MAP-FLA requires only a forward recursion, so that it can operate in continuous mode. In this case, adaptivity to channel variations can be embedded (see Section 16.2.2) because of the short and fixed decision delay. Both the memory and computational requirements, however, grow exponentially with the decision delay so that this parameter should be kept to a minimum. For this reason the FBA may have smaller computational requirements than the FLA.

The most relevant features of the three algorithms are summarized in Table 16.1. Here L_c is the sequence length, and L_{fl} and K are the decision delay for the MAP-FLA and VA, respectively. Usually the inequality

$$L_q \leq \frac{K}{L_{fl}} \leq L_c \tag{16.1.126}$$

is satisfied.

As already mentioned, a novel version of the fixed-lag MAPSD, known as *optimum soft-output algorithm* (OSA), is derived in [94]. It is an improved version because it generates optimum soft outputs with only one forward recursion, but the number of quantities to be stored and recursively updated increases linearly, rather than exponentially, with the decision delay. In the same paper, a *suboptimum soft-output algorithm* (SSA) is also proposed to overcome some disadvantages of the MAP algorithms. The SSA does not require knowledge of the noise variance and the computations it requires are in the logarithmic domain. For this reason, *add-compare-select* (ACS) is the main operation, as in the VA.

Approximations and modifications of the MAP-FBA have also been proposed and include the so-called Log MAP and the Max-Log-MAP algorithms [91], [12], [97]. A *soft-ouput* VA (SOVA) has been also proposed as an alternative to MAP-FBAs [97], [10].

Finally, it is worth pointing out that MAP algorithms provide optimum soft decision metrics for decoding convolutional codes on fading channels. As shown in the previous equations, optimum soft-decision metrics require channel state information. Some interesting considerations on applications to FF fading are illustrated in [98].

Bayesian decision feedback equalization

The relationship of *Bayesian decision-feedback equalizers* (BDFEs) to fixed-lag MAPSDs is analogous to the relationship of DDFSDs or RSSDs to MLSDs: both use decision feedback to reduce the optimal structure's complexity [99]. However, in appearance and operation, Bayesian DFEs are closer to DFEs, but with the DFE's feedforward and feedback tap coefficients replaced by a nonlinear structure [100].

The BDFE treats detection as a nonlinear classification problem [101]. In detecting a symbol, a fixed-length vector of received samples in the vicinity of the current symbol identifies a point in a multidimensional space. The space is divided into M nonlinear regions, one per possible decision. These regions are each the union of smaller decision regions, one per ISI combination. The equalizer makes a decision by determining which decision region the received sample vector lies in. Computational savings are achieved in BDFE over the full MAPSD implementation, since past decisions reduce the number of allowed ISI combinations and, thus, the number of smaller decision regions that exist [100]. A block approach is presented in [102].

Neural networks are effective nonlinear identification algorithms. Several models have been applied to BDFE [101], [103], [104] to achieve the equalization of nonlinearly distorted signals [105] and to suppress adjacent channel interference (ACI) [103] and cochannel interference (CCI) [104]. A promising technique for doubly spread channels is pursued in [106], [100], [107], [108]. In [106], the LMS algorithm is used to adapt a complex radial basis function network in a GSM-like channel.

16.2 EQUALIZATION WHEN THE CIR IS ESTIMATED

The equalization structures of Section 16.1 were derived assuming a known channel. In practice, channel estimation is required. The scope of this section is twofold. First, we show some estimation tools and we discuss blind equalization. Second, we illustrate equalization techniques incorporating channel estimation strategies.

16.2.1 Estimation tools

Principle of per-survivor processing

In practical systems, the channel impulse response $\bar{\mathbf{q}}_k \triangleq [\bar{q}_{k,0}, \dots, \bar{q}_{k,L_q\eta-1}]^T$ at the kth sampling epoch (see (16.1.4), for instance) is not known a priori and must be estimated to compute the equalizer metrics. A common approach is based on data-aided parameter estimation techniques in which the aiding data sequence is generated in a decision-directed mode from *tentative low-delay decisions* at the decoder output [109], [33]. Let the tentative decision on the data symbol at the epoch k be denoted by $\tilde{c}_{k-d-1}$, where d denotes the decoding delay of the tentative decisions. Based on the sequence $\tilde{\mathbf{c}}_{k-d} \triangleq [\tilde{c}_i, i = ..., k-d-2, k-d-1]^T$ and the received signal vector $\mathcal{R}_k$, a data-aided parameter estimator provides the detector with an estimate $\hat{\mathbf{q}}_k$ of the unknown channel $\bar{\mathbf{q}}_k$ as

$$\hat{\mathbf{q}}_k = \mathbf{g}\left(\mathcal{R}_k, \tilde{\mathbf{c}}_{k-d}\right). \tag{16.2.1}$$

Here, $\mathcal{R}_k \triangleq [r_0, r_1, ..., r_k]^T$ and $\mathbf{g}(\cdot)$ denotes the functional dependence of the estimate $\hat{\mathbf{q}}_k$ on the received signal and the sequence of tentative decisions. It is worth noting that a decision delay is inherent in the estimate $\hat{\mathbf{q}}_k$ with respect to the true parameter vector $\bar{\mathbf{q}}_k$.

An alternative to this approach is the application of *per-survivor* estimation [110], [111] of the unknown channel. We can give a formal description by defining the channel symbol sequence associated to the survivor of the state x_k as $\tilde{\mathbf{c}}_{k-1}(x_k)$ in a decoder based on the Viterbi algorithm (see Subsection 16.1.1). Per-survivor estimates $\{\breve{\mathbf{q}}_k(x_k)\}$ of the unknown channel vector $\bar{\mathbf{q}}_k$ based on the data-aided estimator $\mathbf{g}(\cdot)$ in (16.2.1) and the channel symbols associated to each surviving path can be defined as

$$\breve{\mathbf{q}}_k(x_k) = \mathbf{g}\left(\mathcal{R}_k, \tilde{\mathbf{c}}_{k-1}(x_k)\right). \tag{16.2.2}$$

This channel estimate is employed in the evaluation of the branch metrics for all the state transitions going out of state x_k, as illustrated in Figure 16.14.

A heuristic justification for this approximation can be given as follows. If the ignorance of the CIR prevents us from calculating a decision metric in a precise form, estimates of CIR based on the multiple data hypotheses (i.e., on the survivor paths) are evaluated. If a particular survivor represents the correct choice of data, the corresponding estimate is evaluated properly. At each decoding step, however, we do not know which survivor, if any, represents correct data decisions. Then, we extend each survivor by using the channel estimates based on its associated data sequence, i.e., on the best data sequence available.

Different forms of the function $\mathbf{g}(\cdot)$ can be employed in practical applications. In the following discussion we concentrate on the *least mean square* (LMS) and *recursive least mean square* (RLS) algorithms [112]. In addition, we mainly focus on a slowly time-varying channel with baud-rate sampling (i.e., $\eta = 1$) for simplicity.[6]

LMS and RLS

The LMS algorithm results from applying the *stochastic gradient algorithm* to the minimization of the MSE [112]

$$J(\tilde{\mathbf{q}}) \triangleq E\left\{|e_k|^2\right\} \tag{16.2.3}$$

[6]In white noise, estimation of the fractionally spaced channel impulse response reduces to η parallel estimations of η baud-rate spaced channel impulse response [113].

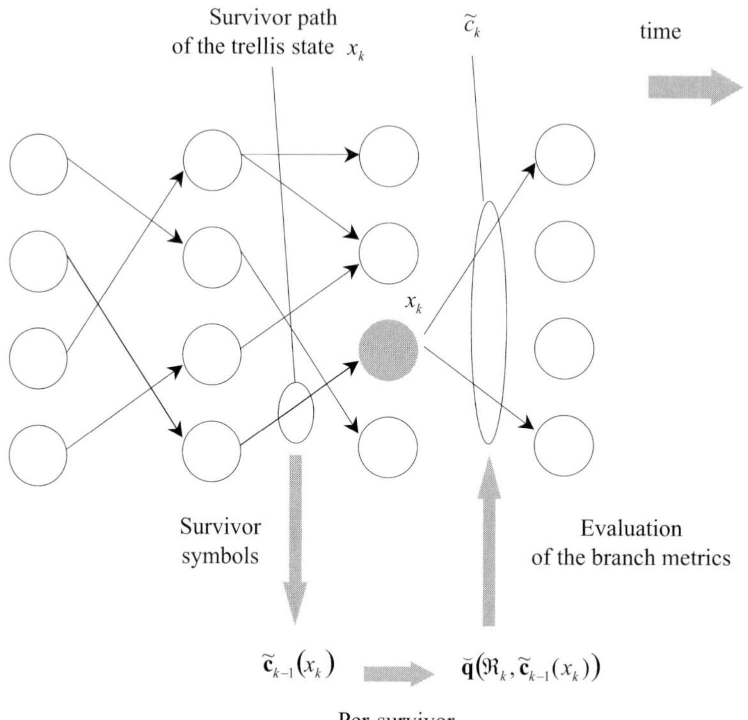

Figure 16.14. Per-survivor channel estimation

between the channel output and its estimated version, where e_k, the estimation error, is defined as

$$e_k \triangleq r_k - \sum_{m=k-L_q+1}^{k} c_m \, \tilde{q}_{k-m} \tag{16.2.4}$$

and $\tilde{\mathbf{q}} \triangleq [\tilde{q}_0, \tilde{q}_1, ..., \tilde{q}_{L_q-1}]^T$ is the estimate of the CIR $\bar{\mathbf{q}} \triangleq [\bar{q}_0, \bar{q}_1, ..., \bar{q}_{L_q-1}]^T$. Taking into account the decision delay d, the LMS algorithm is summarized by the coefficient update equation [112]

$$\tilde{\mathbf{q}}\,[k+1] = \tilde{\mathbf{q}}\,[k] + \mu \, e_{k-d} \, \tilde{\mathbf{c}}_{k-d}^{*} \tag{16.2.5}$$

where $d = 1$ for PSP. Here, $\tilde{\mathbf{q}}\,[k]$ is the estimate of $\bar{\mathbf{q}}$ at the kth step, d is the data decision delay, μ is the so-called *step-size*, $\tilde{\mathbf{c}}_{k-d} \triangleq \left[\tilde{c}_{k-d}, \tilde{c}_{k-d-1}, ..., \tilde{c}_{k-d-L_q+1}\right]^T$ is a vector of data decisions, and $\{e_{k-d}\}$ is the error sequence.

It is well known that the step-size parameter μ controls both the rate of adaptation and the stability of the LMS [112]. For stability, $0 < \mu < 2/\lambda_{\max}$, where $\lambda_{\max}$ is the largest eigenvalue of the received signal covariance matrix. A large value for μ, just below the upper limit, provides rapid convergence, but it also introduces large fluctuations during the steady state operation. Such fluctuations represent a form of self-noise whose variance increases with μ. Therefore, the choice of μ represents a trade-off between rapid convergence and a small self-noise variance during steady state operation.

The LMS is well known for its simplicity but does not provide fast convergence. Its slow convergence rate is due to the fact that there is only one parameter, namely, μ, controlling the rate of adaptation. Generally speaking, a faster converging algorithm is obtained if we employ the RLS criterion [112] for the adjustment of the channel estimator coefficients. In this case, the target at the kth step minimizes the sum of exponentially weighted squared errors, i.e.,

$$J_k(\tilde{\mathbf{q}}) \triangleq \sum_{n=0}^{k} w^{k-n} E\left\{|e_n|^2\right\} \tag{16.2.6}$$

where the parameter w, representing an exponential *weighting factor*, provides a fading memory in the estimation of the channel coefficients and is selected to be in the range $0 < w < 1$. The RLS algorithm can be expressed as

$$\tilde{\mathbf{q}}[k+1] = \tilde{\mathbf{q}}[k] + \mathbf{P}_k\, \tilde{\mathbf{c}}_{k-d}^{*}\, e_{k-d}. \tag{16.2.7}$$

Here, $\mathbf{P}_k$ is an $L_q \times L_q$ square matrix and is updated according to

$$\mathbf{P}_k = \frac{1}{w}\left[\mathbf{P}_{k-1} - \frac{1}{w + \tilde{\mathbf{c}}_{k-d}^{T}\mathbf{P}_{k-1}\tilde{\mathbf{c}}_{k-d}^{T}}\mathbf{P}_{k-1}\tilde{\mathbf{c}}_{k-d}^{*}\tilde{\mathbf{c}}_{k-d}^{T}\mathbf{P}_{k-1}\right]. \tag{16.2.8}$$

It can be shown that $\mathbf{P}_k$ is the inverse of the data autocorrelation matrix

$$\mathbf{R}_k = \sum_{n=0}^{k} w^{k-n}\tilde{\mathbf{c}}_{k-d}^{*}\tilde{\mathbf{c}}_{k-d}^{T}. \tag{16.2.9}$$

At the first iteration of the RLS algorithm, $\mathbf{P}_0$ can be selected to be proportional to the identity matrix.

The RLS algorithm offers a substantially faster convergence with respect to the LMS algorithm. However, the recursive update equation (16.2.8) has poor numerical properties. For this reason, other algorithms with an improved numerical stability have been derived. They are based on the square root factorization of $\mathbf{P}_k$, i.e.,

$$\mathbf{P}_k = \mathbf{S}_k\, \mathbf{S}_k^{T} \tag{16.2.10}$$

where $\mathbf{S}_k$ is a lower triangular matrix. Such algorithms are known as *square root RLS algorithms* [112]. They update the matrix $\mathbf{S}_k$ directly without computing the matrix $\mathbf{P}_k$ explicitly and have a computational complexity proportional to L_q^2.

Another type of RLS algorithm, known as *fast RLS algorithm* [79], [114] has been devised with a computational complexity proportional to L_q.

The channel estimator based on the RLS algorithm can be also implemented as a *lattice structure* [79], [115]. The convergence rate is identical to that of the RLS algorithm given above. However, the computational complexity for the RLS lattice structure is proportional to L_q, but with a larger proportionality constant compared to the fast RLS algorithm.

Let us now comment on the application of the LMS and RLS to the estimation and tracking of nonstationary channels. In this case, both the LMS and the RLS algorithms produce a delayed estimate of the CIR. For instance, in the kth symbol interval, (16.2.8) generates an estimate of the CIR at the $(k-d)$th epoch, i.e., $\bar{\mathbf{q}}_{k-d} \triangleq [\bar{q}_{k-d,0}, \bar{q}_{k-d+1,1}, ..., \bar{q}_{k-d-L_q+1,L_q-1}]^T$ (if $\eta = 1$). Therefore, a channel predictor must be included in the receiver structure if a CIR estimate at the present epoch is desired [116]. The effect of the decision delay on the performance of the LMS algorithm is investigated in [117].

With nonstationary channels, the tracking capability of the LMS algorithm is generally worse than that of the RLS. However, the matrix determining the tracking behavior is the autocorrelation matrix of the channel estimation algorithm, namely $E\left\{\tilde{\mathbf{c}}_{k-d}\,\tilde{\mathbf{c}}_{k-d}^H\right\}$. If the transmitted data symbols are assumed i.i.d., the eigenvalue spread of this autocorrelation matrix is unity; hence, the LMS algorithm can track the channel variations as quickly as does the RLS algorithm [118]. Further information on the stationary and nonstationary learning characteristics of the LMS and RLS algorithms can be found in [119], [120], [121], [122]. Considerable work to improve the RLS algorithm by explicitly modelling the time variation of the channel taps with additional parameters is illustrated in [123], [124]. In fast fading, marked improvements are available.

Pilot-based channel estimation

Pilot signals allow for channel estimation prior to data detection. In the time-invariant channel, only a single training sequence is required; in the FF fading channel, a single pilot tone or a sequence of pilot symbols suffices (details are given in Section 16.2.4). In the DS channel, a single training sequence cannot track the time-varying channel [125], nor can a pilot symbol sequence efficiently measure frequency selectivity [126], [127], since the adjacent data symbols overlap the pilot symbols and obscure the channel information in the known pilot symbol. More general methods are needed so as to maintain orthogonality between the pilot and data-bearing signals at the receiver and to allow the pilot information to be extracted before detection [128], [127], [7]. In general, a *comb* of pilot tones is required to characterize the channel in frequency as well as in time [7].

In many systems, the channel is slowly time varying but highly frequency selective. Then, the method of [129], [130], [131] is superior, where training sequences of pseudorandom symbols are transmitted periodically. The equalizer estimates the channel response from each. By interpolating [129], [131] or Wiener filtering [130] between training sequences, the channel can be estimated over the whole transmission. The sequences are usually several times the length of the received pulse and must be spaced at the Nyquist rate for the channel's Doppler spread. Thus, throughput is low except for small Doppler spread. For faster fading, very short training sequences can be used, as when isolated pilot symbols are employed [7].

Blind techniques and CMA

An equalizer can compensate for channel distortion even in the absence of known symbols by using one of a number of *blind* deconvolution algorithms. These can be classified into four classes: algorithms explicitly using higher order statistics (HOS), subspace algorithms, Bussgang algorithms, and joint data detection and channel estimation techniques [132]. This last class is thoroughly studied in Section 16.2.2 and is further discussed later in this section, so only the first three classes are briefly described here. These remaining algorithms are employed to adapt a linear equalizer's coefficients, but there are also preliminary extensions to DFEs.

HOS: HOS or *cumulants* of the symbol-spaced received samples $\{r_{m\eta}\}$ are estimated by time averaging and then compared with the known cumulants of the input constellation. The mismatch implicitly provides information about the channel and can be used for equalizer adaptation [133], [134], [135], [136], [137], [138], [139], [140]. If symbol-spaced samples are employed, the received sample sequence $\{r_{m\eta}\}$ is stationary, and so second order cumulants

$$\mathrm{cum}_{2,r}\left(\Delta\eta\right) \triangleq E\left\{r^*_{m\eta}\, r_{(m+\Delta)\eta}\right\} \tag{16.2.11}$$

can characterize only the amplitude response of the channel, not its phase. Therefore, higher-order statistics, such as fourth-order cumulants

$$\begin{aligned}\mathrm{cum}_{4,r}\left(\Delta_1\eta, \Delta_2\eta, \Delta_3\eta\right) \triangleq\; & E\left\{r^*_{m\eta}\, r_{(m+\Delta_1)\eta}\, r^*_{(m+\Delta_2)\eta}\, r_{(m+\Delta_3)\eta}\right\} \\ & - E\left\{r_{m\eta}\, r^*_{(m+\Delta_1)\eta}\right\} E\left\{r_{(m+\Delta_2)\eta}\, r^*_{(m+\Delta_3)\eta}\right\} \\ & - E\left\{r_{m\eta}\, r_{(m+\Delta_2)\eta}\right\} E\left\{r^*_{(m+\Delta_1)\eta}\, r^*_{(m+\Delta_3)\eta}\right\} \\ & - E\left\{r_{m\eta}\, r^*_{(m+\Delta_3)\eta}\right\} E\left\{r_{(m+\Delta_2)\eta}\, r^*_{(m+\Delta_1)\eta}\right\}\end{aligned} \tag{16.2.12}$$

are needed. Accordingly, the acquisition and tracking performance of HOS-based algorithms is usually inadequate for wireless channels because an unduly large number of received samples is needed before the HOS estimates are reliable [141], [142], [143]. By comparison, Bussgang algorithms implicitly exploit higher-order statistics more quickly and simply [141].

Subspace methods: By using all the fractionally spaced received samples, subspace methods are able to determine (under general conditions [144], [145]) both the amplitude and phase responses of the channel from second-order statistics only [146], [142], [147]. Symbol-spaced samples are also sufficient if multiple diversity sources are available instead. The central idea is that the noiseless received samples equal the sum of scaled (by the constellation points) and time-shifted (by multiples of the symbol period, T) received pulses. In time-invariant channels, these received pulses are the OCIR q_k, from (14.2.20), so we write

$$z_k = \sum_{m=0}^{L_c-1} c_m \, q_{k-m\eta}. \tag{16.2.13}$$

Therefore, the vector $\mathbf{z}_{m_0} \triangleq \left(z_{m_0\eta},...,z_{(m_0+L_w)\eta-1}\right)$ of received samples (for some start m_0 and duration L_w, in symbol periods) occupies only a subspace spanned by the time-shifted received pulses within that window, $\{q_{k-m\eta}\}$, $m = m_0 - L_q + 1, ..., m_0 + L_w - 1$. This property leads to the observation that the matrices $(1/2)\,E\left\{\mathbf{r}_{m_0}\mathbf{r}_{m_0}^H\right\}$ and $(1/2)\,E\{\mathbf{r}_{m_0}\mathbf{r}_{m_0+1}^H\}$, where $\mathbf{r}_{m_0} \triangleq \left(r_{m_0\eta},...,r_{(m_0+L_w)\eta-1}\right)$, uniquely identify the OCIR $\{q_k\}$. Unfortunately, these methods presume that the duration L_q of the received pulses is known, and they are quite nonrobust when there is a mismatch between the assumed and actual value.

Bussgang algorithms: Bussgang algorithms employ a gradient descent algorithm that attempts to restore some property of the input data that was destroyed by the frequency selective channel [148], [149], [150], [151]. Let us consider a linear equalizer with contents $\mathbf{r}_m$, coefficients $\mathbf{f}_m$ and quantizer input $\tilde{R}_m$ defined by (16.1.44), (16.1.41), and (16.1.64) respectively. Then, its update equation is

$$\mathbf{f}_{m+1} = \mathbf{f}_m - \mu \, \mathbf{r}_m \, e^*\left(\tilde{R}_m\right) \tag{16.2.14}$$

where μ is the stepsize (a small positive constant) and the error factor $e(.)$ is a nonlinear, memoryless mapping that characterizes each Bussgang variant. There are three important variants: (a) decision-directed equalization, (b) the Sato algorithm [148], and (c) the *constant modulus algorithm* (CMA) [149].

The decision-directed algorithm is simply the LMS algorithm with decisions in place of training data, as

$$e\left(\tilde{R}_m\right) \triangleq \tilde{R}_m - Q\left(\tilde{R}_m\right) \tag{16.2.15}$$

where $Q\,(.)$ is the nearest neighbor quantizer. Given a reliable (ideally, *correct*) decision, the decision-directed algorithm minimizes the residual mean-square error between the input and output of the quantizer, and thus it is well suited to tracking. It is unsuitable for acquisition since typically many incorrect decisions are fed back initially and the equalizer may converge to a closed-eye setting (i.e., the cost function has many local minima) [152]. Common practice is to use another blind algorithm until the eye diagram is opened; then, the operation is switched to decision-directed mode.

Sato's algorithm pioneered the field of blind equalization. Given a real constellation and real received samples, it uses

$$e\left(\tilde{R}_m\right) \triangleq \tilde{R}_m - \text{signum}\left(\tilde{R}_m\right). \tag{16.2.16}$$

Essentially, the symbols are quantized as pseudobinary, and this increases the decision reliability when large constellations are employed. It is demonstrably less likely to converge to a local minimum than is the decision-directed version. However, it is known that it can get hung-up when initializing a blind DFE [153], [154].

The algorithm of choice is the CMA [154]. It attempts to ensure that the quantizer input lies as close as possible to a circle with radius $\sqrt{C}$ by minimizing

$$\epsilon_{CMA} \triangleq E\left\{\left(\left|\tilde{R}_m\right|^2 - C\right)^2\right\} \tag{16.2.17}$$

with

$$C \triangleq \frac{E\left\{\left|\tilde{R}_m\right|^4\right\}}{E\left\{\left|\tilde{R}_m\right|^2\right\}}. \tag{16.2.18}$$

It can be used for real/complex signals and constant/nonconstant envelope modulations. In recursive form, the error factor is calculated as

$$e\left(\tilde{R}_m\right) = \tilde{R}_m\left[\left|\tilde{R}_m\right|^2 - C\right]. \tag{16.2.19}$$

Given diversity (through fractional sampling, for instance), a finite-length equalizer adapted by the CMA can be shown to be *globally convergent*, i.e., to converge to the optimum setting whatever its initialization.

Other work on blind equalization includes [155], [156].

16.2.2 Equalizers employing channel estimates

In this section we consider different algorithms for joint channel estimation and equalization, namely, adaptive MLSDs, PSP-MLSDs and MAPBD/MAPSDs. Receiver design requires careful selection from this wide set of equalization structures to achieve a satisfying trade-off between complexity and performance.

Adaptive MLSDs

The MLSD detectors illustrated in Section 16.1 were derived under the assumption of known CIR. In practice, channel estimation is required and *adaptive* MLSD strategies provide the simplest structures. In fact, in this case, a *single* channel estimator operates jointly with the equalization algorithm, as illustrated in Figure 16.15. In this structure, the transmitted symbol sequence is detected by the estimated channel impulse response so that the known channel impulse response $\{q_{i,k-i\eta}\}$ in (16.1.4) is replaced by its estimate $\{\hat{q}_{i,k-i\eta}\}$. Then, the branch metrics of an adaptive LMLSD based on the VA are given by

$$\lambda\left(x_m, x_{m+1}\right) = \sum_{k=m\eta}^{(m+1)\eta-1} \left| r_k - \sum_{i=m-L_q+1}^{m} \tilde{c}_i \, \hat{q}_{i,k-i\eta} \right|^2 . \tag{16.2.20}$$

The channel estimate is updated according to the detected sequence provided by MLSD [109], [3]. If a matched filter is used as a front end, it too must be updated [3]. In practical applications, data transmission normally starts with a training sequence to initialize the channel estimator. The receiver is then switched to a decision-directed mode, where tentative decisions from the survivor sequence with best metric [157] are fed back to the estimator from the Viterbi processor after a certain delay [158], [159], [20]. The Viterbi decision delay, K, is on the order of five times the channel memory, L_q, [79], and, consequently, the channel estimator has only outdated information available, and the estimate suffers a lag error [160], [157]. For time-varying channels, this lag error must be traded off against the accuracy of the tentative decisions, so that the tentative decision delay d can be chosen to be shorter than the Viterbi decision delay [109]. We can also diminish the lag error by predicting the channel estimate [27], [116], [161]. In this case, a channel predictor is inserted at the channel estimator output, as shown in Figure 16.15.

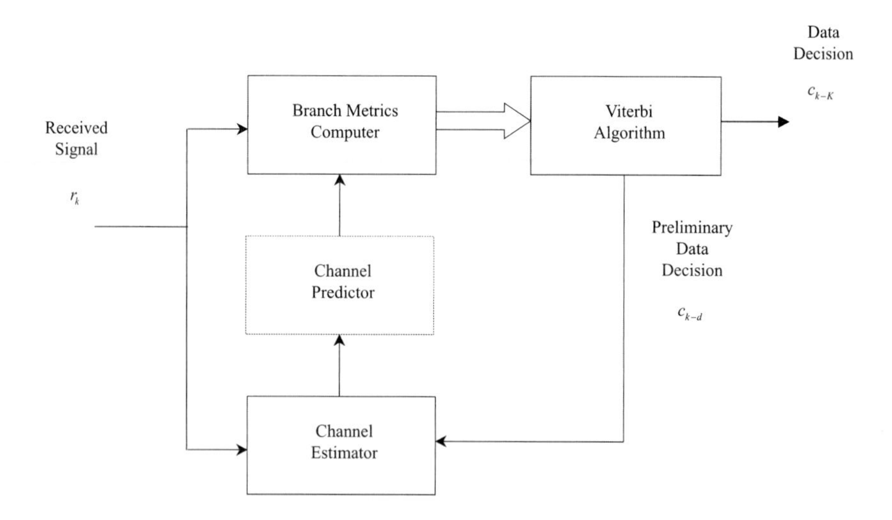

Figure 16.15. Block diagram of a conventional adaptive MLSD. A channel predictor can be inserted to mitigate the lag effects due to channel estimation with tentative decisions.

As with linear equalizers, a fast-acquiring algorithm for channel estimation is desired [27]. For tracking, the low-complexity LMS algorithm is invariably favored both in adaptive MLSD and in adaptive PSP MLSD [3], [160], [109], [162], [20] (see also the following subsection). The channel estimator inputs are the MLSD's tentative decisions. As mentioned in Section 16.2.1, if the input data correlation is low, the tracking ability of the LMS [119], [163], [120], [121], [122] and RLS [118], [120], [164], [165], [132] algorithms approximately match. The BER performance of the LMS-adaptive MLSD with prediction is analyzed in [116], where it is shown that, even in the presence of channel prediction, an error floor occurs with fast fading. Thus, adaptive MLSD is suited only to slowly fading channels.

Adaptive MLSD can be applied to GSM-like systems, as shown in [158], [159], where their error performance is simulated.

The performance improvement offered by trellis-coded-modulation over DS channels is addressed in [20], [166], where the need for effective, low-complexity equalizers compatible with interleaving is identified.

Adaptive MLSD is compared with other equalization strategies in [167], [69], [166], [168].

PSP-MLSDs

The tentative decision delay of adaptive MLSD is unsatisfactory for time-varying channels and to avoid this, *per survivor processing* (PSP) [111] can be applied [160], [169], [5], [162], [170], [171], [172], [87], [173]. Then, a channel estimator is associated with each survivor of an MLSD with an M^{L_q-1}-state trellis, and the corresponding CIR estimate is updated with no lag by use of the survivor's hypothesized symbols [111]. Therefore, the detection metric (16.2.20) becomes

$$\lambda\left(x_m, x_{m+1}\right) = \sum_{k=m\eta}^{(m+1)\eta-1} \left| r_k - \sum_{i=m-L_q+1}^{m} \tilde{c}_i \, \breve{q}_{i,k-i\eta}\left(x_m\right) \right|^2 \tag{16.2.21}$$

where $\{\breve{q}_{i,k-i\eta}\left(x_m\right)\}$ represents the CIR estimate associated with the state x_m at the mth decoding step. A generic block diagram of a PSP-based MLSD is shown in Figure 16.16.

Adaptive PSP MLSD is motivated by the inadequacy of adaptive MLSD in time-varying channels, but it is never truly optimal [174] since, with time-varying channels, the decoding trellis of the ML strategy never becomes a finite-state trellis, as described in Section 16.3.1. LMS [169], [160], [5], [162], [111] or RLS [175] algorithms can employed for channel tracking, but LMS is usually preferred, as explained above. Error floors can still occur, however, even with relatively slow fading. We can appreciably improve the RLS algorithm performance in fast fading by explicitly modelling the time variation of the channel taps with additional parameters, as illustrated in [123], [124].

The performance improves as the quality of the channel estimate gets better. With FF fading, accurate channel prediction can be obtained with PSK and CPM

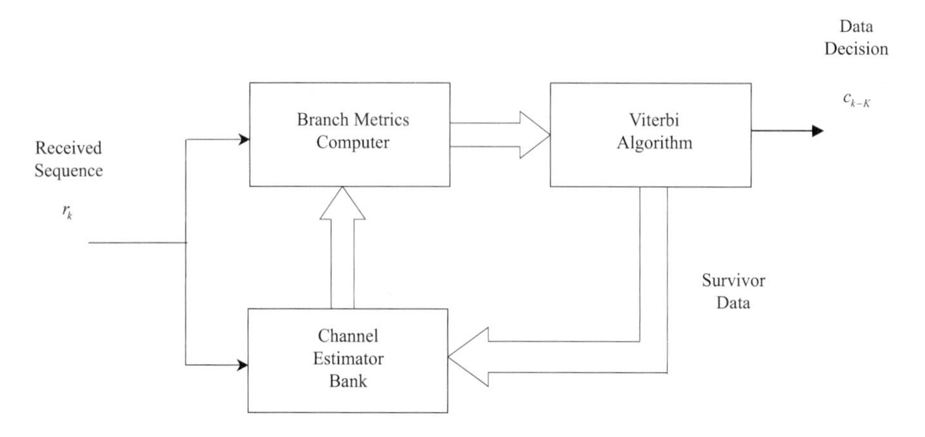

Figure 16.16. Block diagram of a PSP-MLSD

signals resorting to the geometrical approach illustrated in [87], [173]. In this case, a linearly time-selective channel model[7] is adopted [176], that is, the fading distortion is approximated by a straight line over the duration of the transmitter pulse for linearly modulated signals and over the signalling interval for CPM signals, thus allowing for linear changes in the fading with time. The implicit time diversity can then be extracted by two matched filters (for a linearly modulated signal) [87] or two filter banks [173] (for CPM signals). The matched filter outputs provide accurate information about the fading process. Both types of information can be exploited to produce improved per-survivor channel predictions.

Adaptive MLSD and adaptive PSP-MLSD are compared in typical GSM scenarios in [157]. For the nonfading but Doppler-shifted channel, the additional complexity of adaptive PSP MLSD is found to be unwarranted, although in [160], it outperformed adaptive MLSD.

Dai and Shwedyck [177] propose the use of an *autoregressive moving average* (ARMA) channel model of the CIR and a bank of Kalman filters (KFs) for channel tracking. Unfortunately, the complexity of this PSP-KF receiver structure can be prohibitive, especially for long CIRs, due to the filtering complexity. Approximations to the PSP-KF scheme are proposed by Rollins and Simmons in [175]. Complexity reduction is achieved by means of a prediction-feedback mechanism exploiting the parallel structure resulting in the ARMA model when certain model parameters are neglected. The resulting *reduced complexity Kalman filter* (RCKF) receiver provides significant performance improvement over a PSP receiver with RLS channel estimation and only a moderate degradation with respect to the KF-PSP.

At the receiver startup, a PSP-based MLSD can estimate the channel blindly. In particular, fast startup in FF fading is provided by the blind algorithms proposed

[7]This model is expressed by (14.2.57) with $N_{TS} = 2$.

in [87], [173]. Blind acquisition properties of PSP-MLSDs with frequency-selective channels are discussed in [178].

Finally, it is worth remembering the following:

1. Array processing can be incorporated in PSP-based MLSDs. Some error performance results are shown in [179].

2. *Generalized* PSP algorithms, retaining multiple (instead of one) survivors per state, have been proposed. They are investigated in [169].

3. As a performance-complexity trade-off, the number of channel estimators in a PSP-based detector can be reduced [110].

4. An MLSD algorithm employing *per-branch processing* (PBP) embedded in a Viterbi algorithm for joint estimation of the channel CIR and the symbol timing is described in [170]. It employs a bank of M^{L_q}-extended Kalman filters (EKFs).

Recently, the theory of adaptive PSP and PBP-MLSDs has been studied in a framework based on the *expectation* and *maximization* (EM) algorithm. In this context, a generalized ML sequence detection and estimation (GMLSE) criterion has been derived following the EM approach and it has been shown that per-survivor and per-branch processing methods emerge naturally from GMLSE [180].

MAPBD/MAPSDs

As shown in Sections 16.1.6 and 16.1.7, two types of MAP algorithms have been developed for practical applications: one, dubbed MAP-FBA, performs forward and backward recursions [181]; the other, denoted as MAP-FLA, performs only a forward recursion [93]. In time-varying environments, the channel parameters are initially estimated by means of a training sequence and afterwards are adaptively tracked using the decisions of the detection algorithm. Unfortunately, MAP-FBAs cannot deliver reliable decisions to the channel estimator until both the forward and the backward recursions have been completed. For this reason it is reasonable to say that MAP-FBAs are not suitable for time-variant channels when a *single* channel estimator, fed by tentative decisions, is employed in the receiver structure [94].

On the contrary, in this case, a MAP-FLA can be used, as shown in [182], [183], [184], where adaptive MAP receiver structures are described. They recursively deliver reliable hard decisions and employ a *single* Kalman-type channel estimator, as illustrated in Figure 16.17. Such a channel estimator is not fed with tentative data decisions, but with the *a posteriori probabilities* (APPs) of the states of the ISI channel. A training sequence is employed for receiver startup. Simulation results show that the use of soft statistics in channel estimators improves their channel-tracking capabilities and that the proposed MAPSD outperforms conventional adaptive MLSDs on time-varying channels. This approach can be also applied to sparse channels [183] in a way that the receiver complexity does not depend on the overall memory of the channel, but on its few non-zero taps.

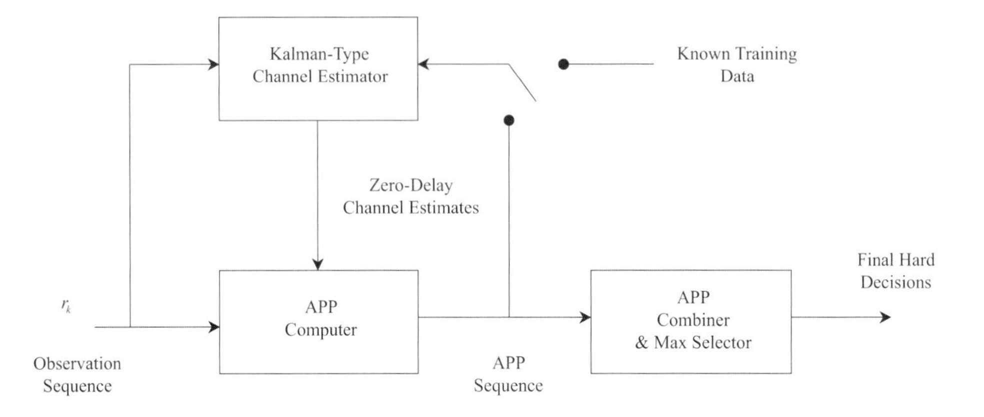

Figure 16.17. Block diagram of an adaptive MAP-FA with single-channel estimator

Channel tracking performance and, consequently, error performance can be improved by the use of multiple channel Kalman estimators, as illustrated in [155], where a *blind* MAP equalization algorithm is derived. A parallel bank of M^{L_q} conditional channel estimates, each corresponding to a different subsequence of length L_q, is updated using a forward MAP strategy. Although the number of channel estimates required for blind MAPSDs is M times larger than needed in blind PSP-MLSDs, it should be noted that the former algorithms exhibit superior performance over the latter ones with fast fading.

If the symbol timing is unknown [185], we can still use the last approach by including an additional ISI term in the receiver memory to account for this uncertainty. This approach leads to an M^{L_q+1}-state MAPSD algorithm employing a bank of *extended* KFs (EKFs) and jointly estimating the channel coefficients and the symbol timing, as shown in Figure 16.18. During the receiver training with a known preamble sequence, an auxiliary EKF is employed to estimate the channel statistics. At the end of the training period, the auxiliary filter is employed to appropriately initialize both the parallel bank of EKFs and the MAP subsequence metrics. For the remaining symbols in the data block, the MAP algorithm makes symbol-by-symbol decisions, whereas the fading channel is tracked by the filter bank. As in [155], different suboptimal approaches can be employed to lower the complexity of the EKF-based MAPSD algorithm, namely: metric pruning to reduce the number of filter and metric computations, the use of LMS filters in place of EKFs, or channel memory truncation combined with decision feedback.

Adaptive versions of the optimum soft-output and the suboptimum soft-output algorithms [94] have been proposed in [186]. Conventional adaptive and per-survivor processing extensions of the above-mentioned algorithms are illustrated and com-

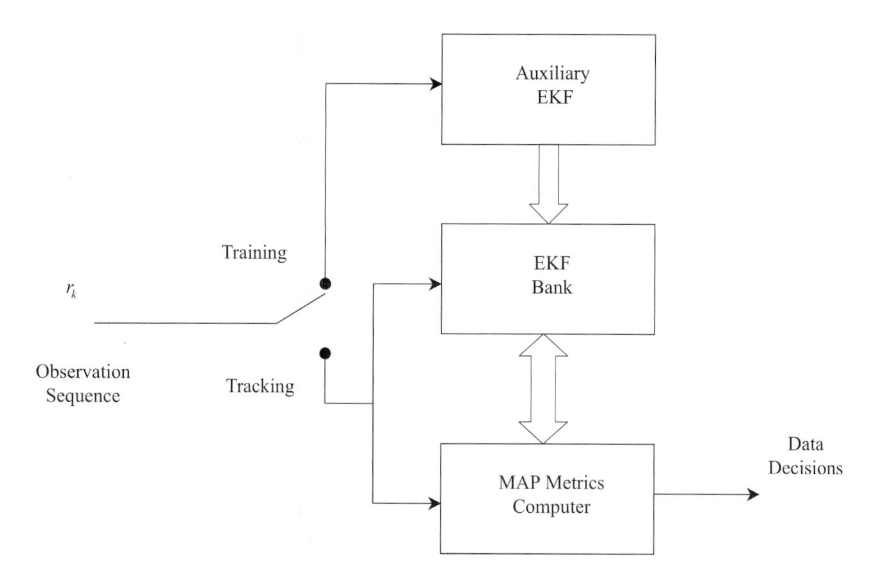

Figure 16.18. Dual-mode EKF-based MAPSD [185]

pared with the performance attained by use of a PSP-MLSD as a first stage in a receiver for an interleaved coded system operating over a DS channel. One of the most significant results is that PSP-SSA has a computational complexity comparable to PSP-MLSD but offers superior performance.

All the algorithms described above belong to the class of MAP-FLAs. Nonetheless, MAP-FBAs can be employed on time-varying channels [187], [188]. In particular, with fading channels the channel estimation process can be embedded in both the forward and the backward recursions, reasoning as the per-survivor receiver. These ideas have been exploited in [188] to devise a fractionally spaced MAP equalizer for DS channels. The forward-backward algorithm operates on an expanded state trellis and accomplishes *per-state* joint MMSE channel estimation and soft data detection.

Finally, it is worth noting that a MAP receiver for FF fading channels has been developed in [189]. The detection algorithm employs a bank of matched stochastic nonlinear filters and is robust against mistuning of receiver parameters.

16.2.3 Inverse filtering

Adaptive LE

The coefficients of a LE's transversal filter (see Subsection 16.1.4) must be adapted to the channel. This adaptation normally occurs as an *acquisition* phase followed by a *tracking* phase (e.g., [190], [191], [167], [192]). During acquisition, either a blind algorithm, such as CMA, or one of the LMS, RLS or LS algorithms in conjunction with a training sequence is used. The speed of acquisition of the CMA and the LMS algorithms is poor, so their use is largely restricted to broadcast signals or to the case when a long training interval is available, such as in frequency-division multiple-access (FDMA) systems. In time-division, multiple-access (TDMA) systems, a relatively short training sequence is usually provided and therefore the RLS or LS algorithms are appropriate. It is also possible to employ the LMS algorithm to train the equalizer, as long as the training data are repeated multiple times.[8] Given a steadily diminishing step-size, the coefficients should converge to their LS solution.

The tracking phase invariably uses decision-directed LMS adaptation, and therefore the acquisition phase should terminate only when sufficient reliability is attained, such as a SER below $5 \cdot 10^{-2}$. If this error rate cannot be achieved, because of a poor SNR, then the decisions after the *decoder*, when re-encoded, may be reliable enough. Certainly the channel code should be exploited where possible [62], [193]. The additional delay requires a reduced step size for stability [194]. Otherwise, adaptation through a blind algorithm should continue (and indeed if and when the SER ever exceeds $5 \cdot 10^{-2}$ again, such as when the channel enters a fade). The SER can be estimated either from the degree of clustering of the quantizer input or, more directly, by comparing the LE's decisions with the re-encoded decoder's decisions.

As already stated, the RLS algorithm is markedly more complicated than the LMS algorithm, yet its tracking performance is not compellingly better [112], [195], [159]. Thus, the RLS algorithm is little used for tracking.

Adaptive DFE

Like an LE, DFEs (see Section 16.1.3) are normally adapted through the decision-directed LMS algorithm. Acquisition with LMS algorithm and a training sequence is also a simple extension of the LE case. However, blind DFE acquisition is much more complicated, since initially a high proportion of erroneous decisions are fed back. One method is to set the DFE's feedback filter's coefficients to zero and adapt the DFE's feedforward filter only during acquisition. Once sufficient reliability is achieved (say, an SER of 0.2), then the feedback filter is adapted also. However, in highly frequency-selective channels with marginal SNRs, the LE may never achieve the desired reliability, whereas a DFE could. Probably the best available scheme for blind DFE acquisition is presented in [196]. An IIR LE is adapted with the CMA algorithm, and once adequately reliable decisions are available, the feedback portion of the IIR filter becomes the FBF's coefficients.

The performance degradation due to imperfect adaptation is considered in [197], [198]. The LMS algorithm's rate of acquisition and tracking is degraded when the input is correlated [199], such as by a delay spread channel. In this case, it is reasonable, albeit highly computationally intensive, to estimate the time-varying

[8]Therefore, the received samples after the training sequence must be buffered, and the receiver must be able to catch up afterwards, perhaps during a pause in transmission or perhaps by operating faster than the symbol rate.

channel impulse response and from that calculate the DFE's coefficients [200], [201], [202].

As with RSSD, the best performance is achieved when the cascade of overall channel and FFF is minimum phase, yet this is difficult to ensure in a time-varying channel. Reversing time may convert a channel with a undesirably large precursor into one that can be equalized more easily [203], [192], [204].

16.2.4 Equalization strategies employing reference-based channel estimators with FF fading

As shown in Section 16.1.5, the use of coherent detection with MRC or ECG on slow FF fading channels requires knowledge of the fading distortion in each symbol interval. In practice, coherent detection is possible if a reference (or sounding) signal [205], [128] is transmitted with the information bearing signal. An accurate phase reference cannot be generated by a Phase-Locked-Loop (PLL) because a PLL cannot track the rapid phase changes of the channel fading [206]. A coherent reference can be made available to the receiver by transmission of a time-continuous sounding signal (*pilot tone*) [207], [208], [209], [210], [211], [212], [213], [214], [215], [216] or by transmission of a sequence of known symbols (*pilot symbols*) interspersed with the data symbols [216], [217], [218], [219], [220], [221]. Several pilot tone techniques have been proposed as follows:

a) The technique described in [207], which consists of sending a continuous wave sounding signal together with a data BPSK signal. The data and the sounding signals can be separated since they are kept orthogonal in phase;

b) The *transparent-tone-in-band* (TTIB) technique [211], where the baseband spectrum is split into two segments. The segment in the upper frequency band is translated up in frequency by an amount equal to the "notch" width, and a reference pilot tone is added at the center of the resulting notch.

c) The *tone calibration technique* (TCT) [209] creates a spectral null in the data signal by means of a zero DC encoding technique (e.g., Manchester coding [19]) and inserts a pilot tone in the null. The TCT scheme is illustrated in Figure 16.19, together with the baseband spectrum of the transmitted signal.

d) The *dual-pilot tone calibration* (DPTC) technique [208], where two pilots are symmetrically located outside the data spectrum near the band edges. DPTC provides better bandwidth efficiency than does TCT, at the price of increased sensitivity of the pilots to frequency shifts.

Pilot-tone techniques lead to robust and simple receiver structures, as evidenced by Figure 16.19. The pilot tone can be separated by relatively simple circuitry from the received signal. The use of coherent detection substantially lowers the error floor level of the receiver. The main disadvantage of pilot tone techniques (and also of the pilot symbol techniques) is that a fraction of the transmitted power is wasted in transmitting reference signals.

Simpler transmitter and receiver processing is achieved by pilot symbol-assisted modulation (PSAM) [216], [217], [218], [219], [220], [221], although frame synchro-

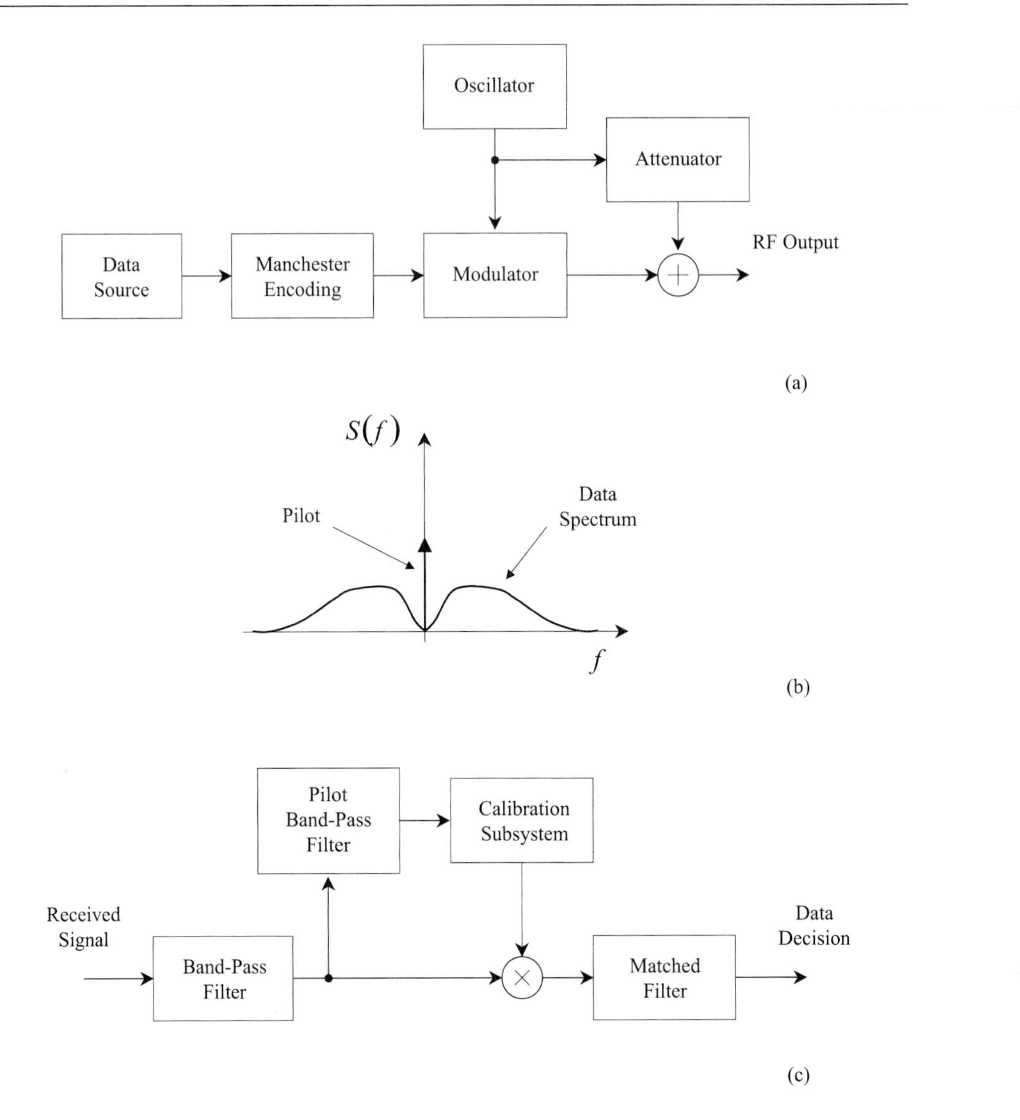

Figure 16.19. Block diagram of the TCT transmitter (a) and receiver (c); power density spectrum of the baseband transmitted signal (b)

nization is required at the receiver. In PSAM transmission, the transmitter periodically sends known symbols, from which the receiver derives its amplitude and phase reference. The PSAM transmitter and receiver schemes are shown in Figure 16.20, together with the transmitted data format. Here, the data-symbol rate is equal to

$$R_s = \frac{K-1}{KT} \tag{16.2.22}$$

$1/KT$ being the *pilot symbol rate*.[9] Like pilot-tone modulation, PSAM suppresses the error floor and offers the further advantage of enabling multilevel modulation without requiring a change of the transmitted pulse shape or of the peak-to-average power ratio. A comparison of PSAM with TTIB [216] has shown it offers substantially better energy efficiency for any practical power amplifier.

Finally, we note that reference-based techniques for coherent detection were first proposed for linearly modulated signals. Recently, Ho and Kim [222] have shown that a pilot symbol-assisted-detection strategy can be implemented for CPM signals.

16.3 EQUALIZATION WHEN THE CIR IS AVERAGED-OVER

As shown in Section 15.4, if the channel statistics are known, different formulations are possible for the ML detection metrics [223], [224]. Whatever the formulation, the optimal detection strategy requires (at least conceptually) the evaluation of the optimal metric (15.4.2) for all M^{L_c} possible data sequences and the selection of the data sequence having minimum metric. One approximation to this brute-force strategy is developed in [225]. Here, the sequences are divided into short blocks, with some overlap at the block edges. Within each block is a relatively small number of data subsequences, and so metrics for each may be feasibly calculated. The subsequence with best metric is detected, then the next block is processed. Symbols that overlap into the next block are treated as exactly known, with values given by the decisions from the previous block. The short block technique is somewhat suboptimal for symbols on the block boundaries.

Most of the ML sequence detectors employing known channel statistics can be derived from (or can be related to) the innovation-based formulation (15.4.21) of the optimal metric, as shown in the following subsection. First, innovation-based MLSDs are developed and are related to adaptive MLSDs. Second, it is shown how the same ideas can be applied in the design of MAPSDs. Finally, other equalization/detection strategies for FF fading channels are considered.

16.3.1 Innovation-based receivers

Innovation-based MSLDs

When the CIR is averaged over, the MLSD metric can be expressed as (see Section 15.4.1)

$$\Lambda(\tilde{\mathbf{c}}) = -\log \prod_{k=0}^{N_z-1} p\left(r_k \middle| \mathcal{R}_{k-1}, \tilde{\mathbf{c}}\right) = -\sum_{k=0}^{N_z-1} \log p\left(r_k \middle| \mathcal{R}_{k-1}, \tilde{\mathbf{c}}\right) \tag{16.3.1}$$

[9] The pilot symbol rate should be at least $2B_{D_{MAX}}$, $B_{D_{MAX}}$ being the largest value of the Doppler bandwidth B_D.

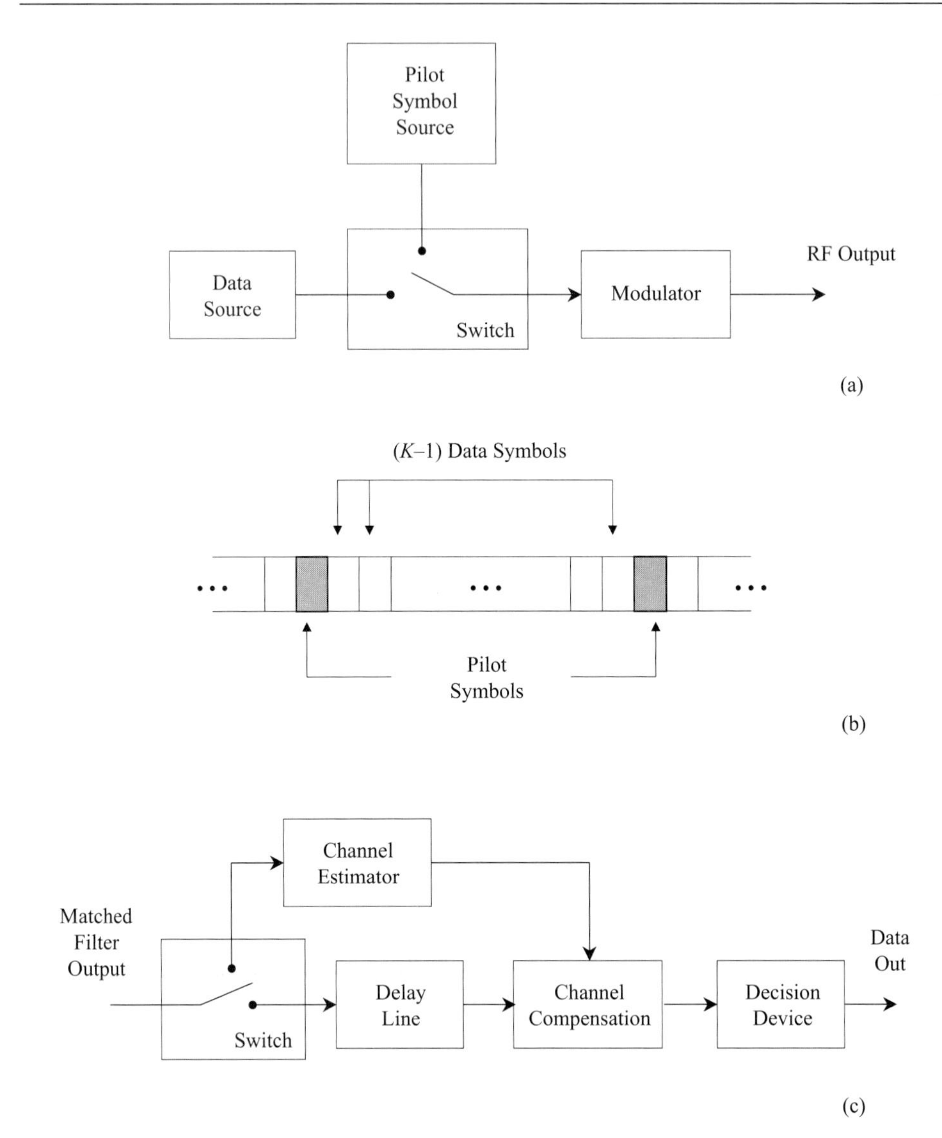

Figure 16.20. PSAM transmitter (a) and receiver (c); location of the pilot symbols in the transmitted symbol sequence

or, equivalently, as

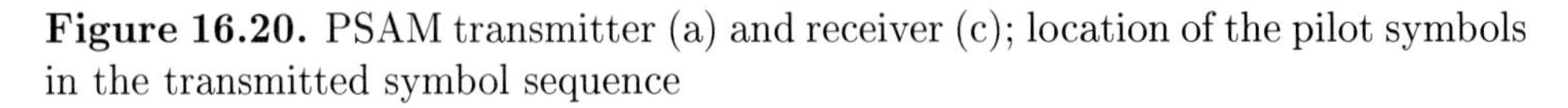

$$\Lambda(\tilde{\mathbf{c}}) = \sum_{k=0}^{N_z-1} \left| \frac{r_k - \eta_{k|\mathcal{R},\tilde{\mathbf{c}}}}{\sqrt{2\,\sigma^2_{k|\mathcal{R},\tilde{\mathbf{c}}}}} \right|^2 + \log\left(2\pi\,\sigma^2_{k|\mathcal{R},\tilde{\mathbf{c}}}\right) \tag{16.3.2}$$

where $\eta_{k|\mathcal{R},\tilde{\mathbf{c}}} \triangleq E\{r_k | \mathcal{R}_{k-1}, \tilde{\mathbf{c}}\}$ and $\sigma^2_{k|\mathcal{R},\tilde{\mathbf{c}}} \triangleq (1/2)\, E\{|r_k - \eta_{k|\mathcal{R},\tilde{\mathbf{c}}}|^2\}$. The MLSD applies the decision strategy

$$\hat{\mathbf{c}} = \arg\min_{\tilde{\mathbf{c}}} \Lambda(\tilde{\mathbf{c}}) \tag{16.3.3}$$

so that, in general, a search over the set $\{\tilde{\mathbf{c}}\}$ of M^{L_c} sequences must be accomplished. This operation can be diagrammatically represented as looking for the path with minimum metric in a trellis having number of states exponentially increasing with time, as illustrated in Figure 16.21 (a). Let us suppose now that the densities $\{p\,(r_k | \mathcal{R}_{k-1}, \tilde{\mathbf{c}})\}$ in (16.3.1) depend only on a finite history of the data, i.e.,

$$p\,(r_k | \mathcal{R}_{k-1}, \tilde{\mathbf{c}}) = p\left(r_k | \mathcal{R}_{k-1}, \tilde{\mathbf{c}}^k_{k-L_m}\right), k = 1, 2, ..., N_z - 1 \tag{16.3.4}$$

for a proper value of the parameter L_m, with

$$\tilde{\mathbf{c}}^k_{k-L_m} \triangleq [\tilde{c}_{k-L_m}, \tilde{c}_{k-L_m+1}, \; ..., \tilde{c}_{k-1}]^T \tag{16.3.5}$$

Then, the trellis of Figure 16.21 (a) *folds* the M^{L_m}-state trellis of Figure 16.21 (b) and, as with ISI channel, the search for the optimal path in the trellis can be carried out by means of the Viterbi algorithm [226]. The equality (16.3.4) is known as the *folding condition* [174], and the parameter L_m is called the *channel memory.* It is worth noting that such condition is equivalent to the following couple of equalities (see (15.4.20))

$$\eta_{k|\mathcal{R},\tilde{\mathbf{c}}} = \eta_{k|\mathcal{R},\tilde{\mathbf{c}}^k_{k-L_m}} \tag{16.3.6}$$

$$\sigma^2_{k|\mathcal{R},\tilde{\mathbf{c}}} = \sigma^2_{k|\mathcal{R},\tilde{\mathbf{c}}^k_{k-L_m}} \tag{16.3.7}$$

for $k = 1, 2, ..., N_z - 1$.

In [227], [67], [226], [172], [228], it is claimed that, with FF fading, folding occurs if one of the following conditions holds:

A.1—the sequence

$$x_k \triangleq \alpha_k + n_k \tag{16.3.8}$$

of fading plus noise samples is an *autoregressive* (AR) process of finite order L_m [227], [67], [172], [228];

A.2—the channel has finite coherence time. This occurs if the autocovariance function $C_a\,(\tau)$ of the fading distortion has finite support, i.e.,

$$C_a\,(kT) = 0 \tag{16.3.9}$$

for $k > L_m$ [226].

With DS fading channels [226], [228], [229], [177], finite time dispersion is required together with one of the previous assumptions for all the taps in the delay

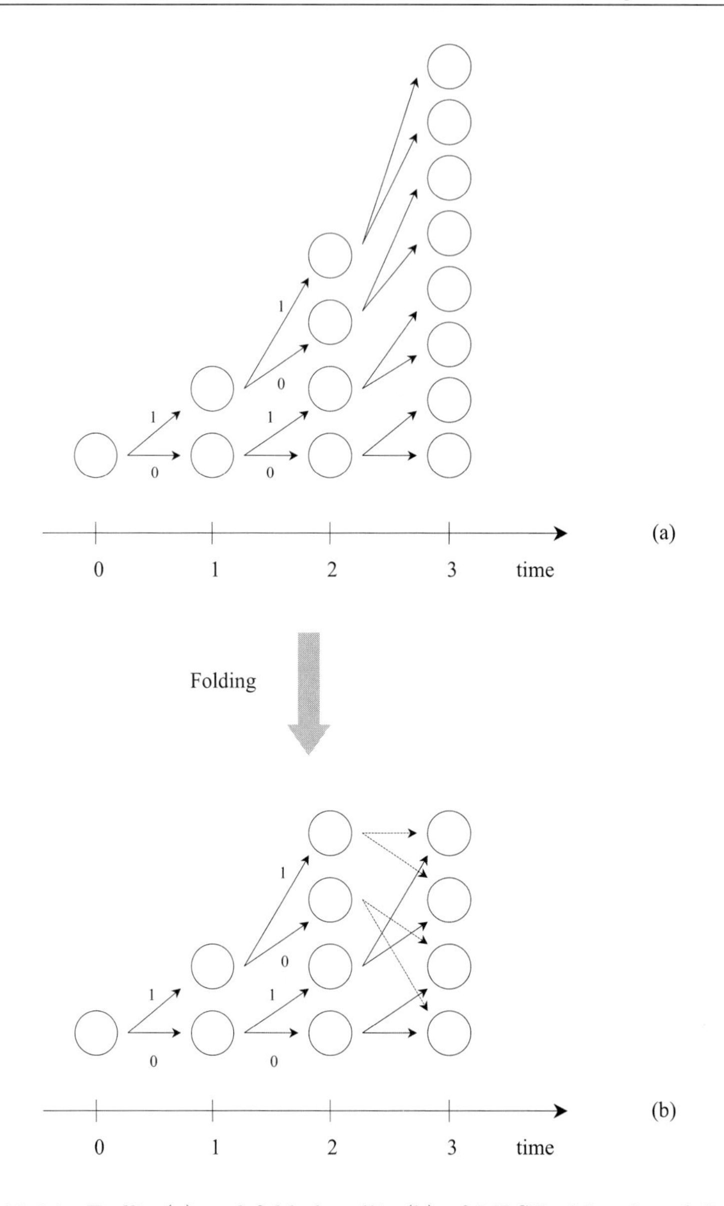

Figure 16.21. Trellis (a) and folded trellis (b) of MLSD. $M = 2$ and $L_m = 2$ are assumed.

line model of the channel [226], [228], or, alternatively, an *autoregressive moving average* (ARMA) model of the CIR vector [177]. For instance, in [228] it is shown that, if the memory associated with the channel time variations is $L_{m\alpha}$ symbol intervals and the channel delay spread is L_{ds} symbol intervals, a trellis with M^{L_m}

states is required with

$$L_m = L_{ds} + L_{m\alpha} - 1. \tag{16.3.10}$$

With finite memory channels the relevant quantities $\{\eta_{k|\mathcal{R},\tilde{\mathbf{c}}^k_{k-L_m}}\}$ (16.3.6) (and, in some instances, $\{\sigma^2_{k|\mathcal{R},\tilde{\mathbf{c}}^k_{k-L_m}}\}$ (16.3.7)) can be evaluated by means of finite-length estimation filters, i.e., time-invariant Wiener predictors [227], [226], [228] or Kalman filters [230], [231], [232], [177].

Innovation-based receivers provide good error performance in fast fading at the price of large complexity. In practice, significant complexity reduction can be achieved. The number of symbols constituting each trellis state can be reduced from L_m to Q (where Q should be properly selected [172]). By predicting the fading process with L_m-tap optimal filters, we can use the survivor path symbols in place of the missing state symbols [172]. This approach represents a form of decision feedback and has been exploited in [24], [25], [26], to reduce the complexity of the ML receiver [2] for frequency-selective channels.

In [174], however, Chugg proved that the folding condition is never met for our model of a DS (or, in particular, FF) channel, and, consequently, all the equalizers mentioned above are examples of *forced folding*. In other words, they are a suboptimal approximation to the optimal estimator-correlator receiver structure.

Let us now comment on the interpretation of innovation-based MLSDs and concentrate, for simplicity, on the case of FF fading channel with PSK or CPM signals. In these cases the contribution of the quantities $\{\sigma^2_{k|\mathcal{R},\tilde{\mathbf{c}}^k_{k-L_m}}\}$ is usually neglected in (16.3.1) [227], [172], [233] so that the optimal metric becomes

$$\Lambda(\tilde{\mathbf{c}}) = \sum_{k=0}^{N_z-1} \left| r_k - \eta_{k|\mathcal{R},\tilde{\mathbf{c}}^k_{k-L_m}} \right|^2 \tag{16.3.11}$$

As shown in Section 15.4.5, the conditional mean $\eta_{k|\mathcal{R},\tilde{\mathbf{c}}_k(L_m)}$ can be rewritten as

$$\eta_{k|\mathcal{R},\tilde{\mathbf{c}}^k_{k-L_m}} = \tilde{d}_k \, \tilde{\alpha}\left\{ k|\, \mathcal{R}_{k-1}, \tilde{\mathbf{c}}^k_{k-L_m} \right\} \tag{16.3.12}$$

where $\tilde{\alpha}\left\{ k|\, \mathcal{R}_{k-1}, \tilde{\mathbf{c}}^k_{k-L_m} \right\}$ is the MMSE one-step prediction of the fading sample α_k, based on $\mathcal{R}_{k-1}$ and assuming that the sequence $\tilde{\mathbf{c}}^k_{k-L_m}$ has been transmitted, and $\tilde{d}_k$ is the kth useful signal sample corresponding to the transmission of the trial sequence $\tilde{\mathbf{c}}$. For instance, in [172] and [227], $\eta_{k|\mathcal{R},\tilde{\mathbf{c}}^k_{k-L_m}}$ is evaluated as

$$\eta_{k|\mathcal{R},\tilde{\mathbf{c}}^k_{k-L_m}} \triangleq \tilde{c}_k \sum_{l=1}^{L_a} p_k \, r_{k-l} \, \tilde{d}^*_{k-l} \tag{16.3.13}$$

where $\{p_k, \;\; k = 1, 2, ..., L_a\}$ denote the coefficients of the one-step MMSE linear predictor (with $L_a \geq L_m$ taps) for the sequence $\{x_k\}$ (16.3.8) of fading plus noise samples. Thus, a VA-based receiver implicitly evaluates multiple channel estimates, i.e., as many channel estimates (carrier references) as the number of trellis states.[10] Each fading estimate is evaluated conditioned on the channel symbols corresponding to a survivor path in the trellis. Then, an innovation-based MLSD can be interpreted as an adaptive MLSD with per-survivor processing [110], provided that proper filtering is employed for channel estimation. Following this approach, VA-based algorithms can be heuristically derived, as in [234]. A similar interpretation of PSP-MLSDs can be provided in the DS case.

MAPSD

Some of the ideas developed for innovation-based MLSD also apply to MAPSD. Here, we concentrate on the FF case, for simplicity, and, in particular, we assume linear modulation and baud-rate sampling ($\eta = 1$) at the receiver side [235]. Then, following the same approach as Section 16.1.7, the L_{fl}-lag symbol-by-symbol MAP strategy can be expressed as

$$\hat{c}_m = \arg\max_{\tilde{c}_m} P\left(\tilde{c}_m | \mathbf{r}_0^{L_{fl}+m}\right) \tag{16.3.14}$$

where the conditional probability $P(\tilde{c}_m | \mathbf{r}_0^{L_{fl}+m})$ is given by

$$P\left(\tilde{c}_m | \mathbf{r}_0^{L_{fl}+m}\right) = \frac{\sum_{\tilde{\mathbf{c}}_{m+1}^{m+1+L_f}, \tilde{\mathbf{c}}_0^{m-1}} p\left(\mathbf{r}_0^{L_{fl}+m} | \tilde{\mathbf{c}}_0^{L_{fl}+m}\right) P\left(\tilde{\mathbf{c}}_0^{L_{fl}+m}\right)}{\sum_{\tilde{\mathbf{c}}_0^{m+1+L_f}} p\left(\mathbf{r}_0^{L_{fl}+m} | \tilde{\mathbf{c}}_0^{L_{fl}+m}\right) P\left(\tilde{\mathbf{c}}_0^{L_{fl}+m}\right)}. \tag{16.3.15}$$

The quantities $\{p(\mathbf{r}_0^{L_{fl}+m} | \tilde{\mathbf{c}}_0^{m+L_{fl}}) P(\tilde{\mathbf{c}}_0^{L_{fl}+m})\}$ represent a set of $M^{m+L_{fl}}$ sufficient statistics for making a MAP decision on c_m with lag L_{fl}. In principle, the computation of such statistics can be carried out by means of a recursive algorithm [93]. In fact, if we assume that the same quantities at the previous step, i.e., $\{p(\mathbf{r}_0^{L_{fl}+m-1} | \tilde{\mathbf{c}}_0^{L_{fl}+m-1}) P(\tilde{\mathbf{c}}_0^{L_{fl}+m-1})\}$, are known and that the sample $r_{m+L_{fl}}$ has

[10] An unambiguous phase reference can be computed only if the transmitted signal contains known features, such as a training sequence, pilot symbols, or a pilot tone. Rotationally invariant codes make an absolute phase reference unnecessary.

been observed, the product $p(\mathbf{r}_0^{L_{fl}+m}|\tilde{\mathbf{c}}_0^{m+L_{fl}})\,P(\tilde{\mathbf{c}}_0^{L_{fl}+m})$ can be also expressed as

$$
\begin{aligned}
&p\left(\mathbf{r}_0^{L_{fl}+m}|\tilde{\mathbf{c}}_0^{L_{fl}+m}\right)\,P\left(\tilde{\mathbf{c}}_0^{L_{fl}+m}\right)\\
&= p\left(\mathbf{r}_0^{L_{fl}+m-1}, r_{m+L_{fl}}|\tilde{\mathbf{c}}_0^{L_{fl}+m-1}, \tilde{c}_{L_{fl}+m}\right)\,P\left(\tilde{\mathbf{c}}_0^{L_{fl}+m-1}, \tilde{c}_{L_{fl}+m}\right)\\
&= p\left(r_{L_{fl}+m}|\mathbf{r}_0^{L_{fl}+m-1}, \tilde{\mathbf{c}}_0^{L_{fl}+m-1}, \tilde{c}_{L_{fl}+m}\right)\,p\left(\mathbf{r}_0^{L_{fl}+m-1}|\tilde{\mathbf{c}}_0^{L_{fl}+m-1}, \tilde{c}_{L_{fl}+m}\right)\\
&\cdot P\left(\tilde{\mathbf{c}}_0^{L_{fl}+m-1}\right)\,P\left(\tilde{c}_{L_{fl}+m}\right)\\
&= P\left(\tilde{c}_{L_{fl}+m}\right)\,p\left(r_{L_{fl}+m}|\mathbf{r}_0^{L_{fl}+m-1}, \tilde{\mathbf{c}}_0^{L_{fl}+m}\right)\\
&\cdot\left[p\left(\mathbf{r}_0^{L_{fl}+m-1}|\tilde{\mathbf{c}}_0^{L_{fl}+m-1}\right)\,P\left(\tilde{\mathbf{c}}_0^{L_{fl}+m-1}\right)\right]
\end{aligned}
\tag{16.3.16}
$$

because the components of $\mathbf{r}_0^{L_{fl}+m-1}$, conditioned on the transmitted data sequence, are independent random variables and do not depend on $c_{L_{fl}+m}$. The last relation provides the desired recursive formula because the quantity in square brackets is the sufficient statistics from the previous update. Using the innovation representation of (15.4.20), the conditional pdf $p(r_{L_{fl}+m}|\mathbf{r}_0^{L_{fl}+m-1}, \tilde{\mathbf{c}}_0^{L_{fl}+m})$ can be written as

$$
p\left(r_{L_{fl}+m}|\mathbf{r}_0^{L_{fl}+m-1}, \tilde{\mathbf{c}}_0^{L_{fl}+m}\right) = \frac{1}{2\pi\sigma^2_{L_{fl}+m|\mathbf{r},\tilde{\mathbf{c}}}}\exp\left[-\frac{\left|r_k - \eta_{L_{fl}+m|\mathbf{r},\tilde{\mathbf{c}}}\right|^2}{2\,\sigma^2_{L_{fl}+m|\mathbf{r},\tilde{\mathbf{c}}}}\right]
\tag{16.3.17}
$$

where

$$
\eta_{L_{fl}+m|\mathbf{r},\tilde{\mathbf{c}}} \triangleq E\left\{r_{L_{fl}+m}|\mathbf{r}_0^{L_{fl}+m-1}, \tilde{\mathbf{c}}_0^{L_{fl}+m}\right\}
\tag{16.3.18}
$$

$$
\sigma^2_{L_{fl}+m|\mathbf{r},\tilde{\mathbf{c}}} \triangleq \frac{1}{2}E\left\{\left|r_{L_{fl}+m} - \eta_{L_{fl}+m|\mathbf{r},\tilde{\mathbf{c}}}\right|^2 |\mathbf{r}_0^{L_{fl}+m-1}, \tilde{\mathbf{c}}_0^{L_{fl}+m}\right\}.
\tag{16.3.19}
$$

Here, the conditional mean $\eta_{L_{fl}+m|\mathbf{r},\tilde{\mathbf{c}}}$ is given by [235]

$$
\eta_{L_{fl}+m|\mathbf{r},\tilde{\mathbf{c}}} = \mathbf{a}\left(L_{fl}+m-1; \tilde{\mathbf{c}}_0^{L_{fl}+m}\right)\left(\tilde{\mathbf{c}}_0^{L_{fl}+m-1}\right)^T
\tag{16.3.20}
$$

where $\mathbf{a}(L_{fl}+m-1; \tilde{\mathbf{c}}_0^{L_{fl}+m})$ is a vector of conditional forward prediction coefficients and $2\,\sigma^2_{L_{fl}+m|\mathbf{r},\tilde{\mathbf{c}}}$ is the corresponding error prediction variance. The conditional mean $\eta_{L_{fl}+m|\mathbf{r},\tilde{\mathbf{c}}}$ is the output of a length $(L_{fl}+m-1)$ linear prediction filter whose coefficients depend on the trial sequence $\tilde{\mathbf{c}}_0^{L_{fl}+m}$. It is also worth noting that predicting $r_{L_{fl}+m}$ by $\eta_{L_{fl}+m|\mathbf{r},\tilde{\mathbf{c}}}$ is loosely equivalent to estimating the channel in a per-survivor fashion. This shows, once again, that MAP algorithms with CIR averaged-over can be related to adaptive MAP ones.

The recursive equation (16.3.16) provides a simple method for updating the decision statistics. However, at time $(L_{fl}+m)$, it generates M new statistics for each one of those produced in the previous symbol interval. Then, the computational complexity of the MAP-FLA (16.3.14)—(16.3.17) increases as $(L_{fl}+m) \cdot M^{L_{fl}+m}$ with time. This is in contrast with the MAP-FLA for ISI channels [93] requiring a fixed number of computations per symbol. This substantial difference can be related to the channel memory, which is fixed with static, ISI channels but is infinite with time-varying channels. Then, in the last case, the quantity $p\left(r_{L_{fl}+m}|\mathbf{r}_0^{L_{fl}+m-1}, \tilde{\mathbf{c}}_0^{L_{fl}+m}\right)$ in (16.3.16) should be evaluated for any possible trial sequence $\tilde{\mathbf{c}}_0^{L_{fl}+m}$, i.e., for any possible path in the state trellis of Figure 16.21 (a). As shown in the previous paragraph, a rigorous simplification of the algorithm is possible if the folding condition applies. In [235], folding is forced, assuming that the fading distortion has finite coherence time. The corresponding algorithm operates on a fixed length observation vector made of the last N received samples and makes use of decision feedback. The complexity is still large, and further reductions are achieved by resorting to thresholding techniques. These allow us to discard unlikely paths but entail an average reduction of the computational load so that instantaneous reduction can be still significant.

The approach illustrated in [235] for FF fading can be extended to frequency-selective fading channels: analytical details are given in [236], [237], [238].

The ideas employed to develop a MAP-FLA can also be applied to design MAP forward-backward algorithms. A detection algorithm belonging to this class is illustrated in [92]. It applies to CPM and PSK signals transmitted over FF channels. To avoid recursions with increasing computational burden, trellis folding is forced under the assumption that the fading plus noise process $\{x_k\}$ (16.3.8) can be accurately approximated by an AR model of order L. Thus, the MAP decoder operates a forward and a backward recursion on a M^L-state trellis. The evaluation of the state transition probabilities (needed for the evaluation of the symbol a posteriori probabilities) involves a form of *per-state* channel, estimation-based linear predictors of order L. The performance results for this MAP demodulator feeding a convolutional decoder show that soft decisions provide a substantial improvement over differential detection and remain robust at fast Doppler rates when the differential detector is useless. Extensions of this approach to DS channels are possible, but at the price of large complexity.

16.3.2 Other equalization strategies with FF fading

Estimating the channel statistics is often a feasible task with FF fading channels. For this reason, a large number of equalization structures exploiting the knowledge of the channel statistics have been developed. In addition, a few simple detectors not relying on the use of channel statistics have been derived. In the following discussion, we illustrate some of these solutions, namely, the *block equalizers*, the *decision-feedback equalizers*, and some *one-shot detectors*.

Block equalizers

In detection of a long data sequence, the sequence of received samples can be partitioned into blocks of length N and a block ML algorithm can be employed at the receiver. Block detectors can be roughly divided into two classes:

1. Multiple-symbol ML detectors [239], [240], [241] and [242]

2. ML detectors employing the EM algorithm [243], [244], [245]

Multiple-symbol ML detectors were proposed by Ho and Fung in [241] for block detection of differentially encoded M-PSK sequences transmitted over Rayleigh-fading channels. This work showed that ML detection can be interpreted as a multiple-symbol differential detector [240] and that it provides an appreciable reduction of the error floor in fast fading with respect to a conventional differential receiver. The main drawbacks with respect to a conventional differential receiver are (a) a complexity increase as M^{L_c} metrics $\{p(\tilde{\mathbf{c}}|\mathbf{r})\}$ (one for each possible data sequence $\tilde{\mathbf{c}}$ of length L_c) must be computed and compared, and (b) the receiver must estimate the second-order channel statistics. Multiple-symbol differential detection algorithms have been also investigated in [239], [246], [242]. The work of [246] has provided an interesting interpretation of the ML block-detection algorithm for QAM signals and has investigated the error performance of suboptimal algorithms in the presence of coding and diversity.

The EM algorithm [247] was proposed in [243], [245] as a solution to the problem of ML estimation of linearly modulated data sequences. Its application leads to a two-step iterative procedure embedding a Kalman filter for channel estimation. The EM receiver needs a startup estimate of the fading channel over the whole received block. This estimate can be easily obtained with the PSAM technique. It has been shown [245] that the algorithm converges to the ML data solution in two or three iterations.

Decision-feedback equalizers

The class of suboptimal decision-feedback receivers offers an interesting complexity-performance trade-off [248], [249], [250], [230], [251], [252], [231], [253], [235], [254] and [255]. Such detectors are based on the idea that in order to detect the kth symbol c_k coherently, an estimate of the fading distortion sample α_k is required. If the data decisions on previous symbols are reliable, they can be used to remove the modulation from the corresponding received signal samples and to predict α_k. A general scheme for these equalizers is shown in Figure 16.22. Decision-feedback leads to a reduction in the number of channel predictors from many, as in PSP-MLSD, to one. The single-channel estimate can be computed by a Wiener predictor [248], [249], [253], [254], by a Kalman filter [230], [251], [252], or by an extended Kalman filter [231]. A drawback of these receivers is that a periodic refresh of their memory with a string of known symbols is required to prevent receiver runaway, i.e., a loss of channel tracking [249], and to solve the phase ambiguity problem.

The error performance of a decision-feedback receiver can be improved (with increased detection latency) if the two-stage architecture of [255] is used, as illustrated in Figure 16.22. In this case, the first stage consists of a symbol-by-symbol detector with an MMSE channel estimator. The data decisions of the first stage are delivered to the second stage, which generates an improved channel estimate by means of an optimal smoother. Finally, this estimate is used to produce new (more reliable) data decisions. A similar architecture has been proposed by Kam in [256].

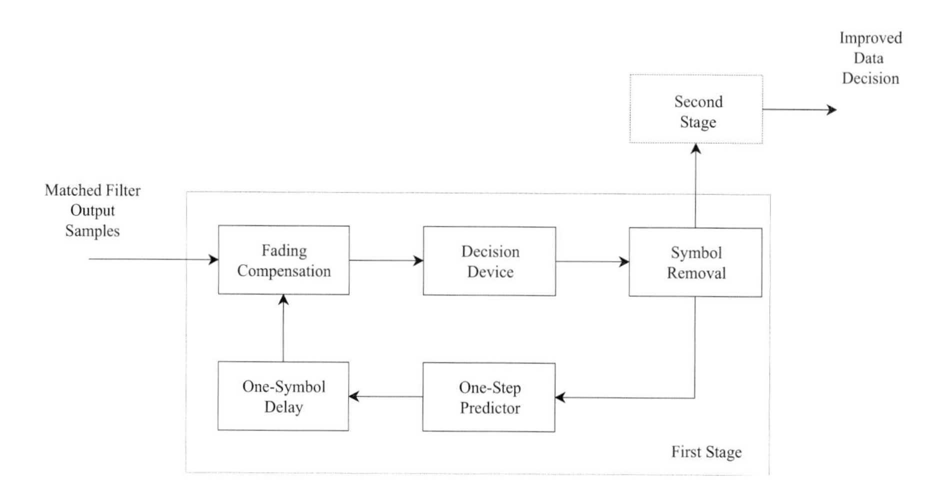

Figure 16.22. One-stage and two-stage decision feedback receiver

ML one-shot detectors

When the observation interval is limited to one or two symbol intervals, averaging over CIR produces simple ML detectors, accomplishing sequence detection on a symbol-by-symbol basis. Strictly speaking, these are not equalizers, since they make no attempt to estimate the channel. However, they are widely used in wireless transmission because they allow detection of a signal in the absence of an explicit channel estimate and offer the advantage of simplicity. Moreover, their analysis provides a basis for approaching that of more complex equalization structures. Classic examples resulting from this approach are the *differential detector* (DD) for M-ary PSK and the *matched filter & envelope detector* (MFED) for energy detection of FSK signals [75]. Both structures are optimal under the assumption of slow fading [77], i.e., if the fading can be approximated as a stepwise function, that does not change appreciably in any signalling interval for an FSK signal or over the signalling pulse duration for the linearly modulated signal of (14.2.1).

An analysis of the error performance of DDs and MFEDs on fading channels can be found in [257], [258], [259], [260], [261], [262], [263], [264] and in [265], [266], [267], [257], [86], respectively. The analysis shows that DDs and MFEDs suffer from two drawbacks: (a) there is a *signal-to-noise-ratio* (SNR) loss with respect to coherent detection; (b) if the fading is fast (changes appreciably in a symbol interval), the detector error performance will exhibit an error floor [257], [268], largely due to the quick phase changes the signal experiences during deep fades [76]. Improved DDs and MFEDs have been derived in [251], [252], [264],[269] and in [86], respectively. They exploit a couple of receive filters [86], [264], or multiple received samples per symbol (multisampling[11]) [251], [252], [269] in order to exploit the *implicit time diversity* of the channel [81]. This results in a low error floor in fast fading at the price, however, of acquiring a more refined knowledge of the channel noise and fading statistics.

Other noncoherent detectors are available for CPM signals and comprise differential detectors and discriminators. An analysis of their error performance is provided in [271], [272], [273] and [242] for the differential receivers, and in [65], [274], [275], [276] and [277] for the discriminators.

Bibliography

[1] G. D. Forney, "The Viterbi algorithm," *Proc. IEEE*, vol. 61, pp. 268–278, Mar. 1973.

[2] G. D. Forney, "Maximum-likelihood sequence estimation of digital sequences in the presence of intersymbol interference," *IEEE Trans. Inf. Theory*, vol. 18, pp. 363–378, May 1971.

[3] G. Ungerboeck, "Adaptive maximum-likelihood receiver for carrier-modulated data-transmission systems," *IEEE Trans. Comm.*, vol. 22, pp. 624–636, May 1974.

[4] L. C. Barbosa, "Maximum likelihood sequence estimators: A geometric view," *IEEE Trans. Inf. Theory*, vol. 35, pp. 419–427, Mar. 1989.

[5] K. M. Chugg and A. Polydoros, "MLSE for an unknown channel—Part I: Optimality considerations," *IEEE Trans. Comm.*, vol. 44, pp. 836–846, July 1996.

[6] B. D. Hart and D. P. Taylor, "Maximum-likelihood synchronization, equalization, and sequence estimation for unknown time-varying frequency-selective Rician channels," *IEEE Trans. Comm.*, vol. 46, pp. 211–221, Feb. 1998.

[7] B. D. Hart and D. P. Taylor, "Extended MLSE receiver for the frequency-flat, fast fading channel," *IEEE Trans. Vehicular Techn.*, vol. 46, pp. 381–389, May 1997.

[8] B. D. Hart and D. P. Taylor, "Extended MLSE diversity receiver for the time- and frequency-selective channel," *IEEE Trans. Comm.*, vol. 45, pp. 322–333, March 1997.

[11] Theoretical considerations of multisample processing in optimal detection can be found in [270].

[9] G. E. Bottomley and S. Chennakeshu, "Unification of MLSE receivers and extension to time-varying channels," *IEEE Trans. Comm.*, vol. 46, pp. 464–471, Apr. 1998.

[10] J. Hagenauer and P. Hoeher, "A Viterbi algorithm with soft-decision outputs and Its applications," in *Proc. of GLOBECOM '89, IEEE Global Telecommun. Conf.*, (Dallas, TX, USA), pp. 1680–1686, 1989.

[11] P. Hoeher, "Advances in soft-output decoding," in *Proc. of GLOBECOM '93, IEEE Global Telecommun. Conf.*, (Houston, TX, USA), pp. 793–797, 1993.

[12] M. P. C. Fossorier, F. Burkert, L. Shu, and J. Hagenauer, "On the Equivalence Between SOVA and Max-log-MAP Decodings," *IEEE Comm. Letters*, vol. 2, pp. 137–139, May 1998.

[13] S. Ariyavisitakul and L. J. Greenstein, "Reduced-complexity equalization techniques for broadband wireless channels," *IEEE J. Sel. Areas Comm.*, vol. 15, pp. 5–15, Jan. 1997.

[14] J. B. Anderson, "Sequential coding algorithms: A survey and cost analysis," *IEEE Trans. Comm.*, vol. 32, pp. 169–176, Feb. 1984.

[15] A. P. Clark, S. N. Abdullah, S. G. Jaysinghe, and K. H. Sun, "Pseudobinary and pseudoquaternary detection processes for linearly distorted multilevel QAM signals," *IEEE Trans. Comm.*, vol. 44, pp. 127–129, Feb. 1988.

[16] T. M. Aulin, "Breadth-first maximum likelihood sequence detection: Basics," *IEEE Trans. Comm.*, vol. 47, pp. 208–216, Feb. 1999.

[17] A. Baier and D. G. Heinrich, "Performance of M algorithm MLSE equalizer in frequency selective fading mobile radio channels," in *Conf. Rec. ICC'89*, (Boston, MA, USA), pp. 281–285, June 1989.

[18] P. Jung, "Performance evaluation of a novel M-detector for coherent receiver antenna diversity in a GSM-type mobile radio system," *IEEE J. Sel. Areas Comm.*, vol. 13, pp. 80–88, Jan. 1995.

[19] R. Mehlan and H. Meyr, "Soft output M-algorithm equalizer and trellis-coded modulation for mobile radio channels," in *Proc. IEEE 42nd Vehicular Technology Conf, VTC-92*, (Denver, CO, USA), pp. 582–591, 1992.

[20] E. Katz and G. L. Stuber, "Sequential sequence estimation for trellis-coded modulation on multipath fading ISI channels," *IEEE Trans. Comm.*, vol. 43, pp. 2883–2885, Dec. 1995.

[21] W. P. Chou and P. J. McLane, "16-state nonlinear equalizer for IS-54 digital cellular channels," *IEEE Trans. Vehicular Techn.*, vol. 45, pp. 13–25, Feb. 1996.

[22] G. Benelli, A. Garzelli, and F. Salvi, "Simplified Viterbi processors for the GSM Pan-European cellular communication system," *IEEE Trans. Vehicular Techn.*, vol. 43, pp. 870–878, Nov. 1994.

[23] P. J. McLane, "A residual intersymbol interference error bound for truncated-state Viterbi detectors," *IEEE Trans. Inf. Theory*, vol. 26, pp. 549–553, Sep. 1980.

[24] A. D. Hallen and C. Heegard, "Delayed decision-feedback sequence estimation," *IEEE Trans. Comm.*, vol. 37, pp. 428–436, May 1989.

[25] M. V. Eyuboglu and S. U. H. Qureshi, "Reduced-state sequence estimation with set partitioning and decision feedback," *IEEE Trans. Comm.*, vol. 36, pp. 13–20, Jan. 1988.

[26] M. V. Eyuboglu and S. U. H. Qureshi, "Reduced-state sequence estimation for coded modulation on intersymbol interference channels," *IEEE J. Sel. Areas Comm.*, vol. 7, pp. 989–995, Aug. 1989.

[27] J. Wu and A. H. Aghvami, "A new adaptive equalizer with channel estimator for mobile radio communications," *IEEE Trans. Vehicular Techn.*, vol. 45, pp. 467–474, Aug. 1996.

[28] H. C. Guren and N. Holte, "Decision feedback sequence estimation for continuous phase modulation on a linear multipath channel," *IEEE Trans. Comm.*, vol. 41, pp. 280–284, Feb. 1993.

[29] G. Ungerboeck, "Channel coding with Multilevel/Phase signals," *IEEE Trans. Inf. Theory*, vol. 28, pp. 55–67, 1982.

[30] R. E. Kamel and Y. Bar-Ness, "Reduced-complexity sequence estimation using state partitioning," *IEEE Trans. Comm.*, vol. 44, pp. 1057–1063, Sep. 1996.

[31] W.-H. Sheen and G. L. Stuber, "Error probability for maximum likelihood sequence estimation of trellis-coded modulation on ISI channels," *IEEE Trans. Comm.*, vol. 42, pp. 1427–1430, Feb./Mar./Apr. 1994.

[32] C. T. Beare, "The choice of the desired impulse response in combined linear-Viterbi algorithm equalizers," *IEEE Trans. Comm.*, vol. 26, pp. 1301–1307, Aug. 1978.

[33] S. U. H. Qureshi and E. E. Newhall, "An adaptive receiver for data transmission Over time-dispersive channels," *IEEE Trans. Inf. Theory*, vol. 19, pp. 448–457, Jul. 1973.

[34] D. D. Falconer and F. R. Magee, "Adaptive channel memory truncation for maximum-likelihood sequence estimation," *Bell Syst. Tech. J*, vol. 52, pp. 1541–1562, Nov. 1973.

[35] W. U. Lee and F. Hill, "A maximum-likelihood sequence estimator with decision-feedback equalisation," *IEEE Trans. Comm.*, vol. 25, pp. 971–979, Sep. 1977.

[36] Y. Gu and T. Le-Ngoc, "Adaptive combined DFE/MLSE techniques for ISI channels," *IEEE Trans. Comm.*, vol. 44, pp. 847–857, July 1996.

[37] K. Wesolowski, "An efficient DFE & ML suboptimum receiver for data transmission over dispersive channels using two-dimensional signal constellation," *IEEE Trans. Comm.*, vol. 35, pp. 337–339, Mar. 1987.

[38] J. Cheung and R. Steele, "Soft-decision feedback equalizer for continuous phase modulated signals in wideband mobile radio channels," *IEEE Trans. Comm.*, vol. 42, pp. 1628–1638, Feb./Mar./Apr. 1994.

[39] S. U. H. Qureshi, "Adaptive equalization," *Proc. IEEE*, vol. 73, pp. 1349–1387, Sep. 1985.

[40] M. Austin, "Decision feedback equalization for digital communication over dispersive channels," *MIT Research Laboratory of Electronics Technical Report 461*, Aug. 1967.

[41] P. Monsen, "Feedback equalization for fading dispersive channels," *IEEE Trans. Inf. Theory*, vol. 17, pp. 56–64, Jan. 1971.

[42] C. A. Belfiore and J. H. Park, "Decision feedback equalization," *Proc. IEEE*, vol. 67, pp. 1143–1156, Aug. 1979.

[43] R. D. Gitlin and S. B. Weinstein, "Fractionally-spaced equalization: An improved digital transversal equalizer," *Bell Syst. Tech. J.*, vol. 60, pp. 856–864, Feb. 1981.

[44] G. Ungerboeck, "Fractional tap-spacing equaliser and consequences for clock recovery in data modems," *IEEE Trans. Comm.*, vol. 24, pp. 856–864, Aug. 1976.

[45] J. Salz, "Optimum mean-square decision feedback equalization," *Bell Syst. Tech. J.*, vol. 52, pp. 1341–1373, Oct. 1973.

[46] D. D. Falconer and G. J. Foschini, "Theory of minimum mean-square error QAM systems Employing decision feedback equalization," *Bell Syst. Tech. J.*, vol. 52, pp. 1821–1849, Dec. 1973.

[47] J. J. O'Reilly and A. M. de Oliveria Duarte, "Error propagation in decision feedback receivers," *IEE Proc. F. Commun., Radar and Signal Process.*, vol. 132, pp. 561–566, Dec. 1985.

[48] D. L. Duttweiler, J. E. Mazo, and D. G. Messerschmitt, "An upper bound on the error probability in decision-feedback equalization," *IEEE Trans. Inf. Theory*, vol. 20, pp. 490–497, July 1974.

[49] M. R. Aaron and M. K. Simon, "Approximation of the error probability in a regenerative repeater with quantized feedback," *Bell Syst. Tech. J.*, pp. 845–1847, Dec. 1966.

[50] A. M. de Oliveria Duarte and J. J. O'Reilly, "Simplified technique for bounding error statistics for DFB receivers," *IEE Proc. F. Commun., Radar and Signal Proc.*, vol. 132, pp. 567–575, Dec. 1985.

[51] R. A. Kennedy and B. D. O. Anderson, "Tight bounds on the error probabilities of decision feedback equalizers," *IEEE Trans. Comm.*, vol. 35, pp. 1022–1028, Oct. 1987.

[52] J. C. Cartledge, "Outage performance of QAM digital radio using adaptive equalization and switched spaced diversity reception," *IEEE Trans. Comm.*, vol. 35, pp. 166–171, Feb. 1987.

[53] P. Balaban and J. Salz, "Optimum diversity combining and equalization in digital data transmission with applications to cellular mobile radio—Part I: Theoretical considerations," *IEEE Trans. Comm.*, vol. 40, pp. 885–894, May 1992.

[54] P. Balaban and J. Salz, "Optimum diversity combining and equalization in digital transmission with applications to cellular mobile radio—Part II: Theoretical considerations," *IEEE Trans. Comm.*, vol. 40, pp. 885–894, 1992.

[55] P. Monsen, "Adaptive equalization of the slow fading channel," *IEEE Trans. Comm.*, vol. 22, pp. 1064–1075, Aug. 1974.

[56] A. Cantoni and P. Butler, "Stability of decision feedback inverses," *IEEE Trans. Comm.*, vol. 24, pp. 970–977, Sep. 1976.

[57] R. A. Kennedy and B. D. O. Anderson, "Error recovery of decision feedback equalizers on exponential impulse response channels," *IEEE Trans. Comm.*, vol. 35, pp. 846–848, Aug. 1987.

[58] R. A. Kennedy and B. D. O. Anderson, "Recovery times of decision feedback equalizers on noiseless channels," *IEEE Trans. Comm.*, vol. 35, pp. 1012–1021, Aug. 1987.

[59] R. A. Kennedy, B. D. O. Anderson, and R. Bitmead, "Channels leading to rapid error recovery for decision feedback equalisers," *IEEE Trans. Comm.*, vol. 37, pp. 1126–1135, Nov. 1989.

[60] N. C. Beaulieu, "Bounds on recovery times of decision feedback equalizers," *IEEE Trans. Comm.*, vol. 42, pp. 2786–2794, Oct. 1994.

[61] R. Agusti and F. Cassadevall, "Performance of fractioned and nonfractioned equalizers with high-level QAM," *IEEE Trans. Comm.*, vol. 5, pp. 476–483, Apr. 1987.

[62] M. V. Eyuboglu, "Detection of coded modulation signals on linear, severely distorted channels using decision-feedback noise prediction with interleaving," *IEEE Trans. Comm.*, vol. 36, pp. 401–409, Apr. 1988.

[63] E. Dahlman and B. Gudmundson, "Performance Improvement in Decision Feedback Equalizers by Using Soft Decision," *Electronics Letters*, vol. 24, pp. 1084–1085, Aug. 1988.

[64] R. Kennedy and Z. Ding, "Design and Optimization of Nonlinear Mapping in Decision Feedback Equalization," in *Proc. of the 35th IEEE Conf. on Decision and Control*, (Kobe, Japan), pp. 1888–1889, 1996.

[65] S. Elnoubi, H. Badr, and E. A. Youssef, "BER improvement of PRCPM in mobile radio channels with discriminator detection using decision feedback equalization," *IEEE Trans. Vehicular Techn.*, vol. 40, pp. 694–699, Nov. 1991.

[66] L. Bin, "Decision feedback detection of minimum shift keying," *IEEE Trans. Comm.*, vol. 44, pp. 1073–1076, Sept. 1996.

[67] D. Boundreau and J. H. Lodge, "Adaptive equalization of CPM signals transmitted over fast Rayleigh flat-fading channels," *IEEE Trans. Vehicular Techn.*, vol. 44, pp. 404–413, August 1995.

[68] G. K. Kaleh, "Channel equalization for block transmission systems," *IEEE J. Sel. Areas Comm.*, vol. 13, pp. 110–121, Jan. 1995.

[69] S. N. Crozier, D. D. Falconer, and S. A. Mahmoud, "Reduced complexity short-block data detection techniques for fading time-dispersive channels," *IEEE Trans. Vehicular Techn.*, vol. 41, pp. 255–265, Aug. 1992.

[70] I. J. Fevrier, S. B. Gelfand, and M. P. Fitz, "Reduced complexity decision feedback equalization for multipath channels with large delay spreads," *IEEE Trans. Comm.*, vol. 47, pp. 927–937, June 1999.

[71] A. B. Marcovitz, "On inverses and quasi-inverses of linear time-varying discrete systems," *J. Franklin Inst.*, vol. 272, pp. 23–44, 1961.

[72] F. Ling and S. U. H. Qureshi, "Convergence and steady state behaviour of a phase-splitting fractionally spaced equalizer," *IEEE Trans. Comm.*, vol. 38, pp. 418–425, 1990.

[73] M. Barton and D. W. Tufts, "A suboptimum linear receiver based on a parametric channel model," *IEEE Trans. Comm.*, vol. 39, pp. 1328–1334, 1991.

[74] S. B. Gelfand, C. S. Ravishankar, and E. J. Delp, "Tree-structured piecewise linear adaptive equalisation," *IEEE Trans. Comm.*, vol. 41, pp. 70–82, Jan. 1993.

[75] M. Schwartz, W. R. Bennett, and S. Stein, *Communication Systems and Techniques*. New York: McGraw-Hill, 1966.

[76] S. Stein, "Fading channel issues in system engineering," *IEEE J. Select. Areas Comm.*, vol. 5, pp. 68–89, Feb. 1987.

[77] J. K. Cavers, "On the validity of the slow and moderate fading models for matched filter detection of Rayleigh fading signals," *Can. J. of Elect. & Comp. Eng*, vol. 17, pp. 183–189, 1992.

[78] D. G. Brennan, "Linear diversity combining techniques," *Proc. IRE*, vol. 47, pp. 1075–1102, June 1959.

[79] J. G. Proakis, *Digital Communications*. New York: McGraw-Hill, 2nd ed., 1989.

[80] T. Eng, N. Kong, and L. B. Milstein, "Comparison of diversity combining techniques for Rayleigh-fading channels," *IEEE Trans. Comm.*, vol. 44, pp. 1117–1129, Sep. 1996.

[81] R. S. Kennedy, *Fading Dispersive Communication Channels*. New York: Wiley-Interscience, 1969.

[82] J. E. Mazo, "Exact matched filter bound for two-beam Rayleigh fading," *IEEE Trans. Comm.*, vol. 39, pp. 1027–1030, July 1991.

[83] R. Price and P. E. Green, "A communication technique for multipath channels," *Proc. IRE*, vol. 46, pp. 555–570, Mar. 1958.

[84] E. J. Baghdady, "Novel techniques for counteracting multipath interference effects in receiving systems," *IEEE J. Sel. Areas Comm.*, vol. 5, pp. 274–285, Feb. 1987.

[85] U. Hansson and T. M. Aulin, "Reduced complexity decision feedback equalization for multipath channels with large delay spreads," *IEEE Trans. Comm.*, vol. 47, pp. 874–883, June 1999.

[86] G. M. Vitetta, U. Mengali, and D. P. Taylor, "Optimal noncoherent detection of FSK signals transmitted Over linearly time-selective Rayleigh fading channels," *IEEE Trans. Comm.*, vol. 45, pp. 1417–1425, Nov. 1997.

[87] G. M. Vitetta, D. P. Taylor, and U. Mengali, "Double-filtering receivers for PSK signals transmitted over Rayleigh frequency-flat fading channels," *IEEE Trans. Comm.*, vol. 44, pp. 686–695, June 1996.

[88] J. Erfanian, S. Pasupathy, and G. Gulak, "Reduced complexity symbol detectors with parallel structures for ISI channels," *IEEE Trans. Comm.*, vol. 42, pp. 1661–1671, Feb./Mar./Apr. 1994.

[89] R. J. McEliece, "On the BCJR Trellis for Linear Block Codes," *IEEE Trans. Inf. Theory*, vol. 42, pp. 1072–1092, July 1996.

[90] V. Franz and J. B. Anderson, "Concatenated decoding with a reduced-search BCJR algorithm," *IEEE J. Sel. Areas Comm.*, vol. 16, pp. 186–195, Feb. 1998.

[91] P. Robertson, E. Villebrun, and P. Hoeher, "A Comparison of optimal and sub-optimal MAP decoding algorithms Operating in the log domain," in *Proc. IEEE Int. Conf. on Commun. (ICC'95)*, (Seattle, WA, USA), pp. 1009–1013, 1998.

[92] M. J. Gertsman and J. H. Lodge, "Symbol-by-symbol MAP demodulation of CPM and PSK signals on Rayleigh flat-fading channels," *IEEE Trans. Comm.*, vol. 45, pp. 788–799, July 1997.

[93] K. Abend and B. D. Fritchman, "Statistical detection for communication channels with intersymbol interference," *Proc. IEEE*, vol. 58, pp. 779–785, May 1970.

[94] Y. Li, B. Vucetic, and Y. Sato, "Optimum soft-output detection for channels with intersymbol interference," *IEEE Trans. Inf. Theory*, vol. 41, pp. 704–713, May 1995.

[95] K. Abend, T. J. Hartley, B. D. Fritchman, and C. Gumacos, "On optimum receivers for channels having memory," *IEEE Trans. Inf. Theory*, vol. 14, pp. 152–157, Nov. 1968.

[96] J. F. Hayes, T. M. Cover, and J. B. Riera, "Optimal sequence detection and optimal symbol-by-symbol detection: Similar algorithms," *IEEE Trans. Comm.*, vol. 30, pp. 152–157, Jan. 1982.

[97] J. Hagenauer, E. Offer, and L. Papke, "Iterative decoding of binary block and convolutional codes," *IEEE Trans. Inf. Theory*, vol. 42, pp. 429–425, Mar. 1998.

[98] M. Rahnema and Y. Antia, "Optimum soft decision decoding with channel state information in the presence of fading," *IEEE Comm. Mag.*, vol. 41, pp. 110–111, July 1997.

[99] G.-K. Lee, S. B. Gelfand, and M. P. Fitz, "Bayesian decision feedback techniques for deconvolution," *IEEE J. Sel. Areas Comm.*, vol. 13, pp. 155–166, Jan. 1995.

[100] S. Chen, B. Mulgrew, and S. McLaughlin, "Adaptive Bayesian equalizer with decision feedback," *IEEE Trans. Sig. Proc.*, vol. 41, pp. 2918–2926, Sep. 1993.

[101] G. J. Gibson, S. Siu, and C. F. N. Cowan, "The application of nonlinear structures to the reconstruction of binary signals," *IEEE Trans. Sig. Proc.*, vol. 39, pp. 1877–1884, Aug. 1991.

[102] D. Williamson, R. A. Kennedy, and G. W. Pulford, "Block decision feedback equalization," *IEEE Trans. Comm.*, vol. 40, pp. 255–264, Feb. 1992.

[103] Z.-J. Xiang and G.-G. Bi, "A new lattice polynomial perceptron and its applications to frequency-selective fading channel equalization and ACI suppression," *IEEE Trans. Comm.*, vol. 44, pp. 761–767, July 1996.

[104] Z. Xiang, G. Bi, and T. Le-Ngoc, "Polynomial perceptrons and Their applications to fading channel equalization and co-channel interference suppression," *IEEE Sig. Proc.*, vol. 32, pp. 2470–2480, Sep. 1994.

[105] P.-R. Chang and B.-C. Wang, "Adaptive decision feedback equalization for digital satellite channels using multilayer neural networks," *IEEE J. Sel. Areas Comm.*, vol. 13, pp. 316–324, Feb. 1995.

[106] S. Cheng, S. McLaughlin, B. Mulgrew, and P. M. Grant, "Adaptive Bayesian decision feedback equalizer for dispersive mobile radio channels," *IEEE Trans. Comm.*, vol. 43, pp. 1937–1945, May 1995.

[107] I. Cha and S. A. Kassam, "Channel equalization using adaptive complex radial basis function networks," *IEEE J. Sel. Areas Comm.*, vol. 13, pp. 122–131, Jan. 1995.

[108] S. Chen and B. Mulgrew, "A clustering technique for digital communications channel equalization using radial basis function networks," *IEEE Trans. Neural Net.*, vol. 4, pp. 570–579, July 1993.

[109] F. R. Magee and J. G. Proakis, "Adaptive maximum-likelihood sequence estimation for digital signalling in the presence of intersymbol interference," *IEEE Trans. Inf. Theory*, vol. 19, pp. 120–124, Jan. 1973.

[110] R. Raheli, G. Marino, and P. Castoldi, "Per-survivor processing and tentative decisions: What Is in between?," *IEEE Trans. Comm.*, vol. 44, pp. 127–129, Feb. 1996.

[111] R. Raheli, A. Polydoros, and C.-K. Tzou, "Per-survivor processing: A general approach to MLSE in uncertain environments," *IEEE Trans. Comm.*, vol. 43, pp. 354–364, Feb./Mar./Apr. 1995.

[112] S. Haykin, *Adaptive Filter Theory.* Englewood Cliffs, NJ: Prentice-Hall, 1986.

[113] S. N. Crozier, D. D. Falconer, and S. A. Mahmoud, "Least Sum of Squared Errors (LSSE) Channel Estimation," *IEE Proc.-F*, vol. 138, pp. 371–378, Aug. 1991.

[114] D. D. Falconer and L. Ljung, "Application of fast Kalman estimation to adaptive equalization," *IEEE Trans. Comm.*, vol. 26, pp. 1439–1446, Oct. 1978.

[115] E. H. Satorius and J. D. Pack, "Application of least squares lattice algorithms to adaptive equalization," *IEEE Trans. Comm.*, vol. 136, pp. 136–142, Feb. 1981.

[116] M.-C. Chiu and C.-C. Chao, "Analysis of LMS-adaptive MLSE equalization on multipath fading channels," *IEEE Trans. Comm.*, vol. 44, pp. 1684–1692, Dec. 1996.

[117] G. Long, F. Ling, and J. G. Proakis, "The LMS algorithm with delayed coefficient adaptation," *IEEE Trans. Sig. Proc.*, vol. 37, pp. 1397–1405, Sep. 1989.

[118] E. Eleftheriou and D. D. Falconer, "Tracking properties and steady-state performance of RLS adaptive filter algorithms," *IEEE Trans. Acoust. Speech and Signal Proc.*, vol. 34, pp. 1097–1109, Oct. 1986.

[119] B. Widrow, J. McCool, M. Larimore, and C. Johnson, "Stationary and nonstationary learning characteristics of the LMS adaptive filter," *Proc. IEEE*, vol. 64, pp. 1151–1162, Aug. 1976.

[120] F. Ling and J. G. Proakis, "Nonstationary learning characteristics of least squares adaptive estimation algorithms," in *Proc. IEEE ICASSP'84*, (San Diego, CA, USA), Mar. 1984.

[121] J. W. M. Bergmans, "Tracking capabilities of the LMS adaptive filter in the presence of gain variation," *IEEE Acoust. Speech and Signal Proc.*, vol. 38, pp. 712–714, Apr. 1990.

[122] O. Macchi, "Optimization of adaptive identification for time-varying filters," *IEEE Trans. Auto. Contr.*, vol. 31, pp. 283–287, Mar. 1996.

[123] D. K. Borah and B. D. Hart, "Frequency-selective fading channel estimation with a polynomial time-varying channel model," *IEEE Trans. Comm.*, vol. 47, pp. 862–871, June 1999.

[124] D. K. Borah and B. D. Hart, "Receiver structures for time-varying frequency-selective fading channels," *IEEE J. Sel. Areas Comm.*, vol. 17, pp. 1863–1875, Nov. 1999.

[125] R. A. Ziegler and J. M. Cioffi, "Estimation of time-varying digital radio channels," *IEEE Trans. Vehicular Techn.*, vol. 41, pp. 134–151, May 1992.

[126] J. K. Cavers, "Pilot symbol assisted modulation and differential detection in fading and delay spread," *IEEE Trans. Comm.*, vol. 43, pp. 2206–2212, July 1995.

[127] D.-L. Liu and K. Feher, "Pilot-symbol aided coherent M-ary PSK in frequency-selective fast Rayleigh fading channel," *IEEE Trans. Comm.*, vol. 42, pp. 54–62, Jan. 1994.

[128] G. D. Hingorani and J. C. Hancock, "A transmitted-reference system for communication in random or unknown channels," *IEEE Trans. Comm. Systems*, vol. 13, pp. 293–301, Sep. 1965.

[129] N. W. K. Lo, D. D. Falconer, and U. H. Sheikh, "Adaptive equalization and diversity combining for mobile radio using interpolated channel estimates," *IEEE Trans. Vehicular Techn.*, vol. 40, pp. 636–645, Aug. 1991.

[130] S. A. Fechtel and H. Meyr, "Optimal parametric feedforward estimation of frequency-selective fading radio channels," *IEEE Trans. Comm.*, vol. 42, pp. 1639–1650, Feb./Mar./Apr. 1994.

[131] T. Eyceoz and A. Duel-Hallen, "Simplified block adaptive diversity equalizer for cellular mobile radio," *IEEE Comm. Letters*, vol. 1, pp. 15–19, Jan. 1987.

[132] S. Haykin, A. H. Sayed, J. R. Zeidler, P. Yee, and P. C. Wei, "Adaptive tracking of linear time-variant systems by extended RLS algorithms," *IEEE Trans. Sig. Proc.*, vol. 45, pp. 1118–1127, May 1997.

[133] O. Shalvi and E. Weinstein, "New criteria for blind deconvolution of nonminimum phase systems (channels)," *IEEE Trans. Inf. Theory*, vol. 36, pp. 312–321, Mar. 1990.

[134] B. Porat and B. Friedlander, "Blind equalization of digital communication channels using high-order moments," *IEEE Trans. Sig. Proc.*, vol. 39, pp. 522–526, Feb. 1991.

[135] F.-C. Zheng, S. McLaughlin, and B. Mulgrew, "Blind equalization of nonminimum phase channels: Higher order cumulant based algorithm," *IEEE Trans. Sig. Proc.*, vol. 41, pp. 681–691, Feb. 1993.

[136] O. Shalvi and E. Weinstein, "Super-exponential methods for blind deconvolution," *IEEE Trans. Inf. Theory*, vol. 39, pp. 504–519, Mar. 1993.

[137] J. K. Tugnait, "Blind equalization of digital communication channel impulse response," *IEEE Trans. Comm.*, vol. 44, pp. 607–1616, Feb./Mar./Apr. 1994.

[138] J. K. Tugnait, "Blind equalization and estimation of digital communication FIR channels using cumulant matching," *IEEE Trans. Comm.*, vol. 43, pp. 1240–1245, Feb./Mar./Apr. 1995.

[139] F. B. Ueng and Y. T. Su, "Adaptive blind equalization using second- and higher order statistics," *IEEE J. Sel. Areas Comm.*, vol. 13, pp. 132–140, Jan. 1995.

[140] M. K. Tsatsanis and G. B. Giannakis, "Equalization of rapidly fading channels: Self-recovering methods," *IEEE Trans. Comm.*, vol. 44, pp. 619–630, May 1996.

[141] E. A. Haykin, *Blind Deconvolution.* Englewood Cliffs, NJ: Prentice-Hall, 1994.

[142] L. Tong, "Blind sequence estimation," *IEEE Trans. Comm.*, vol. 43, pp. 2986–2994, Dec. 1995.

[143] J. L. Valenzuela, A. Valdovinos, and F. J. Casadevall, "Performance of blind equalization with higher order statistics in indoor radio environments," *IEEE Trans. Vehicular Techn.*, vol. 46, pp. 369–374, May 1997.

[144] J. K. Tugnait, "On blind identifiability of multipath channels using fractional sampling and second-order cyclostationary statistics," *IEEE Trans. Inf. Theory*, vol. 41, pp. 308–311, Jan. 1995.

[145] J. K. Tugnait, "Blind equalization and estimation of FIR communications channels using fractional sampling," *IEEE Trans. Comm.*, vol. 44, pp. 324–336, Mar. 1996.

[146] L. Tong, G. Xu, and T. Kailath, "Blind identification and equalization based on second-order statistics: A time domain approach," *IEEE Trans. Inf. Theory*, vol. 40, pp. 340–349, Mar. 1994.

[147] L. Tong and S. Perreau, "Multichannel blind equalization: From subspace to maximum likelihood methods," *IEEE Proc.*, vol. 10, pp. 1951–1968, Oct. 1998.

[148] Y. Sato, "A method of self-recovering equalization for multilevel amplitude -modulation systems," *IEEE Trans. Communications*, vol. 23, pp. 679–682, June 1975.

[149] D. N. Godard, "Self-recovering equalization and carrier tracking in two-dimensional data communication systems," *IEEE Trans. Comm.*, vol. 28, pp. 1867–1875, Nov. 1980.

[150] A. Benveniste, M. Goursat, and G. Ruget, "Robust identification of a nonminimum phase system: Blind adjustment of a linear equalizer in data communications," *IEEE Trans. Auto. Contr.*, vol. 25, pp. 385–399, June 1980.

[151] S. Bellini, "Bussgang techniques for blind equalization," in *Proc. Globecom'86*, pp. 1634–1640, Dec. 1986.

[152] Z. Ding and R. A. Kennedy, "On the Whereabouts of Local Minima for Blind Adaptive Equalizers," *IEEE Trans. Circ. Syst.*, vol. 39, pp. 119–123, Feb. 1992.

[153] R. A. Kennedy, B. D. O. Anderson, and R. R. Bitmead, "Blind Adaptation of Decision Feedback Equalizers: Gross Convergence Properties," *Int. J. of Adapt. Contr. and Sig. Proc.*, vol. 7, pp. 497–524, Nov. 1993.

[154] R. Johnson, P. Schniter, T. J. Endres, J. D. Behm, D. R. Brown, and R. A. Casas, "Blind equalization using the constant modulus criterion: A review," *IEEE Proc.*, vol. 10, pp. 1927–1950, Oct. 1998.

[155] R. A. Iltis, J. J. Shynk, and K. Giridhar, "Bayesian algorithms for blind equalization using parallel adaptive filtering," *IEEE Trans. Comm.*, vol. 42, pp. 1017–1032, Feb./Mar./Apr. 1994.

[156] G.-K. Lee, S. B. Gelfand, and M. P. Fitz, "Bayesian techniques for blind deconvolution," *IEEE Trans. Comm.*, vol. 44, pp. 826–835, July 1996.

[157] G. Castellini, E. D. Re, and L. Perucci, "A continuously adaptive MLSE receiver for mobile communications: Algorithm and performance," *IEEE Trans. Comm.*, vol. 45, pp. 80–89, Jan. 1997.

[158] R. D'Avella, L. Moreno, and M. Sant'Agostino, "An adaptive MLSE receiver for TDMA digital mobile radio," *IEEE J. Sel. Areas Comm.*, vol. 7, pp. 122–129, Jan. 1989.

[159] G. D'Aria, R. Piermarini, and V. Zingarelli, "Fast adaptive equalizers for narrow-band TDMA mobile radio," *IEEE Trans. Vehicular Techn.*, vol. 40, pp. 392–404, May 1991.

[160] H. Kubo, K. Murakami, and T. Fujino, "An adaptive maximum-likelihood sequence estimator for fast time-varying intersymbol interference channels," *IEEE Trans. Comm.*, vol. 42, pp. 1972–1880, Feb./Mar./Apr. 1994.

[161] E. Dahlman, "New adaptive Viterbi detector for fast-fading mobile radio channels," *Electron. Letters*, vol. 26, pp. 1572–1573, Sep. 1990.

[162] K. M. Chugg and A. Polydoros, "MLSE for an unknown channel—Part II: Tracking performance," *IEEE Trans. Comm.*, vol. 44, pp. 949–958, Aug. 1996.

[163] W. Gardner, "Learning characteristics of stochastic-gradient-descent algorithms: A general study, analysis and critique," *Sig. Proc.*, vol. 6, pp. 112–133, Apr. 1984.

[164] J. Lin, J. G. Proakis, F. Ling, and H. Lev-Ari, "Optimal tracking of time-varying channels: A frequency domain approach for known and new algorithms," *IEEE J. Sel. Areas Comm.*, vol. 13, pp. 141–154, Jan. 1995.

[165] B. Toplis and S. Pasupathy, "Tracking improvements in fast RLS algorithms using a variable forgetting factor," *IEEE Acoust. Speech and Signal Proc.*, vol. 36, pp. 206–227, Feb. 1988.

[166] K. A. Hamied and G. L. Stuber, "Performance of trellis-coded modulation for equalized multipath fading ISI channels," *IEEE Trans. Vehicular Techn.*, vol. 44, pp. 50–58, Feb. 1995.

[167] J. G. Proakis, "Adaptive equalization for TDMA digital mobile radio," *IEEE Trans. Vehicular Techn.*, vol. 40, pp. 333–341, May 1991.

[168] D. D. Falconer, A. U. H. Sheikh, E. Eleftheriou, and M. Tobis, "Comparison of DFE and MLSE receiver performance on HF channels," *IEEE Trans. Comm.*, vol. 33, pp. 484–486, May 1985.

[169] N. Seshadri, "Joint data and channel estimation using blind trellis search techniques," *IEEE Trans. Comm.*, vol. 42, pp. 1000–1011, Feb./Mar./Apr. 1994.

[170] R. A. Iltis, "A Bayesian maximum-likelihood sequence estimation algorithm for A priori unknown channels and symbol timing," *IEEE J. Sel. Areas Comm.*, vol. 10, pp. 579–588, Apr. 1992.

[171] M. Erkurt and J. G. Proakis, "Joint data detection and channel estimation for rapidly fading channels," in *Proc. IEEE Globecom Conf.*, (Orlando, FL, USA), Dec. 1992.

[172] G. M. Vitetta and D. P. Taylor, "Maximum likelihood decoding of uncoded and coded PSK signal sequences transmitted over Rayleigh flat-fading channels," *IEEE Trans. Comm.*, vol. 43, pp. 2750–2758, Nov. 1995.

[173] G. M. Vitetta, U. Mengali, and D. P. Taylor, "Blind detection of CPM signals transmitted over frequency-flat fading channels," *IEEE Trans. Vehicular Techn.*, vol. 44, pp. 961–968, Nov. 1998.

[174] K. M. Chugg, "The condition for the applicability of the Viterbi algorithm with implications for fading channel MLSD," *IEEE Trans. Comm.*, vol. 46, pp. 1112–1116, Sep. 1998.

[175] M. E. Rollins and S. J. Simmons, "Simplified per-survivor Kalman processing in fast frequency-selective fading channels," *IEEE Trans. Comm.*, vol. 45, pp. 544–553, May 1997.

[176] P. Bello, "Characterization of Randomly Time-Variant Linear Channels," *IEEE Trans. Comm.*, vol. 11, pp. 360–393, 1963.

[177] Q. Dai and E. Shwedyk, "Detection of bandlimited signals Over frequency-selective Rayleigh fading channels," *IEEE Trans. Comm.*, vol. 42, pp. 941–950, 1994.

[178] K. M. Chugg, "Blind acquisition characteristics of PSP-based sequence detectors," *IEEE J. Sel. Areas Comm.*, vol. 16, pp. 1518–1529, Oct. 1998.

[179] G. Paparisto and K. M. Chugg, "PSP array processing for multipath fading channels," *IEEE Trans. Comm.*, vol. 47, pp. 504–507, Apr. 1995.

[180] H. Zamiti-Jafarian and S. Pasupathy, "Adaptive mlsde Using the em algorithm," *IEEE Trans. Comm.*, vol. 47, pp. 1181–1193, Aug. 1999.

[181] R. W. Chang and J. C. Hancock, "On receiver structures for channel having memory," *IEEE Trans. Inf. Theory*, vol. 12, pp. 463–468, Oct. 1966.

[182] E. Baccarelli, R. Cusani, and S. Galli, "A novel adaptive receiver with enhanced channel tracking capability for TDMA-based mobile radio communications," *IEEE J. Sel. Areas Comm.*, vol. 16, pp. 1630–1639, Dec. 1998.

[183] R. Cusani and J. Mattila, "Equalization of digital radio channels with large multipath delay for cellular land mobile applications," *IEEE Trans. Comm.*, vol. 47, pp. 348–351, Mar. 1999.

[184] E. Baccarelli and R. Cusani, "Combined channel estimation and data detection using soft statistics for frequency-selective fast-fading digital links," *IEEE Trans. Comm.*, vol. 46, pp. 424–427, Apr. 1998.

[185] K. Giridhar, J. J. Shynk, A. R. Iltis, and A. Mahur, "Adaptive MAPSD algorithms for symbol and timing recovery of mobile radio TDMA signals," *IEEE Trans. Comm.*, vol. 44, pp. 927–978, Aug. 1996.

[186] A. Anastasopoulos and A. Polydoros, "Adaptive soft-decision algorithms for mobile fading channels," *Eur. Trans. Telecomm.*, vol. 9, pp. 183–190, Mar./Apr. 1998.

[187] I. Bar-David and A. Elia, "Augumented APP (A^2P^2) for A posteriori probability calculation and channel parameter tracking," *IEEE Comm. Letters*, vol. 3, pp. 18–20, Jan. 1999.

[188] L. Davis, I. Collings, and P. Hoeher, "Joint MAP equalization and channel estimation for frequency selective and frequency flat fading channels," *submitted to the IEEE Trans. Comm.*

[189] F. N. Nunes and J. M. N. Leitao, "A nonlinear filtering approach to estimation and detection in mobile communications," *IEEE J. Sel. Areas Comm.*, vol. 16, pp. 1649–1659, Dec. 1998.

[190] J. G. Proakis and J. H. Miller, "An adaptive receiver for digital signalling through channels with intersymbol interference," *IEEE Trans. Inf. Theory*, vol. 15, pp. 484–497, July 1969.

[191] J. M. Perl, A. Shpigel, and A. Reichman, "Adaptive receiver for digital communication Over HF channels," *IEEE J. Sel. Areas Comm.*, vol. 5, pp. 304–308, Feb. 1987.

[192] Y.-J. Liu, M. Wallace, and J. W. Ketchum, "A soft output bidirectional decision feedback equalization technique for TDMA cellular radio," *IEEE J. Sel. Areas Comm.*, vol. 11, pp. 1034–1045, Sept. 1993.

[193] S. J. Nowlan and G. E. Hinton, "A soft decision-directed LMS algorithm for blind equalization," *IEEE Trans. Comm.*, vol. 41, pp. 275–279, Feb. 1993.

[194] S.-S. Ahn, "Almost-sure Convergence of the DLMS Algorithm," *J. of the Korean Inst. of Telem. and Electr.*, vol. 32, pp. 62–70, Sep. 1995.

[195] E. Eleftheriou, "Adaptive equalization techniques for HF channels," *IEEE J. Sel. Areas. Comm.*, vol. 5, pp. 238–247, Feb. 1987.

[196] T. J. Endres, C. H. Strolle, S. N. Hulyalkar, T. A. Schaffer, A. Shah, M. Gittings, C. Hollowell, A. Bhaskaran, J. Roletter, and B. Paratore, "Carrier independent blind initialization of a DFE using CMA," in *Proc. 1999 2nd IEEE Dig.*

Sig. Proc. Workshop on Sig. Proc. Advances in Wireless Commun. (SPAWC'99), (Annapolis, MD, USA), pp. 239–242, May 1999.

[197] M. Stojnanovic, J. Proakis, and J. Catipovic, "Analysis of the impact of channel estimation errors on the performance of a decision feedback equalizer in fading multipath channels," *IEEE Trans. Comm.*, vol. 43, pp. 877–885, Feb./March/Apr. 1995.

[198] A. Wautier, J.-C. Dany, C. Mourot, and V. Kumar, "A new method for predicting the channel estimate influence on performance of TDMA mobile radio systems," *IEEE Trans. Vehicular Techn.*, vol. 44, pp. 594–602, Aug. 1995.

[199] J. A. Bingham, "Improved methods of accelerating the convergence of adaptive equalizers for partial-response signals," *IEEE Trans. Comm.*, vol. 35, pp. 277–260, Mar. 1987.

[200] I. Lee and J. M. Cioffi, "A fast computation algorithm for the decision feedback equalizer," *IEEE Trans. Comm.*, vol. 43, pp. 2742–2749, Nov. 1995.

[201] N. Al-Dhahir and J. M. Cioffi, "Fast computation of channel-estimate based equalizers in packet data transmission," *IEEE Trans. Signal Proc.*, vol. 43, pp. 2462–2473, Nov. 1995.

[202] B. Farhang-Boroujeny, "Channel Equalization Via Channel Identification: Algorithms and Simulation Results for Rapidly Fading HF Channels," *IEEE Trans. Comm.*, vol. 44, pp. 1409–1412, Nov. 1996.

[203] S. Ariyavisitakul, "A decision feedback equalizer with time-reversal structure," *IEEE J. Sel. Areas Comm.*, vol. 10, pp. 599–613, Apr. 1992.

[204] T. Nagayasu, S. Sampei, and Y. Kamio, "Complexity reduction and performance improvement of a decision feedback equalizer for 16QAM in land mobile communications," *IEEE Trans. Vehicular Techn.*, vol. 44, pp. 570–578, Aug. 1995.

[205] C. K. Rushforth, "Transmitted-reference techniques for random or unknown channels," *IEEE Trans. Inf. Theory*, vol. 9, pp. 39–42, Jan. 1964.

[206] W. J. Weber, "Performance of phase-locked loops in the presence of fading communication channels," *IEEE Trans. Comm.*, vol. 24, pp. 487–499, May 1976.

[207] M. Yokoyama, "BPSK system with sounder to combat Rayleigh fading in mobile radio communication," *IEEE Trans. Vehicular Techn.*, vol. 34, pp. 35–40, Feb. 1985.

[208] M. K. Simon, "Dual-pilot tone technique," *IEEE Trans. Vehicular Techn.*, vol. 35, pp. 63–70, May 1986.

[209] F. Davarian, "Mobile digital communication via tone calibration," *IEEE Trans. Vehicular Techn.*, vol. 36, pp. 55–62, May 1987.

[210] J. P. McGeehan and A. J. Bateman, "Phase-locked transparent tone-in-band (TTIB): A new spectrum configuration particularly suited to the transmission of data over SSB mobile radio networks," *IEEE Trans. Comm.*, vol. 32, pp. 81–87, Jan. 1984.

[211] A. Bateman, "Feedforward transparent tone-in-band: Its implementations and applications," *IEEE Trans. Vehicular Technoloy*, vol. 39, pp. 235–243, Aug. 1990.

[212] J. K. Cavers, "Performance of tone calibration with frequency offset and imperfect pilot filter," *IEEE Trans. Vehicular Techn.*, vol. 40, pp. 426–434, May 1991.

[213] M. Fitz, "Further results in the unified analysis of digital communication systems," *IEEE Trans. Comm.*, vol. 40, pp. 521–532, 1992.

[214] M. P. Fitz, "A dual-tone reference digital demodulator for mobile digital communications," *IEEE Trans. Vehicular Techn.*, vol. 42, pp. 156–165, May 1993.

[215] J. H. Lodge, M. L. Moher, and S. N. Crozier, "A comparison of data modulation techniques for land mobile satellite channel," *IEEE Trans. Vehicular Techn.*, vol. 36, pp. 28–34, Feb. 1987.

[216] J. K. Cavers and M. Liao, "A comparison of pilot tone and pilot symbol techniques for digital mobile communication," in *IEEE Globecom'92 Rec*, pp. 915–921, 1992.

[217] J. K. Cavers, "An analysis of pilot symbol assisted modulation for Rayleigh fading channels," *IEEE Trans. Vehicular Techn.*, vol. 40, pp. 686–693, Nov. 1991.

[218] J. K. Cavers and J. Varaldi, "Cochannel interference and pilot symbol assisted modulation," *IEEE Trans. Vehicular Techn.*, vol. 42, pp. 407–413, Nov. 1993.

[219] S. Sampei and T. Sunaga, "Rayleigh fading compensation for QAM in land mobile radio communications," *IEEE Trans. Vehicular Techn.*, vol. 42, pp. 137–146, May 1993.

[220] T. Sunaga and S. Sampei, "Performance of multi-level QAM with post-detection maximal ratio combining space diversity for digital land-mobile radio communication," *IEEE Trans. Vehicular Techn.*, vol. 42, pp. 294–301, Aug. 1993.

[221] M. L. Moher and J. H. Lodge, "TCMP—a modulation and coding strategy for Rician fading channels," *IEEE J. Select. Areas Comm.*, vol. 7, pp. 1347–1355, Dec. 1989.

[222] P. Ho and J. H. Kim, "On pilot symbol assisted detection of CPM schemes operating in fast fading channels," *IEEE Trans. Comm.*, vol. 44, pp. 337–347, Mar. 1996.

[223] T. Kailath, "Correlation detection of signals perturbed by a random channel," *IRE Trans. Inf. Theory*, vol. 6, pp. 361–366, June 1960.

[224] H. J. Scudder, "Adaptive communication receivers," *IEEE Trans. Inf. Theory*, vol. 11, pp. 167–174, Apr. 1965.

[225] D. W. Matolak and S. G. Wilson, "Detection for a statistically known, time-varying dispersive channel," *IEEE Trans. Comm.*, vol. 44, pp. 1673–1683, Dec. 1996.

[226] R. E. Morley and D. L. Snyder, "Maximum likelihood sequence estimation for randomly dispersive channels," *IEEE Trans. Comm.*, vol. 27, pp. 833–839, June 1979.

[227] J. H. Lodge and M. J. Moher, "Maximum likelihood sequence estimation of CPM signals transmitted over Rayleigh flat-fading channels," *IEEE Trans. Comm.*, vol. 38, pp. 787–794, June 1990.

[228] X. Yu and S. Pasupathy, "Innovations-based MLSE for Rayleigh fading channels," *IEEE Trans. Comm*, vol. 43, pp. 1534–1544, Feb./Mar./Apr. 1995.

[229] X. Yu and S. Pasupathy, "Error performance of innovations-based MLSE for Rayleigh fading channels," *IEEE Trans. Vehicular Techn.*, vol. 45, pp. 631–642, Nov. 1996.

[230] R. Haeb and H. Meyr, "A systematic approach to carrier recovery and detection of digitally phase modulated signals on fading channels," *IEEE Trans. Comm.*, vol. 37, pp. 748–754, July 1989.

[231] A. Aghamohammadi, H. Meyr, and G. Ascheid, "Adaptive synchronization and channel parameter estimation using an extended Kalman filter," *IEEE Trans. Comm.*, vol. 37, pp. 1212–1219, Nov. 1989.

[232] A. Aghamohammadi, H. Meyr, and G. Ascheid, "A new method for phase synchronization and automatic gain control of linearly modulated signals on frequency-flat fading channels," *IEEE Trans. Comm.*, vol. 39, pp. 25–29, Jan. 1991.

[233] G. M. Vitetta and D. P. Taylor, "Multi-sampling receivers for uncoded and coded PSK signal sequences transmitted over Rayleigh frequency-flat fading channels," *IEEE Trans. Comm.*, vol. 44, pp. 130–133, Feb. 1996.

[234] A. N. D'Andrea, A. Diglio, and U. Mengali, "Symbol-aided channel estimation with nonselective Rayleigh fading channels," *IEEE Trans. Vehicular Techn.*, vol. 44, pp. 41–49, Jan. 1995.

[235] J. P. Seymour and M. P. Fitz, "Near optimal symbol-by-symbol detection schemes for flat Rayleigh fading," *IEEE Trans. Comm.*, vol. 43, pp. 1525–1533, Feb./Mar./Apr. 1995.

[236] Y. Zhang, M. P. Fitz, and S. B. Gelfand, "Optimal and Near-Optimal Joint Channel and Data Estimation for Frequency-Selective Rayleigh Fading Channels," in *Proc. of the Thirty-third Annual Allerton Conference on Communication, Control, and Computing*, (University of Illinois, Urbana-Champaign, IL, USA), pp. 618–627, 1999.

[237] G.-K. Lee, S. B. Gelfand, and M. P. Fitz, "Bayesian techniques for equalization of rapidly fading frequency selective channels," *Int. Journ. Wireless Inform. Networks*, vol. 2, pp. 41–53, Jan. 1995.

[238] B. D. Hart and S.Pasupathy, "Innovations-Based map detection for time varying, frequency selective channels," *accepted for publication in IEEE Trans. Comm.*, 2000.

[239] D. Divsalar and M. K. Simon, "Maximum likelihood differential detection of uncoded and trellis coded amplitude phase modulation over AWGN and fading channels—metrics and performance," *IEEE Trans. Comm.*, vol. 42, pp. 76–89, Jan. 1994.

[240] M. Samiuddin and K. H. Biyari, "A comparative study of higher-order differential phase shift keying schemes over AWGN and Rayleigh fading channels," *Int. Journ. Wireless Inform. Networks*, vol. 2, pp. 183–196, 1995.

[241] P. Ho and D. Fung, "Error performance of multiple-symbol differential detection of PSK signals transmitted over correlated Rayleigh fading channels," *IEEE Trans. Comm.*, vol. 40, pp. 25–29, October 1992.

[242] I. Korn, "M-ary CPFSK-DPD with L-diversity maximum ratio combining in Rician fast-fading channels," *IEEE Trans. Vehicular Techn.*, vol. 45, pp. 613–621, 1996.

[243] K. H. Chang, W. Yuan, and C. N. Gheorgiades, "Block-by-block channel and sequence estimation for ISI/Fading channels," in *Proc. 7th Tyrrhenian Workshop on Digital Communications*, pp. 153–170, Sep. 1995.

[244] J. C. Han and C. N. Georghiades, "Sequence estimation in the presence of random parameters via the EM algorithm," *IEEE Trans. Comm.*, vol. 45, pp. 300–308, Mar. 1997.

[245] J. C. Han and C. N. Georghiades, "Pilot symbol initiated optimal decoder for the land mobile fading channel," in *Conf. Rec. Globecom'95*, (Singapore), pp. 42–47, Nov. 1995.

[246] D. Makraris, P. T. Mathiopoulos, and D. P. Bouras, "Optimal decoding of coded PSK and QAM signals in correlated fast fading channels and AWGN: A combined envelope, multiple differential and coherent detection approach," *IEEE Trans. Comm.*, vol. 42, pp. 63–74, Jan. 1994.

[247] T. K. Moon, "The expectation-maximization algorithm," *IEEE Sig. Proc. Mag.*, vol. 13, pp. 47–63, Nov. 1996.

[248] P. Y. Kam and C. H. Teh, "Reception of PSK signals over fading channels via quadrature amplitude estimation," *IEEE Trans. Comm.*, vol. 31, pp. 1024–1027, August 1983.

[249] P. Y. Kam and C. H. Teh, "An adaptive receiver with memory for slowly fading channels," *IEEE Trans. Comm.*, vol. 32, pp. 654–659, June 1984.

[250] P. Y. Kam, "Generalized quadratic receivers for orthogonal signals over the Gaussian channel with unknown phase/fading," *IEEE Trans. Comm.*, vol. 43, pp. 2050–2058, June 1995.

[251] J. H. Painter and L. R. Wilson, "Simulation results for the decision-directed MAP receiver for M-ary signals in multiplicative and additive Gaussian noise," *IEEE Trans. Comm.*, vol. 22, pp. 649–660, May 1974.

[252] J. H. Painter and S. C. Gupta, "Recursive ideal observer detection of known M-ary signals in multiplicative and additive Gaussian noise," *IEEE Trans. Comm.*, vol. 21, pp. 948–953, Aug. 1973.

[253] P. K. Varshney and A. H. Haddad, "A receiver with memory for fading channels," *IEEE Trans. Comm.*, vol. 26, pp. 278–283, Feb. 1978.

[254] R. Schober, W. H. Gerstacker, and J. B. Huber, "Decision-feedback differential detection of MDPSK for flat Rayleigh fading channels," *IEEE Trans. Comm.*, vol. 47, pp. 1025–1035, July 1999.

[255] J. P. Seymour and M. P. Fitz, "Two-stage carrier synchronization techniques for non-selective fading," *IEEE Trans. Vehicular Techn.*, vol. 44, pp. 103–110, Feb. 1995.

[256] P. Y. Kam and H. M. Ching, "Sequence estimation over the slow nonselective Rayleigh fading channel with diversity reception and Its application to Viterbi decoding," *IEEE J. Sel. Areas Comm.*, vol. 10, pp. 562–570, April 1992.

[257] P. A. Bello and B. Nelin, "Predetection diversity combining with selectively fading channels," *IEEE Trans. Comm.*, vol. 10, pp. 32–42, 1962.

[258] W. D. Lindsey, "Error probability for incoherent diversity reception," *IEEE Trans. Inf. Theory*, vol. 11, pp. 491–499, Oct. 1965.

[259] L. J. Mason, "Error Probability Evaluation for Systems Employing Differential Detection in a Rician Fast Fading Environment and Gaussian Noise," *IEEE Trans. Comm.*, vol. 35, pp. 39–46, 1987.

[260] L. J. Mason, "An error probability formula for M-ary DPSK in fast Rician fading and Gaussian noise," *IEEE Trans. Comm.*, vol. 35, pp. 976–978, Jan. 1987.

[261] H. Salwen, "Differential phase-shift keying performance under time-selective multipath fading," *IEEE Trans. Comm.*, vol. 23, pp. 383–385, March 1975.

[262] N. M. Y. Miqagaki and T. Namekawa, "Error rate performance of M-ary DPSK systems in Satellite/Aircraft communications," *Proc. IEEE ICC'79*, pp. 34.6.1–34.6.6, 1979.

[263] A. Neul, "Bit error rate for 4-DPSK in fast Rician fading and Gaussian noise," *IEEE Trans. Comm.*, vol. 37, pp. 1385–1387, Dec. 1989.

[264] G. M. Vitetta, U. Mengali, and D. P. Taylor, "Double-filter differential detection of PSK signals transmitted over linearly time-selective Rayleigh fading channels," *IEEE Trans. Comm.*, vol. 47, pp. 239–247, Feb. 1999.

[265] G. L. Turin, "On optimal diversity reception," *IRE Trans. Inf. Theory*, vol. 7, pp. 154–167, July 1961.

[266] J. N. Pierce, "Theoretical diversity improvement in frequency-shift keying," *Proc. IRE*, vol. 46, pp. 903–910, May 1958.

[267] P. M. Hahn, "Theoretical diversity improvement in multiple frequency shift keying," *IRE Trans. Comm. Systems*, vol. 10, pp. 177–184, June 1962.

[268] I. Korn, "Differential phase shift keying in two-path Rayleigh channel with adjacent channel interference," *IEEE Trans. Vehicular Techn.*, vol. 40, pp. 461–471, 1991.

[269] W. C. Dam and D. P. Taylor, "An adaptive maximum likelihood receiver for correlated Rayleigh-fading channels," *IEEE Trans. Comm.*, vol. 42, pp. 2684–2692, Sep. 1994.

[270] L. Andriot, G. Tziritas, and G. Jourdain, "Discrete realization for receivers. detecting signals over random dispersive channels. Part II: Doppler-spread channel," *Sig. Proc.*, vol. 9, pp. 89–100, Sept. 1985.

[271] S. Elnoubi, "Probability of error analysis of digital partial response continuous phase modulation with noncoherent detection in mobile radio channels," *IEEE Trans. Vehicular Techn.*, vol. 38, pp. 19–30, February 1989.

[272] I. Korn, "Error probability of M-ary FSK with differential phase detection in satellite mobile channel," *IEEE Trans. Vehicular Techn.*, vol. 38, pp. 76–85, May 1989.

[273] I. Korn, "GMSK with differential phase detection in the satellite mobile channel," *IEEE Trans. Comm.*, vol. 38, pp. 1980–1986, Nov. 1990.

[274] S. Elnoubi, "Analysis of GMSK with discriminator detection in mobile radio channels," *IEEE Trans. Vehicular Techn.*, vol. 35, pp. 71–76, May 1986.

[275] I. Korn, "M-ary Frequency Shift Keying with Limiter-Discriminator-Integrator Detector in Satellite Mobile Channel with Narrow-Band Receiver Filter," *IEEE Trans. Comm.*, vol. 38, pp. 1771–1778, 1990.

[276] I. Korn, "GMSK with limiter discriminator detection in satellite mobile channel," *IEEE Trans. Comm.*, vol. 39, pp. 94–101, Jan. 1991.

[277] D. K. Asano and S. Pasupathy, "Improved post-detection processing for limiter-discriminator detection of CPM in a Rayleigh, fast fading channel," *IEEE Trans. Vehicular Techn.*, vol. 44, pp. 729–734, Nov. 1995.

Part IV

Orthogonal Frequency Division Multiplexing

Thomas May and Hermann Rohling

Chapter 17

INTRODUCTION

In typical radio channels, multipath propagation occurs because of many reflections of the transmitted signal, see Part I, Chapter 3. The various propagation paths are characterized by different delays, and this leads to a time-dispersive behavior of the channel. Intersymbol interference (ISI) is caused and has to be taken care of in radio transmission systems. The measures that have to be taken depend on the data rate to be transmitted or, equivalently, on the bandwidth processed by the transmission system.

If the data rate is low and the symbol duration is large as compared to the maximum delay of the channel, it can be possible to cope with the resulting ISI without any equalization. This is the situation in cordless phones with a distance range of only several hundred meters.

As the distance range or the data rate of the system increases, ISI becomes more severe and channel equalization has to be provided. For example, in the GSM system, ISI over five symbols is equalized. For this purpose, a channel estimation is required and filter coefficients must be calculated. A powerful equalizer can also cope with ISI over a longer symbol sequence, but somewhere a limit is given by the computational complexity of the equalization. If, e.g., a data rate of 10 Mbit/s is to be transmitted over a channel with a maximum delay of 10 μs, then ISI extends over 100 symbols and the calculation of the corresponding number of filter coefficients would be too complex.

In situations where a high data rate is to be transmitted over a channel with a relatively large maximum delay, an alternative approach is given by the OFDM (orthogonal frequency division multiplexing) transmission technique. The idea of OFDM is to distribute the high-rate data stream to many low-rate data streams that are transmitted in a parallel way over many subchannels. Thus, in each subchannel the symbol duration is low as compared to the maximum delay of the channel, and ISI can be handled.

The basic principles of OFDM have already been proposed in several publications in the 1960s. However, these ideas could not be implemented efficiently, since powerful semiconductor devices were not available at that time. Today, even relatively complex OFDM transmission systems with high data rates are technically feasible, and such systems can be taken advantage of in frequency-selective radio channels.

In a classical FDM system, narrowband signals are generated independently, assigned to various frequency bands, transmitted, and separated by filters at the receiver. The new aspects of OFDM are that the various signals are generated jointly by a fast Fourier transform (FFT) and that their spectra overlap. As a result, generating the signal is simplified and the bandwidth efficiency of the system is improved.

It is interesting to note that as early as 1961 a code division multiplexing scheme was proposed where sine and cosine functions were used as orthogonal signals [1]. The resulting signal could already be compared with an OFDM signal. However, the fact that this system was identical to a frequency division multiplex was not important for the proposal, and the benefits in frequency-selective channels were not recognized.

Since 1966, FDM systems with overlapping spectra were proposed in several publications [2],[3],[4]. The next step was the proposal to realize an FDM system with the discrete Fourier transform (DFT) [5]. Finally, in 1971, Weinstein and Ebert proposed a complete OFDM system [6], which included generating the signal with an FFT and adding a guard interval in the case of multipath channels. This system is referred to in the following discussion.

In further development, OFDM was discussed for channels with both flat and frequency-selective fading [7]. In [8], OFDM was proposed for broadcast applications and mobile reception. Meanwhile, the OFDM transmission technique is a part of the European DAB and DVB-T broadcasting standards [9]. It has also been chosen as the transmission technique for wireless local area networks in the HIPERLAN 2 standard and in the 5 GHz extension of the IEEE 802.11 standard.

The most important advantage of the OFDM transmission technique as compared to single-carrier systems is obtained in frequency-selective channels. The signal processing in the receiver is rather simple in this case: after transmission over the radio channel, the orthogonality of the OFDM subcarriers is maintained and the channel interference effect is reduced to a multiplication of each subcarrier by a complex transfer factor. Therefore, equalizing the signal is very simple, whereas equalization may not be feasible in the case of conventional single-carrier transmission covering the same bandwidth. This trade-off is depicted in Figure 17.1. It must be mentioned, however, that in [10] a single-carrier system with frequency-domain equalization, which also copes with large delays, was proposed.

In the following sections, the OFDM transmission technique and its application in time-dispersive channels are described. The topics of synchronization, channel estimation, modulation, channel coding, and amplitude limitation of the signal are covered. Furthermore, the extension to OFDM-CDM is briefly discussed.

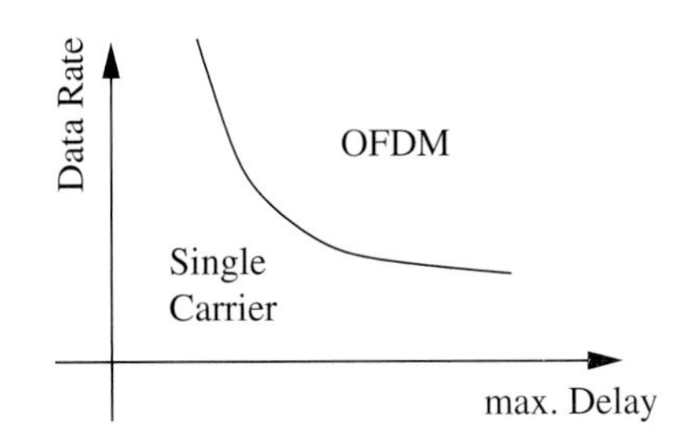

Figure 17.1. OFDM vs. single-carrier transmission

Bibliography

[1] G. A. Franco and G. Lachs, "An orthogonal coding technique for communications," in *IRE International Convention Record*, vol. 9, pp. 126–133, 1961.

[2] R. W. Chang, "Synthesis of band-limited orthogonal signals for multichannel data transmission," *Bell Sys. Techn. Journal*, vol. 45, Dec 1966.

[3] B. Saltzberg, "Performance of an efficient parallel data transmission system," *IEEE Transactions on Communications*, vol. 15, pp. 805–811, Dec 1967.

[4] M. S. Zimmerman and A. L. Kirsch, "The AN/GSC-10 (KATHRYN) variable rate data modem for HF radio," *IEEE Trans. Comm.*, vol. 15, pp. 197–203, 1967.

[5] S. Darlington, "On digital single-sideband modulators," *IEEE Transactions on Circuit Theory*, vol. 17, pp. 409–414, Aug 1970.

[6] S. B. Weinstein and P. M. Ebert, "Data transmission by frequency-division multiplexing using the discrete Fourier transform," *IEEE Transactions on Communications*, vol. 19, pp. 628–634, 1971.

[7] L. J. Cimini, "Analysis and simulation of a digital mobile channel using orthogonal frequency division multiplexing," *IEEE Trans. on Comm.*, vol. 33, pp. 665–675, July 1985.

[8] M. Alard and R. Lassalle, "Principles of modulation and channel coding for digital broadcasting for mobile receivers," Tech. Rep. 224, Aug. 1987.

[9] T. de Couasnon and etal, "OFDMfor digital TV broadcasting," *Signal Processing*, vol. 39, pp. 1–32, 1994.

[10] H. Sari, G. Karam, and I. Jeanclaude, "Channel equalization and carrier synchronization in OFDM systems," in *Proc. 6th Tirrenia International Workshop on Digital Communications*, (Tirrenia), Sept. 1993.

Chapter 18

THE OFDM TRANSMISSION TECHNIQUE

This chapter provides an analytical description of the basic OFDM transmission technique. It also describes how the signal is generated at the transmitter and evaluated at the receiver.

18.1 TRANSMITTER

An OFDM signal as it is described in [1], [2] consists of N subcarriers spaced by the frequency distance Δf. Thus, the total system bandwidth B is divided into N equidistant subchannels. All subcarriers are mutually orthogonal within a time interval of length $T_S = 1/\Delta f$. The kth subcarrier signal is described analytically by the function $\tilde{g}_k(t)$, $k = 0, \ldots, N-1$.

$$\tilde{g}_k(t) = \begin{cases} e^{j2\pi k\Delta f t} & \forall t \in [0, T_S] \\ 0 & \forall t \notin [0, T_S]. \end{cases} \tag{18.1.1}$$

Since the system bandwidth B is subdivided into N narrowband subchannels, the OFDM block duration T_S is N times as large as in the case of a single-carrier transmission system covering the same bandwidth. Typically, for a given system bandwidth, the number of subcarriers is chosen such that the symbol duration is large compared to the maximum delay of the channel. This subcarrier signal $\tilde{g}_k(t)$ is extended by a cyclic prefix (called guard interval) with the length T_G yielding the following signal:

$$g_k(t) = \begin{cases} e^{j2\pi k\Delta f t} & \forall t \in [-T_G, T_S] \\ 0 & \forall t \notin [-T_G, T_S]. \end{cases} \tag{18.1.2}$$

The guard interval is added to the subcarrier signal in order to avoid ISI, which occurs in multipath channels; see Figure 18.1. At the receiver, the guard interval is removed and only the time interval $[0, T_S]$ is evaluated. From this point of view,

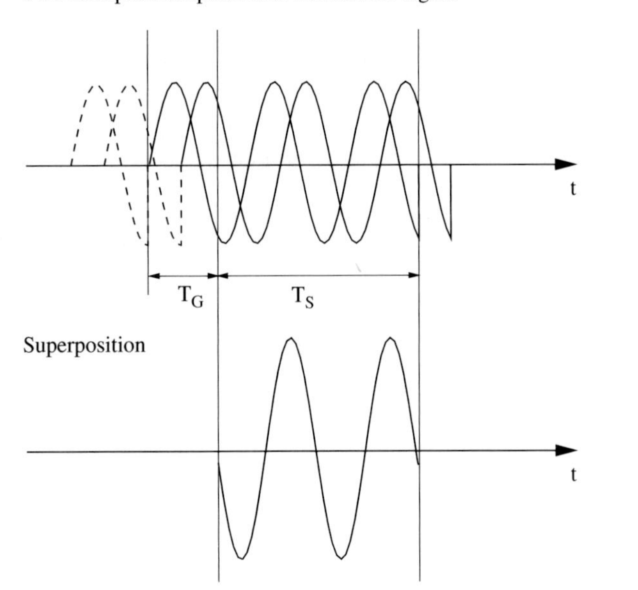

Figure 18.1. An OFDM subcarrier signal transmitted over a multipath channel

the guard interval is a pure system overhead. The total OFDM block duration is $T = T_S + T_G$.

An important advantage of the OFDM transmission technique is that ISI, which occurs in all multipath channels, can be reduced considerably. If the guard interval length T_G is larger than the maximal delay in the radio channel, no ISI occurs at all and the orthogonality of the subcarriers is not affected, as shown in Figure 18.1. Interference with the previously transmitted information only appears within the guard interval, whereas in the evaluated time interval, the multipath channel only changes the amplitude and the phase of the subcarrier signal.

Each subcarrier can be modulated independently with the complex modulation symbol $S_{n,k}$, where the subscript n refers to the time interval and k to the number of the subcarrier in the considered OFDM block. Thus, within the symbol duration T, the following signal of the n-th OFDM block is formed:

$$s_n(t) = \frac{1}{\sqrt{N}} \sum_{k=0}^{N-1} S_{n,k}\, g_k(t - nT). \tag{18.1.3}$$

The total continuous-time signal consisting of all OFDM blocks is

$$s(t) = \frac{1}{\sqrt{N}} \sum_{n=0}^{\infty} \sum_{k=0}^{N-1} S_{n,k}\, g_k(t - nT). \tag{18.1.4}$$

Thus, a rectangular pulse shaping is applied for each subcarrier. In literature, different pulse shaping has also been considered. The effect of this modification is discussed later. Due to the rectangular pulse shaping, the spectra of the subcarriers are sinc-functions, e.g., for the kth subcarrier:

$$G_k(f) = T\operatorname{sinc}\left[\pi T(f - k\Delta f)\right] \qquad \text{where} \tag{18.1.5}$$

$$\operatorname{sinc}(x) = \frac{\sin(x)}{x}. \tag{18.1.6}$$

The spectra of the subcarriers overlap, but the subcarrier signals are mutually orthogonal and the modulation symbols $S_{n,k}$ can be recovered by a correlation technique:

$$\begin{aligned} \langle g_k, g_l \rangle &= \int_0^{T_S} g_k(t)\overline{g_l(t)}dt \\ &= T_S \delta_{k,l} \end{aligned} \tag{18.1.7}$$

$$S_{n,k} = \frac{\sqrt{N}}{T_S} \left\langle s_n(t), \overline{g_k(t - nT)} \right\rangle \tag{18.1.8}$$

where $\overline{g_k(t)}$ is the conjugate of $g_k(t)$. In practical applications, the OFDM signal $s_n(t)$ is generated in a first step as a discrete-time signal in the digital signal processing part of the transmitter. Since the bandwidth of an OFDM system is $B = N\Delta f$, the signal must be sampled with the sampling time $\Delta t = 1/B = 1/N\Delta f$. The samples of the signal are written as $s_{n,i}$, $i = 0, 1, \ldots, N-1$ and can be calculated as

$$s_{n,i} = \frac{1}{\sqrt{N}} \sum_{k=0}^{N-1} S_{n,k} e^{j2\pi ik/N}. \tag{18.1.9}$$

This equation describes exactly the inverse discrete Fourier transform (IDFT), which is typically implemented as an IFFT. After the IFFT, further signal processing can be applied to avoid out-of-band radiation. Out-of-band radiation can occur if amplitude peaks of the OFDM signal are limited by the power amplifier and can occur to the side lobes of the subcarrier spectra. Appropriate signal processing methods are discussed in Chapters 23 and 24. Finally, the signal is D/A-converted and transmitted. The process flow is illustrated in Figure 18.2.

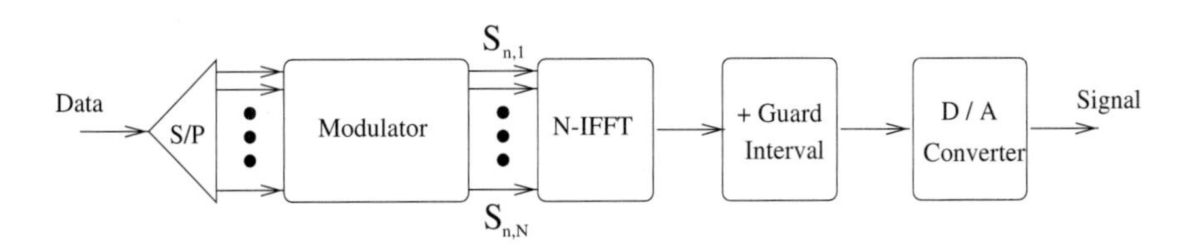

Figure 18.2. Structure of an OFDM transmitter

18.2 RECEIVER

The guard interval length T_G is assumed to be chosen to be larger than the maximal multipath delay of the channel. Furthermore, a time-invariant channel is considered in a first step. Thus, within the time interval that is evaluated by the receiver, the contributions to the received signal of all propagation paths add up to the original modulated subcarrier signals, each multiplied by an individual complex channel transfer factor. This is depicted in Figure 18.1 for a single subcarrier. Thus, the subcarrier orthogonality is not affected at the output of the radio channel.

The same assumption is a good approximation if we consider a time-variant channel where the coherence time is large compared to the symbol duration T. Therefore, the received signal $r_n(t)$ can be separated into the orthogonal subcarrier signals by a correlation technique according to (18.1.8):

$$R_{n,k} = \frac{\sqrt{N}}{T_S} \left\langle r_n(t), \overline{g_k(t - nT)} \right\rangle . \tag{18.2.1}$$

Equivalently, the correlation at the receiver can be implemented as a discrete Fourier transform (DFT) or an FFT, respectively:

$$R_{n,k} = \frac{1}{\sqrt{N}} \sum_{i=0}^{N-1} r_{n,i} \mathrm{e}^{-\mathrm{j}2\pi ik/N} \tag{18.2.2}$$

where $r_{n,i}$ is the ith sample of the received signal $r_n(t)$ and $R_{n,k}$ is the received complex modulation symbol of the kth subcarrier. The FFT and IFFT algorithms can be implemented very efficiently.

If the symbol duration T is chosen to be much smaller than the coherence time of the channel, then the transfer function of the radio channel $H(f,t)$ can be considered constant within the duration of each modulation symbol $S_{n,k}$. In this case, the effect of the radio channel is only a multiplication of each subcarrier signal $g_k(t)$ by a complex transfer factor $H_{n,k} = H(k\Delta f, nT)$. As a result, the received modulation symbol $R_{n,k}$ after the FFT is

$$R_{n,k} = H_{n,k} S_{n,k} + N_{n,k}, \tag{18.2.3}$$

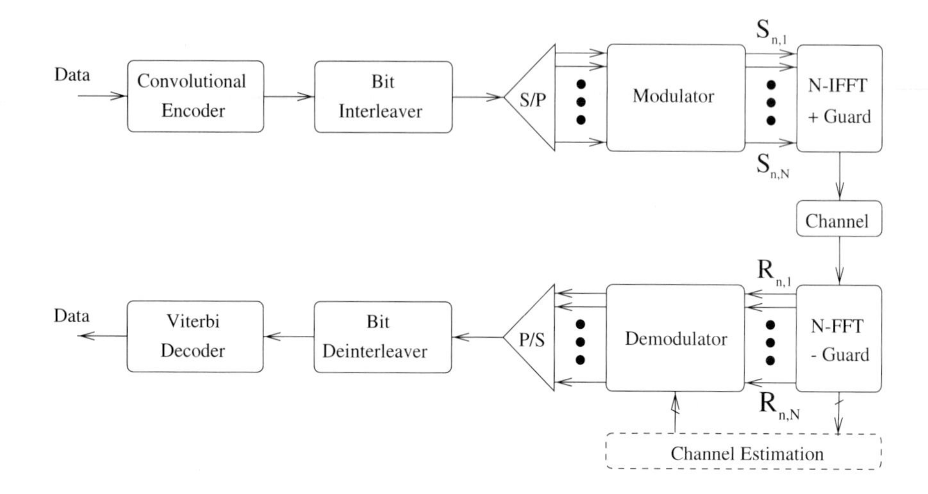

Figure 18.3. Block diagram of an OFDM transmission system with convolutional channel coding

where $N_{n,k}$ is additive noise of the channel. This equation shows the most important advantage of applying the OFDM transmission technique. Even in a channel with a large maximum delay that would result in severe ISI in the case of a single-carrier system, we can ensure that each transmitted modulation symbol is only modified by a complex transfer factor and additive noise and that no ISI occurs at all.

Equation (18.2.3) implies perfect synchronization at the receiver, which is an important topic in OFDM transmission systems, since time and frequency synchronization errors disturb the orthogonality of the subcarriers and therefore reduce the signal-to-noise ratio considerably. Another aspect that is not taken into account in (18.2.3) is the signal distortion due to amplifier nonlinearities. Unfortunately, an OFDM signal has a high peak-to-average power ratio, so the power amplifier usually has to be operated at a large input backoff, mainly in order to avoid out-of-band interference. If, e.g., the interference in the adjacent frequency bands is to be lower than 40 dB below the power density in the OFDM band, then an input backoff of more than 7.5 dB is required. However, the amplitude range of OFDM signals can be reduced by means of coding or signal processing so that a lower input backoff is possible.

A simplified block diagram of an OFDM transmission system with convolutional channel coding is depicted in Figure 18.3. Channel coding is very important for OFDM systems because of the flat-fading behavior of the subcarriers.

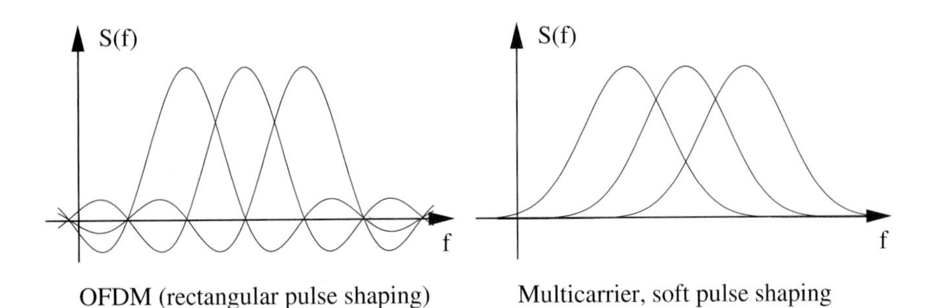

Figure 18.4. Spectra of OFDM and of a multicarrier system with soft pulse shaping

18.3 MULTICARRIER SYSTEM WITH SOFT IMPULSE SHAPING

Usually, in OFDM systems a rectangular pulse shaping is applied. The important advantage of this choice is that the different subcarrier signals are orthogonal and that therefore interchannel interference (ICI) does not occur. However, two disadvantages of this rectangular pulse shaping are related to the sinc shape of the corresponding subcarrier spectra. First, the side lobes of the subcarrier spectra decay slowly with increasing frequency distance and produce out-of-band interference. Second, if frequency synchronization errors are not negligible, each subcarrier causes interference on each other subcarrier [3].

These disadvantages can be avoided if pulse shaping filters are applied. The advantages of different pulse shaping formats with respect to robustness against frequency synchronization errors and against fading effects have been analyzed in [4], [5], [6], [7], [8], [9], [10].

In [7] Gaussian functions are used for pulse shaping. In this case, the subcarrier spectra are also Gaussian functions and decay rapidly. In such a system, however, both ISI and ICI over a few modulation symbols occur. Therefore, equalization must be performed. Figure 18.4 shows the spectra of an OFDM system with rectangular pulse shaping and a multicarrier system with soft pulse shaping. In [7], equalization is performed with a Viterbi equalizer. It is shown that multicarrier systems with soft impulse shaping are more robust against synchronization errors. Furthermore, a diversity gain can be obtained because of the equalization.

In [4], new functions for pulse shaping are constructed; they are longer than one symbol duration and consequently overlap in time direction. The corresponding subcarrier spectra overlap in frequency direction, too. These functions are orthogonal and cause neither ISI nor ICI, provided that synchronization is perfect. Furthermore, the functions are optimized such that minimum out-of-band interference is produced. Indeed, it is shown that out-of-band interference is significantly lower with this impulse shaping. However, since the guard interval technique cannot be applied, ISI does occur if the signal is transmitted over a multipath channel, so an equalizer is necessary.

Pulse shaping can also be implemented in the frequency domain, as described in [10]. In [10], a guard interval is used. Thus, the spectral shaping of the transmitted signal is destroyed and the advantage of lower out-of-band interference is lost. After an appropriate windowing at the receiver, the spectral shaping is restored and so the advantage of an increased robustness against synchronization errors is maintained. Due to the guard interval, ISI is avoided. ICI, which is caused by the pulse shaping, still requires an equalizer.

In [8], a multicarrier system with soft-impulse shaping is proposed in which no equalization is applied. Instead, the subcarriers are separated by a larger frequency distance to reduce ICI at the cost of a lower bandwidth efficiency. It is shown that soft-impulse shaping makes the system more robust against frequency synchronization errors but less robust against large delays.

The cited contributions show that with soft-impulse shaping, an advantage can be obtained with respect to frequency synchronization. On the other hand, this advantage is not large, and a rather good frequency synchronization is still required. An equalizer must be implemented additionally to remove ICI and ISI; this addition requires a channel estimation. The diversity gain that is obtained by the equalizer is an advantage only in the case of uncoded transmission. In a coded system, a larger diversity gain is obtained by the code, so that this advantage of soft impulse shaping vanishes. Furthermore, channel coding with soft-decision decoding which should be applied is more difficult to implement in a system with an equalizer.

Thus, it appears that the additional effort of channel estimation and equalization is not justified by the resulting advantage. In a usual OFDM system, no channel estimation is necessary if differential modulation is applied. With respect to the reduction of out-of-band interference, it seems to be more promising to apply filtering for this purpose.

Bibliography

[1] S. B. Weinstein and P. M. Ebert, "Data transmission by frequency-division multiplexing using the discrete Fourier transform," *IEEE Transactions on Communications*, vol. 19, pp. 628–634, 1971.

[2] M. Alard and R. Lassalle, "Principles of modulation and channel coding for digital broadcasting for mobile receivers," Tech. Rep. 224, Aug. 1987.

[3] T. Pollet, M. V. Bladel, and M. Moeneclaey, "BER sensitivity of OFDM systems to carrier frequency offset and Wiener phase noise," *IEEE Transactions on Communications*, vol. 43, pp. 191–193, Feb.-Apr. 1995.

[4] A. Vahlin and N. Holte, "Optimal finite duration pulses for OFDM," *IEEE Transactions on Communications*, vol. 44, pp. 10–14, Jan. 1996.

[5] R. Haas and J. C. Belfiore, "A time-frequency well-localized pulse for multiple carrier transmission," *Wireless Personal Communications*, vol. 5, pp. 1–18, 1997.

[6] W. Kozek and A. F. Molisch, "Nonorthogonal pulseshapes for multicarrier communications in doubly dispersive channels," *IEEE J. Selected Areas Communications*, vol. 16, pp. 1579–1589, 1998.

[7] K. Matheus and K.-D. Kammeyer, "Optimal design of a multicarrier system with soft impulse shaping including equalization in time and frequency direction," in *Proc. Globecom '97*, (Phoenix), Nov. 1997.

[8] W. Rhee, J. C. Chuang, and L. Cimini, "Performance comparison of OFDM and multitone with polyphase filter bank for wireless communications," in *Proc. IEEE VTC '98*, (Ottawa, Canada), pp. 768–772, May 1998.

[9] P. K. Remvik, N. Holte, and A. Vahlin, "Fading and carrier frequency offset robustness for different pulse shaping filters in ofdm .," in *IEEE-Proceedings of Vehicular Technology Conference*, pp. 777–781, 1999.

[10] S. B. Slimane, "OFDM schemes with non-overlapping time waveforms," in *Proc. IEEE VTC '98*, (Ottawa, Canada), pp. 2067–2071, May 1998.

Chapter 19

SYNCHRONIZATION

In an OFDM transmission system, synchronization must be accomplished with respect to the frame structure, the OFDM block structure, and carrier frequency. The frequency synchronization has to be more exact than in single carrier systems [1], since a frequency offset results in ICI over all subcarriers. Especially in TDMA communication systems with short transmission frames, it is important to achieve a fast synchronization in order to use the transmission capacity efficiently.

19.1 FRAME AND TIMING SYNCHRONIZATION

The first task to be performed is frame synchronization. A simple approach to this problem is insertion of a zero block at the beginning of the transmission frame [2], [3], [4]. In this zero block, no signal is transmitted. Based on this block, the receiver can detect the beginning of a frame. An appropriate synchronization scheme is described in [5]. This concept can also be used to obtain a rough timing synchronization.

Different approaches of timing synchronization make use of training blocks or periodic signals that are used instead of or in addition to the zero block. In [6], the proposal is made to use a chirp signal as a training burst that can be detected by a correlation technique. Similarly, a timing synchronization can be based on periodic signals. In [4], an additional zero block and a reference block are transmitted at the beginning of a frame. This block contains a periodically repeated training sequence and is evaluated for timing and frequency acquisition. These training blocks are evaluated by a correlation technique [7].

Once a rough timing synchronization is established, a more exact synchronization can be obtained by evaluation of the guard interval of the OFDM blocks; the interval must be longer than the maximum delay of the channel in this case [8], [9], [10]. This synchronization is also accomplished with a correlation where the following correlation metric is maximized [4]:

$$\Lambda(j) = \sum_{i=0}^{N_G-1} r_{n,j-i} \cdot r^*_{n,j-i-N}, \qquad (19.1.1)$$

where N_G is the number of samples in the guard interval. Results in [8] and [4] indicate that a good synchronization can be obtained for SNR values above 10 dB in the AWGN channel and 15 dB in a fading radio channel.

19.2 CARRIER FREQUENCY SYNCHRONIZATION

A frequency synchronization error f_{off} leads to ICI over all subcarriers. The resulting interference power for each subcarrier is [11]

$$P_I = \sum_{k \neq l} \text{sinc}^2[\pi(k - l - f_{off}/\Delta f)]. \tag{19.2.1}$$

For a frequency offset of more than $0.1\,\Delta f$, the interference power is larger than 15 dB below the signal power. Thus, the frequency synchronization has to be very exact.

The frequency offset f_{off} can be written as

$$f_{off} = n\Delta f + f'_{off}, \tag{19.2.2}$$

that is, it consists of an integer part n and a remaining offset f'_{off}. The carrier synchronization schemes known from literature mostly consist of two stages. In one stage, the integer part of the frequency offset is compensated for and in the other stage, f'_{off}, which is smaller than half the subcarrier distance, is removed. The frequency offset is compensated by multiplying the time signal by $\exp(-\text{j}2\pi f_{off}t)$.

The methods described in [12], [11], [3], [13], [4] make use of pilot signals to measure the frequency offset.

In [13], three OFDM blocks are transmitted for frequency synchronization. First, two identical pseudorandom binary sequences are transmitted. The phase difference of modulation symbols received on the same subcarrier in these two OFDM blocks is evaluated to determine f'_{off}. This idea also appears in [12]. Then, another OFDM block, which is differentially modulated with another pseudorandom binary sequence, is transmitted. By means of correlation the receiver can detect the integer part n of the frequency offset with this sequence.

In [11], pilot symbols are inserted in two successive OFDM symbols at a number of uniformly spaced subcarriers. The pairs of pilot symbols transmitted over the same subcarrier are differentially modulated with a pseudorandom sequence. This can be used to obtain a rough frequency synchronization using a correlation. In a second step, the synchronization can be improved by measurement of the phase difference of the same pairs of pilot symbols. Thus, frequency synchronization is achieved with much less overhead than in [13].

Frequency and timing synchronization is performed on the basis of a single OFDM block (pilot block) in [3]. Additionally, a rough timing synchronization is achieved with a zero block. In the pilot block, only a part of the subcarriers is selected according to a pseudorandom sequence and is transmitted. At the receiver, the frequency offset is estimated roughly by a correlation so that the transmitted pilot subcarriers can be identified. In a second step, pairs of pilot subcarriers are evaluated to determine the timing and frequency offsets more exactly, and the average is taken from the results.

References [14], [15], [16] have proposed a frequency synchronization without using pilot signals.

In [14], the guard interval determines the phase offset, exploiting the fact that a frequency offset leads to a phase rotation between the samples $r_{n,i}$ in the guard interval and $r_{n,i+N}$. However, more than 50 OFDM blocks are needed for acquisition.

In [15], the frequency synchronization is accomplished in the following way. The OFDM time signal is oversampled by a factor of two. With every second sample, the FFT is calculated, yielding the sequence z_k^1; with the other samples, the sequence z_k^2 is also calculated by an FFT. The receiver calculates the values

$$c_k = z_k^2 \cdot (z_k^1 \cdot \mathrm{e}^{\mathrm{j}\pi k/N})^*. \tag{19.2.3}$$

The sign pattern of the sequence c_k determines the integer part of the frequency offset, and the remaining offset can also be measured on the basis of the c_k. In this case, averaging over many OFDM blocks is also necessary for a stable synchronization.

In communication systems with short transmission frames, it is important that the synchronization be performed as fast as possible. Therefore, pilot signals must be used. If a system with continuous transmission is considered, a synchronization scheme without pilot signals may be acceptable, although acquisition takes much more time.

Bibliography

[1] T. Pollet, M. V. Bladel, and M. Moeneclaey, "BER sensitivity of OFDM systems to carrier frequency offset and Wiener phase noise," *IEEE Transactions on Communications*, vol. 43, pp. 191–193, Feb.-Apr. 1995.

[2] K. Fazel *et al.*, "A concept of digital terrestrial television broadcasting," *special issue "Multi-Carrier Communications," Wireless Personal Communications*, vol. 2, pp. 9–27, 1995.

[3] H. Nogami and T. Nagshima, "A frequency and timing period acquisition technique for OFDM systems," in *Proc. IEEE PIMRC '95*, pp. 1010–1015, 1995.

[4] T. Keller and L. Hanzo, "Orthogonal frequency division multiplex synchronization techniques for wireless local area networks," in *Proc. PIMRC '96*, (Piscataway, NJ), pp. 963–967, IEEE, 1996.

[5] W. D. Warner and C. Leung, "OFDM/FM frame synchronization for mobile radio data communication," *IEEE Transactions on Vehicular Techn.*, vol. 42, pp. 302–313, August 1993.

[6] L. Hazy and M. El-Tanany, "Synchronization of OFDM systems over frequency-selective fading channels," in *Proc. IEEE VTC '97*, (Phoenix, USA), pp. 2094–2098, 5.-7. May 1997.

[7] T. M. Schmidl and D. C. Cox, "Robust frequency and timing synchronization for ofdm," *IEEE Trans. Comm.*, vol. 45, pp. 1613–1621, 1997.

[8] J.-J. V. de Beek, M. Sandell, M. Isaksson, and P. O. Börjesson, "Low complex frame synchronization in OFDM systems," in *Proc. International Conference on Universal Personal Communications (ICUPC)*, (Tokyo, Japan), pp. 982–986, Nov. 1995.

[9] M. Speth, F. Claßen, and H. Meyr, "Frame synchronization of OFDM systems in frequency selective fading channels," in *Proc. VTC '97*, (Phoenix, USA), pp. 1807–1811, 5.-7. May 1997.

[10] J. J. vandeBeek, M. Sandell, and P. O. Borjesson, "Ml estimation of time and frequency offset in ofdm systems," *IEEE Trans. Signal Processing*, vol. 45, pp. 1800–1805, 1997.

[11] F. Claßen and H. Meyr, "Frequency synchronization algorithms for OFDM systems suitable for communication over frequency-selective fading channels," in *Proc. VTC '94*, (Stockholm, Sweden), pp. 1655–1659, June 1994.

[12] P. H. Moose, "A technique for orthogonal frequency division multiplexing frequency offset correction," *IEEE Transactions on Communications*, vol. 42, pp. 1590–1598, Oct. 1994.

[13] S. Hara, M. Mouri, M. Okada, and N. Morinaga, "Transmission performance analysis of multi-carrier modulation in frequency selective fast Rayleigh fading channels," *Wireless Personal Communications*, vol. 2, pp. 335–356, Feb. 1996.

[14] F. Daffara and O. Adami, "A new frequency detector for orthogonal multicarrier transmission techniques," in *Proc. IEEE VTC '95*, (Chicago, USA), pp. 804–809, 1995.

[15] J. S. Oh, Y. M. Chung, and S. U. Lee, "A carrier synchronization technique for OFDM on the frequency-selective fading environment," in *Proc. VTC '96*, pp. 1574–1578, 1996.

[16] M. Okada, M. Mouri, S. Hara, S. Komaki, and N. Morinaga, "A maximum likelihood symbol timing, symbol period and frequency offset estimator for orthogonal multi-carrier modulation signals," in *Proc. IEEE ICT '96*, pp. 596–601, 1996.

Chapter 20

MODULATION

As shown in (18.1.3), each subcarrier can be modulated independently by a complex modulation symbol. Generally, this complex value is generated by mapping a sequence of m bits to one out of $M = 2^m$ points in the constellation diagram of the applied modulation scheme. In the case of differential modulation this mapping is followed by a differential encoding; see Figure 20.1.

20.1 WITHOUT DIFFERENTIAL ENCODING

If no differential encoding is applied, the data bits to be transmitted are directly mapped to the modulation symbols $S_{n,k} \in \Psi$. Examples of this modulation technique are M-PSK and M-QAM. For M-PSK, the set of modulation symbols is $\Psi = \{e^{j2\pi i/M} | i = 0, \ldots, M-1\}$.

QAM schemes are applied in the terrestrial transmission of DVB. In this application, a modification of QAM, termed multiresolution QAM (MR-QAM), has been analyzed. In [2], both MR-QAM and MR-DAPSK, which is the corresponding technique for differential amplitude and phase modulation, are described.

20.2 WITH DIFFERENTIAL ENCODING

If differential encoding is applied, the transmitted information is contained in the quotient of two successive modulation symbols. In this case, the data bits to be transmitted are not directly mapped to the modulation symbols $S_{n,k}$, but rather to the quotient $B_{n,k}$ of two successively transmitted modulation symbols. This technique can be applied either in time or in frequency direction.

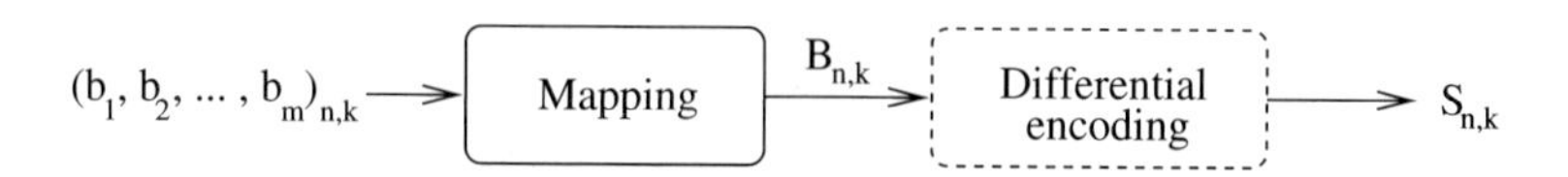

Figure 20.1. Block diagram of differential subcarrier modulation. Reprinted from [1], copyright IEEE.

The operation of differential encoding in time direction can be described analytically by the following multiplication:

$$S_{n,k} = S_{n-1,k} \cdot B_{n,k}. \tag{20.2.1}$$

For differential encoding in time direction, it is important that the coherence time of the channel be large compared to the symbol duration to ensure that the channel transfer factors $H_{n-1,k}$ and $H_{n,k}$ are approximately equal.

Differential encoding can alternatively be performed in frequency direction as described in (20.2.2).

$$S_{n,k} = S_{n,k-1} \cdot B_{n,k}. \tag{20.2.2}$$

In this case, the coherence bandwidth of the channel has to be large compared to the subcarrier spacing.

Depending on the channel and on the design of the OFDM system differential, encoding in time or frequency direction may be more suitable. Since differential encoding in a time direction requires an entire OFDM block for the transmission of reference symbols whereas differential encoding in a frequency direction requires only a single reference symbol per OFDM block, the latter technique can have an advantage with respect to the system overhead. This advantage applies especially to communication systems where only a few OFDM blocks or even only a single block is transmitted per channel access.

It is also possible to use differential encoding in both time and frequency directions. The first transmitted OFDM block can be differentially encoded in a frequency direction and all subsequent blocks in a time direction. Thus, only a single reference symbol is required at the beginning of a transmission. In the following discussion, only differential encoding in the time direction is considered.

In the case of M-DPSK, for example, $B_{n,k} \in \Psi = \{e^{j2\pi i/M} | i = 0, \ldots, M-1\}$, the constellation of $B_{n,k}$ is the same as that of $S_{n,k}$. However, M-DPSK has a poor performance if M is large. To increase the bandwidth efficiency, this scheme can be extended to a differential amplitude and phase modulation (M-DAPSK), which shows a substantial performance improvement over M-DPSK for $M \geq 16$.

DAPSK can be described as a differentially encoded APSK, the signal-space diagram of which is defined by the signal set

$$\Psi = \{a^A \cdot e^{j\Delta\varphi \cdot P} | A \in \{0, \ldots N_a - 1\}, P \in \{0, \ldots, N_p - 1\}\}. \tag{20.2.3}$$

As an example, a 64-APSK signal-space diagram with $N_a = 4$ and $N_p = 16$ is depicted in Figure 20.2. Note that the amplitudes are spaced by a factor a. The mapping of the m input bits is done separately for amplitude and phase, using m_a and m_p bits, respectively. To minimize the bit-error rate, Gray mapping is used.

The number of amplitude states $N_a = 2^{m_a}$, the number of phases per amplitude state $N_p = 2^{m_p}$, and the amplitude ratio a are free parameters that must be

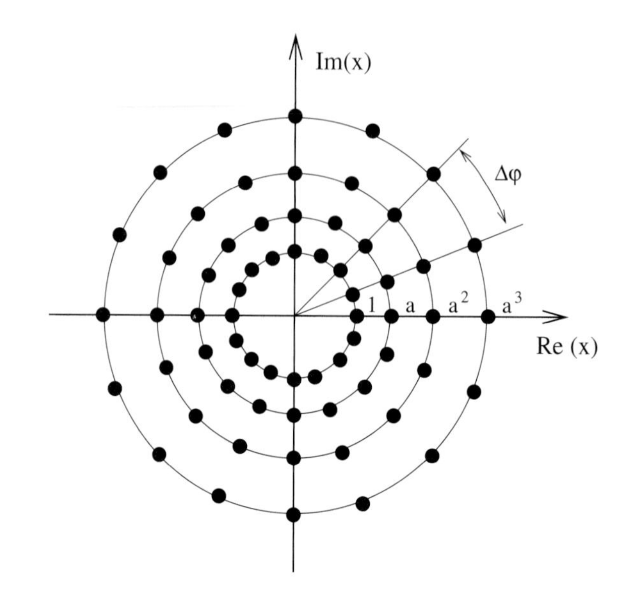

Figure 20.2. 64–(D)APSK signal space diagram. Reprinted from [3], copyright IEEE.

	N_a	N_p	a (inc. dem)	a(quasi-coh. dem.)
M¡16	1	M	-	-
M=16	2	8	2.0	1.8
M=32	2	16	1.6	1.45
M=64	4	16	1.4	1.38
M=128	4	32	1.3	1.21

Table 20.1. Optimal DAPSK modulation parameters for quasi-coherent and non-coherent demodulation. Reprinted from [1], copyright IEEE.

optimized depending on the number of signal states $M = N_a \cdot N_p = 2^m$ and the demodulation method (quasi-coherent or noncoherent). The optimized values for different numbers of bits per symbol are given in Table 20.1.

Differential encoding of the APSK symbols $Q_{n,k}$ can be performed as follows:

$$S_{n,k} = Q_{n,k} \odot S_{n-1,k} \tag{20.2.4}$$

$$= a^{[A(Q_{n,k})+A(S_{n-1,k})] \bmod N_a} \cdot e^{j\frac{2\pi}{N_p}[P(Q_{n,k})+P(S_{n-1,k})]}. \tag{20.2.5}$$

To ensure that $S_{n,k} \in \Psi$, the amplitude exponents $A(Q_{n,k})$ and $A(S_{n-1,k})$ are added modulo the number N_a of amplitude circles. This addition is not explicitly necessary for the phase, because of its inherent periodicity. Alternatively, the modulo operation can be integrated into the amplitude mapping, see Figure 20.3, so

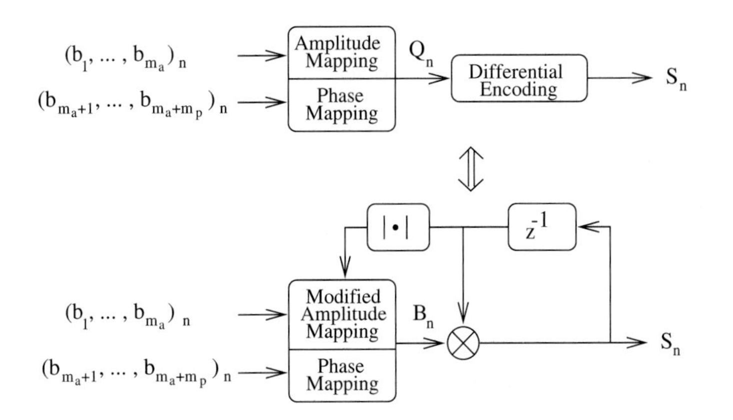

Figure 20.3. DAPSK modulation. Reprinted from [3], copyright IEEE.

that differential encoding can again be described by a multiplication according to (20.2.1):

$$S_{n,k} = B_{n,k} \cdot S_{n-1,k}, \tag{20.2.6}$$

where

$$B_{n,k} \in \Psi' = \left\{ a^{A'} \cdot e^{j\Delta\varphi P} \,\middle|\, \begin{array}{l} A' \in \{-N_{a+1}, \ldots, N_a - 1\} \\ P \in \{0, \ldots, N_p - 1\} \end{array} \right\}. \tag{20.2.7}$$

As an example, this modified mapping is given in Table 20.2 for $m_a = 2$.

$\lvert B_{n,k}\rvert$		$(b_1 b_2)_{n,k}$ 00	01	11	10
$\lvert S_{n-1,k}\rvert$	1	1	a	a^2	a^3
	a^1	1	a	a^2	$1/a$
	a^2	1	a	$1/a^2$	$1/a$
	a^3	1	$1/a^3$	$1/a^2$	$1/a$

Table 20.2. Modified amplitude mapping in case of four amplitude circles. Reprinted from [1], copyright IEEE.

Bibliography

[1] H. Rohling, T. May, K. Brueninghaus, and R. Grnheid, "Broadband ofdm radio transmission for multimedia applications," *Proc. IEEE*, vol. 87, pp. 1778 –1789, 1999.

[2] V. Engels and H. Rohling, "Multi-resolution 64-DAPSK modulation in a hierarchical COFDM transmission system," *IEEE Transactions on Broadcasting*, vol. 44, pp. 139–149, March 1998.

[3] H. Rohling, K. Brueninghaus, and T. May, "High rate ofdm modem with quasi-coherent dapsk," in *Proc. ICT '97, Melbournce, April 1997*, pp. 211 –216, IEEE, 1997.

Chapter 21

DEMODULATION

This chapter describes how the transmitted data bits can be recovered from the received symbol sequence $R_{n,k}$. This activity could more exactly be termed hard output demodulation. In case of channel coding, soft-decision decoding is usually preferred; that is, a soft-output demodulation is required. This topic is discussed in more detail in Chapter 22.

Depending on the type of modulation (differential or nondifferential), there are different ways to demodulate the received complex sequence, as discussed in detail in this chapter.

21.1 COHERENT DEMODULATION OF NONDIFFERENTIAL MODULATION

Coherent demodulation has to be applied if a nondifferential modulation scheme is used in the transmitter. Originally, the term coherent means that the mixer in the receiver is synchronized in frequency and phase with the carrier frequency of the received signal. The term coherent applied to an OFDM transmission system means that each subcarrier must be synchronized, or the phase offset must be known. If any kind of amplitude modulation like M-QAM is applied, then the attenuation of each subcarrier has to be known, too. To generate this information in the receiver, we must perform a channel estimation to provide estimates $\widehat{H}_{n,k}$ for the channel transfer factors. The decision is based on the quotient

$$D_{n,k}^{c} = \frac{R_{n,k}}{\widehat{H}_{n,k}} = S_{n,k} + \frac{N_{n,k}}{\widehat{H}_{n,k}} \Rightarrow \widehat{S}_{n,k} = \mathrm{dec}\{D_{n,k}^{c}\}. \qquad (21.1.1)$$

For $D_{n,k}^{c}$ the receiver makes a decision according to given thresholds, as is expressed by the operator $\mathrm{dec}\{\cdot\}$.

21.1.1 Channel estimation

The channel estimation can be performed on the basis of known symbols (pilot symbols) which are included in the OFDM signal before transmission [1], [2], [3]. In

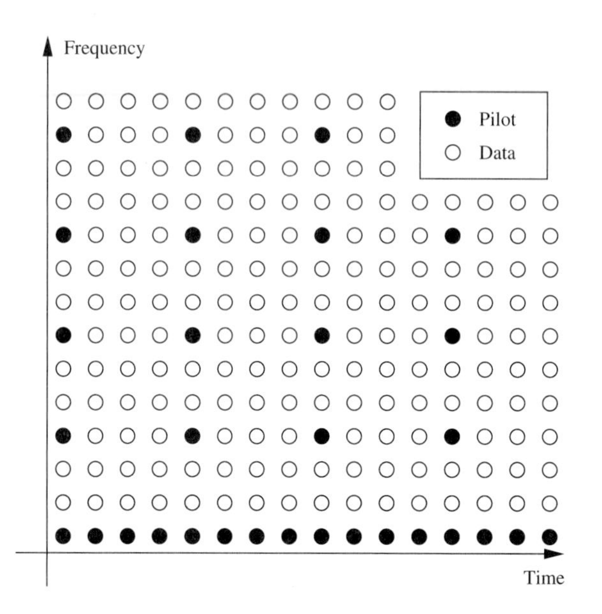

Figure 21.1. Example of a pilot grid for OFDM

the first step, the receiver extracts the transfer factors at those times and frequencies at which pilot symbols have been transmitted. Then, the transfer factors between the positions of pilot symbols can be interpolated by means of filtering. If the transfer function is filtered in time direction with a given density of pilot symbols, then the maximal Doppler frequency of the channel must be sufficiently small, according to the sampling theorem. In the same way, filtering in the frequency direction means that the maximal delay of the channel must be sufficiently small.

Figure 21.1 illustrates an example of a pilot grid. To obtain good estimates at the border of the OFDM blocks, too, the first and the last subcarrier are only modulated with pilot symbols. The other pilot symbols are spaced by $n_f = 3$ subcarriers in frequency direction and by $n_t = 4$ modulation symbols in time direction. References [2], [4], show that the conditions

$$n_f < \frac{1}{\tau_{max}\Delta f} \tag{21.1.2}$$

$$n_t < \frac{1}{2f_{D,max}T} \tag{21.1.3}$$

must be fulfilled according to the sampling theorem. However, to get a reasonable noise reduction by filtering, the pilot symbol spacing should be smaller than half the values given above [4], so that the channel transfer function is oversampled.

The receiver determines the transfer factors at the locations of pilot symbols and determines the other transfer factors by means of filtering. Ideally, a two-dimensional Wiener filter could be applied for this purpose. This filter would be designed according to the autocorrelation function of the channel transfer function.

At the locations of pilot symbols the receiver measures the complex values

$$P_{n,k} = \frac{R_{n,k}}{S_{n,k}}, \tag{21.1.4}$$

which are written as a vector p. The channel transfer factors $H_{n,k}$ to be estimated are stored in the vector h. Furthermore, R_{pp} denotes the auto-covariance matrix of p and R_{hp} the cross-covariance matrix between h and p. The estimates of the transfer factors are calculated by

$$h = R_{hp} R_{pp}^{-1} p. \tag{21.1.5}$$

To reduce the computational complexity associated with the channel estimation, two one-dimensional Wiener filters can be used instead. In this case, the transfer function is first filtered in frequency direction, and then the result is filtered in time direction. Reference [2] analyzed channel estimation with several filters. For a fixed calculation complexity, the best results were obtained with two one-dimensional filters.

With respect to a receiver with ideal knowledge of the transfer factors, the BER curves typically degrade by approximately 1.5 dB due to this channel estimation if the channel statistics are roughly known. If the assumptions concerning maximal delay and Doppler shift are not met, then larger degradations occur.

21.2 NONCOHERENT DEMODULATION OF DIFFERENTIAL MODULATION

If differential encoding is used in the transmitter, the demodulation can be performed either noncoherently (nc) or quasi-coherently (qc). With noncoherent demodulation, the decision is based on the quotient of two successive symbols

$$D^{nc}_{n,k} = \frac{R_{n,k}}{R_{n-1,k}} = \frac{S_{n-1,k} \cdot B_{n,k} \cdot H_{n,k} + N_{n,k}}{S_{n-1,k} \cdot H_{n-1,k} + N_{n-1,k}} \tag{21.2.1}$$

$$\widehat{B}_{n,k} = \text{dec}\{D^{nc}_{n,k}\}. \tag{21.2.2}$$

If the coherence time of the channel is large enough, successive channel transfer factors are strongly correlated so that $H_{n,k} \approx H_{n-1,k}$ and therefore cancel out in (21.2.1) (if the noise influence is neglected). However, $D^{nc}_{n,k}$ is affected by twice the noise power of $D^{c}_{n,k}$, leading to a higher bit-error rate than that of coherent demodulation with perfect channel estimation. The advantage of noncoherent demodulation is that no channel estimation has to be performed. Thus, the computation complexity in the receiver is relatively low.

Demodulation of DAPSK symbols requires thresholds to be defined for the decision variable $D^{nc}_{n,k}$ noncoherent demodulation. Since the bits are independently mapped to amplitude and phase, it is sufficient to define independent amplitude and phase thresholds.

In case of noncoherent demodulation, the following threshold is optimum for the phase:

$$T_p^{nc} = \Delta\varphi/2 + i\Delta\varphi \quad i = 0, \ldots, N_p - 1. \tag{21.2.3}$$

With respect to the amplitude, the optimal thresholds are hard to calculate because of the division of two successively received symbols (21.2.1). Furthermore, they depend on the signal-to-noise ratio. However, there exists a good approximation, which is given by the geometric average of two adjacent valid amplitudes [5]:

$$T_a^{nc} \approx a^j \cdot \sqrt{a} \quad j = 0 \ldots N_a - 2. \tag{21.2.4}$$

21.3 QUASI-COHERENT DEMODULATION

Quasi-coherent demodulation is another way to demodulate differential modulation. As with coherent demodulation, the channel influence is removed in a first step and differential decoding takes place as a second step. Due to differential encoding, there is no need to determine the channel phase exactly but only up to an ambiguity of $\pm n \cdot 2\pi/N_p$ rad, where N_p is the number of phases. In this case, no pilot symbols are required to estimate the unknown channel transfer factors [6].

Apart from differential decoding, the processing in a quasi-coherent receiver is similar to that in a coherent one, i.e.,

$$\widehat{B}_{n,k} = \frac{\widehat{S}_{n,k}}{\widehat{S}_{n-1,k}} = \frac{\mathrm{dec}(R_{n,k}/\widetilde{H}_{n,k})}{\mathrm{dec}(R_{n-1,k}/\widetilde{H}_{n-1,k})} = \frac{\mathrm{dec}(D^c_{n,k})}{\mathrm{dec}(D^c_{n-1,k})}. \tag{21.3.1}$$

In case of an incorrect decision, this error influences two successive symbols due to differential encoding. Therefore, the error probability with quasi-coherent demodulation is approximately twice as high as with coherent detection, provided that the error rate is relatively small ($< 10\%$). Doubling the error rate with respect to coherent demodulation results in a smaller SNR loss than does noncoherent demodulation (at least in case of perfect channel estimation).

For quasi-coherent demodulation of DPSK and DAPSK, the following thresholds are used:

$$T_p^{qc} = \Delta\varphi/2 + i\Delta\varphi \quad i = 0, \ldots, N_p - 1 \tag{21.3.2}$$

$$T_a^{qc} = a^j \cdot \frac{1+a}{2} \quad j = 0, \ldots, N_a - 2 \tag{21.3.3}$$

Bibliography

[1] P. Höher, "TCM on frequency–selective land–mobile fading channels," in *Proc. of the 5th International Workshop on Digital Communications*, (Amsterdam), Elsevier, 1991.

[2] M. Sandell, *Design and Analysis of Estimators for Multicarrier Modulation and Ultrasonic Imaging.* PhD thesis, Lulea University of Technology, Lulea, Sweden, Sept. 1996.

[3] Y. G. Li, L. C. Cimini, and N. R. Sollenberger, "Robust channel estimation for ofdm systems with rapid dispersive fading channels," *IEEE Trans. Comm.*, vol. 46, pp. 902–915, 1998.

[4] S. Kaiser, *Multi-Carrier CDMA Mobile Radio Systems — Analysis and Optimization of Detection, Decoding and Channel Estimation.* PhD thesis, Düsseldorf, Germany, 1998.

[5] V. Engels and H. Rohling, "Multilevel differential modulation techniques (64-DAPSK) for multicarrier transmission systems," *European Transactions on Telecommunications*, vol. 6, pp. 663–640, Nov., Dec. 1995.

[6] A. J. Viterbi, "Nonlinear estimation of PSK-modulated carrier phase with application to burst digital transmission," *IEEE Trans. Commun.*, vol. IT-29, pp. 543–551, July 1983.

Chapter 22

CHANNEL CODING

As explained, the radio channel attenuates each OFDM subcarrier by a complex transfer factor $H_{n,k}$. If the channel is a multipath channel with many propagation paths and without a line-of-sight (LOS) path, then the complex transfer factors are Gaussian distributed according to the central limit theorem. Consequently, the amplitudes of the transfer factors are Rayleigh distributed.

This means that even at a very high average signal-to-noise ratio (SNR) there are always some subcarriers that are strongly attenuated and have a rather low SNR. Thus, many bit errors occur on these subcarriers. The situation is similar to a narrowband fading channel, which is not surprising, since an OFDM transmission system consists of many narrowband fading channels.

For this reason, channel coding is an important topic for OFDM systems. Without channel coding the typical flat-fading curve is obtained for the bit-error rate (BER). With channel coding, large SNR gains are achieved. In this situation, the code exploits the frequency diversity provided by the system bandwidth. The additional gain that can be realized with soft-decision instead of hard-decision decoding is very large.

Therefore, convolutional codes are a reasonable choice because soft-decision decoding can easily be performed by the Viterbi algorithm, which is usually applied to decode convolutional codes. Furthermore, the code rate of convolutional codes can be adjusted in a flexible way by puncturing the code. Code puncturing can be applied to adapt the error-correcting capability of the code to the channel state and to the desired level of error protection.

We can improve the performance of simple convolutional codes by concatenating two or more codes. In the DVB-T standard, for example, a punctured convolutional code is concatenated with an outer Reed-Solomon code.

Another example is the parallel concatenation of two recursive systematic convolutional codes, well known as turbo codes [1]. For a serial concatenation of convolutional codes with turbo decoding, an even better performance has been obtained in [2]. These concepts can also be applied reasonably to OFDM systems.

22.1 SOFT OUTPUT DEMODULATION

To apply soft-decision decoding, the decoder needs metric increments for the received modulation symbols instead of hard decisions. This means the demodulator has to provide soft outputs $\lambda_{n,k}^{\nu}$ for each symbol $R_{n,k}$, where the superscript ν enumerates the (possibly transmitted) modulation symbols in the set Ψ.

22.1.1 Coherent

As explained, the received modulation symbol of the kth OFDM subcarrier in the nth time interval is

$$R_{n,k} = H_{n,k}S_{n,k} + N_{n,k}, \tag{22.1.1}$$

where $S_{n,k}$ is the transmitted symbol, $H_{n,k}$ is the complex-channel transfer factor of the considered subcarrier, and $N_{n,k}$ is additive white Gaussian noise with the noise power $2\sigma^2$. The probability density function (pdf) of $R_{n,k}$ is a complex Gaussian distribution

$$p\left(R_{n,k}|S_{n,k}\right) = \frac{1}{2\pi\sigma^2}\mathrm{e}^{-\frac{|R_{n,k}-H_{n,k}S_{n,k}|^2}{2\sigma^2}}. \tag{22.1.2}$$

A maximum likelihood sequence estimator would have to choose one out of all possibly transmitted symbol sequences μ

$$\langle S_{n,k}\rangle^{(\mu)} = \langle S_{n,k}(\mu)\rangle\,. \tag{22.1.3}$$

The sequences which possibly have been transmitted are defined by the channel coding scheme. $S_{n,k}(\mu)$ is the modulation symbol of subcarrier k in the nth OFDM block if the sequence μ has been transmitted. The sequence estimator determines an estimate $\widehat{\langle S_{n,k}\rangle}$ according to the following criterion:

$$\widehat{\langle S_{n,k}\rangle} = \arg\max_{\mu} P\left(\langle R_{n,k}\rangle \,|\, \langle S_{n,k}\rangle^{(\mu)}\right) \tag{22.1.4}$$

$$= \arg\max_{\mu} \prod_{k} p\left(R_{n,k}|S_{n,k}(\mu)\right) \tag{22.1.5}$$

$$= \arg\min_{\mu} \sum_{k} \left|R_{n,k} - H_{n,k}S_{n,k}(\mu)\right|^2. \tag{22.1.6}$$

The summands in (22.1.6) can be calculated by the demodulator for each possibly transmitted modulation symbol $S_{n,k}^{\nu}$ with $\nu = 1, \ldots, M$ and passed to the decoder as metric increments

$$\lambda_{n,k}^{\nu} = \left| R_{n,k} - H_{n,k} S_{n,k}^{\nu} \right|^2 \tag{22.1.7}$$

for further evaluation and final decision. The decoder estimates the symbol sequence or the corresponding bit sequence on the basis of the given metric increments according to (22.1.6).

This metric, consisting of squared Euclidean distances as metric increments, can be used with all coherent demodulation techniques because it is based on the assumptions that only the transmitted modulation states are statistically independent and that the received modulation states are Gaussian distributed due to the AWGN.

22.1.2 Noncoherent

In the case of differential modulation techniques, soft outputs are calculated in a similar way. Decision variables that are chosen should have a pdf that can be approximated by a Gaussian distribution, and the variance of this distribution is determined. Then, the demodulator generates a metric consisting of squared distances with these appropriate decision variables weighted according to the determined variance.

For higher-level M-DPSK ($M \geq 8$), the pdf of the phase difference $\varphi_{n,k} = \arg D_{n,k}^{nc}$ of successive symbols can already be approximated by a Gaussian distribution

$$p\left(\varphi_{n,k} | \psi_{n,k}\right) \approx \frac{1}{\sqrt{2\pi\sigma_{\varphi}^2}} \mathrm{e}^{-\frac{|\varphi_{n,k} - \psi_{n,k}|^2}{2\sigma_{\varphi}^2}}, \tag{22.1.8}$$

where $\psi_{n,k}$ is the transmitted and $\varphi_{n,k}$ the received phase difference. This results in the following decision criterion of the Viterbi algorithm for M-DPSK signals [3]:

$$\widehat{\langle \psi_{n,k} \rangle} = \arg\min_{\mu} \sum_{k} \left| H_{n,k} \right|^2 \left| \varphi_{n,k} - \psi_{n,k}(\mu) \right|^2 . \tag{22.1.9}$$

The amplitude information of DAPSK is contained in the quotient $B_{n,k}$ of the amplitudes of two successive modulation symbols—this is the usual decision variable. To generate a metric for the Viterbi algorithm, we use the new decision variable $W_{n,k}$ instead of $B_{n,k}$. $W_{n,k}$ is approximately Gaussian distributed again.

$$W_{n,k} = \ln\left|\frac{R_{n,k}}{R_{n-1,k}}\right|, \quad V_{n,k} = \ln\left|\frac{S_{n,k}}{S_{n-1,k}}\right| \tag{22.1.10}$$

$$p\left(W_{n,k}|V_{n,k}\right) \approx \frac{1}{\sqrt{2\pi\sigma_w^2}} \mathrm{e}^{-\frac{\left|W_{n,k}-V_{n,k}\right|^2}{2\sigma_w^2}} \tag{22.1.11}$$

$$\sigma_w^2 = \sigma_\varphi^2 = \frac{\sigma^2}{\left|H_{n,k}S_{n,k}\right|^2} + \frac{\sigma^2}{\left|H_{n-1,k}S_{n-1,k}\right|^2} \tag{22.1.12}$$

Using the decision variables $W_{n,k}$ and $\varphi_{n,k}$, we can also give a metric with squared distances for DAPSK modulation [3]. The transmitted symbols $B_{n,k}$ are described by $V_{n,k}$ and $\psi_{n,k}$.

$$\widehat{\langle B_{n,k}\rangle} = \arg\min_{\mu} \sum_k d_{n,k}^2(\mu) RI_{n,k}^2 \tag{22.1.13}$$

$$d_{n,k}(\mu) = \left(W_{n,k} - V_{n,k}(\mu)\right)^2 + \left(\varphi_{n,k} - \psi_{n,k}(\mu)\right)^2 \tag{22.1.14}$$

$$RI_{n,k}^2 = \frac{1}{\left|1/R_{n,k}\right|^2 + \left|1/R_{n-1,k}\right|^2} \tag{22.1.15}$$

Again, each summand in (22.1.13) is the metric increment for the corresponding symbol $B_{n,k}$. For all possible symbols $B_{n,k}^{\nu}$, metric increments can be calculated as

$$\lambda_{n,k}^{\nu} = (d_{n,k}^{\nu})^2 RI_{n,k}^2 \tag{22.1.16}$$

by the demodulator. By the term $d_{n,k}^{\nu}$ the position of the received symbols in the constellation diagram is considered. $RI_{n,k}$ is information about the reliability of the subcarrier (Reliability Information) since it is mainly determined by the channel transfer factor of the kth subcarrier.

22.1.3 Quasi-coherent

In the case of quasi-coherent detection, we evaluate the two successively received modulation symbols separately in order to calculate the metric increments. The Viterbi decoder uses the following metric:

$$\begin{aligned}\widehat{\langle B_{n,k}\rangle} &= \arg\max_{\mu} \prod_k p\left(R_{n,k}, R_{n-1,k}|B_{n,k}(\mu)\right) \\ &\approx \arg\min_{\mu} \lambda_{n,k}(\mu)\end{aligned} \tag{22.1.17}$$

where for all possible $B_{n,k}^{\nu}$ metric increments, $\lambda_{n,k}^{\nu}$ can be calculated as follows:

$$\lambda_{n,k}^{\nu} = \min_{\frac{S^{\kappa}}{S^{\rho}} = B_{n,k}^{\nu}} \left[|R_{n,k} - H_{n,k} S^{\kappa}|^2 + |R_{n-1,k} - H_{n-1,k} S^{\rho}|^2 \right]. \quad (22.1.18)$$

Again, all possible metric increments $\lambda_{n,k}^{\nu}$ can be calculated by the demodulator and passed to the decoder.

22.2 BIT METRIC

As described above, the demodulator calculates metric increments $\lambda_{n,k}^{\nu}$ for each modulation symbol $S_{n,k}^{\nu}$ or $B_{n,k}^{\nu}$. In the case of convolutional codes, the task of the decoder is to decide for the bit sequence $\langle b_{n,k,j} \rangle$ that leads to the minimal metric. The subscript $j = 1, \ldots, m$ enumerates the bits that are transmitted by a modulation symbol. If the Viterbi algorithm is to be used for decoding, metric increments $\lambda_{n,k,j}$ for each bit are required instead of metric increments for each symbol, in order to make the following decision:

$$\widehat{\langle b_{n,k,j} \rangle} = \arg\max_{\mu} P\left(\langle R_{n,k} \rangle \mid \langle b_{n,k,j} \rangle^{(\mu)} \right) \quad (22.2.1)$$

$$\approx \arg\min_{\mu} \sum_{k,j} \lambda_{n,k,j}(\mu). \quad (22.2.2)$$

A metric increment $\lambda_{n,k,j}^{\kappa}$ for the assumption that $b_{n,k,j} = \kappa \in \{-1, 1\}$ can be obtained by

$$\lambda_{n,k,j}^{\kappa} = \min_{S_{n,k}^{\nu} | b_{n,k,j} = \kappa} \lambda_{n,k}^{\nu}. \quad (22.2.3)$$

In the case of differential modulation, $B_{n,k}^{\nu}$ replaces $S_{n,k}^{\nu}$ in (22.2.3). The Viterbi decoder chooses the bit sequence $\langle b_{n,k,j} \rangle$ that minimizes the sum in (22.2.2). For all bits $b_{n,k,j}$ of a subcarrier k, the corresponding metric increments are determined on the basis of the modulation symbol with the highest conditional probability $P(R_{n,k}|S_{n,k})$ and by which the assumed bit (-1 or 1) is transferred.

This way of calculating and using a bit metric for the Viterbi algorithm only approximates an MLSE. For an optimal decision, the metric increments for all bits in the bit sequence which are associated with the same subcarrier would have to refer to the same modulation symbol.

22.3 MULTILEVEL CODING

In the discussion above, we considered modulation and channel coding separately. A joint optimization of coding and modulation can achieve a better performance by taking advantage of the fact that different error events occur with very different

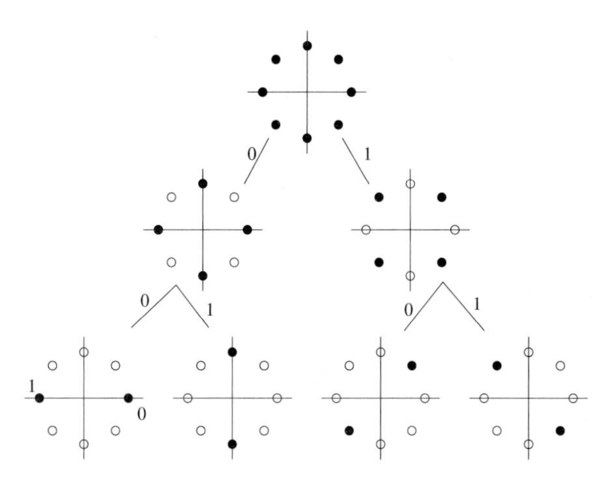

Figure 22.1. Set partitioning for 8–PSK. Reprinted from [7], copyright IEEE.

probabilities. A general approach of combining coding and modulation is multilevel coding [4].

Another approach to a joint optimization of coding and modulation is the concept of trellis-coded modulation (TCM) [5]. Several publications have proposed that TCM be applied for channel coding in OFDM transmission systems. In the discussion about channel coding for narrowband fading channels, it has been recognized, however, that TCM performs worse than simple convolutional codes in this situation [6]. In this section, only the multilevel coding concept is explained.

If a modulation scheme with m bits per symbol is used, the data bits are encoded into m sequences of l bits by m different encoders. These bit sequences are mapped to l modulation symbols.

The set of all modulation symbols is partitioned into two subsets such that the error probability within each subset is minimized. In the case of nondifferential modulation, this means that the Euclidean distance of the modulation symbols in each subset must be maximized. Subsequently, each of the subsets is recursively partitioned again until the subsets contain only a single modulation symbol. A well-known example for this set partitioning is given in Figure 22.1 for 8-PSK.

Based on this set partitioning, the data bits are mapped to modulation symbols in the following way. One data bit is taken from each of the m coded sequences. The first data bit selects one of the two subsets of the first partitioning level. Each of the following data bits selects one of the two subsets of the next partitioning level until the modulation symbol to be transmitted is defined after m selections.

The receiver evaluates the transmitted modulation symbols and subsequently determines the coded bit sequences. Provided that the receiver is able to decode without errors the first bit sequence that corresponds to the first partitioning level, this information can be used to evaluate the second partitioning level and so on. As the error probability within the subsets decreases from level to level, the codes need less redundancy to correct errors.

For differential modulation, the set of all possible quotients $B_{n,k} = S_{n,k}/S_{n-1,k}$ has to be partitioned so that the error probability within each subset is maximized. In the case of M-DPSK, that leads to the same set partitioning which is known for M-PSK. For M-DAPSK, the set partitioning is discussed in the following paragraphs.

We assume that the M-DAPSK symbol B_1 with logarithm of amplitude V_1 and phase difference ψ_1 is transmitted, that the symbol B_2 with V_2 and ψ_2 belongs to the same subset as B_1, and that the decisions variables that the receiver calculates from the received symbol B' are $W = V_1 + \Delta V$ and $\varphi = \psi_1 + \Delta\psi$.

The probability density functions of W and φ are approximately Gaussian as shown above. The approximated joint pdf of W and φ is

$$\begin{aligned} p(W, \varphi | V_1, \psi_1) &\approx \frac{1}{2\pi\sigma_w^2} \mathrm{e}^{-\frac{1}{2\sigma_w^2}\left(|W-V_1|^2 + |\varphi-\psi_1|^2\right)} \\ &= \frac{1}{2\pi\sigma_w^2} \mathrm{e}^{-\frac{1}{2\pi\sigma_w^2}\left(|\Delta V|^2 + |\Delta\psi|^2\right)}. \end{aligned} \tag{22.3.1}$$

The receiver decides incorrectly if

$$p(W, \varphi | V_2, \psi_2) > p(W, \varphi | V_1, \psi_1) \tag{22.3.2}$$

$$\Rightarrow (W - V_2)^2 + (\varphi - \psi_2)^2 < (W - V_1)^2 + (\varphi - \psi_1)^2. \tag{22.3.3}$$

The decision variables W and φ make the situation similar to the partitioning of a nondifferential modulation scheme like M-QAM, where the Euclidean distance within the subsets is maximized. Here, the error probability within a subset is minimized if the smallest distance $d_{i,j}$ between two symbols B_i and B_j in the subset is maximized.

$$d_{i,j}^2 = (V_i - V_j)^2 + (\psi_i - \psi_j)^2 \tag{22.3.4}$$

The distance measure $d_{i,j}$ is nothing else but a Euclidean distance referring to the coordinates V and ψ.

The optimal set partitioning is given for 16-, 32-, and 64–DAPSK in Figures 22.2 – 22.4. The partitioning diagrams refer to the Q symbols (see Section 20.2) that represent the information to be transmitted.

22.4 PERFORMANCE OF CODED OFDM SYSTEMS

22.4.1 System parameters

The baseband part of a general OFDM transmission system with equalization and channel coding was shown in Figure 18.3. In this section, we consider OFDM

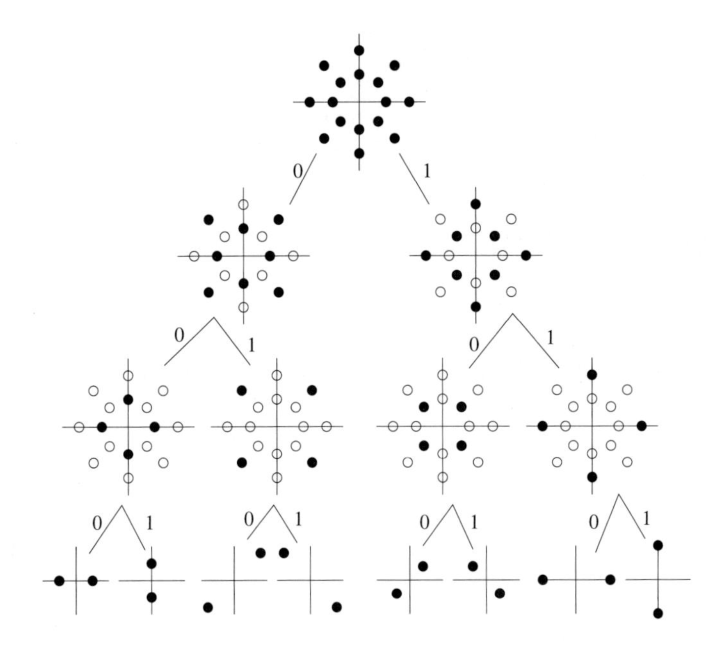

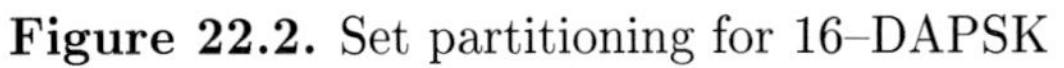

Figure 22.2. Set partitioning for 16–DAPSK

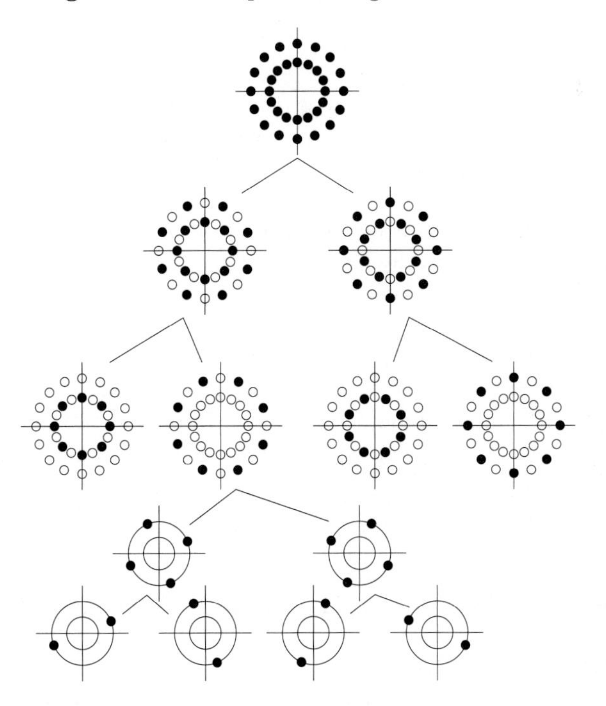

Figure 22.3. Set partitioning of 32–DAPSK

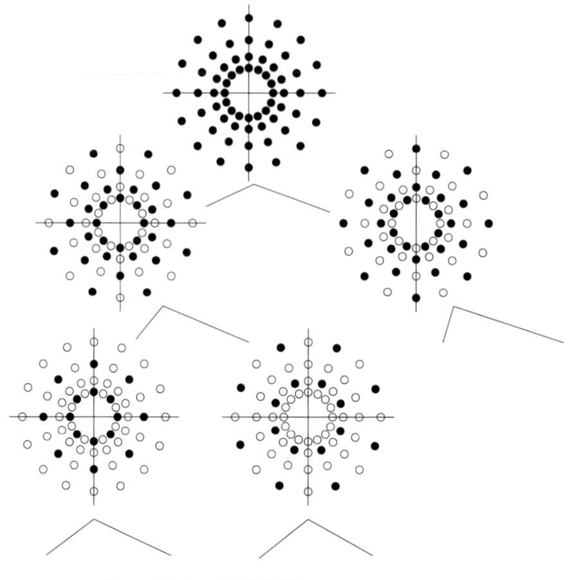

Figure 22.4. Set partitioning of 64–DAPSK. Reprinted from [7], copyright IEEE.

systems with 256 subcarriers, a total bandwidth of 7.16 MHz, and a guard interval of 7 μs. The net data rates are 18 and 27 Mbit/s for 16-QAM/DAPSK and 64-QAM/DAPSK, respectively.

For each OFDM block, the sequence of data bits to be transmitted is encoded by a convolutional code with memory 6 from [8]. The coded bits are interleaved by a block interleaver to avoid burst errors that would affect the performance of a convolutional code.

The modulation takes place in the frequency domain as a mapping of the coded and interleaved data bits to a sequence of modulation symbols with or without differential encoding. Differential modulation is performed in the time domain so that each modulation symbol $S_{n,k}$ refers to the previously transmitted symbol $S_{n-1,k}$ of the same OFDM subcarrier.

This signal is transmitted over the frequency-selective radio channel. The channel has been simulated according to the WSSUS channel model with the power delay profile "Typical Urban" from [9] (maximum delay $\tau = 7\mu$s), not time-varying.

If a nondifferential modulation scheme (PSK or QAM) is used, an ideal knowledge of the channel transfer factors is assumed. A practical transmission system would have to use some of the subcarriers for the transmission of pilot symbols for the channel estimation. Doing so would cause both an additional overhead and a system degradation, as compared to the ideal channel estimation, of approximately

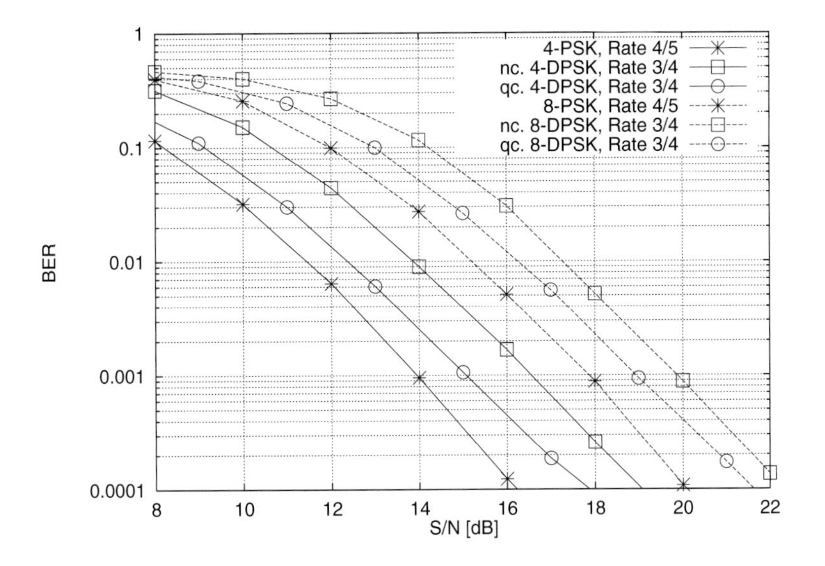

Figure 22.5. Comparison of 4- and 8-DPSK and -PSK, 6% pilot symbols assumed for PSK

1.5 dB.

If differential modulation is considered, pilot signals are not necessary, so less redundancy has to be transmitted in this case. To make the situation comparable with systems that employ nondifferential modulation, we use a convolutional code with increased redundancy instead of the pilot symbols. The code rates are chosen in such a way that the same user data rate is transmitted in the systems that are compared. Such a comparison with a fixed data rate seems to be reasonable. For example, if 6% of the subcarriers are needed for the transmission of pilot signals in the case of M-QAM as a nondifferential modulation scheme, the code rate $R = 4/5$ can be used for M-QAM, and the code rate $R = 3/4$ for M-DAPSK. If 11% pilot symbols are required, the code rates are $R = 3/4$ for M-QAM or -PSK, and $R = 2/3$ for M-DAPSK or -DPSK. The performance of these systems can be compared reasonably.

22.4.2 Results

In Figure 22.5, 4- and 8-PSK and -DPSK are compared with the assumption that 6% pilot symbols are necessary for the channel estimation. For the case that 11% pilot symbols are required, Figure 22.6 shows simulation results. The amount of pilot symbols actually required for the channel estimation depends on the coherence time and coherence bandwidth of the channel and on the OFDM system parameters.

Simulation results for 16-DAPSK and -QAM are shown in Figure 22.7, where the code rates are chosen such that the data rate is the same if 6% pilot symbols

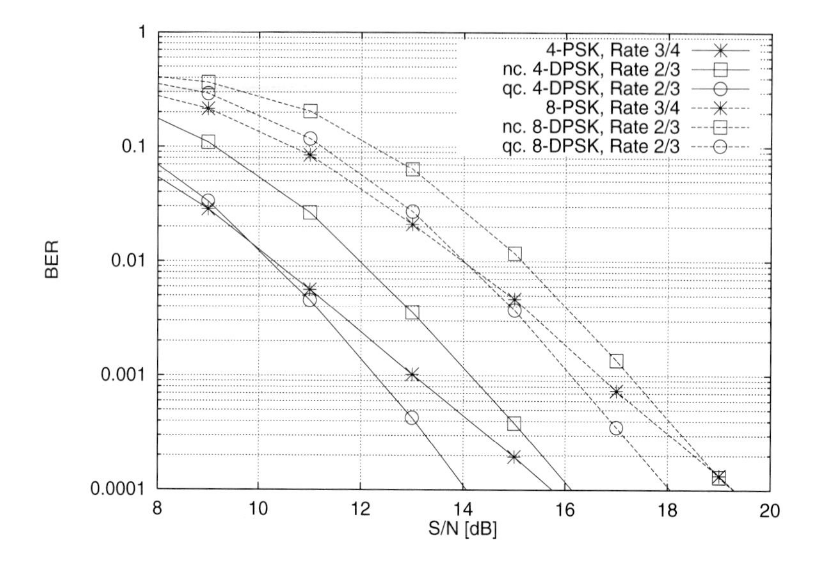

Figure 22.6. Comparison of 4- and 8-DPSK and -PSK, 11% pilot symbols assumed for PSK

are used for 16-QAM. In Figure 22.8, the corresponding results for 64-DAPSK and 64-QAM are given.

With multilevel coding, an additional coding gain can be obtained. An example for 16-DAPSK is depicted in Figure 22.9, where a convolutional code with code rate $R = 3/4$ is compared with a multilevel coding schemes based on convolutional codes with the code rates 1/2, 2/3, 5/6, and 11/12 for the four bit levels. The overall code rate of the multilevel code is 0.73.

The main difference between coherent and noncoherent systems is the channel estimation required for the coherent system. This channel estimation is an additional effort that is made in order to achieve a performance gain over the differential systems. However, this gain is not large if the redundancy due to pilot symbols is considered (which can be replaced by additional code redundancy in the case of differential modulation) and if, additionally, a loss of 1.5 dB due to a real channel estimation is taken into account.

The decision on which modulation scheme is to be preferred depends on the properties of the considered radio channel. If only 6% pilot symbols are sufficient for adaptation to the frequency-selective and time-variant channel, the performance of 64-QAM compared with 64-DAPSK as an example may justify the additional effort for channel estimation. But if more pilot symbols are needed, it can be preferable to decide for differential modulation. A large amount of pilot symbols can be required because of the channel behavior. Especially in a communications system with many users, each of them transmitting only short data blocks, the overhead due to the

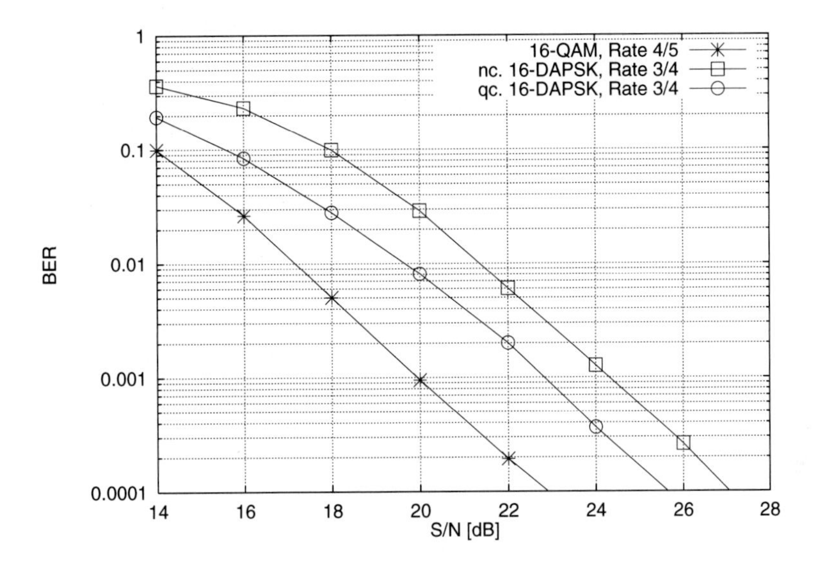

Figure 22.7. Comparison of 16-DAPSK and 16-QAM

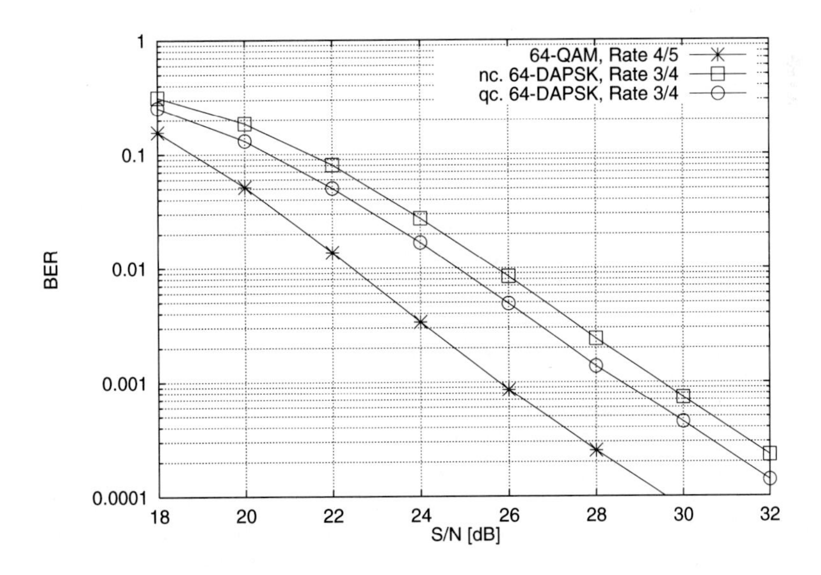

Figure 22.8. Comparison of 64-DAPSK and 64-QAM

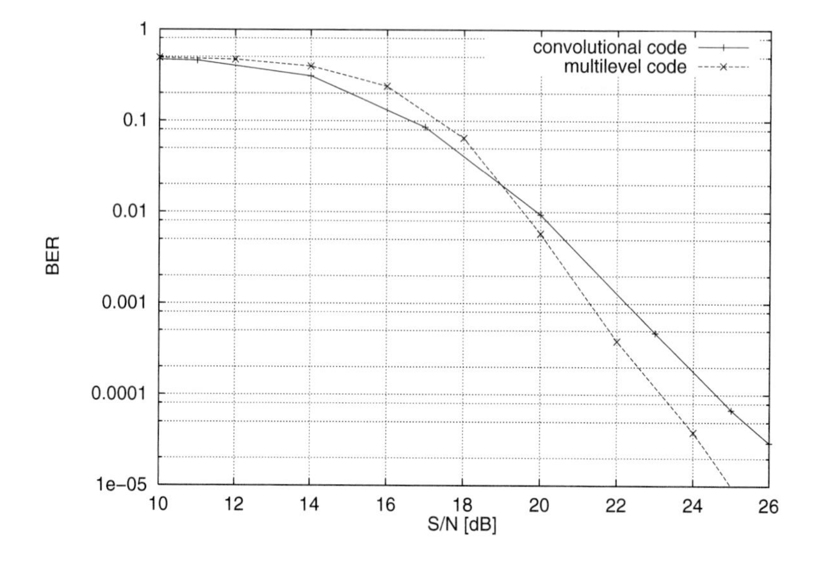

Figure 22.9. 16-DAPSK with convolutional code $R = 3/4$ and with multilevel code with rates 1/2, 2/3, 5/6, 11/12. Reprinted from [7], copyright IEEE.

pilot symbols is high.

The BER performance of the transmission systems can be further improved by multilevel coding. The improvement compared with a simple convolutional code is about 1.5 dB.

Bibliography

[1] C. Berrou, A. Glavieux, and P. Thitimajshima, "Near shannon limit error-correcting coding and decoding: Turbo-codes (1)," in *Proc. IEEE International Conference on Communications (ICC '93)*, (Geneva), 1993.

[2] S. Benedetto and G. Montorsi, "Iterative decoding of serially concatenated convolutional codes," *IEE Electronics Letters*, vol. 32, pp. 1186–1187, 13. June 1996.

[3] T. May, H. Rohling, and V. Engels, "Performance analysis of Viterbi decoding for 64–DAPSK and 64–QAM modulated OFDM signals," *IEEE Transactions on Communications*, vol. 46, pp. 182–190, Feb. 1998.

[4] H. Imai and S. Hirakawa, "A new multilevel coding method using error-correcting codes," *IEEE Transactions on Information Theory*, vol. 23, pp. 371–377, May 1977.

[5] G. Ungerboeck, "Channel coding with multilevel/phase signals," *IEEE Transactions on Information Theory*, vol. 28, pp. 55–67, Jan. 1982.

[6] E. Zehavi, "8-PSK trellis codes for a Rayleigh channel," *IEEE Transactions on Communications*, vol. 40, pp. 873–884, May 1992.

[7] H. Rohling, T. May, K. Brueninghaus, and R. Grnheid, "Broadband ofdm radio transmission for multimedia applications," *Proc. IEEE*, 1999.

[8] Y. Yasuda, K. Kashiki, and Y. Hirata, "High-rate punctured convolutional codes for soft decision Viterbi decoding," *IEEE Transactions on Communications*, vol. 32, pp. 315–319, Mar. 1984.

[9] C. 207, "Digital land mobile radio communications," Tech. Rep. ISBN 92-825-9946-9, Commission of the European Union, Luxembourg, 1989.

Chapter 23

AMPLITUDE LIMITATION OF OFDM SIGNALS

An OFDM signal is the sum of many subcarrier signals that are modulated independently by different modulation symbols. Considering the data to be transmitted as a stochastic process, the amplitude of the OFDM signal is a stochastic process, too. According to the central limit theorem, it obeys a complex Gaussian distribution if the number of subcarriers is large.

Thus, OFDM signals have a very large peak-to-average power ratio. In the transmitter, the maximal output power of the amplifier therefore limits the peak amplitude of the signal, and this effect produces interference both within the OFDM band and in adjacent frequency bands. Furthermore, in OFDM systems a rectangular pulse is usually used for modulation. The corresponding pulse spectrum is a sinc function, which also causes out-of-band interference. These effects, particularly the problem of amplitude limitation of OFDM signals, have been discussed in several publications.

In this chapter, we propose a modification of the OFDM signal to remove the signal peaks that exceed a given amplitude threshold. The out-of-band interference produced by this method can be kept within clearly defined limits while minimum interference within the OFDM band is obtained.

In most of the publications about amplitude limitation of OFDM signals, it is assumed that the limitation can be achieved by predistortion of the signal, that is, the amplifier behaves like an ideal limiter. This means that the signal is amplified linearly up to a maximum input amplitude A_0, and larger amplitudes are limited to A_0. This model is also adopted in this chapter. The input power of the amplifier as compared to the threshold is described by the input backoff $IBO = 10 \log \frac{A_0^2}{P_S}$ where P_S is the average power of the baseband OFDM signal.

23.1 PROPOSALS IN LITERATURE

In the literature, two kinds of approaches are investigated which ensure that the transmitted OFDM signal $s(t)$ does not exceed the amplitude A_0 if a given input backoff is used. The first method makes use of redundancy in such a way that any data sequence leads to an OFDM signal with $|s(t)| \leq A_0$ or that at least the probability of higher amplitude peaks is greatly reduced. These approaches do not result in interference of the OFDM signal [1], [2], [3].

In the second approach, the OFDM signal is manipulated by a correcting function that eliminates the amplitude peaks. The out-of-band interference caused by the correcting function is zero or negligible. However, interference of the OFDM signal itself is tolerated to a certain extent [4], [5], [6]. This concept is described in this section. First, however, some of the redundancy approaches are discussed.

23.1.1 Inserting redundancy

In [1], short block codes are applied to enable a lower input backoff in OFDM systems with four or eight subcarriers. The principle of this idea is to select from the multitude of all possible OFDM blocks those which fulfill the condition $|s(t)| \leq A_0$ at a given input backoff. These "suitable" OFDM blocks are assigned to different data bit sequences by a code. Beyond limiting the amplitude of the resulting OFDM signals, these codes can also be used for error correction at the receiver.

Assuming that in an OFDM system with a large number of subcarriers only a very small fraction of all possible OFDM blocks are selected as suitable, this method still requires little redundancy. If, e.g., this fraction is 10^{-9}, the system can transmit approximately 30 bits less per OFDM block, which could be acceptable. However, it seems to be necessary to find a way of constructing complying codes. If these codes cannot be constructed, the effort for this approach in terms of memory is generally too large. Even for a simple system with 16 subcarriers and QPSK, billions of assignments would have to be stored. For this reason, practical application of this scheme is limited to systems with very few subcarriers.

In [3], algorithms are proposed in which the same data sequence can be represented by several different OFDM blocks. The transmitter generates all possible signals corresponding to a data sequence and chooses the most suitable one for transmission. The receiver must additionally be told which of the signals has been chosen. This can be achieved with little redundancy. If differential modulation is applied between adjacent subcarriers, the receiver does not even need any side information. However, in this case, on several subcarriers reference symbols are transmitted for the differential demodulation. This scheme allows us, e.g., to decrease the input backoff from 12 dB to 9 dB at the same level of out-of-band interference.

A proposal that realizes an OFDM transmission with a constant envelope using 50% redundancy is proposed in [2]. In this scheme, instead of one OFDM block, two blocks are transmitted, calculated from $s(t)$. However, this calculation is nonlinear and causes out-of-band interference. The objective of this approach is not to avoid out-of-band interference but to avoid interference of the OFDM signal.

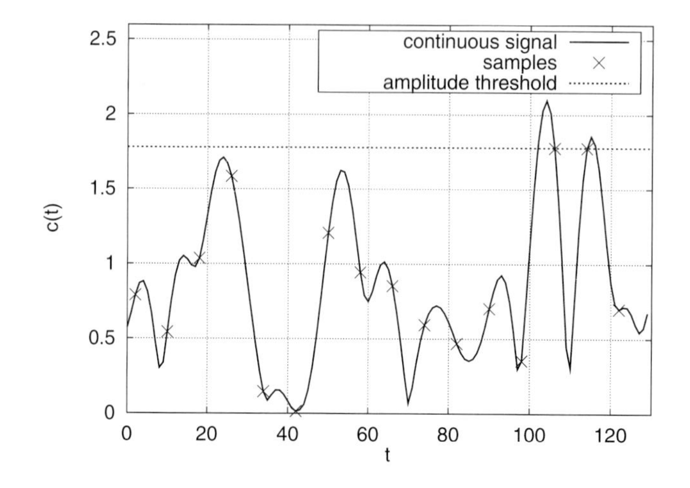

Figure 23.1. The continuous-time OFDM signal exceeds the amplitude threshold A_0 in spite of clipping. Reprinted from [6], copyright IEEE.

23.1.2 Correcting the OFDM signal

The second approach in which the OFDM signal is corrected with suitable functions avoids out-of-band interference but tolerates interference of the OFDM signal itself. In the simplest case, the sampled signal is limited to the amplitude A_0 [7]. This method is termed *clipping*. Clipping does not cause out-of-band interference if $s(t)$ is not oversampled. However, without oversampling, the analog signal after the D/A conversion will exceed the amplitude threshold; see Figure 23.1.

This effect has to be considered. Furthermore, the OFDM signal must be filtered because of the rectangular modulation pulse. For both reasons, oversampling of the signal is necessary. In [5], the proposal is made to apply clipping to the oversampled signal that causes out-of-band interference. This interference is taken care of by a FIR filter that also removes the side lobes of the modulation pulse. However, the filter leads to new amplitude peaks in the signal, but, after all, the peak-to-average power ratio of the signal is reduced by this method.

In [4], the authors corrected OFDM signal by multiplying it with a correcting function $k(t)$. If the signal exceeds the amplitude threshold A_0 at the times t_n, then the corrected signal $c(t)$ is

$$c(t) = s(t)k(t) \quad \text{where} \tag{23.1.1}$$

$$k(t) = 1 - \sum_n A_n g(t - t_n) \tag{23.1.2}$$

$$g(t) = \mathrm{e}^{-t^2/2\sigma^2} \tag{23.1.3}$$

$$A_n = \frac{|s(t_n)| - A_0}{|s(t_n)|}. \tag{23.1.4}$$

Thus, the signal is attenuated by a Gaussian function at all positions where it has high amplitude peaks. The spectrum $C(f)$ of the corrected signal is

$$C(f) = S(f) * K(f) \tag{23.1.5}$$

$$= S(f) - S(f) * \left[\mathrm{e}^{-f^2/2\sigma_f^2} \sum_n B_n \mathrm{e}^{\mathrm{j}2\pi t_n f} \right], \tag{23.1.6}$$

where B_n are constants and σ_f^2 is the variance of the Gaussian function $g(t)$ in the frequency domain. So, the correction broadens the signal spectrum by the width of a Gaussian function with the variance $\sigma_f^2 = 1/2\pi\sigma^2$. The Gaussian function used for the correction should be narrow in the time domain so that only a small interval of the signal is attenuated. It should also be narrow in the frequency domain so that the signal spectrum is broadened as little as possible. If, for an OFDM system with the bandwidth B, a Gaussian function with $\sigma = 5/B$ is chosen, then $\sigma_f = B/10\pi$, which results in an acceptable broadening of the signal spectrum.

Obviously, with this method we can limit the amplitude of the oversampled signal without causing out-of-band interference, except in a narrow frequency band adjacent to the OFDM band. However, if many amplitude peaks have to be corrected, the entire signal is attenuated and the peak-to-average power ratio of the signal cannot be improved beyond a certain figure.

This scheme of correcting the OFDM signal can be realized for any number of subcarriers, and it does not need any redundancy. It causes interference of the OFDM signal, but this is of secondary importance in a fading environment in which OFDM is typically applied. The important task is to avoid out-of-band interference.

23.2 ADDITIVE BAND-LIMITED CORRECTING FUNCTIONS

Each manipulation of an OFDM signal that is performed to reduce amplitude peaks can also be seen as an additive correction. The corrected signal can be written as

$$c(t) = s(t) + k(t). \tag{23.2.1}$$

In the following discussion, we consider correcting functions $k(t)$ where

$$k(t) = \sum_n A_n g(t - t_n) \tag{23.2.2}$$

$$A_n = -\left(|s(t_n)| - A_0\right) \frac{s(t_n)}{|s(t_n)|}, \tag{23.2.3}$$

that is, the correcting function $k(t)$ is composed of the auxiliary function $g(t)$, which must be normalized so that $g(0) = 1$. This correction limits the signal $s(t)$ to A_0 at the positions t_n of amplitude peaks. Of course, the correction could cause the signal to exceed the amplitude threshold at a different position. We ignore this effect now, and later we will see that it is negligible. In the following, we determine an auxiliary function $g(t)$ that produces no out-of-band interference and causes interference of the OFDM signal with minimal power.

The result of this optimization is the following auxiliary function $g(t)$ with the spectrum $G(f)$.

$$g(t) = \mathrm{sinc}(\pi B t) \mathrm{e}^{\mathrm{j}\pi B t} \tag{23.2.4}$$

$$G(f) = \begin{cases} \frac{1}{B} & \text{for } f \in \{0, B\} \\ 0 & \text{for } f \notin \{0, B\} \end{cases} \tag{23.2.5}$$

The spectrum $G(f)$ is band-limited so that no out-of-band interference is caused. It can be shown that correcting a single amplitude peak with this function leads to interference of the OFDM signal with minimal power for the given constraint of band-limitation. Because of the normalization $g(0) = 1$, it is sufficient to show that the power P_g of $g(t)$ is minimum.

$$P_g = \int_{-\infty}^{\infty} |g(t)|^2 dt \tag{23.2.6}$$

$$= \int_0^B |G(f)|^2 \, df = \frac{1}{B} \tag{23.2.7}$$

$$g(0) = \int_0^B G(f) df = 1 \tag{23.2.8}$$

The phase of $G(f)$ is constant. If we chose an auxiliary function $\tilde{g}(t)$ with nonconstant phase of $\tilde{G}(f)$, we could reduce the power $P_{\tilde{g}}$ by setting $\arg \tilde{G}(f) = 0$ and renormalizing $\tilde{g}(t)$. Thus, the phase spectrum of the auxiliary function with minimal power must be constant.

For any band-limited auxiliary function $g'(t) \neq g(t)$ with the spectrum $G'(f)$, $\arg G'(f) = 0$, and $g'(0) = 1$, the power $P_{g'}$ can be calculated according to

$$P_{g'} = \int_0^B (G'(f))^2 \, df \tag{23.2.9}$$

$$= \int_0^B (G(f) + [G'(f) - G(f)])^2 \, df. \tag{23.2.10}$$

The term $G'(f) - G(f)$ can be abbreviated as $\Delta(f)$.

$$P_{g'} = \int_0^B \left(G^2(f) + 2G(f)\Delta(f) + \Delta^2(f)\right) df \tag{23.2.11}$$

$$= P_g + 2G(f) \int_0^B \Delta(f) df + \int_0^B \Delta^2(f) df \tag{23.2.12}$$

$$= P_g + \int_0^B \Delta^2(f) df \tag{23.2.13}$$

$$> P_g. \tag{23.2.14}$$

Therefore, the auxiliary function $g(t)$ can correct a single amplitude peak in the OFDM signal with minimal interference of the signal and without any out-of-band interference. For practical application however, the extent of the sinc function in time must be limited, for example, by windowing.

If the OFDM signal is not oversampled, then the sampled auxiliary function $g(n\Delta t)$ is zero for all $n \neq 0$. The correction scheme is identical with clipping in this case.

The spectrum of the corrected signal is

$$C(f) = S(f) + K(f) \tag{23.2.15}$$

$$= S(f) + \sum_n A_n G(f) e^{j2\pi t_n f} \tag{23.2.16}$$

$$= S(f) + \sum_n \left[\frac{A_n}{N} e^{j2\pi t_n f} \sum_{k=0}^{N-1} \delta(f - k\Delta f) \right]. \tag{23.2.17}$$

Each correction of an amplitude peak causes interference on each subcarrier and the power of the correcting function is distributed evenly to all subcarriers.

This correcting scheme was applied to an OFDM transmission system with 128 subcarriers in a simulation. The signal $s(t)$ was oversampled by a factor of four and normalized so that the signal power is one. Then, the signal was corrected with $k(t)$. For the correction, the amplitude threshold A_0 was set according to the input backoff that had been chosen. After the correction, the signal was limited to the amplitude A_0 to account for the limitation of amplitude peaks which may have remained.

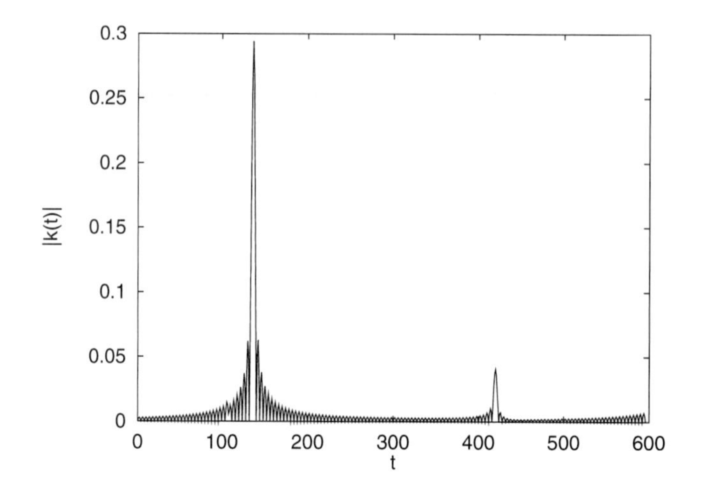

Figure 23.2. Example of a correcting function $k(t)$ for an OFDM signal with 128 subcarriers, oversampling. Reprinted from [6], copyright IEEE.

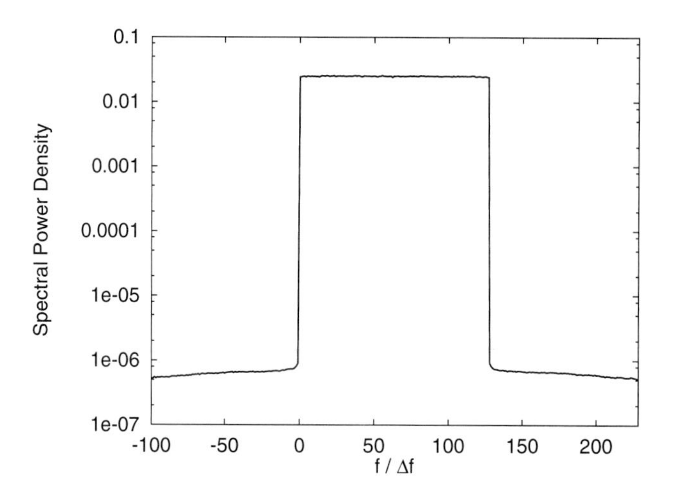

Figure 23.3. Power density spectrum of the interference of the OFDM signal which is caused by correction and limitation at $IBO = 4$ dB. Reprinted from [6], copyright IEEE.

Figure 23.2 shows an example of a correcting function. In Figure 23.3, the power density spectrum of the interference caused by the correction and the limitation of the signal is shown for an input backoff $IBO = 4$ dB. We can see that the interference power is concentrated within the OFDM bandwidth. Despite the correction of

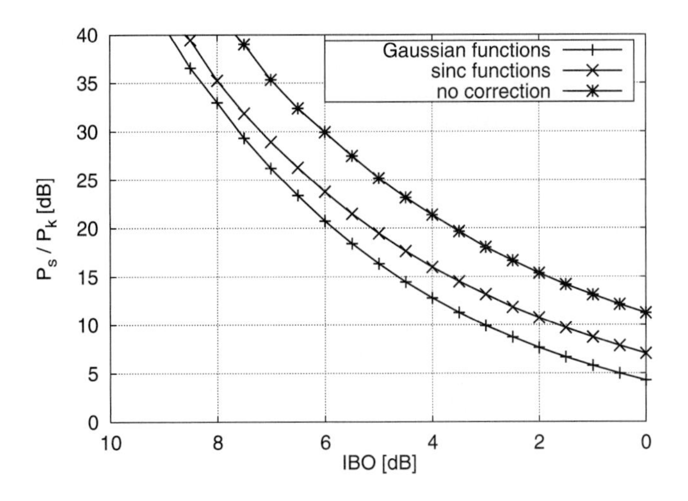

Figure 23.4. Signal-to-interference ratio SIR in the OFDM band in the case of correction with Gaussian functions (multiplicative), with sinc functions (additive) and without any correction. Reprinted from [6], copyright IEEE.

the OFDM signal, the signal still happens to exceed the amplitude threshold A_0. For this reason, the limiter causes out-of-band interference, the power of which is more than 60 dB below the signal power.

The signal-to-interference power ratio SIR can be determined both within the OFDM band and in the adjacent frequency bands. The SIR in the OFDM band is the quotient of the power densities of the signal and of the interference. For the SIR in the adjacent frequency band we denote the quotient of the signal power density in the OFDM band and the average power density of the interference in a frequency band with the bandwidth $B/10$, directly adjacent to the OFDM band.

In Figure 23.4 the resulting SIR in the OFDM band is depicted for correction with Gaussian functions (multiplicative), with sinc functions (additive) and without any correction. In each case, the signal has additionally been limited according to the IBO.

Without any correction, we obtain the least interference power because the signal is modified only where the amplitude threshold of the limiter is exceeded. The correcting functions additionally modify the signal within a certain area around the amplitude peaks.

The SIR in the adjacent frequency bands is shown in Figure 23.5.

The correction of the signal has the effect of an attenuation. This is obvious in the case of the multiplicative correction with Gaussian functions, but the additive correction with sinc functions and the limitation of the signal also reduce the signal power. The effective attenuation of the signal due to the correction and the limitation is displayed in Figure 23.6.

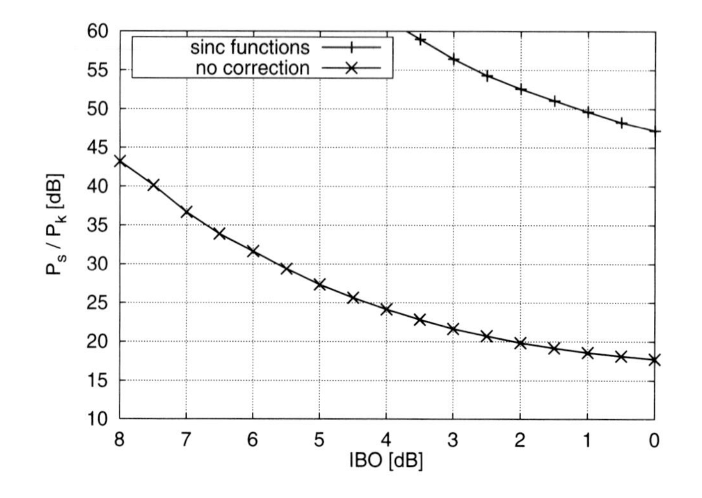

Figure 23.5. Signal-to-interference ratio SIR adjacent to the OFDM band in the case of additive correction with sinc functions and without any correction. Reprinted from [6], copyright IEEE.

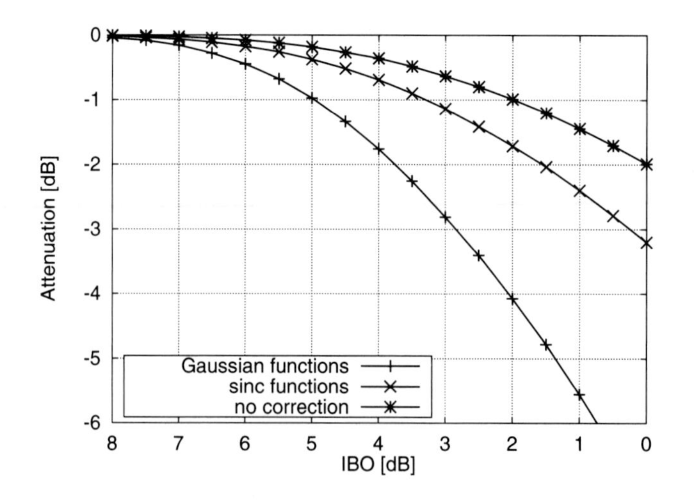

Figure 23.6. Attenuation of the signal due to the correction with Gaussian functions, with sinc functions and with limitation of the signal. Reprinted from [6], copyright IEEE.

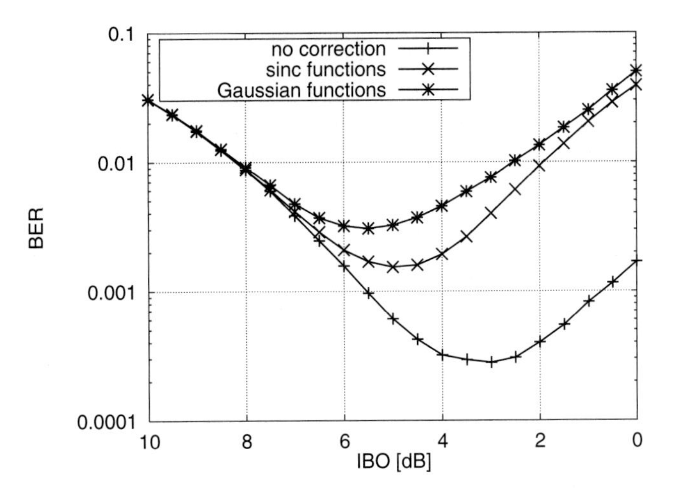

Figure 23.7. Bit-error rate BER as a function of the input backoff IBO of the limiter (amplitude threshold A_0) for the AWGN channel with $A_0^2/N = 18$ dB. Reprinted from [6], copyright IEEE.

Because of this attenuation of the signal, the real input power of the amplifier is lower than the IBO suggests—which the correction of the signal was designed for. Note that IBO refers to the OFDM signal before the correction. The correction with sinc functions causes a lower attenuation than the correction with Gaussian functions, so the same IBO leads to a higher input power of the amplifier in this case.

For the AWGN channel and DQPSK modulation, Figure 23.7 shows the bit-error rate as a function of the input backoff IBO. In the corresponding simulation, the noise power N has been fixed so that $A_0^2/N = 18$ dB. Thus, the signal-to-noise ratio SNR depends on the IBO. Still, the input backoff refers to the OFDM signal before correction.

For a WSSUS radio channel with multipath propagation and consequently with frequency-selective fading, Figure 23.8 shows simulation results with $A_0^2/N = 30$ dB. In this case, the interference due to the correction is less grave because it is attenuated by the radio channel as well, whereas the noise is added to the attenuated signal. Accordingly, a lower input backoff is optimum, unlike the case of the AWGN channel.

23.2.1 Additive correcting functions of finite length

As we saw, the infinite extent of the sinc function in time prohibits its use as a correcting function in practical applications. Additive correcting functions with a finite extent in time can be obtained with methods known from the field of digital filter design.

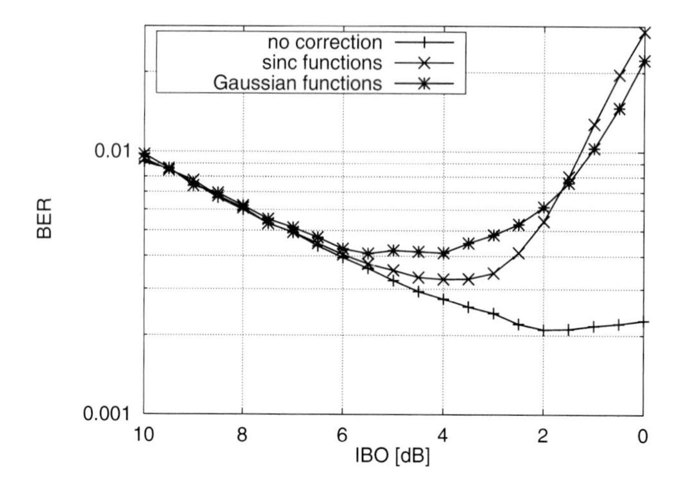

Figure 23.8. Bit error rate BER as a function of the input backoff IBO of the limiter (amplitude threshold A_0) for a channel with frequency-selective Rayleigh fading and $A_0^2/N = 30$ dB. Reprinted from [6], copyright IEEE.

Multiplication of the sinc function with an appropriate windowing function is one possible way to limit the length of the correcting function and hence the number of samples to be added to those of the OFDM signal whenever an amplitude peak is detected. The impact on the power density spectrum of the interference then depends on the shape and the extent of the windowing function used.

However, to minimize the number of nonzero samples of the correcting function while keeping the deviation of the correcting function's power density spectrum from the spectrum of the sinc function within clearly defined limits, a different approach appears to be more suitable. Instead of modifying the sinc function by means of windowing, we can use the coefficients of a specifically designed linear phase Chebyshev filter (equiripple filter) to correct the sampled OFDM signal.

We start with a tolerance scheme for the ripple of the correcting function's amplitude spectrum within the OFDM band, its minimum attenuation outside the OFDM band, and the width of the transient band as depicted in Figure 23.9. Then we use the Parks-McClellan algorithm [9] to obtain the coefficients $g_{Ch}(n)$ of a linear phase Chebyshev filter, the transfer function of which fits within that scheme. The linear phase property ensures a distinct maximum of the correcting function at $n = 0$. Again we assume $g_{Ch}(n)$ to be normalized so that $g_{Ch}(0) = 1$.

Figure 23.10 illustrates such a correcting function; Figure 23.11 shows the corresponding power density spectrum of the resulting interference when we correct the OFDM signal to achieve an input backoff of 4 dB. Although the spectrum of the correcting function itself is designed to have constant ripple outside the OFDM band, we can observe a decay of the interference's spectral power density outside

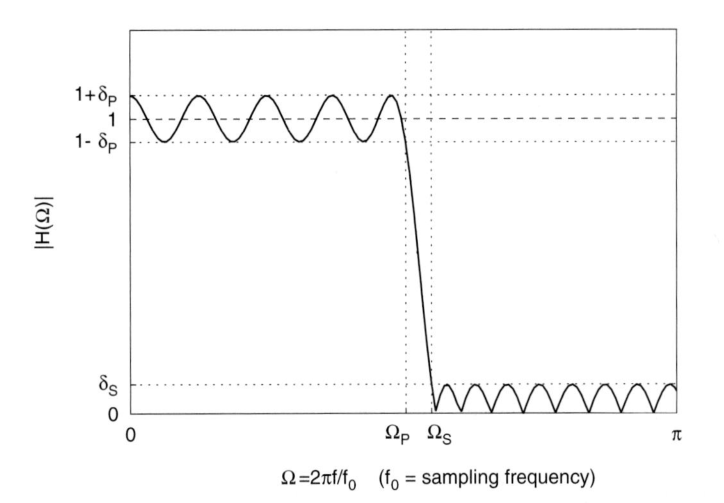

Figure 23.9. Tolerance scheme for the design of equiripple filters and finite length correcting functions, respectively. Reprinted from [8], copyright IEEE.

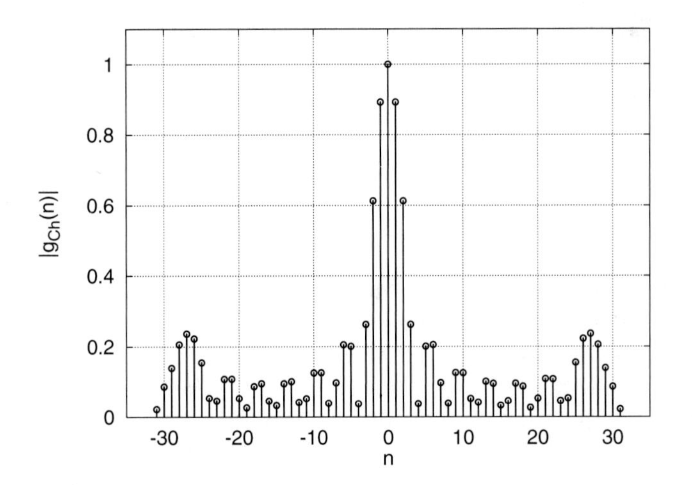

Figure 23.10. Example of a finite length correcting function. (Design parameters: $\delta_P = 0.5$; $\delta_D = -40$ dB; $\Omega_s = 1.1 \cdot \Omega_P$; $2\Omega_P \widehat{=} (N+1)\Delta f$, where Δf denotes the subcarrier spacing and N the number of subcarriers). Reprinted from [8], copyright IEEE.

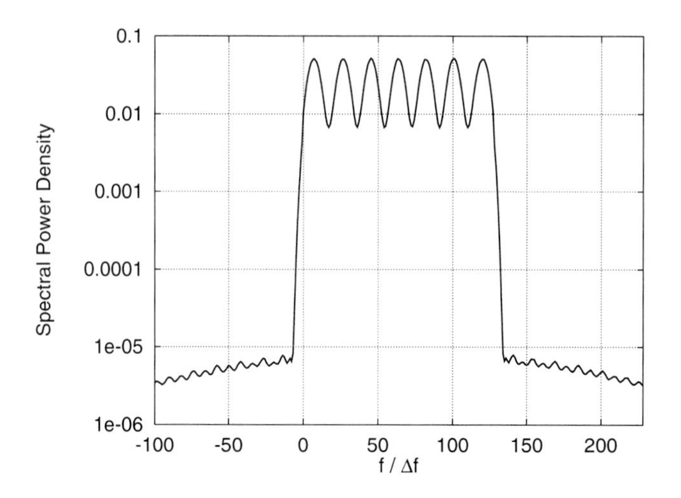

Figure 23.11. Power density spectrum of the interference caused by additive correction (finite length correcting function, design parameters as in Figure 23.10) and limitation at $IBO = 4$ dB. Reprinted from [8], copyright IEEE.

the OFDM band. This decay originates from that amplitude peaks extending over several successive samples so that the instances of the correcting function to be added to the signal are not statistically independent with respect to amplitude, phase, and location on the time axis. This effect increases with the oversampling factor.

Choice of design parameters

The tolerance parameters δ_S, Ω_P, and Ω_S are determined by the maximum allowable amount of interference within the adjacent OFDM band and the frequency gap between adjacent OFDM bands. For our simulations we chose $\delta_S = -40$ dB and $(\Omega_S - \Omega_P) \widehat{=} 0.05 N \Delta f$, thus keeping the power density of the interference in the adjacent OFDM band below -50 dB relative to the spectral power density of the OFDM signal within the OFDM band, assuming that adjacent OFDM bands are spaced by at least 5% of their bandwidth.

The choice of δ_P, however, is independent of the restrictions concerning interband interference but has considerable impact on the correcting function's extent in time and hence on the complexity of the correcting process and on the delay of the OFDM signal. The higher the value of δ_P is, the lower the extent of $g_{Ch}(n)$ in time becomes.

The influence of δ_P on the bit error rate is shown in Figure 23.12 for the AWGN channel and in Figure 23.13 for the frequency-selective Rayleigh fading channel. The increase in bit errors is caused by three facts:

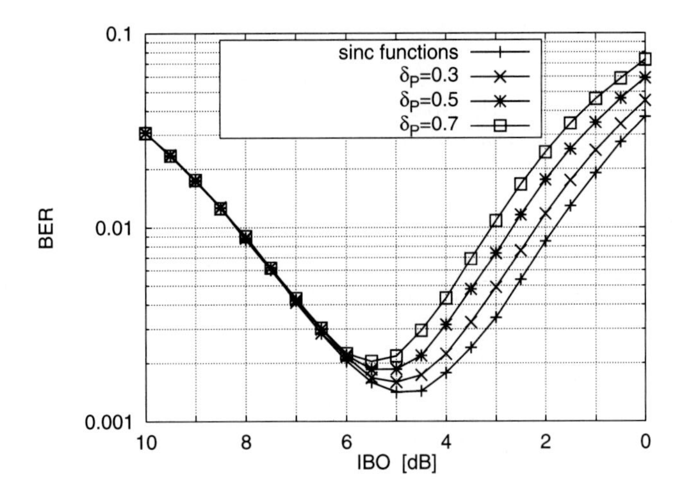

Figure 23.12. Bit-error rate BER as a function of the input backoff IBO for various values of the spectral ripple δ_P of the correcting function within the OFDM band (AWGN channel, $A_0/P_N = 18$ dB). Reprinted from [8], copyright IEEE.

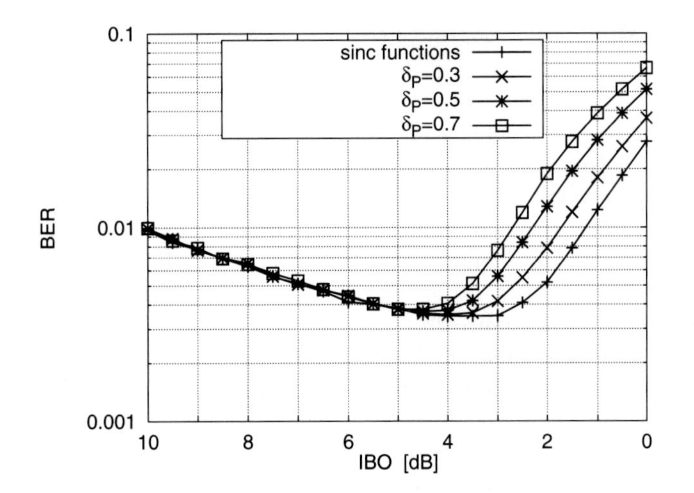

Figure 23.13. Bit-error rate BER as a function of the input backoff IBO for various values of the spectral ripple δ_P of the correcting function within the OFDM band (frequency-selective Rayleigh fading channel, $A_0/P_N = 30$ dB). Reprinted from [8], copyright IEEE.

1. A higher δ_P leads to an increase in the energy of $g_{Ch}(n)$ and hence in the interference power within the OFDM band.

2. The maximum amplitude of the sidelobes of $g_{Ch}(n)$ increases significantly for large values of δ_P, so the addition of the correcting function tends to create additional peaks within the OFDM-signal, some of which will be corrected afterward at the expense of a raise in interference power. Furthermore, the remaining out-of-band interference increases, too, and this increase must be taken into account for a proper choice of δ_P.

3. The SIR strongly varies with the subcarrier frequency, which increases the average bit error probability as compared to a system with equal SIR on all subcarriers.

The effect mentioned in item 3 can be mitigated to some extent if we compensate for the variation of the SIR within the OFDM-band by applying an appropriate amplification or attenuation, respectively, to each subcarrier before the IDFT is carried out.

BER reduction by predistortion

To obtain the amplification factor for each subcarrier in order to minimize the BER for a given value of δ_P, both the channel noise and the interference caused by the peak reduction must be accounted for. Since the I- and Q-components of the interference are not statistically independent, their effect on the bit-error rate differs from that of additive white Gaussian noise. Furthermore, unlike the channel noise, the interference is subject to frequency-selective fading caused by multipath propagation.

It has proved useful to define an equivalent signal-to-noise ratio $(P_S/P_N)_{eq} = f(P_S/P_k)$ that denotes the signal-to-noise ratio leading to the same BER as the respective signal-to-interference ratio (P_S/P_k), assuming the interference to have a constant spectral power density within the OFDM band. For transmission over an AWGN channel and a frequency-selective Rayleigh fading channel, respectively, the equivalent signal-to-noise ratio $(P_S/P_N)_{eq}$ is plotted against the signal-to- interference ratio P_S/P_k in Figure 23.14. To compensate for the spectral ripple of the correcting function, the modulation symbol of each subcarrier $S(n)$, $0 \leq n < N$ is amplified/attenuated in order to satisfy the condition

$$\frac{1}{(P_S/P_N)_n} + \frac{1}{(P_S/P_N)_{eq,n}} = \text{const.}, \qquad (23.2.18)$$

where $(P_S/P_N)_n$ denotes the signal-to noise ratio and $(P_S/P_N)_{eq,n}$ denotes the equivalent signal-to-noise ratio according to the signal-to interference ratio of the nth subcarrier. As seen in (23.2.18), the calculation of the amplification factors requires knowledge about the power of the channel noise. Although with communications applications the user-specific value of P_N can be obtained either by the

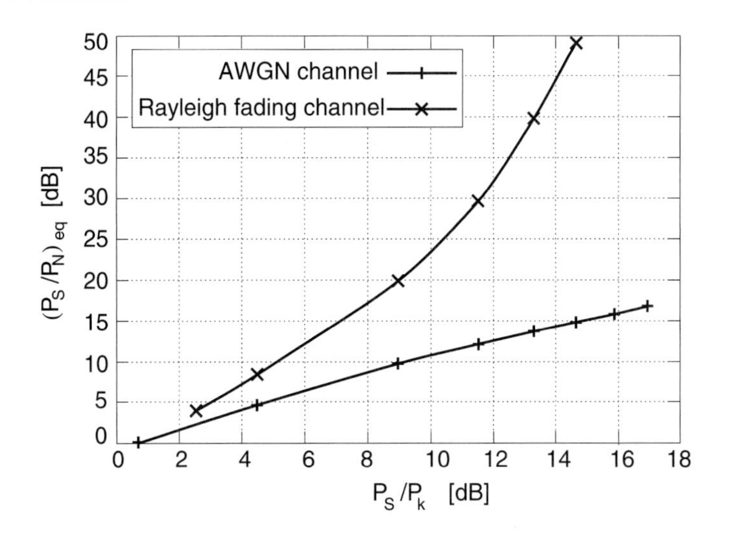

Figure 23.14. Equivalent signal-to-noise ratio $P_S/P_{N,eq}$ as a function of the actual signal-to-interference ratio P_S/P_k

mobile or by the base station, there is no way to obtain P_N in broadcast systems, the more so since each of a possibly high number of mobiles encounters a different value of P_N. In this case, however, calculating the amplification factors with a suitably chosen assumption for P_N can still lead to a significant improvement of the BER over a wide range of SNRs, hence allowing an increase in δ_P with only little performance degradation.

Assuming the accurate value of P_N to be known for the calculation of the amplification factors, we obtain almost equal bit error probability for all subcarriers and a total bit-error rate as shown in Figure 23.15 for the AWGN channel.

23.2.2 Resulting performance gain

Depending on the requirements concerning the spacing of adjacent OFDM bands and the tolerable amount of interband interference, the application of additive correction to the OFDM signal allows a decrease of the input backoff; in most cases, this aproach results in a significant decrease of the BER despite the additional interference within the OFDM band.

Assuming a gap of 5% of the OFDM bandwidth B between adjacent frequency bands, an input backoff of at least 8.9 dB is required in order to keep the spectral power density of the interference in the adjacent frequency band at least 50 dB below the power density of the OFDM signal within the OFDM band if no correction of the OFDM signal is performed.

If additive correction by means of sinc functions or finite-length correcting functions with appropriate predistortion is applied to the OFDM signal, the lowest BER

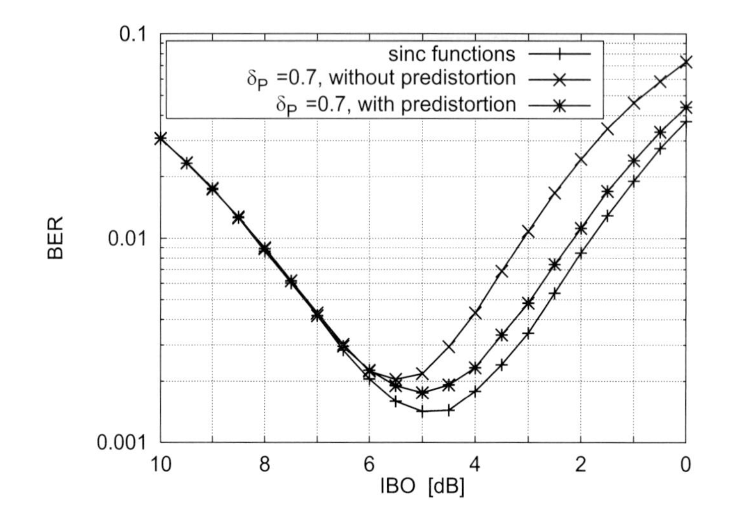

Figure 23.15. Bit-error rate BER as a function of the input-backoff IBO for finite length correcting functions with and withot predistortion as compared to correction with sinc functions. Reprinted from [8], copyright IEEE.

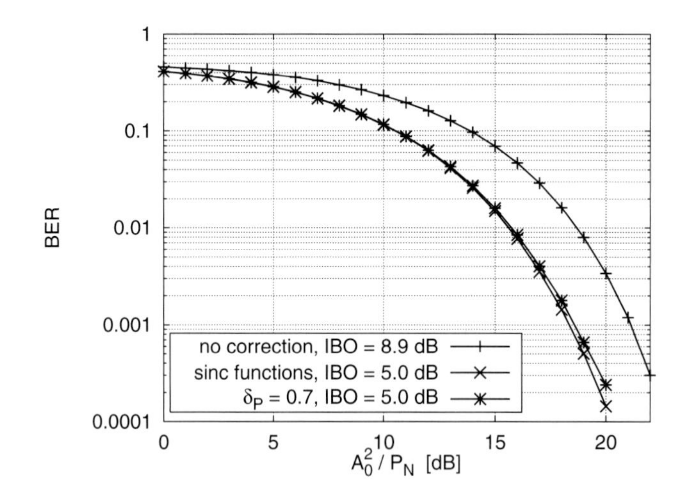

Figure 23.16. Bit-error rate as a function of the A_0^2/P_N ratio with and without additive correction (AWGN channel). Reprinted from [8], copyright IEEE.

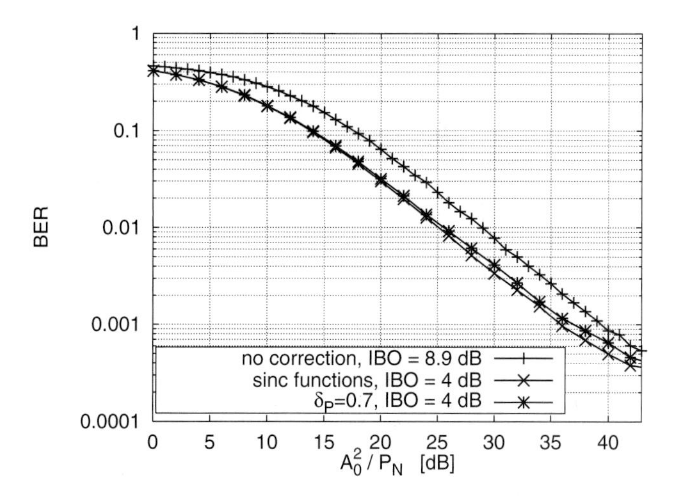

Figure 23.17. Bit-error rate as a function of the A_0^2/P_N ratio with and without additive correction (frequency-selective, Rayleigh-fading channel). Reprinted from [8], copyright IEEE.

is obtained at an input backoff of 5 dB for transmission over an AWGN channel or at an input backoff of 4 dB for transmission over a frequency-selective Rayleigh-fading channel, respectively (see Figure 23.7, Figure 23.8). In Figure 23.16 and Figure 23.17 the bit-error rate is plotted against the A_0^2/P_N ratio. At a BER of 10^{-3} we observe a gain of about 3 dB for the AWGN channel as well as for the frequency-selective Rayleigh-fading channel, respectively, due to the signal correction as compared to a system without signal correction. This means that a less costly power amplifier can be used if additive signal correction is applied, because the limiting threshold A_0 may be decreased by 3 dB as compared to a system without signal correction.

23.2.3 Complexity aspects

The complexity of the additive signal correction and its contribution to the system's latency depend on the correcting function's extent in time. For an OFDM system with $N = 128$ subcarriers and an oversampling factor of $\eta = 4$, the numbers m of non-zero samples of several correcting functions with various combinations of δ_S, and δ_P are given in Table 23.1. The correcting function leading to the results presented in the previous section ($\delta_S = 40$ dB, $\delta_P = 0.7$) has an extent in time equivalent to 8.8% of the duration of an OFDM symbol. This percentage is nearly independent of the oversampling factor and approximately inversely proportional to the relative transient bandwidth $(\Omega_S - \Omega_P) \cdot \eta/2\pi$.

If additive correction is applied to an OFDM signal oversampled by a factor of less than 4, which is desirable for practical applications, then we see an increase in

δ_S [dB]	δ_P	m	$\frac{m}{N\eta}$ [%]	δ_S [dB]	δ_P	m	$\frac{m}{N\eta}$ [%]
-20	0.1	65	12.7	-40	0.1	111	21.7
-20	0.3	39	7.6	-40	0.3	79	15.4
-20	0.5	27	5.3	-40	0.5	61	11.9
-20	0.7	17	3.3	-40	0.7	45	8.8
-30	0.1	89	17.4	-40	0.9	23	4.5
-30	0.3	61	11.9	-50	0.1	135	26.4
-30	0.5	43	8.4	-50	0.3	101	19.7
-30	0.7	29	5.7	-50	0.5	79	15.4
-30	0.9	13	2.5	-50	0.7	63	12.3
				-50	0.9	39	7.6

Table 23.1. Extent of finite length correcting functions (number of coefficients of linear phase Chebyshev filters) for various sets of design parameters ($(\Omega_S - \Omega_P) \mathrel{\widehat{=}} 0.05N\Delta f$, oversampling factor $\eta = 4$)

the amount of signal peaks exceeding the amplitude threshold after the correction because the detection of the maximum signal amplitude and the peak's location on the time axis is less accurate than with higher oversampling factors. This increase leads, in turn, to an increase in the amount of remaining out-of-band power which, depending on the requirements concerning the amount of out-of-band interference, might not be tolerable.

We can compensate this effect by introducing a *correction threshold* B_0 slightly lower than the limiting threshold A_0 of the amplifier. We then perform additive correction with the objective to prevent the amplitude of the sampled OFDM signal from exceeding B_0 rather than A_0. If we choose B_0 properly, the increase in bit-error rate as compared to a system with a higher oversampling factor is almost negligible, and the out-of-band interference can be kept within the desired limits.

Bibliography

[1] A. E. Jones and T. A. Wilkinson, "Combined coding for error control and increased robustness to system nonlinearities in OFDM," in *Proc. IEEE VTC '96*, (Atlanta, USA), pp. 904–908, April 1996.

[2] R. Dinis and A. Gusmao, "CEPB–OFDM: A new technique for multicarrier transmission with saturated power amplifiers," in *Proc. IEEE ICCS '96*, (Singapore), Nov. 1996.

[3] S. H. Müller, R. W. Bäuml, R. F. H. Fischer, and J. B. Huber, "OFDM with reduced peak-to-average power ratio by multiple signal representation," *Annals of Telecommunications*, vol. 52, pp. 58–67, 1997.

[4] M. Pauli and H.-P. Kuchenbecker, "Minimization of the intermodulation distor-

tion of a nonlinearly amplified OFDM signal," *Wireless Personal Communications*, vol. 4, no. 1, pp. 93–101, 1997.

[5] X. Li and L. Cimini, "Effects of clipping and filtering on the performance of OFDM," in *Proc. IEEE VTC '97*, (Phoenix, USA), May 1997.

[6] T. May and H. Rohling, "Reducing the peak-to-average power ratio in OFDM radio transmission systems," in *Proc. VTC '98*, (Ottawa, Canada), 18-21.5. 1998.

[7] R. O'Neill and L. Lopes, "Performance of amplitude limited multitone signals," in *IEEE VTC '94*, pp. 1675–1679, June 1994.

[8] M. Lampe and H. Rohling, "Reducing out-of-band emissions due to non-linearities in OFDM systems," in *Proc. VTC '99*, (Houston, USA), 17.-21.5. 1999.

[9] A. Oppenheim and R. W. Schafer, *Discrete-Time Signal Processing*. Englewood Cliffs, NJ: Prentice Hall, 1989.

Chapter 24

FILTERING OF THE OFDM SIGNAL

In this chapter, we propose a method for suppressing the out-of-band emissions from the rectangular modulation pulse by means of digital filtering. Because filtering requires an additional extension of the guard interval to prohibit intersymbol interference, minimization of the duration of the filter's impulse response is essential in order to keep the overhead low.

Nonlinearities are not the only source of out-of-band emissions in OFDM systems. Because of the rectangular shape of the modulation pulse, the power density spectrum of the OFDM signal can be described as a superposition of sinc^2-functions:

$$\Phi(f) = \sum_{i=0}^{N-1} \operatorname{sinc}^2(\pi T(f - i\Delta f)). \tag{24.0.1}$$

As depicted in Figure 24.1, this superposition leads to a significant spectral power density in the adjacent frequency bands. These out-of-band emissions can be suppressed by means of impulse shaping or, more effectively, by digital filtering of the OFDM signal.

To avoid intersymbol interferences (ISI) from the filtering, the guard interval must be extended by the duration of the filter's impulse response. Hence, the objective for the design of an appropriate FIR filter is to obtain the desired suppression of the out-of-band emissions while keeping the filter's impulse response as short as possible.

The simulation results presented in this chapter were obtained with a *linear phase Chebyshev filter,* which with a given tolerance scheme for the filter's transfer function (see Figure 23.9) requires a relatively low number of coefficients.

24.1 CHOICE OF DESIGN PARAMETERS

For the tolerance parameters δ_S, Ω_P, and Ω_S, considerations similar to those outlined in Section 23.2.1 must be made. Assuming the same restrictions concerning

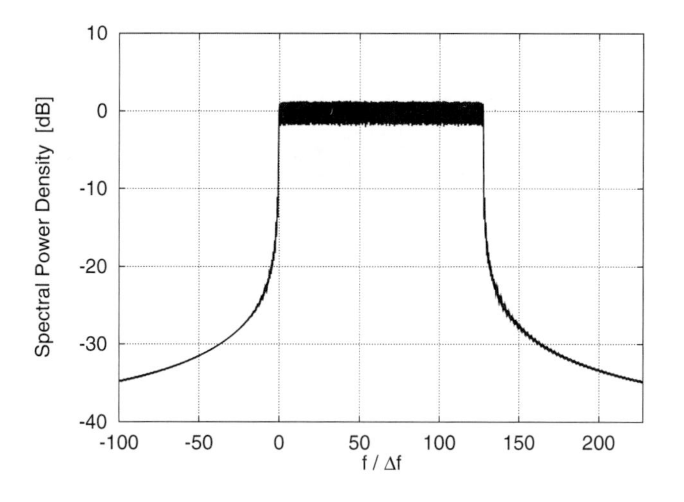

Figure 24.1. Power density spectrum of an OFDM signal with $N = 128$ subcarriers

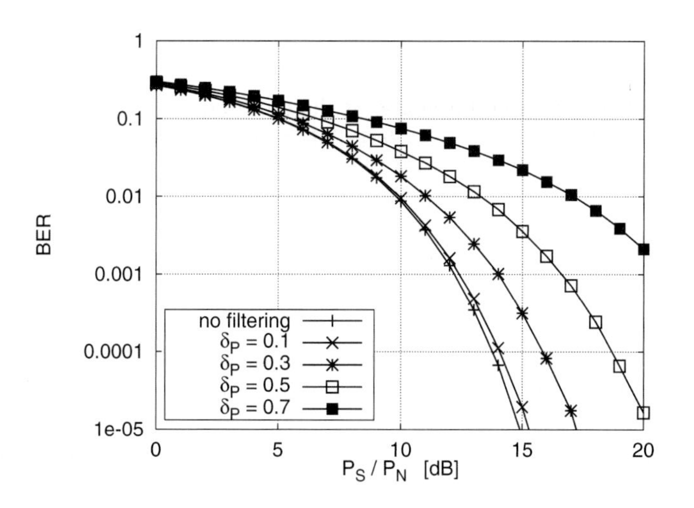

Figure 24.2. Bit-error rate BER as a function of the signal-to-noise ratio for various values of the passband ripple δ_P of the linear phase Chebyshev filter (AWGN channel)

the tolerable amount of out-of-band power as with the additive signal correction, simulations were carried out, with $\delta_S = -30$ dB and $(\Omega_S - \Omega_P) \hat{=} 0.05N\Delta f$. Again the choice of the passband ripple δ_P is independent of those restrictions and thus offers a means of influencing the number of filter coefficients.

Since the passband ripple leads to a subcarrier-specific amplification or attenu-

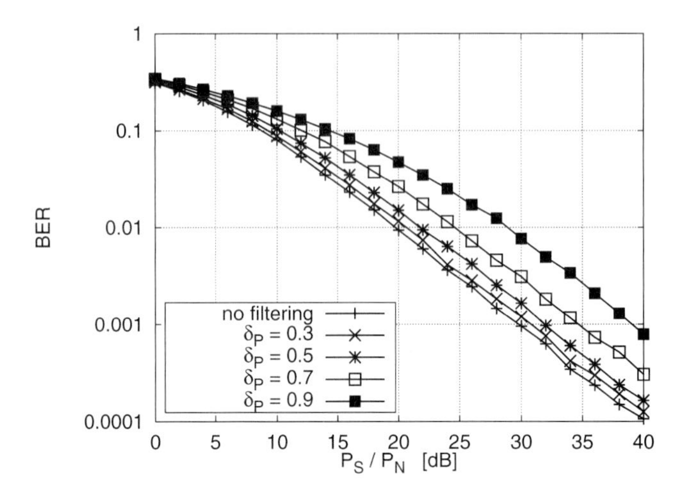

Figure 24.3. Bit-error rate BER as a function of the signal-to-noise ratio for various values of the passband ripple δ_P of the linear phase Chebyshev filter (frequency-selective Rayleigh-fading channel)

ation of the OFDM signal, the bit-error rate BER increases with δ_P as shown in Figure 24.2 for the AWGN channel and in Figure 24.3 for the frequency-selective Rayleigh-fading channel, respectively. This effect can be entirely compensated for by application of an appropriate predistortion of the OFDM signal to equalize the signal-to-noise ratio for all subcarriers.

24.2 BER REDUCTION BY PREDISTORTION

Equal signal-to-noise ratio can be obtained for all subcarriers if each subcarrier is amplified or attenuated according to the equation

$$\tilde{S}_n = \frac{S_n}{H_{Ch}(n\Delta f)} \qquad 0 \leq n < N, \tag{24.2.1}$$

where S_n denotes the complex valued modulation symbol assigned to the nth subcarrier and $H_{Ch}(f)$ denotes the transfer function of the digital filter.

Applying this kind of predistortion leads to an increase in the average power of the OFDM signal before filtering and hence in the spectral power density outside the OFDM band. To meet the requirements concerning the out-of-band emissions, the stop-band attenuation must be increased as compared to a system without predistortion.

For a passband ripple of $\delta_P = 0.7$, Figure 24.4 shows the effect of predistortion on the power density spectrum of the filtered OFDM signal. If both predistortion and filtering are applied to the OFDM signal, the same bit-error rate can be obtained as without filtering, independently of the filter's passband ripple.

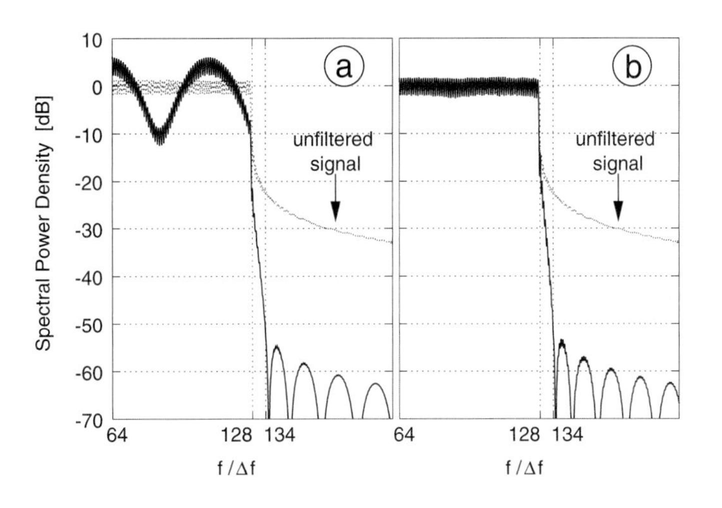

Figure 24.4. Power density spectrum of the OFDM signal after filtering ($N = 128$; $\delta_P = 0.7$; $(\Omega_S - \Omega_P) \widehat{=} 0.05N\Delta f \approx 6\Delta f$); (a) $\delta_S = -30$ dB, no predistortion; (b) $\delta_S = -33$ dB, predistortion

24.3 COMPLEXITY ASPECTS

Since the choice of δ_P does not affect the bit-error rate if appropriate predistortion of the OFDM signal is applied, raising δ_P is an effective means for lowering the number of filter coefficients needed to meet the requirements concerning the out-of-band emissions. This is true even though δ_S needs to be lowered to compensate for the increase in spectral power density outside the OFDM band due to the predistortion.

Table 23.1 gives an overview of the number of filter coefficients for various combinations of δ_S and δ_P. For an oversampling factor of 4, $\delta_P = 0.9$, and $\delta_S = -40$ dB , for example, 23 filter coefficients are required; hence, to avoid intersymbol interference, the guard interval must be extended by 4.5% of the net duration of an OFDM symbol.

For values of δ_P very close to 1 and low values of δ_S, the Parks-McClellan algorithm does not necessarily yield a usable result, so δ_P can be increased up to a certain limit only.

However, given a certain passband ripple δ_P, the number of filter coefficients can be further reduced if the filter design is carried out for a tolerance scheme, as depicted in Figure 24.5. To account for the decay of the spectral power density of the unfiltered signal outside the OFDM band, δ_S is given as a function of frequency. The power density spectrum of the filtered OFDM signal then has constant ripple outside the OFDM band, as shown in Figure 24.6.

To give an example, 21 coefficients are needed for a filter with $(\Omega_S - \Omega_P) \widehat{=} 0.05N\Delta f$, $\delta_P = 0.97$ and constant δ_S of -50 dB. This number can be reduced to 15 if $\delta_S(\Omega)$ is matched to the spectral properties of the unfiltered OFDM signal.

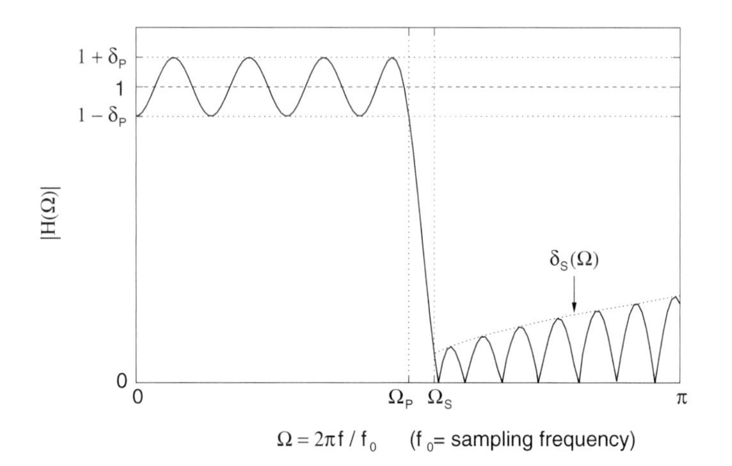

Figure 24.5. Tolerance scheme with frequency-varying minimum stopband attenuation

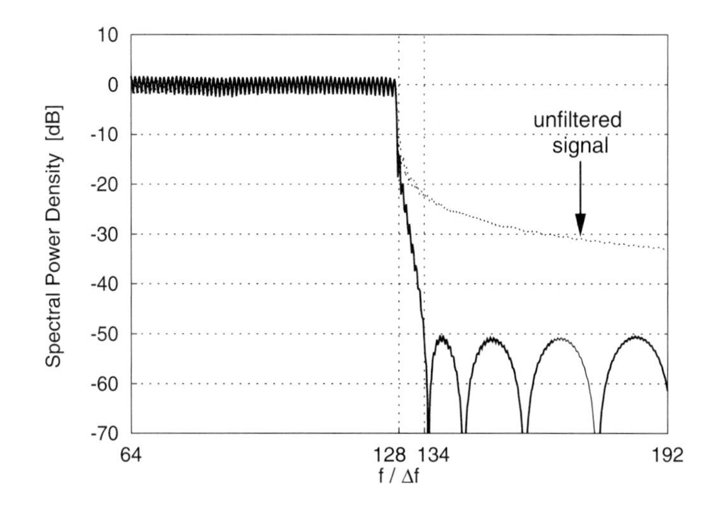

Figure 24.6. Power density spectrum of the OFDM signal after predistortion and filtering (filter designed with tolerance scheme depicted in Figure 24.5. $N = 128$; $\delta_P = 0.7$; $\min(\delta_S(\Omega_S)) = -33$ dB; $(\Omega_S - \Omega_P) \widehat{=} 0.05 N\Delta f \approx 6\Delta f$)

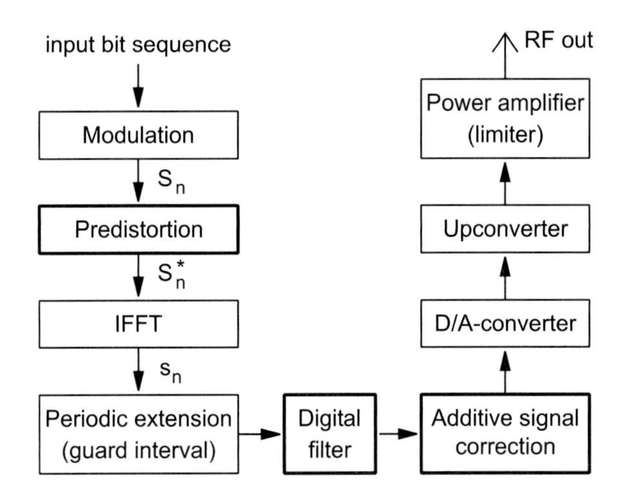

Figure 24.7. OFDM transmitter employing digital filtering and additive signal correction

In some cases, a further reduction of the number of filter coefficients might be obtained if minimum phase Chebyshev filters are used instead of linear phase filters; the latter, for a given tolerance scheme, require the smallest number of coefficients among all types of FIR filters. However, designing minimum phase Chebyshev filters to match a tolerance scheme with frequency-varying minimum stopband attenuation as in Figure 24.5, is not as simple as with linear phase filters. As shown in this section, even with linear phase filters the impulse response and hence the required extension of the guard interval can be reduced to about 3% of the net duration of an OFDM symbol.

24.4 OFDM SYSTEM WITH ADDITIVE CORRECTION AND FILTERING

Figure 24.7 illustrates the block diagram of an OFDM transmitter, incorporating both digital filtering and additive correction of the baseband signal in order to avoid out-of-band interference. We can, in a single operation, perform predistortion that accounts for the spectral ripple of the correcting function and for the passband ripple of the filter's transfer function.

Filtering should be carried out before the signal correction, because filtering the corrected signal instead could create additional amplitude peaks exceeding the limiting threshold.

The structure of an OFDM receiver is depicted in Figure 24.8. Neither the filtering nor the additive signal correction requires any additional signal processing in the receiver as compared to a system without these means of reducing the out-of-band power.

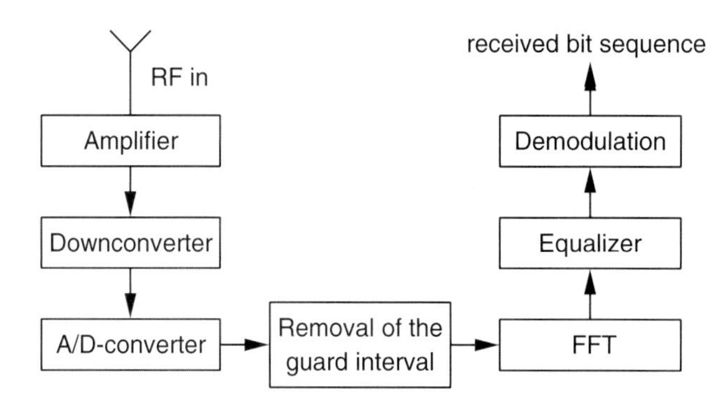

Figure 24.8. General structure of an OFDM receiver

Chapter 25

OFDM-CDM AND SINGLE-CARRIER TRANSMISSION WITH FREQUENCY-DOMAIN EQUALIZATION

The main advantage of the OFDM transmission technique is its ability to transmit a high data rate over a broadband channel with low equalization effort. However, the technique has the disadvantage that the OFDM subcarriers suffer from flat fading, so strong channel coding is required in order to achieve a good performance.

A different approach is given by the combination of OFDM with code division multiplexing (CDM) [1], [2]. In an OFDM-CDM system, each data symbol is spread over several subcarriers before subcarrier modulation and despread in the receiver. Reference [3] showed that the combination of spreading and coding performs better than coding alone at code rates above 1/2. This combination can also be seen as a concatenation of two codes, of course, and the performance of a coded OFDM system could also be improved by applying a more powerful concatenated code instead of, say, a simple convolutional code.

In the following sections, we briefly review OFDM-CDM, then discuss a special case where a Fourier matrix is used for spreading.

25.1 PRINCIPLE OF A MULTICARRIER CDM SYSTEM

With OFDM-CDM, the complex data symbols X_i, $i = 1, \ldots, M$, which can be elements of an arbitrary symbol set, e.g., PSK or QAM, are spread over N subcarriers by means of orthogonal codes. The spreading can be described analytically by a

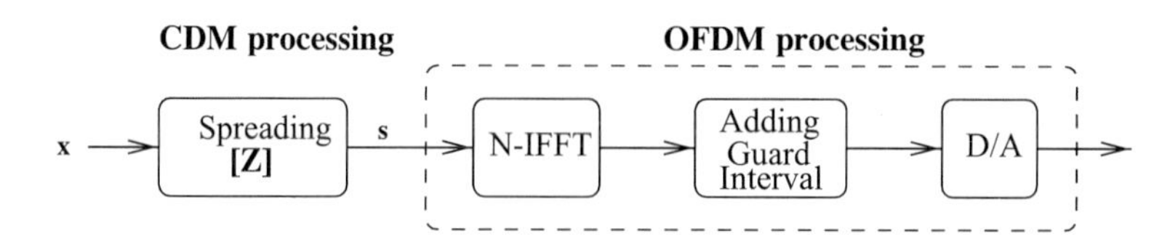

Figure 25.1. Block diagram of a multicarrier CDM transmitter. Reprinted from [11], copyright IEEE.

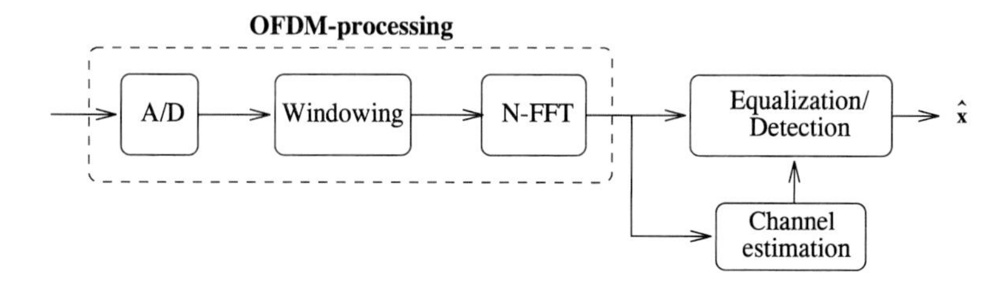

Figure 25.2. Block diagram of a multicarrier CDM receiver. Reprinted from [11], copyright IEEE.

matrix operation

$$\mathbf{s} = [\mathbf{Z}] \cdot \mathbf{x}, \tag{25.1.1}$$

where $\mathbf{x} = (X_1, \ldots, X_M)^T$is the vector of data symbols, $\mathbf{s} = (S_1, \ldots, S_N)^T$ is the vector of modulation symbols, and $[\mathbf{Z}]$ is the code matrix of dimension $N \times M$ ($M \leq N$). Apart from this spreading, the transmitter is identical to a conventional OFDM transmitter, i.e., the resulting signal sequence is IFFT-processed, periodically extended (a guard interval is added), and D/A converted; see Figure 25.1. In the receiver, the inverse processing is performed; see Figure 25.2.

Since the orthogonality of the spreading codes is destroyed in frequency-selective radio channels, equalization must be implemented. In the literature, various kinds of equalizers are discussed [4], [5], [6], [7], [8]. To initialize equalization, the channel transfer function must be known, and therefore a channel estimation must be performed in the receiver, too.

Due to the periodic extension of the signal in the transmitter and windowing in the receiver, the channel influence can be modelled by a cyclic convolution in the time domain or, alternatively, by a multiplication at discrete points in the frequency domain. Moreover, successive symbols do not interfere if the length of the guard interval is longer than the maximal multipath delay. Thus, the transmission of each symbol can be described analytically by a series of matrix multiplications, as depicted in Figure 25.3.

The performance of the overall transmission system does not depend on the specific type of the orthogonal spreading matrix as long as every element of the matrix has the same magnitude—which means that the energy is uniformly spread

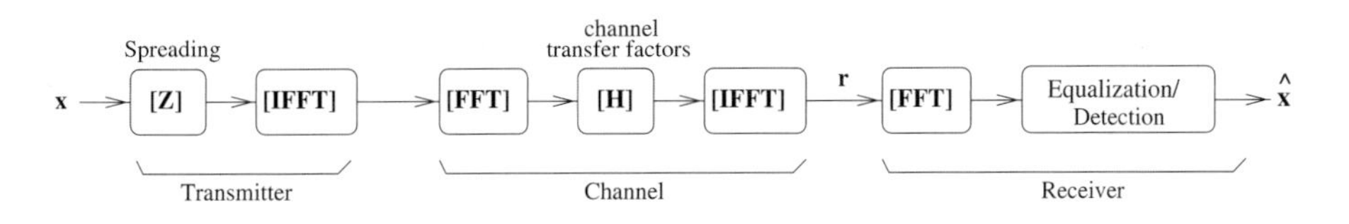

Figure 25.3. Model of a multicarrier CDM system. Reprinted from [11], copyright IEEE.

over all subcarriers [9].

In the literature, Walsh-Hadamard codes are generally used as spreading codes, but in principle, every other orthogonal code matrix can be applied. An interesting special case results if the Fourier matrix is used as an orthogonal code matrix. Considering the case that all available orthogonal codes are used, i.e., the spreading matrix is a square matrix of dimension $N \times N$, the matrix of the Fast Fourier transform meets the above-mentioned demands as well as the Walsh-Hadamard matrix. Since the multicarrier transmission signal is generated by an inverse Fourier transform (IFFT), both operations—spreading and IFFT—cancel each other, resulting in a pure single-carrier system with blockwise processing [10], [11].

Thus, the transmission system degenerates into a linearly modulated, single-carrier system with guard interval, that is,

$$\begin{aligned} s(t) = \textstyle\sum_n X_n \cdot p(t - nT) \quad \text{where} \quad x_n = x_{n-N} \quad \text{if} \\ i(N + N_G) - N_G \leq n < i(N + N_G). \end{aligned} \tag{25.1.2}$$

In (25.1.2), the parameter N_G describes the length of the guard interval in data symbols spaced by the symbol duration T.

25.2 CONSEQUENCES FOR THE TRANSMISSION SYSTEM

25.2.1 Dynamic of the signal envelope

In Chapter 23 we saw how the envelope of an OFDM signal has a high dynamic, which means either that additional signal processing is required or that the power amplifier must operate at a large input backoff. This disadvantage can be overcome if the Fourier matrix is used for spreading. Since the multicarrier CDM system is thereby transformed into a single-carrier system with linear modulation, the amplitude dynamic depends only on the pulse shaping $p(t)$. In connection with second-generation mobile systems, this topic has been analyzed extensively and solutions are well known: in case of binary transmission, for example, application of MSK (minimum shift keying), which uses a sinusoidal pulse, leads to a constant envelope signal. Quaternary transmission can be realized with low-amplitude dynamic by use of offset-QPSK. Low amplitude dynamic is an advantage compared both to an OFDM-CDM system with a Walsh spreading matrix and to a conventional OFDM system without a CDM component.

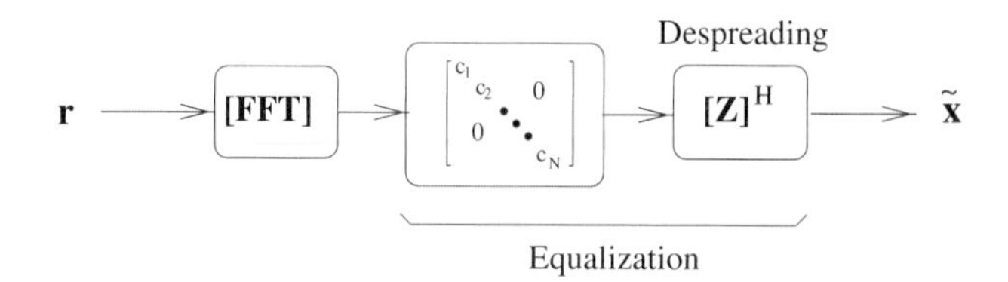

Figure 25.4. Model of an OFDM-CDM receiver with linear equalization. Reprinted from [11], copyright IEEE.

25.2.2 Computation complexity

The complexity of the equivalent single-carrier system is considerably lower than the complexity of a general OFDM-CDM system since neither matrix multiplications nor measures for signal dynamic reduction are required, which makes the single-carrier system very attractive for mobile applications.

Compared to a conventional OFDM system with nondifferential modulation, the single-carrier system requires a similar, overall computation effort. However, in the single-carrier system, the receiver must perform two Fourier transforms instead of one, whereas the transmitter does not perform a Fourier transform at all. Thus, a major part of the computation effort is moved to the receiver. Whether this is desirable depends on the considered system.

25.2.3 Synchronization

To avoid interchannel interferences, both a conventional OFDM system and a general OFDM-CDM system require a precise frequency synchronization. This synchronization is achieved with special synchronization symbols in combination with sophisticated frequency synchronization circuits. The equivalent single-carrier system, however, is less sensitive to frequency errors, thereby allowing simpler and cheaper synchronization circuits.

25.2.4 Performance

Since the single-carrier system, which results from an OFDM-CDM system with Fourier spreading matrix, is a special case of an OFDM-CDM system, the same detection algorithms can be applied and the same performance can be obtained.

Figure 25.4 illustrates an OFDM-CDM receiver with linear equalization [4].

In the case of a Fourier spreading matrix, despreading in the receiver is achieved by an inverse Fast-Fourier matrix, i.e., $[\mathbf{Z}]^H = [\mathbf{IFFT}]$. The overall processing corresponds to a frequency-domain equalization (FDE) since the received signal vector is first Fourier-transformed, multiplied by a diagonal equalization matrix, and then transformed back into the time domain. Such an FDE was originally proposed in [12] to speed up the convergence of the equalizer coefficients in conjunction with the stochastic gradient method. In [13], the FDE was picked up again to compare single-carrier and multicarrier (OFDM) transmission. Recently, the FDE was discussed for use in future mobile communication systems in combination with antenna diversity [14].

In contrast to a conventional single-carrier system, the single-carrier system resulting from OFDM-CDM with Fourier spreading matrix is characterized by a guard interval. The question arises why it should be advantageous to waste some part of the bandwidth. There are two reasons: the first one is the better performance at the same number of equalizer coefficients in the case of linear or decision-feedback equalization. This peformance gain results from the fact that the periodic extension of the data sequence in the transmitter, in connection with the signal that windows the receiver, leads to a cyclic convolution of the data sequence and the discrete-time channel impulse response. If the corresponding discrete-time channel transfer function contains no zeroes, this cyclic convolution can be perfectly inverted in the receiver without requiring more than N(= length of data sequence) coefficients. To achieve the same effect with a conventional single-carrier system, an equalizer with an infinite number of taps is generally required, and this requirement cannot be implemented in a real system. Thus, an equalization error occurs, and limits the performance at high signal-to-noise-ratios.

The superiority of the single-carrier system with guard interval is proved by evaluation of the minimum mean square error J_{min} at the equalizer output [15]:

$$J_{min} = 1 - \epsilon^H [\Gamma]^{-1} \epsilon \tag{25.2.1}$$
$$[\Gamma] = E\{\mathbf{r}^* \mathbf{r}^T\} \qquad \epsilon = E\{\mathbf{r}^* X\}.$$

In (25.2.1), the vector $\mathbf{r}$ describes the input sequence to be filtered and X describes the desired output data, respectively. The difference between both single-carrier systems consists in different covariance matrices $[\Gamma]$ as depicted in Figure 25.5. Because of the periodic extension of the data in the transmitter, the covariance matrix is periodically extended as well, yielding the relation

$$[\Gamma](\text{withguard}) = [\Gamma](\text{withoutguard}) + [\Delta] \tag{25.2.2}$$

with $[\Gamma]$ and $[\Delta]$ being Hermitian matrices.

Using the relationship given in (25.2.1) and (25.2.2), we can show that $J_{min}(\text{withguard}) \leq J_{min}(\text{withoutguard})$.

Figure 25.6 shows simulation results for single-carrier systems with and without guard interval according to the WSSUS channel model. Both systems have a bandwidth of 2 MHz and make use of linear equalization according to the MMSE criterion. The system with guard interval corresponds to an OFDM-CDM system with a Fourier matrix for spreading. It has a block length of $T_N = 32$ μs and a guard interval of $T_G = 6.6$ μs. The number of subcarriers or the number of symbols per block, respectively, is $N = 64$.

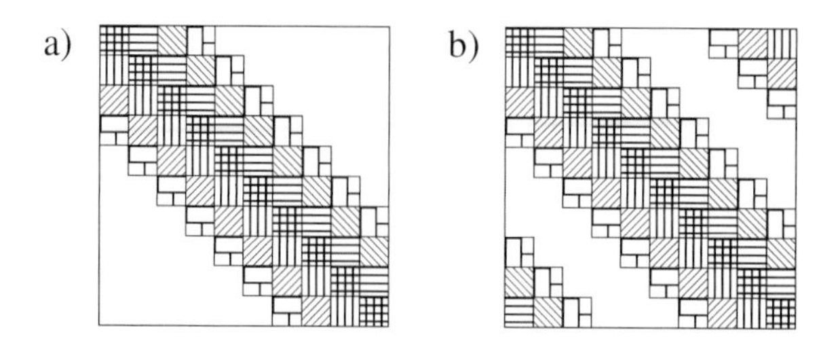

Figure 25.5. Shape of the covariance matrix $[\Gamma]$ in single-carrier systems a) without and b) with periodic extension (guard interval). Reprinted from [11], copyright IEEE.

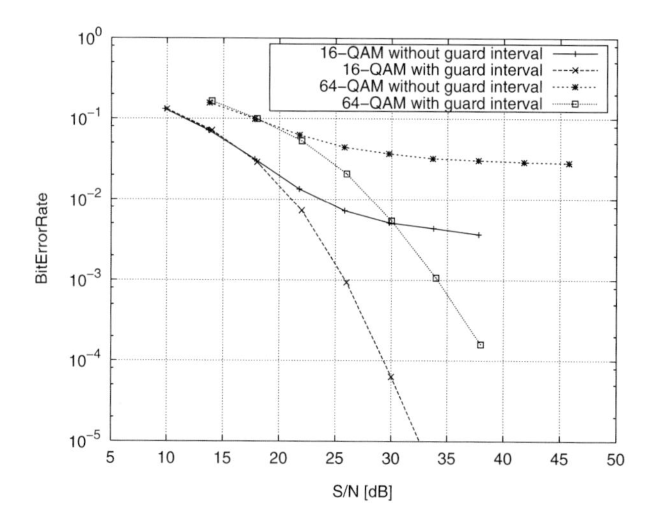

Figure 25.6. Performance comparison of single-carrier systems with and without guard interval (linear equalization). Reprinted from [11], copyright IEEE.

The single-carrier system without guard interval has a modulation pulse with a raised cosine characteristic and the roll-off factor $\beta = 0.2$. The symbol duration is $T_S = 600$ ns.

At low signal-to-noise ratios (SNR), both systems show similar performance; at high SNR values, an error floor can be observed if no guard interval is used. The level of the error floor depends on the type of modulation and may be of minor importance if binary or quaternary modulation is used in combination with channel coding.

The second reason in favor of using a guard interval is the relatively low initialization effort of the equalizer. Since the optimum equalizer coefficients c_i, $i = 1, \ldots, N$, in case of a single-carrier system with guard interval and FDE are given by $k_i^*/(|k_i|^2 + \text{const})$ [14], only two multiplications are required per coeffi-

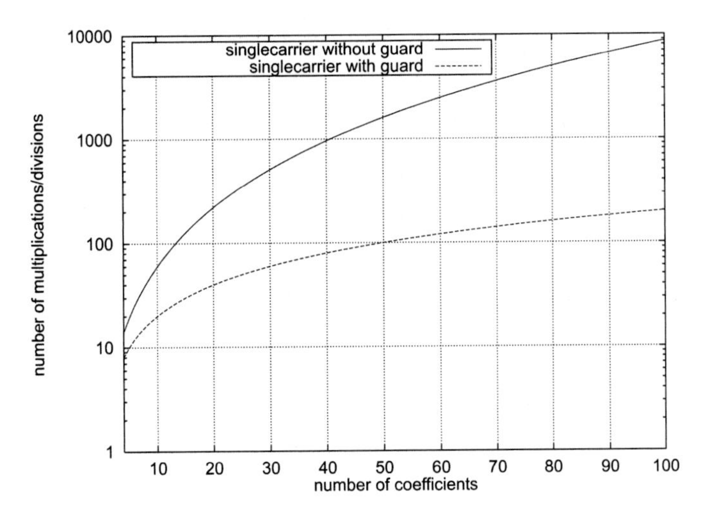

Figure 25.7. Initialization effort of equalizer coefficients assuming that the channel impulse response spans 10% of the equalizer length. Reprinted from [11], copyright IEEE.

cient, and the main effort is to estimate the channel transfer factors k_i. In contrast, initializing the filter coefficients in a single-carrier system without guard interval requires us to solve the set of linear equation $[\Gamma]\mathbf{c} = \epsilon$ (assuming equalization in the time domain). Due to the special form of this set ($[\Gamma]$ is a Hermitian and positive definite band matrix), we can efficiently apply the Cholesky decomposition. However, if the number of filter coefficients is large, i.e., in highly time dispersive channels, the computation complexity of the initialization process is considerably higher than for the system with guard interval; see Figure 25.7.

As a result, adding a guard interval allows even single-carrier systems to cope with very long channel impulse responses, which multicarrier systems were originally designed for, since the computation complexity is lower than for conventional single-carrier systems without guard interval, at least in the case of linear equalization.

25.3 COMPARISON OF THE CONCEPTS

Considering a conventional single carrier system, time-dispersive channels lead to intersymbol interference which makes an equalizer necessary. If the maximum excess delay of the channel is large as compared to the modulation symbol duration, then the effort for channel estimation and equalization can become unacceptable. In this case, a different transmission technique must be applied.

The OFDM transmission technique solves this problem by distributing the data to be transmitted to a multitude of narrowband subchannels. The symbol duration in each subchannel is large, and thus, we can avoid ISI by appending a guard interval. This technique allows transmission of a high data rate over a frequency-selective channel with a low equalization effort. If differential modulation is applied, no channel estimation and equalization is required and a very simple system results.

On the other hand, some drawbacks have to accepted if the OFDM transmission technique is chosen. Synchronization must be more precise compared to single-carrier transmission. Furthermore, OFDM signals are characterized by a large peak-to-average power ratio, which means that the amplifier must operate at an appropriate input backoff. Finally, since the OFDM subchannels are narrowband, strong channel coding must be applied in order to avoid the BER curves that are typical for flat-fading situations. It is clear, however, that for data transmission over fading channels, channel coding is necessary anyway. With respect to these topics, many research results published in the last years have been outlined in the previous chapters.

The OFDM concept has been extended to OFDM-CDM, a transmission technique where the information to be transmitted is spread over several OFDM subcarriers. Arbitrary orthogonal codes can be used for spreading. The effect of this additional measure is that narrowband fading is avoided. This can be seen as a frequency diversity effect. Thus, at high code rates, an OFDM-CDM system performs better than a pure OFDM system. However, equalization and despreading in the receiver lead to a noise amplification. For this reason, the pure OFDM system performs better at low code rates where sufficient frequency diversity is provided by the code. However, the performance of OFDM-CDM systems can be improved with more sophisticated detection methods like iterative despreading and decoding. A major disadvantage of OFDM-CDM systems is the fact that coherent detection is required, so channel estimation and equalization cannot be avoided.

If a Fourier matrix is used for spreading in an OFDM-CDM system, then spreading and the IFFT at the transmitter cancel each other, resulting a single-carrier transmission system with a guard interval results. At the receiver, the signal processing can be interpreted as a frequency-domain equalization, with the additional advantage that the transmitted signal does not have the typical amplitude peaks of an OFDM signal. Thus, it seems to be reasonable to use the Fourier matrix for spreading if spreading is to be applied at all.

Bibliography

[1] V. M. DaSilva and E. S. Sousa in *Proc. IEEE ICUPC '93*, (Ottawa, Canada), pp. 995–999, Oct 1993.

[2] N. Yee, J.-P. Linnartz, and G. Fettweis, "Multi-carrier CDMA in indoor wireless radio networks," in *Proc. IEEE Int. Symp. On Personal, Indoor and Mobile Radio Communications (PIMRC '93)*, Sept. 1993.

[3] S. Kaiser, "Trade-off between channel coding and spreading in multicarrier CDMA systems," in *Proc. ISSSTA '96*, (Mainz, Germany), pp. 1366–1370, 1996.

[4] N. Yee and J. Linnartz, "Wiener filtering of multi-carrier CDMA in Rayleigh fading channel," in *Proc. IEEE Int. Symp. On Personal, Indoor and Mobile Radio Communications (PIMRC '94)*, pp. 1344–1347, 1994.

[5] W. G. Teich, "Detection method for MC-CDMA based on a recurrent neural network structure," *Multi-Carrier Spread Spectrum*, pp. 135–142, 1997.

[6] M. Reinhardt, T. Huschka, and J. Lindner, "Performance of combined equalization and TCM decoding for Rayleigh fading channels," in *Proc. IEEE Int. Symp. On Personal, Indoor and Mobile Radio Communications (PIMRC '96)*, (Taipei, Taiwan), pp. 1092–1096, 1996.

[7] M. Reinhardt *et al.*, "Transformation methods and iterative equalization and decoding for symbol spread transmission," in *4th. Int. Symposium on Comm. Theory & Applications*, (Ambleside, UK), pp. 228–229, 1997.

[8] T. Müller, K. Brüninghaus, and H. Rohling, "Performance of coherent OFDM CDMA for broadband mobile communications," *Wireless Personal Communications*, vol. 2, pp. 295–305, 1996.

[9] D. Rainish, "Diversity transform for fading channels," *IEEE Transactions on Communications*, vol. 44, no. 12, pp. 1653–1661, 1996.

[10] J. Lindner, "Channel coding and modulation for transmission over multipath channels," *Archiv für Elektronik und Übertragungstechnik (AEÜ)*, vol. 49, no. 3, pp. 111– 119, 1995.

[11] K. Brüninghaus and H. Rohling, "Multi-carrier spread spectrum and its relationship to single-carrier transmission," in *Proc. VTC '98*, (Piscataway, N.J.), pp. 2329–2332, IEEE, 1998.

[12] T. Walzman and M. Schwartz, "Automatic equalization using the discrete Fourier domain," *IEEE Transactions on Information Theory*, vol. 19, pp. 59–68, Jan. 1973.

[13] H. Sari, G. Karam, and I. Jeanclaude, "Channel equalization and carrier synchronization in OFDM systems," in *Proc. 6th Tirrenia International Workshop on Digital Communications*, (Tirrenia), Sept. 1993.

[14] G. Kadel, "Diversity and equalization in frequency domain—a robust and flexible receiver technology for broadband mobile communication systems," in *Proc. VTC '97*, (Phoenix, USA), pp. 894–898, 1997.

[15] J. Proakis, *Digital Communications*. Singapore, McGraw-Hill Book Company, 3rd ed., 1995.

Part V

Code Division Multiple Access

Alois M. J. Goiser: Chapters 25-28,
Moe Z. Win and George Chrisikos: Chapter 29,
Savo Glisic: Chapters 30-33

Chapter 26

BASICS OF CODE DIVISION MULTIPLE ACCESS

The spread-spectrum technique is a special communication concept, able to communicate through environments of severe interference. In such terms, it is a robust communication scheme. Although it behaves very differently from conventional communication schemes, it has the potential capability to cope with severe situations. Especially in a multipath environment the spread-spectrum technique offers a unique receiver structure that collects the signal energy from different multipath components. While it collects the signal energy, it is analogous to a rake and therefore it is termed *Rake receiver.*

The Rake receiver is a potential candidate for high-performance communication through multipath channels. In a point-to-point communication system, the Rake receiver exploits the frequency selectivity of the channel with the help of pseudonoise sequences. The next step is to assume a multipath channel and a multiple-access environment. As we see later, a direct-sequence spread-spectrum link in a multiuser environment is interference limited. Therefore, each processing step that reduces the interference enhances the quality of the link or increases the capacity. There are two main interference sources. The interference is primarily caused by other users. This interference can be mitigated with power control and reduced with multiple-user detectors. The other source of interference is introduced by the multipath channel. The Rake receiver can potentially reduce the multipath interference with its unique structure, which mimics the reverse of the channel impulse response. Advanced Rake receiver structures use interference reduction schemes that cope with both interference sources.

First, we give some brief background information about spread-spectrum systems and CDMA technology. Then, we introduce the conventional Rake receiver and study its behavior. Equipped with that knowledge, we focus on advanced Rake receiver structures in CDMA environments.

26.1 INTRODUCTION

For a proper operation of the Rake receiver, all multipath signal components should be synchronized. That includes synchronization of code delay and signal phase for each component. In addition, for MRC operation the channel intensity coefficients should be estimated. Although a joint estimation of these parameters is an optimum solution, due to its complexity this approach might be not a primary choice of practical designers. For this reason we also present simpler algorithms where channel parameters are estimated separately.

For Rake receiver processing in a mobile environment, we need some fundamental understanding of spread-spectrum systems, and we provide a discussion of basics in this chapter. Spread-spectrum technology has moved from military applications to commercial applications. The reason is the robustness in interference-prone environments, especially when the interference is not precisely predictable. Spread-spectrum systems show their inherent interference reduction capabilities in such environments. The unpredictability of interference is a consequence of the global communication concept.

26.2 SPREAD-SPECTRUM CONCEPT

The most popular spread-spectrum technique for commercial applications is the direct-sequence technique. This technique uses a simple shift-register-generated pseudonoise (PN) sequence. The periodic autocorrelation function of a PN sequence is shown in Figure 26.1. The length of the PN sequence is $L_s = 2^n - 1$, and n is the length of the shift register. The relationship between sequence length and data-bit duration T_D is given in (26.2.1) with T_c as chip duration, the fundamental pulse duration in a direct-sequence spread-spectrum system. The transmitter substitutes a positive data-bit within the data-bit duration with the PN sequence itself and for a negative data-bit with the amplitude-inverted copy of the PN sequence. This substitution can be referred to as code shift keying. The PN signal spreads the information signal (data signal) in bandwidth and is referred to as spreading signal. It is important to note that the spreading-sequence must be a PN sequence with its distinct properties [1], [2], [3].

$$T_D = L_s\, T_c = (2^n - 1)\, T_c \tag{26.2.1}$$

The receiver recovers the information (data bits) with the help of the correlation principle and the PN properties of the used spreading signal. The correlation process offers the potential processing gain G_p, which is unique for spread-spectrum systems and equal to the number of chips used in the spreading signal ($G_p = L_s$). For further discussions on spread-spectrum technology, we refer to [4], [5], [6], [1], [7], [8].

If we change our communication topology from point-to-point to point-to- multipoint, we have changed the communication environment from single-link to a multiple-access link. The multiple-access scheme in a spread-spectrum system is

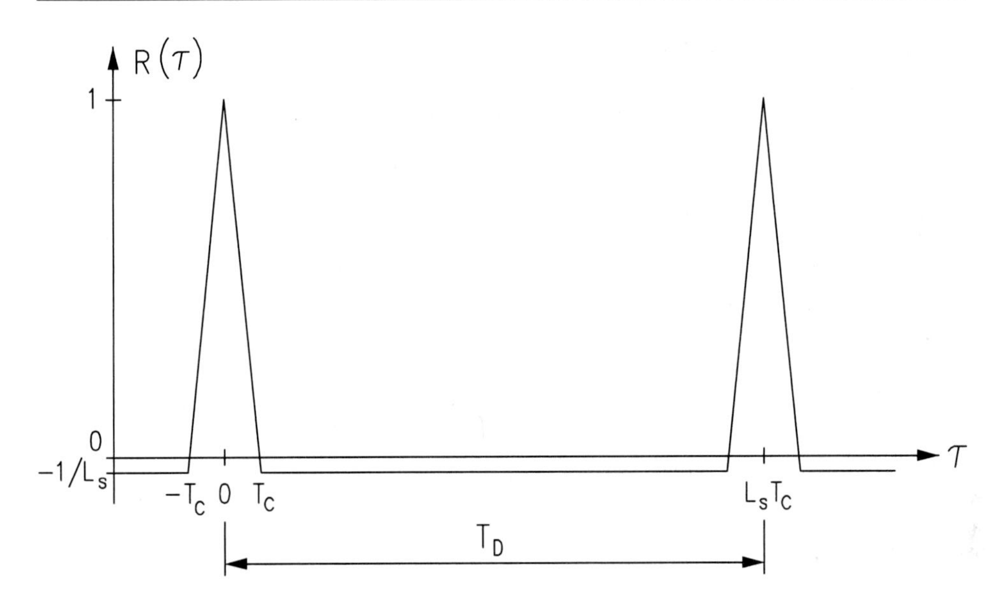

Figure 26.1. Periodic autocorrelation function of a PN-sequence. The sequence length is $L_s = 2^n - 1$, and n is the length of the shift register.

termed *code-division multiple-access* (CDMA). A brief description is given in the next section. For a more detailed discussion, refer to [4], [1], [9], [10], [11].

26.3 CODE-DIVISION MULTIPLE-ACCESS PRINCIPLE

Each access to a common channel needs some form of orthogonality. For frequency-division multiple-access (FDMA), we achieve orthogonality in the frequency domain by selecting nonoverlapping unique frequency bands to each user. We achieve orthogonality in the time domain by selecting nonoverlapping unique time segments to each user; this process is referred to as *time-division multiple access* (TDMA). The spread-spectrum form of multiple access exploits the orthogonality in the code domain and is termed *code-division multiple access* (CDMA).

The multiuser environment in the spread-spectrum case is set up for each user in assigning each user a unique spreading sequence out of a family of orthogonal sequences. Each user in a CDMA network occupies the same channel bandwidth. We can distinguish between synchronous and asynchronous systems. In a synchronous CDMA (S-CDMA) system, each data bit starts at the same time instant, assuming that some global time information is available. The philosophy is explained for synchronous CDMA systems and extended later to asynchronous transmission (A-CDMA).

For simplicity, let's assume we have an isolated cell with one base station in the center and K active users. We assume further that the CDMA cell is designed for K_{nom} users in a certain area. For these assumptions, a preset bit-error rate is not exceeded. To meet future demands, we assume that K' additional orthogonal spreading sequences are available. These assumptions lead to a total of K_{max} orthogonal sequences allowing us to address $K_{max} = K_{nom} + K'$ different users. More precisely, we change from users to information channels to point out the flexibility of CDMA systems. Each available orthogonal sequence corresponds to one information channel. Therefore, a user can accumulate more information channels for a certain service. Details are discussed in Section 26.4.2.

A CDMA system is clearly not a collision avoidance system like FDMA and TDMA. The opposite is true and explains the differences in the behavior of CDMA systems compared to FDMA and TDMA. In general, the collisions at the channel is a disadvantage of a CDMA system and can be mitigated by careful selection of the sequences and power control that is close to perfect.

To discuss the behavior of a CDMA system, we have to investigate the actual signal-to-interference ratio (SIR) corresponding to the other active information channels. For simplicity, we neglect the signal-to-noise ratio (SNR) introduced by the receiver's front end and the background noise. This assumption is realistic if the number of active information channels is large. The point is that the SIR is the figure that determines the bit-error rate, which is the ultimate measure in a digital communication system. Therefore, the performance and the behavior of a CDMA system must be discussed in terms of SIR.

Now we are ready to discuss more detailed mitigation strategies to reduce the effect of the collisions. The number of active information channels is a random number. Connected to that and corresponding to the behavior of a speaking person, the conclusion is that the channel is only occupied during active voice periods. In that context, if the information channel carries a data signal, the data signal behaves like a continuously speaking person.

The main advantage of a CDMA system is the high degree of flexibility. To exemplify flexibility, we present two cases.

In our first case, if the number of active information channels is less than the nominal value, we conclude from the preceding discussion that we can note an enhancement in SIR that is directly reflected in an improvement in bit-error rate. That improvement can be invested in improvement of service quality or in enlargement of cell size. The latter is called *breathing cells.*

If the number of active information channels is temporarily larger than the nominal value, then service quality is degraded for all the active information channels. This situation, called *soft capacity,* is impossible in FDMA and TDMA.

Suppose soft capacity occurs in a multicell environment frequently of the same position; that is, the number of active information channels exceeds the nominal value. The problem can easily be adjusted without complicated frequency planning by addition of another base station to the so-called hot spot.

For our second case, we change to a multipath and multicell environment like a terrestrial cellular mobile radio network. In a multicell environment, we must distinguish between interference corresponding to the own cell and the surrounding cells. In non-spread-spectrum communication systems, performance is severely degraded by the multipath components.

A conventional correlation spread-spectrum receiver has the potential benefit, corresponding to the two valued correlation functions of the PN-like spreading signal, to reduce the interference of the multipath components that are delayed more than a chip duration. This type of receiver is referred to as a *multipath rejection receiver* (MWU). The MWU selects the strongest path and uses it in the correlation process and the subsequent data decision, disregarding the other components of the signal. This behavior reduces the interference and ensures that only the signal component which constructively adds to the detection process is used but at the expense of reduced signal power. In other words, the strategy of the MWU receiver is to avoid a shift from signal power to interference power due to the dispersive characteristic of the channel.

The next step is not to suppress the multipath components but to sum them to maximize the signal power in the correlation process. The necessary receiver structure is the Rake receiver, which combats multipath interference. In the rest of the chapter, we focus on the Rake concept in conjunction with time-dispersive channels. This standard Rake receiver structure is inherently a single-user receiver and ineffective in a near-far situation.

As discussed later, multiuser detection can significantly reduce the interference from unwanted users. Well-known multiuser detector algorithms like the decorrelating receiver and the minimum mean-square error (MMSE) receiver structures offer superior performance in terms of error probability and capacity with a linear increase in complexity [12]. Their superiority is based on the precise knowledge of the desired signal distorted by the channel. This leads to the effort of channel estimation, which increases the complexity and power consumption of the receiver [13]. Therefore, much attention has been paid to suboptimum solutions that reduce the complexity and keep the reduction in performance in affordable limits. The most promising candidates are blind multiuser algorithms that can reduce interference and recover the data bits without inserting training sequences in the data stream. Unfortunately, most of the blind multiuser receptions are computation intensive. Chapter 31 describes the behavior of an adaptive Rake receiver for a frequency-selective fading environment with the assumption that the desired user's signature is known without the use of channel estimates.

26.4 CDMA-NETWORK DESIGN ISSUES

In this section we focus on design issues of direct-sequence, spread-spectrum networks. We discuss the necessity to achieve a trade-off between available bandwidth, necessary data rate, processing gain, and the number of selectively addressable information channels. For bandwidth or power-limited channels in additive white Gaussian noise, we show trade-off for synchronous and asynchronous code-division, multiple-access schemes. Finally, we comment on the blind and assumption-driven computer search to find application-specific sequences.

26.4.1 Assumptions

To start, we design a cluster for K independent, addressable, information channels. An information channel could be a user in a cellular mobile environment (voice), a data channel of a cable modem in a local area network (LAN), or a data channel for a wireless information network (WIN).

The investigated concept is by nature a code-division, multiple-access scheme. We assume a cable modem in an S-CDMA environment with a cluster size corresponding to K_{nom} channels. For this application, the following assumptions are made:

- One cluster supports K_{nom} channels.
- A fixed bit-error rate should be maintained for full coverage (all K_{nom} are active).
- The available channel bandwidth is 3 MHz (roll-off included).
- Each channel has the same data rate.
- The codes are perfectly aligned (synchronized).
- The channel noise is modeled as additive white Gaussian noise (AWGN).

Some remarks about this example: The main interference is the interference from other active users. The bit-error rate is usually low, to assign the channels to voice or data services. We have a certain number of channels available in the cluster. The maximum number of nodes in the cluster corresponds to the available channels. If a particular node needs more channels to increase the data rate for a specific service, then the number of nodes decreases (discrete data rates). We call this situation *dynamic channel assignment.*

26.4.2 Trade-off

The trade-off between available bandwidth (bandwidth of the physical channel B_T), data rate (R_D), processing gain ($G_p = L_s$), the necessary number of addressable channels (K_{nom}), and adjacent and co-channel interference (roll-off) is addressed in this section.

Figure 26.2 shows the relationship between processing gain and symbol rate. Figure 26.3 shows the relationship between modulation index and data rate corresponding to the chosen processing gains (G_p = 64 chips and G_p = 256 chips) in Figure 26.2. These two figures can be read in different ways.

Let's assume that a certain data rate has to be maintained, for instance, $R_D \approx 70$ kbits/s. This assumption is suitable for ISDN channels with a net data

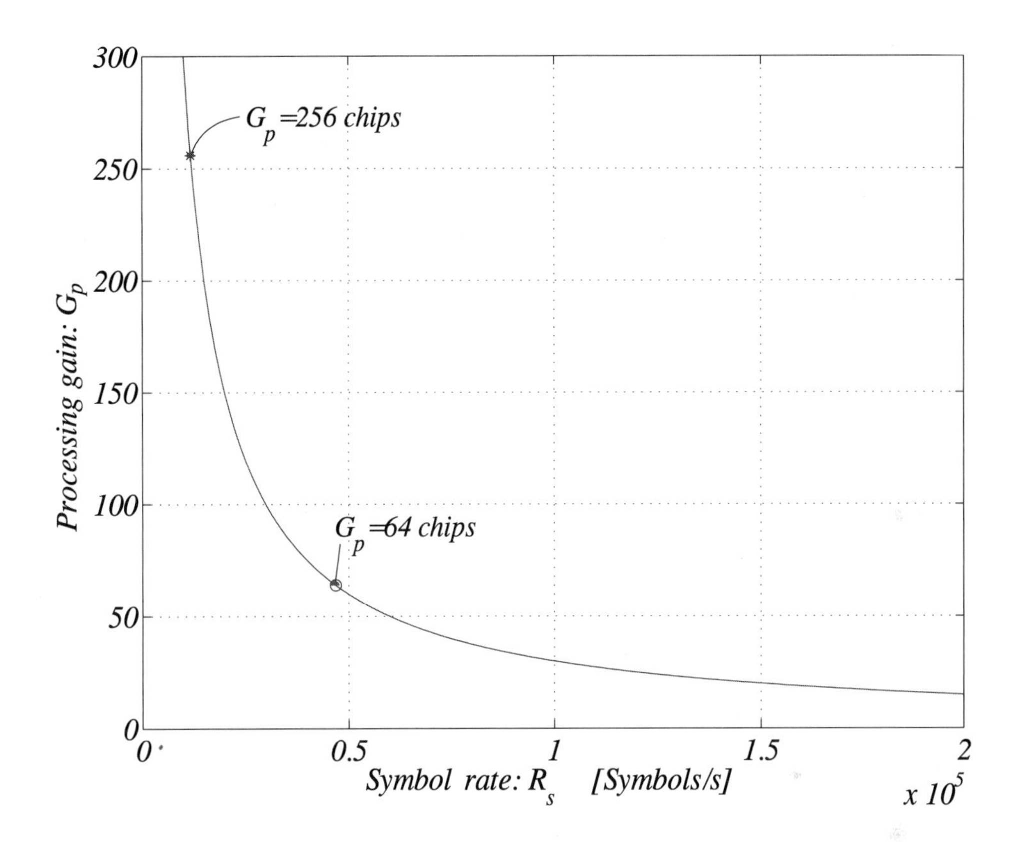

Figure 26.2. Processing gain to symbol rate relation for $B_T = 3$ MHz

rate of 64 kbits/s plus a coding overhead. From Figure 26.3 we see that for a processing gain $G_p = 64$ and modulation index $M = 4$, a suitable data rate is achieved. If the processing gain is suitable with respect to the interference reduction capability, the number of channels, and the achieved bit-error rate, then our design is finished. But if we need a higher processing gain to achieve the necessary number of channels or if the MUI for a S-CDMA system degrades the bit-error rate to a rate of no interest, then we must read the two figures the other way around. We have to fix the necessary processing gain, which corresponds to a certain curve in Figure 26.3 and cross that curve with the necessary data rate (horizontal line). The next modulation index (to higher values) is the choice for the system. Figure 26.3 shows the tendency of an increased processing gain to make the design more complex to maintain a certain data rate.

The trade-off applied to our examples lead to the conclusion that in a S-CDMA design we need a high processing gain (sequence length) to achieve the number of channels and maintain the low bit-error rate. For comparison, we find in [4], [5], [9]

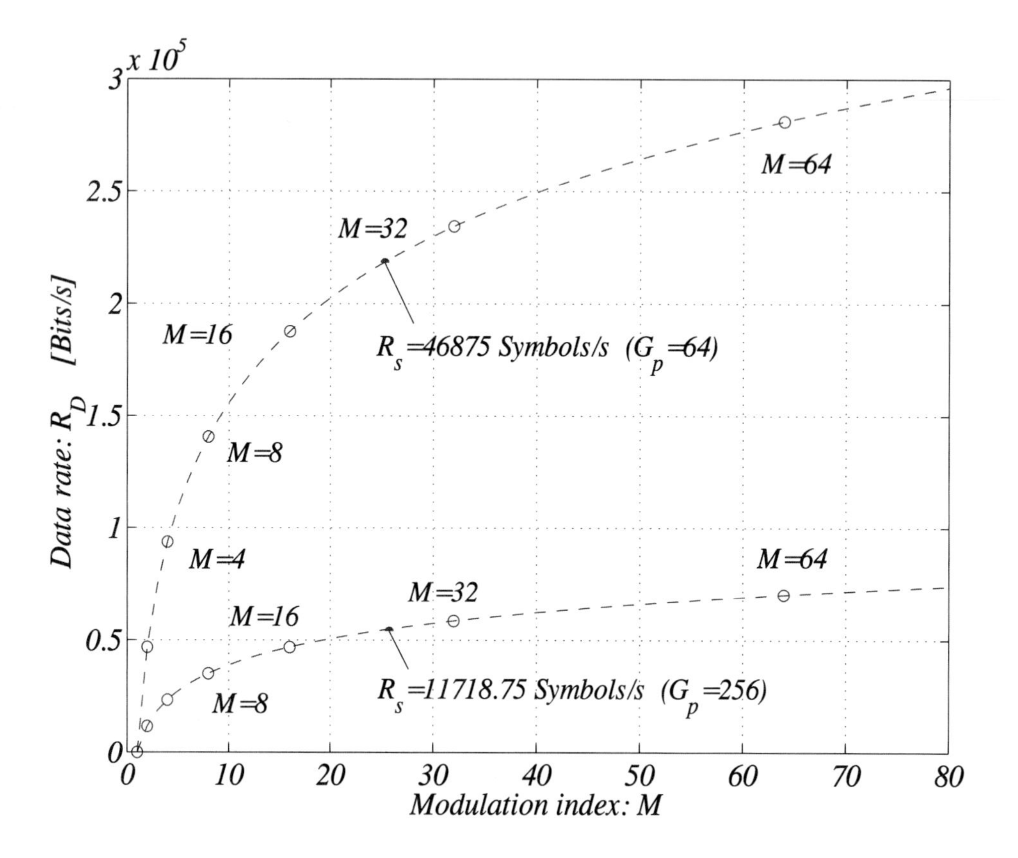

Figure 26.3. Modulation index to data-rate relation for $B_T = 3$ MHz, $G_p = 64$, and $G_p = 256$

that in the A-CDMA design we need the processing gain primarily to mitigate the interference. We conclude that the main issue in the design is the processing gain, regardless of whether S-CDMA or A-CDMA is chosen as system layout.

26.4.3 Code selection

We assume that the properties of the most popular code families are known [14]. The code selection is mainly based on the correlation properties of the sequence. The relevant correlation properties depend on the system design (S-CDMA or A-CDMA). The correlation properties raise the following questions:

Q1: How many sequences are available in the family?

Q2: Must we choose a smaller set of sequence out of a larger set?

To answer Q1: For future changes, the code set should be greater than the nominal value ($K_{max} > K_{nom}$). This code overhead can be used at the expense of a lower bit-error rate in future applications. This answer is highly recommended for a forward-looking design. The codes should be numbered because a figure of merit is predefined. Then, we can assign the codes to the channels in a predefined sequence. The optimum value for K_{nom} is that value when the figure of merit changes significantly to bad values. (This situation must not occur.)

To answer Q2: If more sequences than are necessary are available, we can choose a subset with better probabilities. For a given code set, the bounds are known. If we are dissatisfied with the bounds (in general, correlation bounds are of interest), we can find a smaller set with significantly better properties. The degree of freedom we are exploiting is the reduction of the family members. We can find the members of the smaller set only with a computer search. Now the question is, how to choose the members?

To answer that question: If we have a certain number of appropriate sequences and if, when we check the next sequence, we must decide whether to include or reject it, we face a dilemma. Let's say the investigated sequence collides with only one sequence. Which of those two sequences should we investigate further? If we reject the colliding sequence and include the investigated one, we might find that the next candidate collides with the newly included (investigated) sequence but does not collide with the one we just rejceted.

We can readily see the importance of the starting sequence and the sequence of test for inclusion. Therefore, we must check all possibilities, that is, conduct a *blind search.* From Figure 26.4 we derive all solutions that meet a chosen goodness criterion.

All solutions are included in

$$N = K! \cdot 2^K \tag{26.4.1}$$

trials. For instance, if $K = 10$, we have to check $N = 3.7159 \cdot 10^9$ possibilities to find the number of sequences included in the family for a maximum allowed cross-correlation value. As the numerical example shows for usual values of K, we cannot check all possibilities to find all solutions.

The search strategy can only be to stop at the first occurrence of a hit. We refer to this as the *assumption-driven search strategy.* A hit is defined as the situation in which the necessary amount of sequences are found under the condition of the goodness criterion. The occurrence of this event and the time consumption are not predictable. The influence of the goodness criterion is obvious. Therefore, we should not choose a too-tight cross-correlation bound as figure of merit.

26.4.4 Performance

Due to the power limitation of the channel and S-CDMA-concept, each active channel adjusts its transmitting power to $1/K$ of the total power. The necessary bit-error rate must be established for full coverage, and coverage must include the bit-error

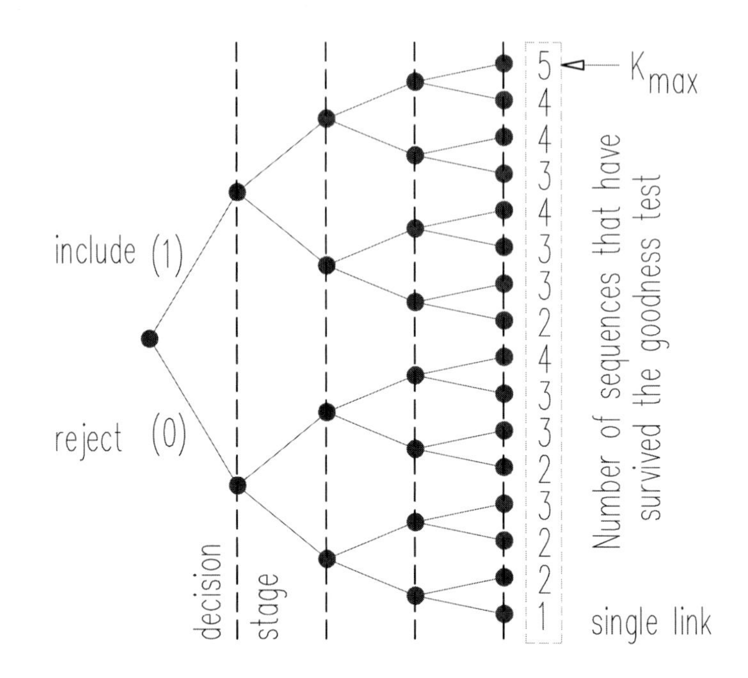

Figure 26.4. Inclusion decision tree

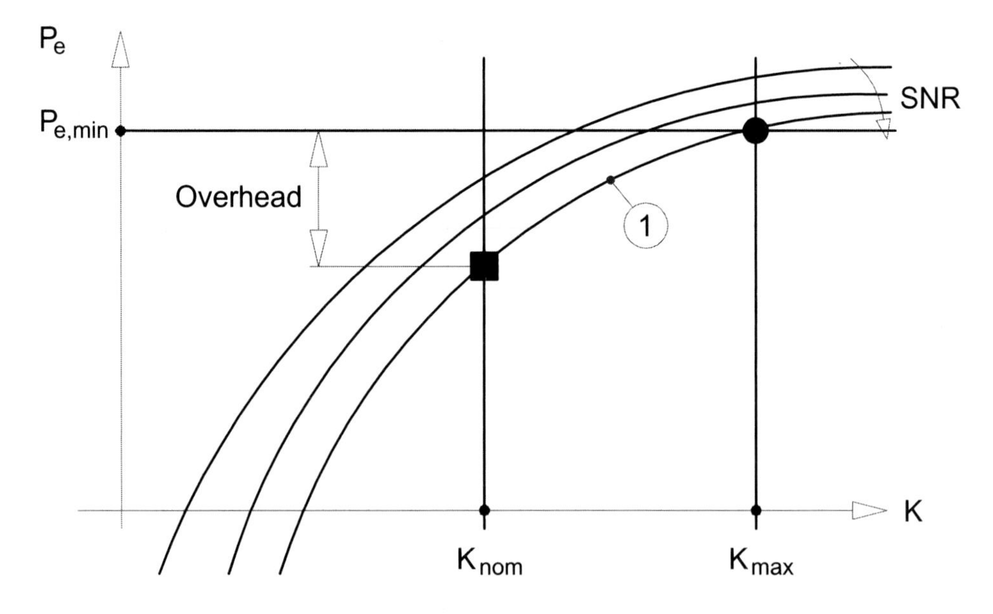

Figure 26.5. Bit-error-rate behavior for S-CDMA

rate for a power-limited channel, where the rate is strictly dependent on the number of active users.

In a bandwidth-limited channel and with a S-CDMA-concept in use, the transmitted power for each channel is the same, and therefore the bit error rate did not depend on the active users if the codes are perfectly balanced.

Figure 26.5 illustrates the behavior of the bit-error rate for S-CDMA in a power-limited channel. Curve 1 is the nominal bit-error rate for the design for various active channels. The square indicates K_{nom}. K_{nom} is assigned a value such that a suitable margin to the necessary bit-error rate P_{min} is achieved. The margin can be used to increase the nominal number of active users for future applications, or it can be used to overcome an unpredictable increase of interference. The behavior of the bit-error rate for S-CDMA in bandwidth-limited channels is dictated by the processing gain and the actual signal-to-interference ratio.

26.4.5 Design improvements

If we have designed the network in the above-mentioned way, we can further improve the performance of the network if we have some additional degrees of freedom. For instance, we can select sequences from a larger set to maintain correlation properties to enhance the probability of data detection or to reduce the mean synchronization time. One possibility to achieve that improvement is to select sequences that are perfect within a correlation window [6]. That means that the correlation is a quasi-perfect correlation within the window $\leq 1/L \approx 0$.

If we add one chip to a sequence with odd-sequence length to get an even sequence length that is a perfectly balanced sequence, we have reduced the co-channel interference to zero in an S-CDMA system, at the expense of a slightly reduced pseudonoise property. This result can be valuable if we have many channels within the cluster and roughly no interference from outside. In that case, where MUI is the dominating interference, perfectly balanced sequences can improve the bit-error rate significantly.

A very attractive improvement can be achieved if interference reduction schemes are applied to the receiver prior to data detection. This topic is out of the scope of this text. But if the mentioned design hints are not successful, we have to use interference reduction schemes. An excellent tutorial can be found in [15]. Figure 26.6 shows that at the expense of complexity, additional interference reduction is possible. The "Lower Limit" indicates perfect multiuser detection for which the transmission system behaves like a single link transmission. Up to the mark K_p, the processing gain is the major parameter that reduces the multiple user interference [16].

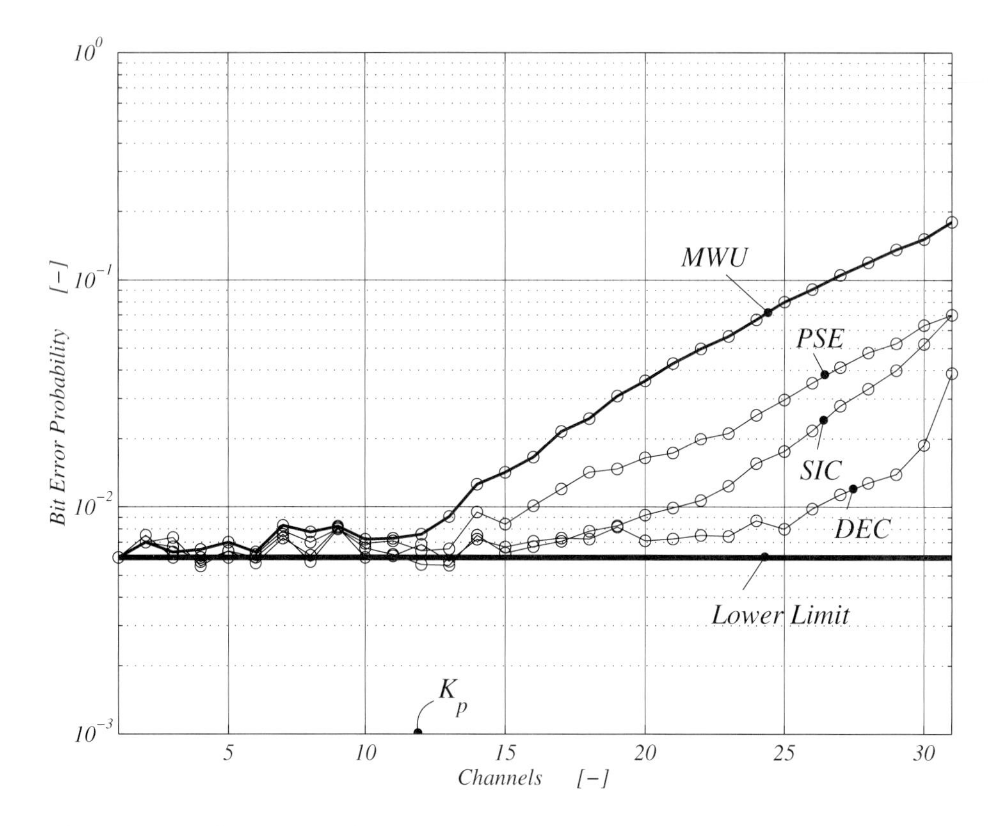

Figure 26.6. Simulation of the bit-error probability versus the active users. SIC: successive interference cancellation detector. DEC: decorrelation detector. PSE: partial sequence estimation detector. MWU: multipath rejection receiver. These receiver structures are described in Chapters 27, 28. Parameters: $G_p = 31$, Gold codes.

Bibliography

[1] R. L. Pickholtz, D. L. Schilling, and L. B. Milstein, "Theory of spread-spectrum communications—a tutorial," *IEEE Transactions on Communications*, vol. 5, pp. 855–884, 1982.

[2] S. Golomb, *Shift Register Sequences.* San Francisco: Holden Day Inc., 1967.

[3] B. Sklar, *Digital Communications.* Englewood Cliffs, Prentice-Hall International, 1982.

[4] P. W. Baier, "Spread-spectrum-technik und CDMA," *telecom praxis*, vol. 5, 1995.

[5] S. Glisic and B. Vucetic, *Spread Spectrum CDMA Systems for Wireless Communications.* Artech House, 1997.

[6] A. Goiser, *Handbuch der Spread-Spectrum Technik.* Wien: Springer, 1998.

[7] R. A. Scholtz, "The origins of spread spectrum communications," *IEEE Transactions on Communication*, vol. COM-30, pp. 822–854, 1982.

[8] R. A. Scholtz, "Notes on spread-spectrum history," *IEEE Transactions on Communication*, vol. 31, pp. 82–84, 1983.

[9] D. Schilling, L. Milstein, and R. Pickholtz, "Spread-spectrum for commercial communications," *IEEE Communications Magazine*, vol. 4, pp. 66–79, 1991.

[10] D. Schilling, L. Milstein, R. Pickholtz, F. Bruno, E. Kanterakis, M. Kullback, V. Ereg, W. Biederman, D. Fishman, and D. Salerno, "Broadband CDMA for personal communications systems," *IEEE Communications Magazine*, vol. 29, pp. 86–93, 1991.

[11] R. A. Scholtz, "The evolution of spread-spectrum multiple-access communications," in *IEEE-Third International Symposium on Spread-Spectrum Techniques and Applications*, (Oulu, Finland), pp. 4–13, 1994.

[12] R. Lupas and S. Verdu, "Linear multiuser detectors for synchronous CDMA channels," *IEEE Transactions on Information Theory*, vol. 35, pp. 123–136, 1989.

[13] H. Liu and G. Xu, "A subspace method for signature waveform estimation in synchronous CDMA systems," *IEEE Transactions on Communications*, vol. 44, pp. 1346–1354, 1996.

[14] H. D. Lüke, *Korrelationssignale.* Berlin, Heidelberg: Springer Verlag, 1992.

[15] L. B. Milstein, "Interference rejection techniques in spread spectrum communications," *Proceedings of the IEEE*, vol. 76, pp. 657–671, 1988.

[16] A. Goiser, "A fast multiuser detector for DS/CDMA-networks using partial sequence estimation," in *IEEE Proceedings Milcom'99*, 1999.

Chapter 27

CONVENTIONAL RAKE RECEIVERS

The first Rake receiver was presented by Price and Green in 1958 [1]. The Rake receiver technology depends on the information that is available and introduced in the algorithm to derive an estimate of the data sequence. The primary question in the design of a Rake receiver for a single link (point-to-point) is whether the channel parameters are available. We refer to a Rake receiver that has the channel information available, mainly on the basis of measurements, as Rake receiver; or, if it is necessary to distinguish between different Rake structures, as a conventional Rake receiver. Rake receivers that make no attempt to derive some information about the channel are referred to as blind Rake receivers. The major advantage of blind Rake receiver structures is that the overall performance of the system is enhanced because there is no need to insert sequences for channel estimation in the data stream or a pilot channel.

For future applications, we have a multiple access scheme (CDMA) in a multipath environment. Sophisticated Rake receiver algorithms have been devised to cope with both interference sources. It is difficult to distinguish the category to which such receivers belong. If we assume that the multipath interference is the basis, we have a Rake receiver with multiple-user detection capabilities. If we take the multiuser detector (MUD) as the basis, we have a MUD with Rake structure. Both fields of interest, Rake and MUD, have fused, and publications can be found in each of the research areas.

27.1 FUNCTIONALITY OF RAKE RECEIVERS

Before we start discussing the interactions of the Rake receiver in different environments, we discuss the theoretical behavior of the Rake receiver under ideal conditions. The results are drawn from a direct-sequence, spread-spectrum system in a multipath environment. The data rate is chosen so that the intersymbol interference is negligible. This choice is directly related to a data-bit duration, which is much longer than the delay spread of the channel. This means that the channel is frequency nonselective with respect to the data rate. The chip rate corresponds to the processing gain much more so than to the data rate, and the chip duration is much smaller than the delay spread, resulting in disturbances of the signal. Due to the spread-spectrum bandwidth, the channel is frequency selective and slowly fading with the capability to resolve the signal. The time resolution corresponds to the bandwidth in use.

The key issue is to adapt the two valued correlation functions of a single MWU receiver to a significant multipath component of the channel. Corresponding to the impulse response of the channel, the basic Rake structure consists of parallel MWU receivers to achieve uncorrelated multipath components. The derived uncorrelated multipath components are subsequently weighted and added to form the data-decision variable. In the derivation of the Rake receiver structure, we use the tapped delay line model described in previous chapters:

$$h(\tau;t) = \sum_{n=0}^{N_m-1} h_n(t)\delta(\tau - nT_c). \tag{27.1.1}$$

The path weights represent uncorrelated Gaussian processes due to the uncorrelated scatterers (WSSUS). Corresponding to the Gaussian assumption, the processes are also statistically independent, reminding us that the Rake receiver employs a kind of diversity reception. The method is based on the independence of the distinct multipath components of the channel. Even if one path suffers from a deep fade, other signal components may still carry enough signal energy so that the information can be recognized in the receiver.

The conventional Rake receiver can be realized in two ways. The first philosophy is to take away the channel-introduced multipath delays. To do that, the received signal passes a delay line before it is multiplied with the reference of the spreading sequence. The other philosophy is to mimic the channel's impulse response for the reference signal. This is implemented as follows: The replica of the spreading sequence passes the delay line and is subsequently multiplied with the received signal. Figure 27.1 shows the structure of the Rake receiver. The adjustable delays are matched to the inverse of the channel's impulse response.

27.2 RAKE BEHAVIOR FOR SINGLE-LINK CDMA COMMUNICATIONS

As we see from Figure 27.1, we need accurate estimates of the delayed components of the signal. The resolution in time and magnitude is important. The received delayed components of the signal correspond to the multipath intensity profile (MIP). The MIP changes with time, corresponding to the topology between transmitter and receiver. To design a Rake receiver we must introduce some terms. The most important figure is the number of significant delays of the signal that arrive at the receiving antenna, and that number depends on the time dispersion of the signal

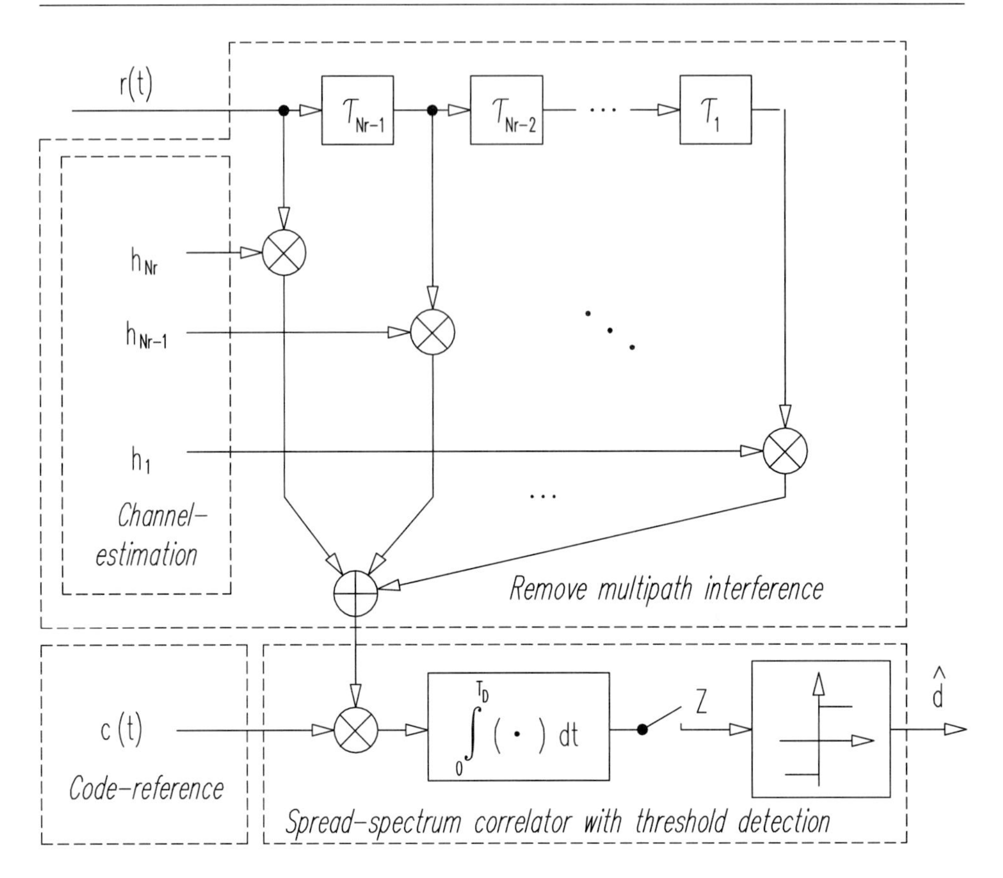

Figure 27.1. Rake receiver model

power.

In addition to the technical reasons for selecting the number of Rake fingers is the economic question. The solution is a compromise and is different for a mobile unit and for a base station, but the actual number has to be omitted from our discussion. For further discussions we assume N Rake fingers.

A pragmatic approach to estimate the time dispersion is to use the output signal of a passive correlator. The reference of the passive correlator is a PN sequence. The stored reference is transmitted through the channel and undergoes all the channel effects before it is received by the Rake antenna. The received and distorted, but unmodulated, orthogonal PN sequence is fed to the passive correlator to derive the delays in multiples of chip durations and the corresponding magnitudes. Simple signal processing selects the most powerful delays and adjusts the delays of the Rake fingers.

The estimated magnitudes are necessary for the maximum ratio combining. In

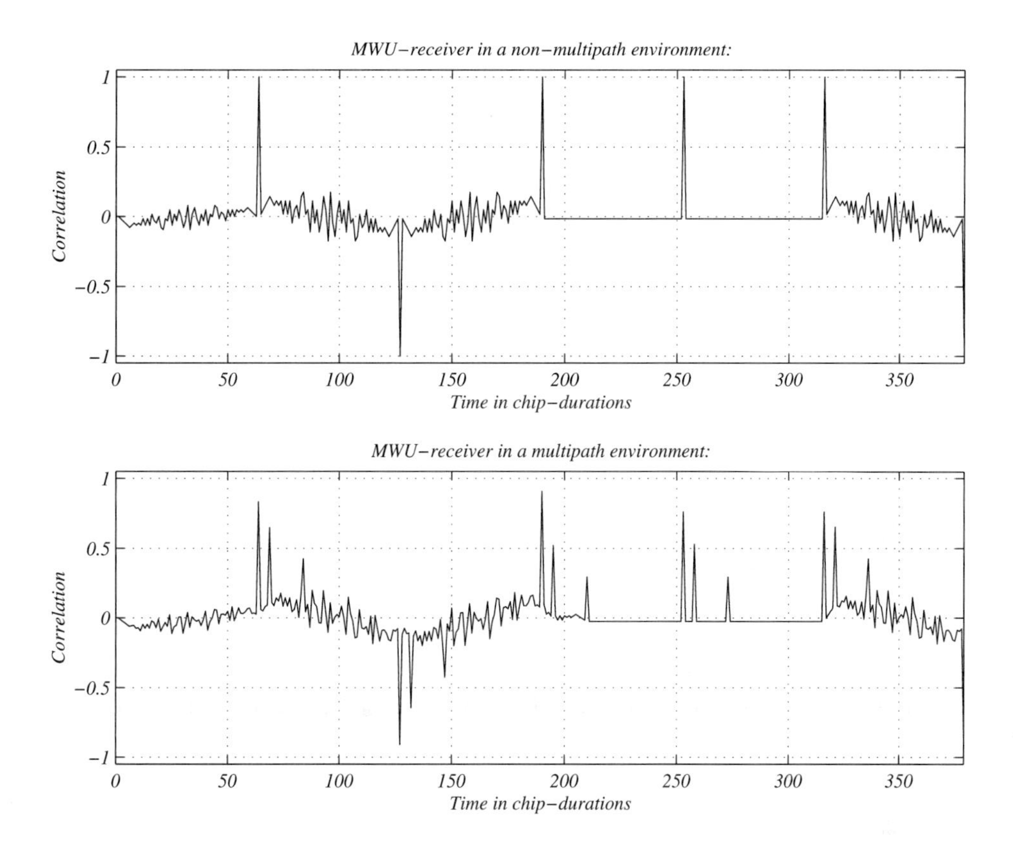

Figure 27.2. Channel behavior for a direct-sequence (PN) signal $\mathbf{d} = [6, 1]$ and data-vector [1,0,1,1,1,0]. Channel: $\tau = [5, 20] \cdot T_c$, $MIP = [0.6, 0.3, 0.1] \cdot$signal power

Figure 27.2 the upper trace shows a received and data-modulated, direct-sequence signal in a nondispersive channel. The data sequence is chosen such that a run of three positive data bits occurs. The mid-data bit shows the well-known two-value correlation peak. The lower trace shows the same data sequence for three signal paths. The mid-data bit shows the profile of the multipath channel.

The performance of the Rake receiver is significantly improved compared to non-spread-spectrum communication systems. The improvement is based on the enhancement of the signal-to-interference ratio. The Rake receiver collects more independent signal components from the multiple signal paths and outperforms all conventional receiver structures. Some quantitative results are given in Chapter 29.

Figure 27.3 visualizes the correlation behavior of the Rake receiver in a multipath channel and compares it to the MWU receiver. The curves are drawn under the assumption that the parameters of the channel are perfectly known and that the spread-spectrum signals of the received signal and the locally generated signal are

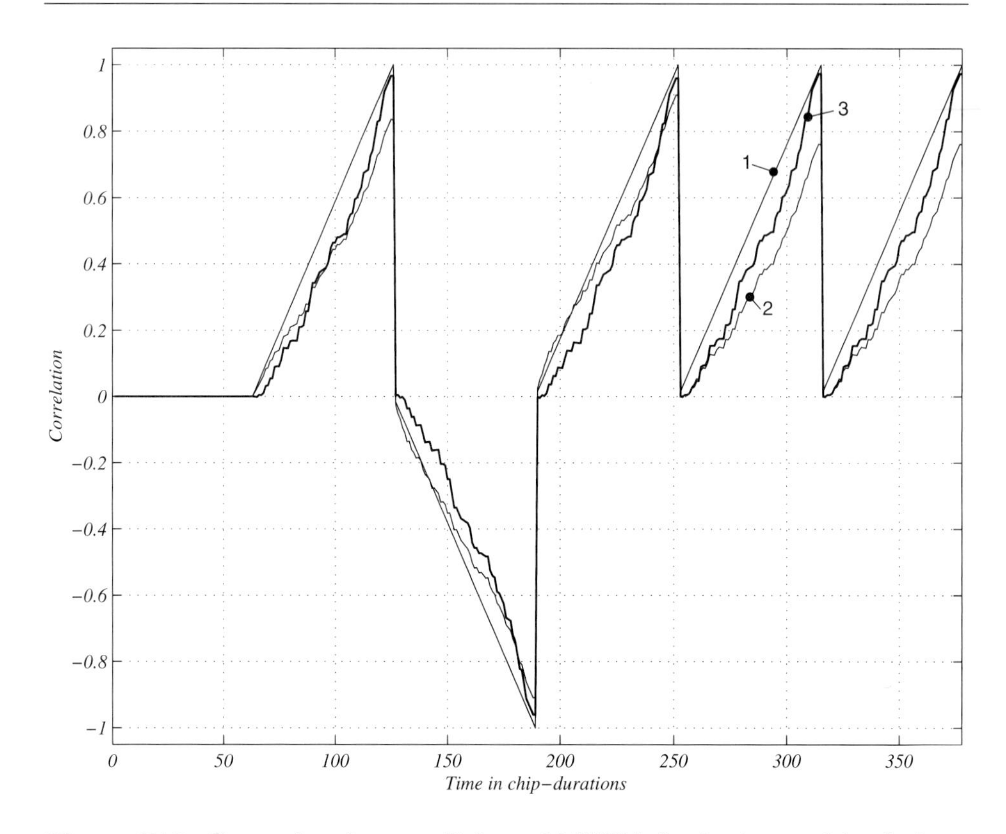

Figure 27.3. Comparison between Rake and MWU behavior in a multipath channel, without any other interference and assuming perfect channel estimates and perfect spread-spectrum synchronization. 1) Ideal correlation; 2) MWU receiver; 3) Rake receiver. The spread-spectrum signal is a direct-sequence signal [6,1] with a length of $L_s = 63$ chips. The initial load was the all one word. The multipath channel characteristic corresponds to that used in Figure 27.2.

perfectly aligned. From the data decision instants $k \cdot L_s$, we can recognize that the energy of the Rake receiver equals the energy of the signal. Additionally, we can recognize that the energy of the MWU receiver is always significantly less than the energy of the Rake receiver. The energy difference depends on the energy distributed to the multipath components.

The following situations influence the performance of the Rake receiver:

- The optimum situation occurs when the preselected number of Rake fingers exactly matches the number of uncorrelated signal paths. This presumes that the magnitudes of the delayed signal components are strong enough to be detected correctly. In that case, we collect all the time-dispersed signal energy in the correlation and data detection process, and the bit error rate is roughly the same as for an AWGN channel.

- The suboptimum situation is when we have more Rake fingers than we need to collect the signal energy. In that situation, we have wasted implementation complexity.

- If we have fewer Rake fingers than we need to collect the energy of the transmitted signal, we degrade the performance of the system, but we are still better than competing systems.

- A very nasty situation arises when we have enough Rake fingers but we cannot detect the delays because of the sensitivity of the detection circuitry. This case occurs when a small number of strong delays occur, many very weak signal delays occur, and the energy in the unresolved multipath components adds up to a significant number.

The performance is estimated by simulations. The assumptions corresponding to the spread-spectrum concept are listed in Table 27.1. The available Rake fingers are selected from the channel estimate in decaying order.

Parameter	Value
Chip duration	$2.5 \cdot 10^{-7}$
Processing gain	255 chips

Table 27.1. Spread-spectrum parameters

In Figure 27.4, we assume that the channel is perfectly estimated by the Rake receiver. A six-ray model for typical urban is assumed. The performance of the Rake receiver increases if more signal paths N_r up to the available channel paths N_m are collected to enhance the overall SIR. Curve 3 in Figure 27.4 corresponds to the perfect Rake ($N_r = N_m$) and is the baseline for the comparison.

If we add more Rake fingers than there are significant signal path components, they introduce more noise power in the decision process and the performance degrades. This situation is illustrated in Figure 27.5.

In the simulation results, the more realistic cases—the channel is not perfectly estimated—are also included. That situation is modeled with an uncertainty region corresponding to rough magnitude estimates and rough delay estimates.

27.3 RAKE BEHAVIOR FOR MULTIPLE-CELL CDMA NETWORKS

The power of the Rake concepts is achieved for multiple-cell CDMA networks. The power is the high degree of flexibility, especially during the handover process.

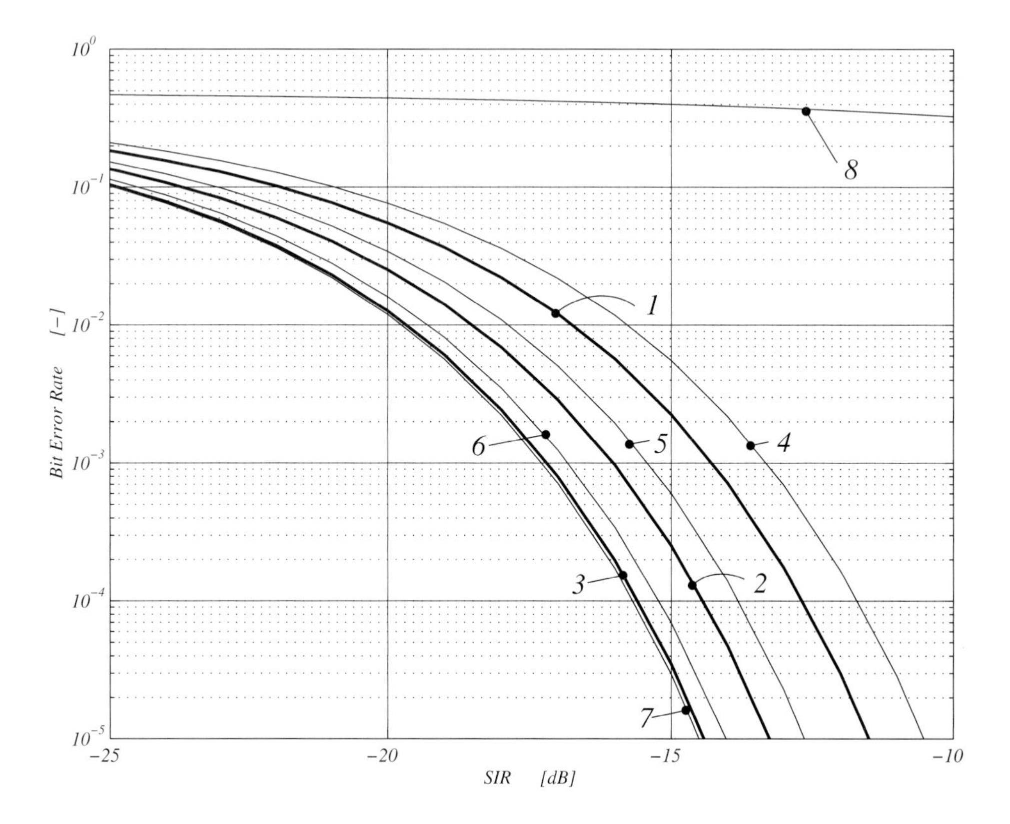

Figure 27.4. Simulated bit error rate for Rake receivers for perfect and imperfect estimated channel parameters (typical urban). Perfect channel estimates: 1) $N_r = 1$; 2) $N_r = 3$; 3) $N_r = N_m = 6$. Imperfect channel estimates: 4) $N_r = 1$; 5) $N_r = 3$; 6) $N_r = N_m = 6$. For comparison: 7) perfect spread-spectrum receiver; 8) non-spread-spectrum system.

Let's assume we have implemented the Rake algorithm in software and are able to reconfigure the Rake receiver structure.

During the handover the Rake receiver in the mobile unit receives its signal from two different cells. This form of diversity is paid for with a decrease in capacity. During handover we have two different multipath channels with different, significant, multipath signal components. The flexibility is based on the collection of all the significant signal paths regardless of which channel or base station they originated from.

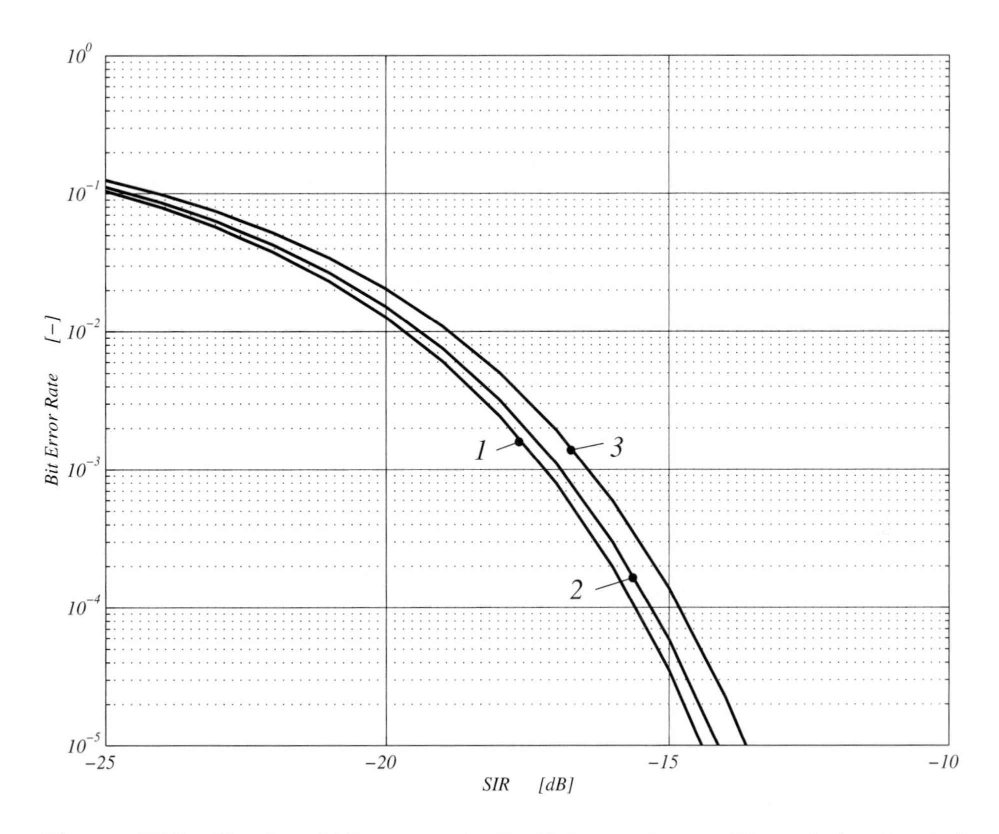

Figure 27.5. Simulated bit error rate for Rake receivers with perfect estimated channel parameters (typical urban). 1) $N_r = N_m = 6$; 2) $N_r = 8$; 3) $N_r = 10$.

Bibliography

[1] R. Price and P. E. Green, "A communication technique for multipath channels," *Proceedings of the IRE*, vol. 46, pp. 555–570, 1958.

Chapter 28

ADVANCED RAKE RECEIVER STRUCTURE

In this chapter we present an adaptive Rake receiver structure with low complexity that operates in a multipath and multiuser (CDMA) environment. The receiver knows the signatures of the active information channels but does not account for information about the different multipath channels, corresponding to the different locations of the users. For simplicity, we assume that each user occupies only one information channel. This assumption allows us to use the information channel and user interchangeably. This type of Rake receiver is called a *D-Rake* receiver [1]; it is so called because of its tendency (a) to mitigate interference as would a decorrelating multiuser receiver and (b) to combine signal components from the particular user. Its extension to antenna array CDMA systems is straightforward.

28.1 PRINCIPLE

The baseband form of the received signal in a CDMA environment is given in (28.1.1). This equation contains $a^{(k)}$ as the signature waveform corrupted by the channel, K as the number of active users, and $D_m^{(k)}$ as the mth data bit of the kth user. T_D represents the duration of a data bit, and $n(t)$ corresponds to additive white noise.

$$r(t) = \sum_{k=1}^{K} \sum_{m=-\infty}^{\infty} D_m^{(k)} \cdot a^{(k)} (t - m\, T_D) + n(t) \tag{28.1.1}$$

As stated in Chapter 27 we can recover the information under ideal conditions (Nyquist pulse-shaping filter, perfect orthogonal CDMA sequences, perfect spread-spectrum synchronization) by correlating the received signal with a local replica of the desired user's signature. For a CDMA system in a multipath environment, each user has its own multipath channel. All the multipath channels form a composite channel response, including the timing offsets, delays, and multipath reflections. We assume that the composite channel response is unknown to the receiver.

In (28.1.2) we use Ψ for the chip shape, L_s represents the length of the direct-sequence signal, $C_n^{(k)}$ is the nth chip within a data bit of the kth user, $h^{(k)}$ is the composite channels impulse response, and T_c represents the chip duration.

$$a^{(k)}(t) = \sum_{n=1}^{L_s} C_n^{(k)} \cdot \Psi\left(t - n\,T_c\right) \cdot h^{(k)}\left(t - n\,T_c\right) \qquad (28.1.2)$$

Further, we assume a quasi-synchronous CDMA system in which all the multipath signals arrive within several chip durations. Equation (28.1.3) is the defining equation for N_d in multiples of the chip duration, which must be significantly less than the data-bit duration.

$$N_d\,T_c = \tau_{delay} \ll T_D \qquad (28.1.3)$$

For a suitably designed CDMA system (28.1.3), we can change to a discrete model and rewrite (28.1.1) as (28.1.4). Equation (28.1.5) is the defining equation for $\mathbf{A}$ and $\mathbf{d}_n$.

$$\mathbf{r}_n = \sum_{k=1}^{K} D_m^{(k)} \cdot \mathbf{a}^{(k)} + \mathbf{n}_n = \left[r_{1,n}, \ldots, r_{L_s-N_d,n}\right]^T \qquad (28.1.4)$$

$$\mathbf{r}_n = \underbrace{\left[\mathbf{a}^{(1)}, \ldots, \mathbf{a}^{(K)}\right]}_{\stackrel{\text{def}}{=}\mathbf{A}} \cdot \underbrace{\begin{bmatrix} D_n^{(1)} \\ \vdots \\ D_n^{(K)} \end{bmatrix}}_{\stackrel{\text{def}}{=}\mathbf{d}_n} + \mathbf{n}_n \qquad (28.1.5)$$

Corresponding to the assumptions of a quasi-synchronous CDMA system and a delay spread significantly less than the data-bit duration, we are close to the intersymbol, interference-free condition for samples within a data-bit duration. In (28.1.5), $\mathbf{d}_n$ is the data-bit vector, $\mathbf{A}$ is the distorted signature waveforms, $\mathbf{a}^{(k)}$ is the distorted signature of the kth user, and $\mathbf{n}_n$ is the noise vector.

From (28.1.2) we can derive a discrete form for the distorted user signature k in (28.1.6) with $\mathbf{C}^{(k)}$ as code matrix and $h_l^{(k)}$ as in the chip-duration, sampled channel response.

$$\mathbf{a}^{(k)} = \begin{bmatrix} c_{N_d}^{(k)} & \cdots & c_1^{(k)} \\ c_{N_d+1}^{(k)} & \cdots & c_2^{(k)} \\ \vdots & \vdots & \vdots \\ c_{L_s}^{(k)} & \cdots & c_{N_d-L_s}^{(k)} \end{bmatrix} \begin{bmatrix} h_1^{(k)} \\ \vdots \\ h_{N_d}^{(k)} \end{bmatrix} \stackrel{\text{def}}{=} \mathbf{C}^{(k)}\,\mathbf{h}^{(k)} \qquad (28.1.6)$$

In (28.1.6) we expressed $\mathbf{a}^{(k)}$ as the product of the code vector delay matrix $\mathbf{C}^{(k)}$ and the sampled channel response. The code vector delay matrix $\mathbf{C}^{(k)}$ is known to the receiver, and the channel parameters are unknown. It is remarkable that the parameter set for the unknown channel is small in size. Reference [2] presented a subspace algorithm to estimate $\mathbf{h}^{(k)}$ and $\mathbf{a}^{(k)}$. When the $\mathbf{a}^{(k)}$ is calculated and no noise is present, we can recover the data-bit vector $\mathbf{d}_n$ perfectly. For the noiseless case, we have to calculate the pseudoinverse $\mathbf{A}^{\dagger}$ of $\mathbf{A}$ to decorrelate the received signal (28.1.7). In the AWGN case, with power σ_n^2, we have to use (28.1.8) for the MMSE (minimum mean-square error) receiver. The basis for the MMSE reception is the minimum output error (MOE) idea. In practice, $\mathbf{a}^{(k)}$ is estimated with the subspace method or training sequences.

$$\mathbf{d}_n = \mathbf{A}^{\dagger} \cdot \mathbf{r}_n \tag{28.1.7}$$

$$\mathbf{d}_n = \left(\mathbf{A}\,\mathbf{A}^H + \sigma_n^2\,\mathbf{I}\right)^{-1}\,\mathbf{a}^{(k)} \tag{28.1.8}$$

Now, we focus on less complex but also blind algorithms for real-time implementation that can recover $\mathbf{d}_n$ from $\mathbf{r}_n$ without any information about the channel. The subspace approach is computation intensive, so we assume that the decorrelation operation is also performed without knowledge of $\mathbf{a}^{(k)}$ and $\mathbf{h}^{(k)}$. If we assume, without loss of generality, that the first user is the desired user, we can rewrite (28.1.2) as (28.1.9). To keep the text readable, we omit the subscript (28.1.10).

$$\mathbf{r}_n = \mathbf{a}^{(1)}\,d_n^{(1)} + \mathbf{u}_n \qquad \text{with: } \mathbf{u}_n = \underbrace{\sum_{k=2}^{K} \mathbf{a}^{(k)}\,d_n^{(k)}}_{\text{MUI}} + \mathbf{n}_n \tag{28.1.9}$$

$$\begin{aligned} \mathbf{r}_n &= \mathbf{a}\,d_n + \mathbf{u}_n \\ &= \mathbf{C}\,\mathbf{h}\,d_n + \mathbf{u}_n \\ &= [\mathbf{c}_1, \ldots, \mathbf{c}_{N_d}] \cdot \begin{bmatrix} h_1^{(1)} \\ \vdots \\ h_{N_d}^{(1)} \end{bmatrix} \cdot d_n + \mathbf{u}_n \end{aligned} \tag{28.1.10}$$

We distinguish between two realizations. First, we introduce a *constant blind reception*, and then we present an *adaptive implementation*.

28.2 CONSTANT BLIND RECEPTION

In the constrained blind reception case, we recover $\mathbf{d}_n$ from $\mathbf{r}_n$ by using all the available signal energy and reducing the interference produced by the other active

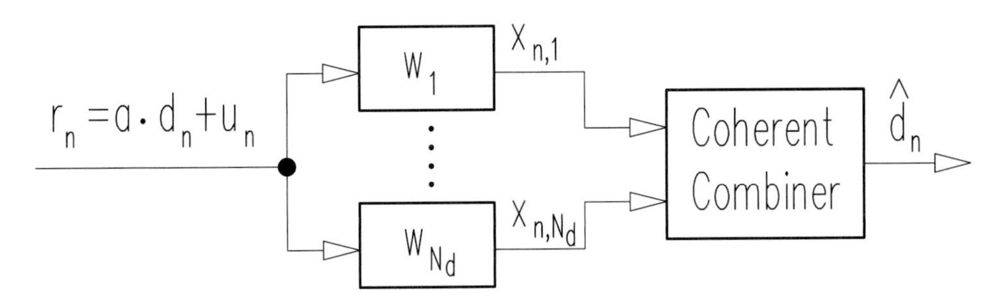

Figure 28.1. D-Rake receiver

users and the noise. This is precisely the idea of the conventional Rake receiver, where the multipath combining coefficients can be estimated from the covariance matrix of the despread signal without the use of training sequences. From (28.1.10) it is evident that the desired signal can be decomposed into a linear combination of signals projected onto a set of delayed code vectors. We exploit this structure to approach optimum estimates for $\mathbf{d}_n$ directly from $\mathbf{r}_n$ without raising the complexity significantly.

Following these ideas, we implement a receiver structure, as shown in Figure 28.1, called a D-Rake. The D-Rake consists of two stages. The first stage performs the adaptive weighting by means of a set of adaptive weight vectors $\mathbf{w}_l$. Stage two combines the different estimates coherently. To be more specific, we construct in the weighting stage a special set of weight vectors to extract the desired signal along individual code vectors while significantly reducing the multiuser interference. Then we constructively combine the extracted signals to achieve near-optimum signal estimation. The extraction operation in the first stage is mathematically represented in (28.2.1), where $\mathbf{1}_l$ is an all-zero vector except for a 1 on the lth position. The output of the lth finger in Figure 28.1 is given in (28.2.2).

$$\mathbf{C}^H \mathbf{w}_l = [0, \ldots, 1, \ldots, 0]^T \stackrel{\text{def}}{=} \mathbf{1}_l, \qquad 1 \le l \le N_d \tag{28.2.1}$$

$$\begin{aligned} x_{n,l} &= \mathbf{w}_l^H \mathbf{r}_n \\ &= \mathbf{w}_l^H (\mathbf{a}\, d_n + \mathbf{u}_n) \\ &= \underbrace{\mathbf{w}_l^H \mathbf{C}}_{\mathbf{1}_l^H} \mathbf{h}\, d_n + \mathbf{w}_l^H \mathbf{u}_n \\ &= h_l\, d_n + \mathbf{w}_l^H \mathbf{u}_n \\ &= h_l\, d_n + e_{n,l} \end{aligned} \tag{28.2.2}$$

In (28.2.2), $e_{n,l}$ denotes the effective interference and noise after filtering. The essence of (28.2.2) is that $\mathbf{w}_l$ emphasizes the desired signal along $\mathbf{c}_l$ and disregards all the other users $\mathbf{c}_x$ with $1 \le x \le N_d, x \ne l$. The N_d weighted vectors extract all signals at different delays. Corresponding to the N_d fingers, the signal power after the first stage is given in (28.2.3).

$$P = \sum_{l=1}^{N_d} |h_l|^2 = ||\mathbf{h}||^2 \tag{28.2.3}$$

The best signal estimate is related to the maximum of the ratio of signal power to the total interference power (MUI, AWGN) at the output of each finger and is defined in (28.2.4). For simplicity, we refer to Γ_l as signal-to-interference ratio without distinguishing between the interference sources.

$$\Gamma_l = \frac{|h_l|^2 \; E\{d_n \, d_n^*\}}{E\left\{e_{n,l} \; e_{n,l}^*\right\}} = \frac{|h_l|^2}{\mathbf{w}_l^H \; \mathbf{R}_{uu} \; \mathbf{w}_l} \tag{28.2.4}$$

Notice, that the output signal power is fixed to $|h_l|^2$. Maximizing the signal-to-interference ratio Γ_l is equivalent to minimizing $E\{e_{n,l} \; e_{n,l}^*\}$.

We derive the MOE receiver with the help of (28.2.5), where $\mathbf{R}_{rr}$ represents the autocovariance matrix of $\mathbf{r}_n$. Under the condition $\mathbf{w}^H\mathbf{a} = 1$, the MOE receiver is equivalent to the MMSE receiver.

$$\mathbf{w}_l = \min_{\mathbf{w}_l} \left\{\mathbf{w}_l^H \; \mathbf{R}_{rr} \; \mathbf{w}_l\right\} \qquad \rightarrow \mathbf{C}^H\mathbf{w}_l = \mathbf{1}_l \tag{28.2.5}$$

Since $\mathbf{a}$ is unknown, each finger in the D-Rake provides only the constrained MOE estimate of the signal. Nevertheless, $\mathbf{w}_l$ is capable of eliminating all the interference and perfectly recovering the signal in the absence of noise. A more compact form that uses vector notation is given in (28.2.6).

$$\begin{aligned} \mathbf{x}_n &\stackrel{\text{def}}{=} \begin{bmatrix} x_{n,1} \\ \vdots \\ x_{n,N_d} \end{bmatrix} \\ &= \mathbf{h}\, d_n + \begin{bmatrix} e_{n,1} \\ \vdots \\ e_{n,N_d} \end{bmatrix} \\ &\stackrel{\text{def}}{=} \mathbf{h}\, d_n + \mathbf{e}_n \end{aligned} \tag{28.2.6}$$

The second stage of the D-Rake coherently combines the outputs of all fingers from the weighting section to further enhance the signal-to-interference ratio.

In general, the optimum combining vector in (28.2.7) requires explicit knowledge of the channel coefficients. But, due to the interference decorrelation and noise reduction in the first stage, the total signal power in $\mathbf{x}_n$ becomes significantly higher than the power of $\mathbf{e}_n$, and we can make the approximation in (28.2.8). This enables us to approximate the optimum combining vector with the principal eigenvector of $\mathbf{R}_{xx}$, which we can obtain blindly by using standard decomposition techniques. The equation in (28.2.8) represents the trade-off between complexity and performance.

$$\mathbf{R}_{xx}^{-1}\,\mathbf{h} \tag{28.2.7}$$

$$\mathbf{R}_{xx} = \mathbf{h}\mathbf{h}^H + \mathbf{R}_{ee} \approx \mathbf{h}\mathbf{h}^H \tag{28.2.8}$$

The D-Rake receiver and the conventional Rake receiver are very similar in structure. To verify that similarity, we set $\mathbf{w}_l = \mathbf{c}$ and reduce the D-Rake to the conventional Rake receiver. However, the D-Rake receiver and the conventional Rake receiver differ significantly in their behavior. The D-Rake is by design a decorrelating receiver. This can easily be seen in (28.2.10) for the case that there is no noise and the parameters have converged to their optimum values and the condition in (28.2.9) is fulfilled.

$$K + N_d - 1 \leq L_s - N_d \tag{28.2.9}$$

$$\mathbf{w}_{l,\mathrm{opt}}^H \left[\mathbf{a}^{(1)}, \mathbf{a}^{(2)}, \ldots, \mathbf{a}^{(N_d)}\right] = \left[h_l^{(1)}, 0, 0, \ldots, 0\right] \tag{28.2.10}$$

If the condition in (28.2.9) holds, then (28.2.11) is fulfilled and the signal can be perfectly recovered. This recovery is impossible for the conventional Rake receiver.

$$x_{n,l} = \mathbf{w}_l^H\,\mathbf{r}_n = h_l\,d_n \tag{28.2.11}$$

In the presence of noise, the D-Rake receiver outperforms the conventional Rake receiver because each finger of the D-Rake behaves like a constraint MOE receiver.

28.3 ADAPTIVE RECEIVER

Now, we change from the constrained blind reception to an adaptive philosophy. This approach enables the D-Rake to accurately track a time-variant channel. The adaptive philosophy is based on estimates from incoming signals for the weight vector in the first stage and the coherent combining coefficients in the second stage.

The weight vectors are adapted through least mean square (LMS) algorithms. The power at the output of each finger is given in (28.3.1). The task is to search for a weight vector $\mathbf{w}_l$ that minimizes P_{MOE} with respect to $\mathbf{C}^H \mathbf{w}_l = \mathbf{1}_l$. The gradient of the cost function is given in (28.3.2). We approximate the autocorrelation matrix with the instantaneous received signal vector $\mathbf{r}_n$ and derive a more user friendly equation for the gradient of the cost function (28.3.3).

$$P_{\mathrm{MOE}} = \mathbf{w}_l^H \, \mathbf{R}_{rr} \, \mathbf{w}_l \tag{28.3.1}$$

$$\nabla_{\mathbf{w}_l} \left(P_{\mathrm{MOE}}\right) = 2 \, \mathbf{R}_{rr} \, \mathbf{w}_l \qquad 1 \le l \le N_d \tag{28.3.2}$$

$$\nabla_{\mathbf{w}_l} \left(P_{\mathrm{MOE}}\right) \approx 2 \, \mathbf{r} \, \mathbf{r}^H \, \mathbf{w}_l \qquad 1 \le l \le N_d \tag{28.3.3}$$

To speed the search, we restrict the search in the constrained subspace. That is, we evaluate all the projections of the gradient of the output energy onto the subspace orthogonal to the code matrix $\mathbf{C}$. In (28.3.4) we define the projection matrix $\mathbf{P}_{\mathbf{C}}^{\perp}$ and arrive at the recursion rule in (28.3.5).

$$\mathbf{P}_{\mathbf{C}}^{\perp} = \mathbf{I} - \mathbf{C}^H \, \left(\mathbf{C} \, \mathbf{C}^H\right)^{-1} \, \mathbf{C} \tag{28.3.4}$$

$$\mathbf{w}_{l,n+1} = \mathbf{w}_{l,n} - \mu \, \mathbf{P}_{\mathbf{C}}^{\perp} \, \mathbf{r}_n \, \mathbf{r}_n^H \, \mathbf{w}_{l,n} = \left(\mathbf{I} - \mu \, \mathbf{P}_{\mathbf{C}}^{\perp} \, \mathbf{r}_n \, \mathbf{r}_n^H\right) \, \mathbf{w}_{l,n} \tag{28.3.5}$$

$$\mathbf{w}_{l,0} = \mathbf{C}^{\dagger} \, \mathbf{1}_l \tag{28.3.6}$$

If we use $\mathbf{w}_{l,0}$ in (28.3.6) as the initial weight vector with $\mathbf{C}^{\dagger}$ as pseudoinverse of $\mathbf{C}$, we recognize that the constraint condition is fulfilled in each iteration step. The factor μ in (28.3.5) has to be chosen as a trade-off between convergence of the algorithm and its steady state error.

For coherent combining in the second stage, we need an estimate of the principal vector of $\mathbf{R}_{xx}$. We can derive it with standard adaptive eigendecomposition techniques. The computational effort is on the order of $O(N_d^3)$ due to the condition $N_d \ll L$, which is negligible compared to the computational effort of the first stage.

The performance of adaptive receiver designs is addressed to misadjustment, convergence, and steady state behavior of the algorithm and signal-to-interference ratio calculations.

The estimate of the performance bound is based on the assumption that the estimation error in the first stage dominates compared to the estimation error in the second stage. We derive the optimum weight vector for the lth finger of the D-Rake using Lagrange multipliers as shown in (28.3.7). It can be easily seen that the weight vector in each finger of the D-Rake converges to the optimum value as the number of observations increases.

$$\mathbf{w}_{l,\text{opt}} = \mathbf{R}_{rr}^{-1}\,\mathbf{C}^H\left(\mathbf{C}\,\mathbf{R}_{rr}^{-1}\,\mathbf{C}^H\right)^{-1}\mathbf{1}_l \tag{28.3.7}$$

The outputs of the first stage of the D-Rake are combined coherently, as shown in (28.3.8). Because both stages can be modeled with a unique weight vector, shown in (28.3.9), we derive the best performance that can be achieved, which corresponds to the maximum signal-to-interference ratio in (28.3.10).

$$\mathbf{x} = \sum_{l=1}^{N_d} h_l^*\,\mathbf{w}_{l,\text{opt}}^H\,\mathbf{r}_n \tag{28.3.8}$$

$$\mathbf{w}_{\text{DRake}} = \sum_{l=1}^{N_d} h_l\,\mathbf{w}_{l,\text{opt}} = \mathbf{R}_{rr}^{-1}\,\mathbf{C}\left(\mathbf{C}^H\,\mathbf{R}_{rr}^{-1}\,\mathbf{C}\right)^{-1}\mathbf{h} \tag{28.3.9}$$

$$\begin{aligned}\Gamma_{\text{DRake}} &= \frac{E\left\{\mathbf{w}_{\text{DRake}}^H\,\mathbf{a}\,d_n\,d_n^*\,\mathbf{a}^H\,\mathbf{w}_{\text{DRake}}\right\}}{E\left\{\mathbf{w}_{\text{DRake}}^H\,\mathbf{u}\,\mathbf{u}^H\,\mathbf{w}_{\text{DRake}}\right\}}\\ &= \frac{||\mathbf{h}||^4}{\mathbf{w}_{\text{DRake}}^H\,\mathbf{R}_{uu}\,\mathbf{w}_{\text{DRake}}}\end{aligned} \tag{28.3.10}$$

For comparison, we calculate the signal-to-interference ratio for the MMSE receiver and the single-finger (or single-arm, SA) receiver. Equations (28.3.11) and (28.3.12) show the corresponding weighting vector for the MMSE receiver and the SA receiver, respectively. With the weighting vectors, we derive the signal-to-interference ratios for the MMSE receiver and the SA receiver in (28.3.13) and (28.3.14), respectively.

$$\mathbf{w}_{\text{MMSE}} = \mathbf{R}_{rr}^{-1}\,\mathbf{a} = \mathbf{R}_{rr}^{-1}\,\mathbf{C}\,\mathbf{h} \tag{28.3.11}$$

$$\mathbf{w}_{\text{SA}} = \mathbf{w}_{1,\text{opt}} \tag{28.3.12}$$

$$\Gamma_{\text{MMSE}} = \frac{|\mathbf{w}_{\text{MMSE}}\,\mathbf{a}|^2}{\mathbf{w}_{\text{MMSE}}^H\,\mathbf{R}_{uu}\,\mathbf{w}_{\text{MMSE}}} \tag{28.3.13}$$

$$\Gamma_{\text{SA}} = \frac{|h_1|^2}{\mathbf{w}_{1,\text{opt}}\,\mathbf{R}_{uu}\,\mathbf{w}_{1,\text{opt}}} \tag{28.3.14}$$

$$\Gamma_{\text{MMSE}} \geq \Gamma_{\text{DRake}} \geq \Gamma_{\text{SA}} \tag{28.3.15}$$

The signal-to-interference ratios for the D-Rake, MMSE, and SA receivers in (28.3.10), (28.3.13), and (28.3.14), respectively, provide an upper bound of the mean-square error (MSE).

When the signal-to-interference ratio is high, the relation in (28.3.15) holds. Especially in the noise-free case, the investigated receivers show zero-forcing behavior if their values have converged to the optimum values and they have the same asymptotic performance. If noise is present, the signal-to-interference ratio for the D-Rake receiver and the SA receiver is upper-bounded by the signal-to-interference ratio of the complex MMSE receiver. The principal vector of $\mathbf{R}_{xx}$ is the optimum combining vector if the signal-to-interference ratio is reasonably high.

28.4 PERFORMANCE

A figure of merit in adaptive filtering is the steady state behavior represented by the excess mean-square error of the adaptive algorithm. We derive the steady state signal-to-interference ratio for the D-Rake receiver by following the standard steps for adaptive LMS algorithms [3], [4]. We define an error $\Delta x_n = x_n - x_{\text{opt}}$ as the difference from the actual x_n value to the optimum value x_{opt}. Especially for the weighting vector in the lth finger of the D-Rake receiver at time instant n, we derive (28.4.1). Equation (28.4.2) is derived from (28.3.5) if we subtract $\mathbf{w}_{l,\text{opt}}$ on both sides.

$$\Delta\mathbf{w}_{l,n} = \mathbf{w}_{l,n} - \mathbf{w}_{l,\text{opt}} \tag{28.4.1}$$

$$\Delta\mathbf{w}_{l,n+1} = \left[\mathbf{I} - \mu\,\mathbf{P}_{\mathbf{C}}^{\perp}\,\mathbf{r}_n\,\mathbf{r}_n^H\right]\,\Delta\mathbf{w}_{l,n} - \mu\,\mathbf{P}_{\mathbf{C}}^{\perp}\,\mathbf{r}_n\,\mathbf{r}_n^H\,\mathbf{w}_{l,\text{opt}} \tag{28.4.2}$$

After some manipulations, following [4], with the extension from a projection vector to a orthogonal projection matrix $\mathbf{P}_{\mathbf{C}}^{\perp}$, we obtain the excess MSE in (28.4.3) with $\epsilon_{l,\min}$ from (28.4.4) as the minimum output energy. While the output signal power is fixed at $|h_l|^2$, the constrained MOE at the output of the lth finger is given in (28.4.5).

$$J_{l,\text{ex}} \approx \epsilon_{l,\min} \frac{\frac{\mu}{2}\cdot\text{tr}\left(\mathbf{P}_{\mathbf{C}}^{\perp}\,\mathbf{R}_{rr}\right)}{1 - \frac{\mu}{2}\cdot\text{tr}\left(\mathbf{P}_{\mathbf{C}}^{\perp}\,\mathbf{R}_{rr}\right)} \tag{28.4.3}$$

$$\epsilon_{l,\min} = \mathbf{w}_{l,\text{opt}}^H \, \mathbf{R}_{rr} \, \mathbf{w}_{l,\text{opt}} = \mathbf{1}_l^H \left(\mathbf{C}^H \, \mathbf{R}_{rr}^{-1} \, \mathbf{C} \right)^{-1} \mathbf{1}_l \tag{28.4.4}$$

$$J_{l,\min} = \epsilon_{l,\min} - |h_l|^2 \tag{28.4.5}$$

With these results we can calculate the MSE and the signal-to-interference ratio for the steady state condition, given in (28.4.6) and (28.4.7), respectively.

$$\text{MSE}_{l,\text{ss}} = J_{l,\min} + J_{l,\text{ex}} \tag{28.4.6}$$

$$\Gamma_{l,\text{ss}} = \frac{|h_l|^2}{J_{l,\min} + J_{l,\text{ex}}} \tag{28.4.7}$$

We obtain the steady state mean-square error at the output of the D-Rake receiver for the assumption that the power of $\mathbf{e}_n$ in $\mathbf{x}_n$ (see (28.2.6)) after the first stage output is small compared to the signal strength $\mathbf{d}_n$. In that case, the combining vector can be approximated with the channel response $\mathbf{h}$. With this idea in mind we can rewrite the output of the lth finger in (28.2.8), in which $e_{l,n}$ denotes the overall error.

$$x_{l,n} = h_l \, d_{1,n} + e_{l,n} \qquad\qquad 1 \leq l \leq N_d \tag{28.4.8}$$

With (28.2.10) we derive the decision variable after the coherent combiner in (28.2.9).

$$E\left\{ e_{l,\infty} \, e_{l,\infty}^* \right\} = J_{l,\min} + J_{l,\text{ex}} = \text{MSE}_{l,\text{ss}} \tag{28.4.9}$$

$$\begin{aligned} Z_n &= \sum_{l=1}^{N_d} h_{l-1}^2 \, d_n + \sum_{l=1}^{N_d} h_l \, e_{l,n} \\ &= ||\mathbf{h}||^2 \, d_n + \sum_{l=1}^{N_d} h_l \, e_{l,n} \end{aligned} \tag{28.4.10}$$

To keep the calculations of the steady state signal-to-interference ratio of the D-Rake simple, we assume that the outputs of the different finger are statistically independent.

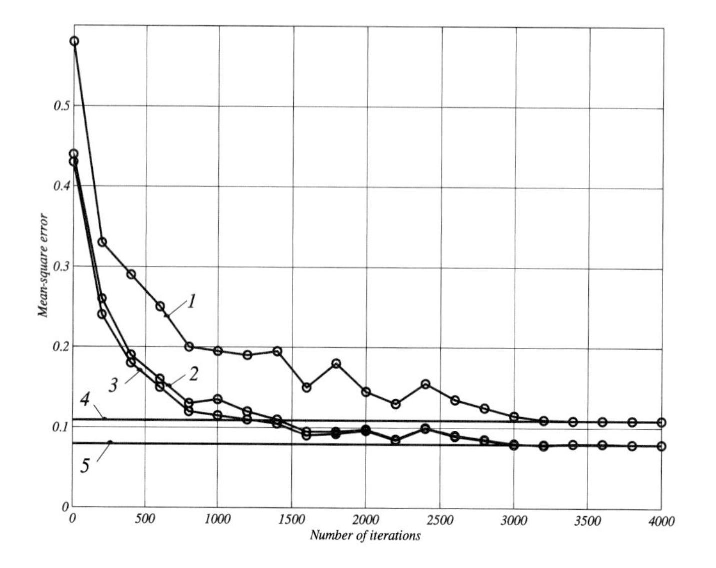

Figure 28.2. Comparison of the mean-square error for the SA receiver (1), D-Rake receiver (2), and the MOE receiver (3). The asymptotic behavior of the SA receiver is shown by line (4), and the asymptotic behavior for the D-Rake is shown by line (5). A total of $K = 10$ active users was chosen; the sequence length is $L_s = 32$. The delay spread in multiples of the chip duration was fixed to $N_d = 3$, and the number of multipath components for each user varies between $1 \leq N_m \leq 10$.

$$\Gamma_{\text{DRake,ss}} = \frac{||\mathbf{h}||^4}{\sum_{l=1}^{N_d} |h_l|^2 \; (J_{l,\min} + J_{l,\text{ex}})} \tag{28.4.11}$$

The ultimate figure in the design process of the D-Rake receiver is the signal-to-interference ratio in (28.2.11). Figure 28.2 shows the mean-square error for the D-Rake receiver and compares it to the SA receiver and the MOE receiver. A total of $K = 10$ active user was chosen, and the sequence length is $L_s = 32$. The delay spread in multiples of the chip duration was fixed to $N_d = 3$, and the number of multipath components for each user varies between $1 \leq N_m \leq 10$. The asymptotic behavior of the D-Rake receiver is the same as the asymptotic behavior of the MOE receiver. The convergent behavior is roughly the same for all investigated receivers. Some slight differences are noticeable in the asymptotic behavior with the SA receiver.

Figure 28.3 compares the mean-square error of the D-Rake receiver, the conventional Rake receiver, and the MOE receiver. The assumptions are the same as for

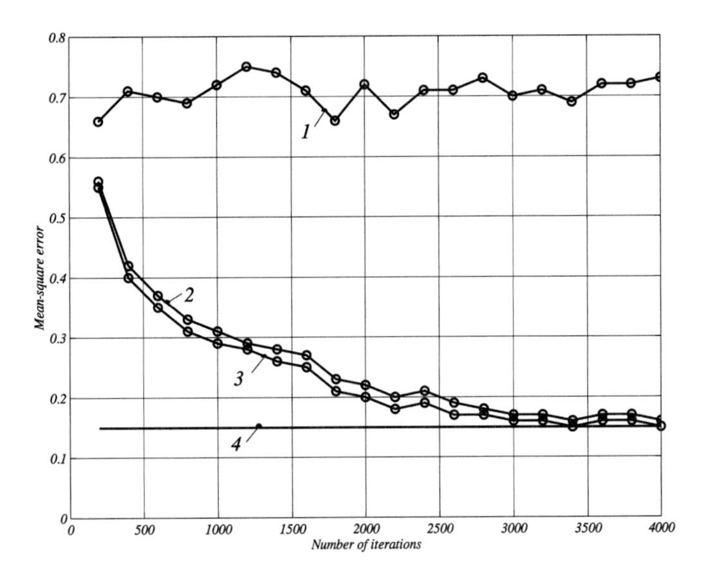

Figure 28.3. Comparison of the mean-square error for the conventional Rake receiver (1), D-Rake receiver (2), and the MOE receiver (3). The asymptotic behavior of the D-Rake is shown by line (4). A total of $K = 15$ active users was chosen, and the sequence length is $L_s = 32$. The delay spread in multiples of the chip duration was fixed to $N_d = 3$, and the number of multipath components for each user varies between $1 \leq N_m \leq 10$.

Figure 28.2, except the number of active users is increased to $K = 15$. This figure points out the potential difference between the conventional Rake receiver and its complex counterparts. The adaptive Rake receivers show their superiority over the conventional Rake. On the other hand, the D-Rake is still close to the MOE receiver. This is surprising because the D-Rake does not have the information about the channel that the MOE has.

The steady state mean-square error after coherent combining for the D-Rake receiver is illustrated in Figure 28.4. The D-Rake behaves as expected. As the number of active users increases, the steady state mean-square error increases linearly.

The conclusion from Figure 28.5 is that the D-Rake has excellent near-far resistance as is inherent in blind multiuser detectors [4].

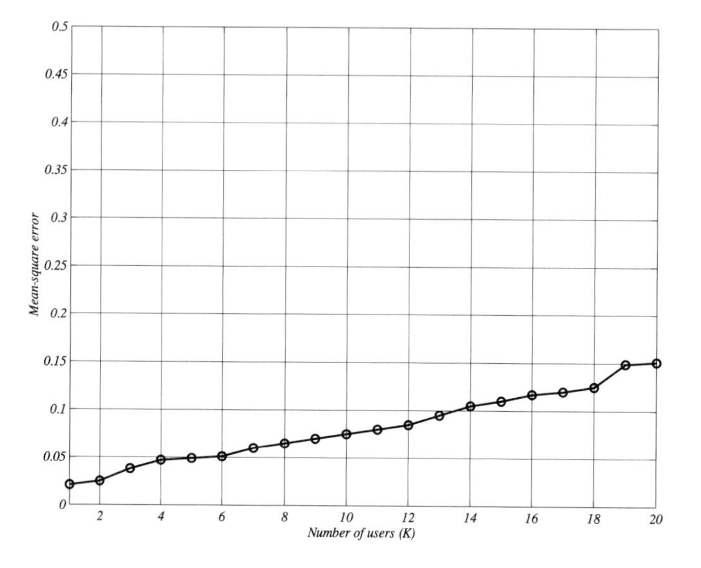

Figure 28.4. Steady state mean-square error after coherent combining for the D-Rake receiver. The sequence length is $L_s = 32$. The delay spread in multiples of the chip duration was fixed to $N_d = 3$, and the number of multipath components for each user varies between $1 \leq N_m \leq 10$.

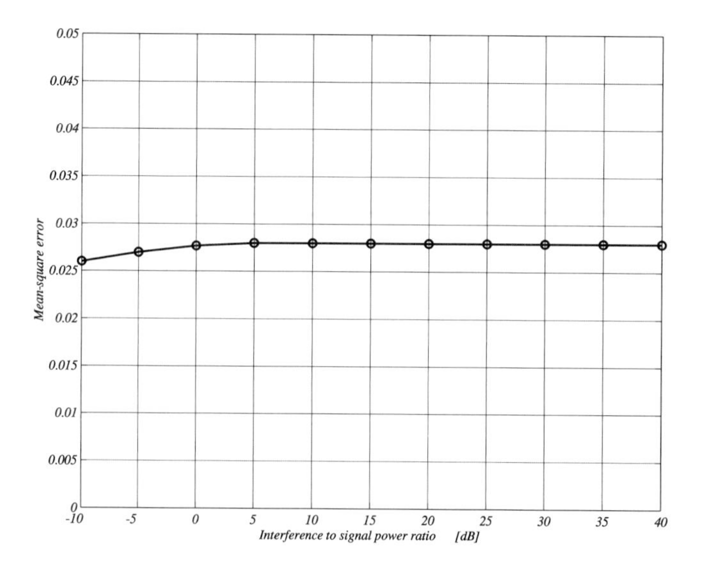

Figure 28.5. Mean-square error versus interference-to-signal ratios for the D-Rake receiver

Bibliography

[1] H. Liu and K. Li, "A decorrelating Rake receiver for CDMA communications over frequency-selective fading channels," *IEEE Transactions on Communications*, vol. 47, pp. 1036–1045, 1999.

[2] H. Liu and G. Xu, "A subspace method for signature waveform estimation in synchronous CDMA systems," *IEEE Transactions on Communications*, vol. 44, pp. 1346–1354, 1996.

[3] S. Haykin, *Adaptive Filter Theory.* Upper Saddle River, NJ: Prentice Hall, 1996.

[4] M. L. Honig, U. Madhow, and S. Verdu, "Blind adaptive multiuser detection," *IEEE Transactions on Information Theory*, vol. 41, pp. 944–960, 1995.

Chapter 29

IMPACT OF SPREADING BANDWIDTH AND SELECTION DIVERSITY ORDER ON RAKE RECEPTION

In this chapter, we develop an analytical framework to quantify the effects of the spreading bandwidth (BW) on spread spectrum systems operating in dense multipath environments in terms of the receiver performance, the receiver complexity, and the multipath channel parameters. The focus of this chapter is to characterize the performance of a Rake receiver tracking the L strongest multipath components in wide-sense stationary uncorrelated scattering (WSSUS) Gaussian channels with frequency-selective fading. Closed form expressions for the mean and the variance of the combiner output signal-to-noise ratio (SNR) as well as analytical symbol error probability (SEP) expressions of the Rake receiver are derived in terms of the number of combined paths, the spreading BW, and the multipath spread of the channel. The proposed problem is made analytically tractable by transforming the physical Rake paths, which are correlated when ordered, into the domain of a "virtual Rake" receiver with independent virtual paths. This results in a simple derivation of the mean, the variance, and the SEP for a given spreading BW and an *arbitrary* number of combined paths.

29.1 INTRODUCTION

Multiple-access systems based on spread spectrum (SS) signaling properties are suitable for dealing with fading multipath channels [1, 2, 3, 4]. These SS multiple access techniques have recently seen significant deployment in wireless communication systems, and they have also been proposed for third-generation wireless access

[5]. One benefit of SS systems is that with a sufficiently wide transmission BW, it is possible to resolve the closely spaced multipath components encountered in the channel. Alternatively, systems using narrowband transmissions perceive most of the closely spaced multipath components as a single faded signal.

29.1.1 The Rake receiver

The "optimum" receiver for detecting signals in a multipath environment, when the observation noise and the undesired interference are modeled as additive white Gaussian noise (AWGN), is a matched filter or a correlation receiver where the reference template signal is the response of the transmission medium to a transmitted signal (composite of the channel *and* the transmitted signal).[1] In practice, this definition leads to a Rake receiver, which is based on optimality theory tempered by some heuristic ideas. Rake receivers resolve the components of a received signal (arriving at different times) and combine them to provide diversity. Discussions on classical Rake receivers can be found in [6, 7, 8, 9] as well as Chapter 27.

One version of the Rake receiver consists of multiple correlators (fingers) where each of the fingers can detect/extract the signal from one of the multipath components created by the channel. The outputs of the fingers are combined to yield the benefits of Rake diversity. The equivalent matched filter version of the receiver involves a matched front-end processor (MFEP) (matched *only* to the transmitted signature waveform) followed by a tapped delay line and a combiner. Multipath components with delays greater than the chip duration (approximately equal to the inverse of the spreading BW) apart are resolved by the MFEP, which is synchronized with the initial path of the received signal. The MFEP output is passed through a tapped delay line filter with $N_\mathrm{r} = T_\mathrm{d}/T_\mathrm{c}$ taps, where T_d is the maximum excess delay from the first arriving path and T_c is the chip duration. The output of the taps provides N_r diversity paths, all of which must be combined for the best possible performance.

We introduce the term *all-Rake (ARake)* receiver to describe the receiver with unlimited resources (taps or correlators) and instant adaptability, so that it can, in principle, combine *all* of the resolved multipath components. For a dense multipath channel with a fixed T_d, the number of resolvable multipath components N_r increases with the spreading BW. However, the number of multipath components that can be utilized in a typical Rake combiner is limited by power consumption issues, design complexity, and the channel estimation [10, 11]. ARake receivers in conjunction with multiuser detection are discussed in Chapter 28.

29.1.2 The selective Rake receiver

Complexity and performance issues have motivated studies of multipath combining receivers that process only a *subset* of the available N_r resolved multipath compo-

[1]Strictly speaking, this "one shot" approach is optimal for detecting the single signal (one symbol) or equivalently intersymbol interference (ISI) free channel. However, the optimality still holds reasonably well as long as ISI is moderate.

nents but achieve better performance than does a single-path (SP) receiver (also called MWU in Chapter 27). This reduced-complexity multipath combining system selects the L best paths (from N_r available diversity paths) and then combines the selected subset of multipath components according to a chosen criterion (see also Chapter 27). We refer to such receivers as *selective Rake (SRake)* receivers [12, 13, 14, 15, 16, 17, 18]. Selecting the "best" paths can be accomplished by selecting the multipath components with the largest signal-to-noise ratio or signal-plus-noise [19, 20]. The selected subset of multipath components can then be combined by equal gain combining or maximal-ratio combining (MRC) [9, 21, 22]. This chapter considers the SRake receiver that selects the L best paths with the largest SNR (from N_r available diversity paths) and combines them by MRC [9, 21].

For a given transmission BW and for a typical power delay profile (PDP), fundamental questions related to the SRake receiver design are:

1. How does the average SNR at the combiner output increase with an increasing number of fingers?

2. What is the behavior of the SNR fluctuation at the combiner output in a multipath fading environment as a function of the number of fingers?

3. What is the impact of the number of combined paths on the SEP performance?

Answers to these questions can provide insights and suggestions regarding the number of fingers to be used in the design of a Rake receiver and the performance improvement versus complexity trade-off. From the receiver complexity point of view, the question can be rephrased: For a fixed complexity receiver operating in an environment with a given PDP, what is the optimal spreading BW to be used?

29.1.3 Previous work

The design of receivers for code division multiple access (CDMA) systems and, in particular, the choice of the number of combined paths is often based on simulations and empirical knowledge [5, 23, 24, 25, 26, 27]. This is partially due to an incomplete theoretical understanding. Results of simulation studies which address the problem regarding the number of Rake fingers vary. In [23] and [24], it is reported that the performance for a 4-finger Rake receiver is about 2 dB better than for a 2-finger Rake receiver at a bit error rate of 10^{-3}. In [26], it is reported that an increase in the number of fingers results in a decrease in performance when turbo coding is considered. Some studies argue that the larger the number of active fingers, the better the performance, and arbitrarily choose a 6-finger Rake implementation [27], whereas some papers simply suggest that an "adequate" number of fingers needs to be used.

Recently, some theoretical and simulation-based efforts have been made to investigate the effects of multipath fading [28, 29, 30, 31, 32]. The effects of Rayleigh fading on direct sequence (DS) CDMA signals with arbitrary pulse shape is quantified in [28]. This analysis was carried out in the frequency domain, and the relationship between the signal power variation and the spreading BW was derived. A similar investigation was made by simulation of the reception of DS-SS signals by means of the geometric propagation model [29]. The results of [28] are evaluated for specific pulse shapes in [30, 31]. In [33, 34], the channel is modeled to consist of several groups of paths, where one dominant path per group is combined in the receiver.

Typical studies on the performance of DS-CDMA systems as a function of the number of Rake fingers assume an ARake model ($L = N_r$) (see, for example, [35, 36]). However, as pointed out in [36], a more realistic model is the Rake receiver tracking the L strongest of N_r paths. In fact, quasi-analytical/experimental analysis of the ultrawide BW SRake receiver (with large N_r) in [37] shows that an acceptable level of SEP performance can be achieved even with $L = 1$ and $L = 2$. This finding indicates that a high diversity order can be achieved even with a selection diversity (SD) or SP receiver. If one were to use an ARake model to analyze such a system, the results would be misleading in that it would suggest that more fingers were required. In [38], only $L = 2$ and $L = 3$ out of N_r paths were analyzed. In [32] the dependence of the signal power variance on the receiver processing in terms of parameters such as chip rate, processing gain, and number of multipath components tracked is investigated for $L = 1$, 2, 4, and 8 through simulations. The bit error rate performance of an SRake receiver was analyzed in [39] for binary differential phase shift-keying modulation.[2] A closely related problem of hybrid selection/maximal-ratio combining (H-S/MRC) receivers in a more general diversity setting was considered in [40, 41, 42, 43, 44, 45, 46, 47, 48, 49].

29.1.4 Analytical framework for selective Rake combining

The principal contributions of this chapter are in discovering a methodology and deriving expressions for the mean and the variance of the combiner output SNR as well as SEP expressions of the SRake receiver for arbitrary L and N_r (i.e., arbitrary spreading BW). Based on our previous work [13, 14, 15, 16, 17, 18], we present an analytical framework to quantify the effects of spreading BW on SS systems operating in dense multipath environments in terms of the receiver complexity and multipath channel parameters. We assume that instantaneous estimation of all possible multipaths is feasible, such as with slow fading. However, SRake combining also offers improvement in fast-fading conditions, and our results serve as bounds on the performance when ideal channel estimates are not available. The mean SNR at the output of the SRake combiner is derived, since it is a commonly accepted performance measure. In order to assess the effectiveness of the SRake receiver in the presence of multipath, the variance of the combiner output SNR and SEP expressions of the SRake receiver are also derived. The proposed problem is made

[2] In [39], the probability density function (pdf) of the sum of the signals with the L strongest path SNR's was obtained as a convolution of the pdfs of the strongest, the second strongest, ..., and the L-th strongest paths. We remark that, in general, the pdf of the sum of the random variables is the convolution of the individual pdfs only if the random variables are independent.

analytically tractable by transformation of the physical Rake paths into the domain of a "virtual Rake" receiver and results in a simple derivation of the mean, the variance, and the SEP for a given spreading BW and an *arbitrary* number of combined paths. The well-known results for the SP receiver and the ARake receiver can be obtained as special cases of the SRake receiver results.

We first derive general expressions and then focus on the special case of constant PDP over an interval. Several researchers have previously considered this type of PDP [50, 51, 52, 36] to study various aspects of DS-CDMA systems. Propagation measurements in urban and suburban environments [53, 54, 55, 56, 57, 58] and mountainous terrain [59] exhibit characteristics supporting such a PDP since they show channels with energy spread over a continuum of arrival times. Thus, the use of this PDP serves as a basic model for analyzing the performance of Rake receivers operating in dense multipath environments. We plot the obtained results and observe the property of diminishing returns with an increase in the number of combined paths.

Second generation cellular systems that are based on CDMA (IS-95) operate at a chip rate of 1.2288 Mchips/s, and the third-generation (e.g., IMT-2000) systems will most probably operate at 3.84 Mchips/s. To support higher bit rates, larger spreading BWs have been proposed [5]. As the third generation activities progress, the designers of receivers will face the question of how many fingers should be included in Rake receivers or how many fingers should be active at any one time. The results of this chapter provide part of the necessary inputs for making those decisions.

29.2 SYSTEM MODEL

The basis for our analysis begins with a description of the signal and channel models as well as the MFEP encountered in Rake receivers. Before we delve into the analysis of the SRake receiver, we review the statistical properties of the MFEP output. This review establishes the relationship between the MFEP output, the PDP, the sequence correlation properties, and the spreading BW.

29.2.1 Signal and channel models

In this section, we introduce the notation for the signal and channel models. The transmitted signal is modeled as[3]

$$\tilde{s}(t) = \Re\left\{s(t)e^{j2\pi f_c t}\right\}, \tag{29.2.1}$$

where $s(t)$ is the equivalent low-pass (ELP) signal transmitted over the time-variant channel, with impulse response

$$\tilde{h}(t,\tau) = \Re\left\{h(t,\tau)e^{j2\pi f_c t}\right\}, \tag{29.2.2}$$

[3]The notation $\Re\{\cdot\}$ denotes the real-part operator.

where $h(t,\tau)$ is the ELP time-variant channel impulse response with t and τ denoting the time and delay variables respectively. The function $h(t,\tau)$ is typically referred to as the input-delay spread function representing the response of the channel at time t due to an impulse applied at time $t-\tau$ [60, 61].[4]

The output of the channel is modeled as

$$\tilde{y}(t) = \Re\left\{y(t)e^{j2\pi f_c t}\right\}, \tag{29.2.3}$$

where $y(t)$ is the ELP output signal. The ELP channel output $y(t)$ can be expressed in terms of the ELP transmitted signal $s(t)$, with energy $2E_s$, in the time domain as

$$y(t) = \int_{-\infty}^{+\infty} h(t,\tau)s(t-\tau)\, d\tau\,. \tag{29.2.4}$$

For a physical channel, $h(t,\tau)$ must have finite support over the positive values of τ, satisfying the causality condition.

The ELP received signal can be modeled as

$$r(t) = y(t) + n(t)\,, \tag{29.2.5}$$

where $n(t)$ is an additive white Gaussian noise process with two-sided power spectral density $2N_0$.[5] The additive noise process $n(t)$ is independent of the channel $h(t,\tau)$ and therefore independent of the process $y(t)$ [see (29.2.4)].

29.2.2 Matched front-end processor

Consider a MFEP with ELP impulse response

$$f_M(t) = \begin{cases} s^*(T_s - t), & 0 < t < T_s \\ 0, & \text{otherwise}, \end{cases} \tag{29.2.6}$$

where T_s is the symbol duration. The MFEP output can be written as[6]

$$r_M(t) = \int_0^{+\infty} r(\alpha)\, f_M(t-\alpha)\, d\alpha \tag{29.2.7}$$

$$= \int_0^{+\infty} r(t-\alpha)\, f_M(\alpha)\, d\alpha\,. \tag{29.2.8}$$

[4] We use the results and nomenclature of the characterization of time-variant channels with a minimal review. An excellent treatment on this subject can be found in [60, 61]; see also Chapter 3.

[5] The term "Gaussian" denotes "ELP complex circular Gaussian."

[6] Since the rest of the chapter deals with the ELP notation, we shall drop the term ELP in the subsequent sections. Here, and throughout the chapter, the range of the integration is determined by the support of the integrand.

Either (29.2.7) or (29.2.8) can be used as a starting point for the derivations to follow. Substituting (29.2.5) into (29.2.8), we can write the MFEP output as[7]

$$r_{\mathrm{M}}(t) = y_{\mathrm{M}}(t) + n_{\mathrm{M}}(t)\,, \tag{29.2.9}$$

where $y_{\mathrm{M}}(t)$ and $n_{\mathrm{M}}(t)$ are the contributions of the signal and noise, respectively, to the MFEP output.

Using (29.2.4), we can write the process $y_{\mathrm{M}}(t)$ as

$$y_{\mathrm{M}}(t) = \int_0^{+\infty}\int_{-\infty}^{+\infty} h(t-\alpha,\tau)s(t-\alpha-\tau)f_{\mathrm{M}}(\alpha)\,d\tau d\alpha\,. \tag{29.2.10}$$

The noise process $n_{\mathrm{M}}(t)$ is given by

$$n_{\mathrm{M}}(t) = \int_0^{+\infty} n(t-\alpha)\,f_{\mathrm{M}}(\alpha)\,d\alpha\,. \tag{29.2.11}$$

29.2.3 Statistical properties of the MFEP output

Statistical properties of $r_{\mathrm{M}}(t)$ can be written in terms of the statistical properties of the random time-variant channel $h(t,\tau)$ and the additive noise $n(t)$. Using the fact that $n(t)$ is a zero-mean process, we can write the mean value of the MFEP output process as

$$\mathbb{E}\{r_{\mathrm{M}}(t)\} = \int_0^{+\infty}\int_{-\infty}^{+\infty} \mathbb{E}\{h(t-\alpha,\tau)\}\,s(t-\alpha-\tau)f_{\mathrm{M}}(\alpha)\,d\tau d\alpha\,. \tag{29.2.12}$$

Note that $r_{\mathrm{M}}(t)$ is zero-mean if the two-dimensional process $h(t,\tau)$ is zero-mean.

The correlation function of $r_{\mathrm{M}}(t)$ is given by

$$R_{\mathrm{r}}(t_1,t_2) = \mathbb{E}\{r_{\mathrm{M}}^*(t_1)r_{\mathrm{M}}(t_2)\}\,. \tag{29.2.13}$$

Since the noise $n(t)$ is zero mean and is independent of the time-variant channel $h(t,\tau)$, $R_{\mathrm{r}}(t_1,t_2)$ becomes

$$R_{\mathrm{r}}(t_1,t_2) = R_{\mathrm{y}}(t_1,t_2) + R_{\mathrm{n}}(t_1,t_2)\,, \tag{29.2.14}$$

where $R_{\mathrm{y}}(t_1,t_2)$ and $R_{\mathrm{n}}(t_1,t_2)$ are correlation functions of $y_{\mathrm{M}}(t)$ and $n_{\mathrm{M}}(t)$, respectively.

We can show, using (29.2.10), that the correlation function of $y_{\mathrm{M}}(t)$ is

$$\begin{aligned} R_{\mathrm{y}}(t_1,t_2) = \int_0^{+\infty}\int_0^{+\infty}\int_{-\infty}^{+\infty}\int_{-\infty}^{+\infty} & R_{\mathrm{h}}(t_1-\alpha_1,t_2-\alpha_2;\tau_1,\tau_2) \\ & s^*(t_1-\alpha_1-\tau_1)f_{\mathrm{M}}^*(\alpha_1)\,s(t_2-\alpha_2-\tau_2)f_{\mathrm{M}}(\alpha_2)\,d\tau_1 d\tau_2\,d\alpha_1 d\alpha_2\,, \end{aligned} \tag{29.2.15}$$

[7] We have carried out the derivations for both representations. The derivation using (29.2.8) as a starting point is presented in this chapter.

where $R_{\rm h}(t_1-\alpha_1, t_2-\alpha_2; \tau_1, \tau_2)$ is defined to be

$$R_{\rm h}(t_1-\alpha_1, t_2-\alpha_2; \tau_1, \tau_2) \triangleq \mathbb{E}\{h^*(t_1-\alpha_1, \tau_1)h(t_2-\alpha_2, \tau_2)\} \,. \tag{29.2.16}$$

For a slowly varying WSSUS channel, the correlation function of the MFEP output is derived in Appendix A of this Part as

$$R_{\rm y}(t_1, t_2) = \begin{cases} \displaystyle\int_{-\infty}^{+\infty} P_{\rm h}(0, \tau)\widetilde{R}_{\rm s}^*(t_1 - T_{\rm s} - \tau)\widetilde{R}_{\rm s}(t_2 - T_{\rm s} - \tau)d\tau & |t_2 - t_1| < \frac{2}{B_{\rm s}} \\ 0, & \text{otherwise,} \end{cases} \tag{29.2.17}$$

where $P_{\rm h}(0; \tau)$ is the PDP, also known as the multipath intensity profile, and

$$\widetilde{R}_{\rm s}(t - T_{\rm s} - \tau) \triangleq \int_0^{+\infty} s(t - \alpha - \tau) f_{\rm M}(\alpha)\, d\alpha \,. \tag{29.2.18}$$

In SS parlance, $\widetilde{R}_{\rm s}(\tau)$ is the periodic time autocorrelation function of the baseband spread signature sequence. For sequence designs as in [62], the function $\widetilde{R}_{\rm s}(\tau)$ possesses small sidelobes and has a narrow peak over the interval $|\tau| < 1/B_{\rm s}$, where $B_{\rm s}$ denotes the spreading BW. The spreading BW is roughly equal to the chip rate $R_{\rm c}$ defined by $R_{\rm c} = 1/T_{\rm c}$, where $T_{\rm c}$ is the chip duration. The symbol time $T_{\rm s}$ is equal to the product of chip duration and the processing gain of the SS system. In deriving (29.2.17), we have assumed that the side-lobes of $\widetilde{R}_{\rm s}(\tau)$ are negligible as presented in Appendix A.

The correlation function of $n_{\rm M}(t)$ can be derived similarly as

$$\begin{aligned} R_{\rm n}(t_1, t_2) &= \int_0^{+\infty}\int_0^{+\infty} \mathbb{E}\{n^*(t_1 - \alpha_1)n(t_2 - \alpha_2)\}\, f_{\rm M}^*(\alpha_1) f_{\rm M}(\alpha_2)\, d\alpha_1 d\alpha_2 \\ &= 2N_0 \int_0^{+\infty} f_{\rm M}^*(\alpha_1) f_{\rm M}(\alpha_1 + t_2 - t_1)\, d\alpha_1 \\ &= 2N_0 \widetilde{R}_{\rm s}(t_2 - t_1)\,. \end{aligned} \tag{29.2.19}$$

In deriving (29.2.19) we have used the property that

$$\mathbb{E}\{n^*(t_1 - \alpha_1)n(t_2 - \alpha_2)\} = 2N_0 \delta_{\rm D}(t_2 - t_1 + \alpha_1 - \alpha_2)\,, \tag{29.2.20}$$

where $\delta_{\rm D}(\cdot)$ is the Dirac delta function.

29.3 THE SELECTIVE RAKE RECEIVER

Let γ_i denote the instantaneous SNR of the MFEP output samples defined by

$$\gamma_i \triangleq \frac{|y_{\rm M}(t_i)|^2}{\mathbb{E}\{|n_{\rm M}(t_i)|^2\}}\,, \tag{29.3.1}$$

where the t_i's are the sampling instants. We model the γ_i's as continuous random variables with pdf $f_{\gamma_i}(x)$ and mean

$$\Gamma_i = \mathbb{E}\{\gamma_i\} = \frac{\mathbb{E}\{|y_{\rm M}(t_i)|^2\}}{\mathbb{E}\{|n_{\rm M}(t_i)|^2\}}. \tag{29.3.2}$$

The instantaneous output SNR of the SRake receiver is

$$\gamma_{\rm SRake} = \sum_{i=1}^{L} \gamma_{(i)} \qquad 1 \le L \le N_{\rm r}, \tag{29.3.3}$$

where $\gamma_{(i)}$ is the ordered γ_i, i.e., $\gamma_{(1)} > \gamma_{(2)} > \cdots > \gamma_{(N_{\rm r})}$, and $N_{\rm r}$ is the number of resolvable multipath components. As pointed out, $N_{\rm r}$ increases with the spreading BW. It is apparent that several multipath combining receivers such as the SP and ARake receivers turn out to be special cases of (29.3.3). Note that the possibility of at least two equal $\gamma_{(i)}$ is excluded, since $\gamma_{(i)} \neq \gamma_{(j)}$ *almost surely* for continuous random variables γ_i.[8]

We model the time-varying channel $h(t,\tau)$ as a two-dimensional complex circular Gaussian process with zero mean. Since $y_{\rm M}(t)$ is a linear transformation of $h(t,\tau)$ [c.f. (29.2.10)], it is also a complex Gaussian process with zero mean and correlation function $R_{\rm y}(t_1,t_2)$. This implies that $|y_{\rm M}(t_i)|$ is Rayleigh distributed and consequently, the pdf of the γ_i is given by

$$f_{\gamma_i}(x) = \begin{cases} \frac{1}{\Gamma_i} e^{-\frac{x}{\Gamma_i}}, & 0 < x < \infty \\ 0, & \text{otherwise}. \end{cases} \tag{29.3.4}$$

For a WSSUS channel with constant PDP, we can show by (29.2.17) that the correlation function of the MFEP output samples is[9]

$$R_{\rm y}(t_1,t_2) = \begin{cases} 4\beta \frac{T_{\rm c}}{T_{\rm d}} E_{\rm s}^2, & |t_2 - t_1| = 0 \\ 4\tilde{\beta} \frac{T_{\rm c}}{T_{\rm d}} E_{\rm s}^2, & |t_2 - t_1| = T_{\rm c} \\ 0, & |t_2 - t_1| = kT_{\rm c} \qquad k = 2, 3, \ldots. \end{cases} \tag{29.3.5}$$

The coefficients β and $\tilde{\beta}$ depend on the specific chip pulse shape. We consider three types of pulse shapes in this chapter, namely, the rectangular, half-sine, and raised cosine pulse shapes. Each pulse shape has finite support over the interval $[0, T_{\rm c})$ and is normalized to constant energy. Such time-limited pulse shapes have been used previously in the literature [64]. The rectangular pulse shape is defined by

$$p_{T_{\rm c}}(t) \triangleq \begin{cases} 1, & 0 \le t < T_{\rm c} \\ 0, & \text{otherwise}. \end{cases} \tag{29.3.6}$$

[8] In our context, the notion of "almost sure" or "almost everywhere" can be stated mathematically as: if $\mathcal{N} = \{\gamma_{(i)} = \gamma_{(j)}\}$, then $\Pr\{\mathcal{N}\} = 0$ [63].

[9] Although we are concerned only with the values of $|t_2 - t_1|$ at integer multiples of $T_{\rm c}$ in (29.3.5), $R_{\rm y}(t_1,t_2)$ is, in general, a continuous function of $|t_2 - t_1|$, as can be seen in (29.2.17).

Similarly, the expressions for the half-sine and raised cosine pulse shapes are defined by

$$p_{T_c}(t) \triangleq \begin{cases} \sqrt{2}\,\sin\left(\frac{\pi t}{T_c}\right), & 0 \leq t < T_c \\ 0, & \text{otherwise}, \end{cases} \tag{29.3.7}$$

and

$$p_{T_c}(t) \triangleq \begin{cases} \sqrt{\frac{2}{3}}\left[1 - \cos\left(\frac{2\pi t}{T_c}\right)\right], & 0 \leq t < T_c \\ 0, & \text{otherwise}, \end{cases} \tag{29.3.8}$$

respectively. The coefficients β and $\tilde{\beta}$ are straightforward to compute and are given by the following expressions:

$$\beta = \begin{cases} \frac{2}{3}, & \text{rectangular pulse} \\ \frac{5}{2\,\pi^2} + \frac{1}{3}, & \text{half-sine pulse} \\ \frac{35}{24\,\pi^2} + \frac{1}{3}, & \text{raised cosine pulse}\ , \end{cases} \tag{29.3.9}$$

and

$$\tilde{\beta} = \begin{cases} \frac{1}{6}, & \text{rectangular pulse} \\ \frac{5}{4\,\pi^2} - \frac{1}{12}, & \text{half-sine pulse} \\ \frac{1}{12} - \frac{35}{48\,\pi^2}, & \text{raised cosine pulse}\ . \end{cases} \tag{29.3.10}$$

Numerically, $\beta = 0.6666, 0.5866, 0.4811$ and $\tilde{\beta} = 0.1666, 4.332 \times 10^{-2}, 9.453 \times 10^{-3}$ for the rectangular, half-sine, and raised cosine pulses, respectively.

The correlation function $R_y(t_1, t_2)$ given in (29.3.5) implies that samples of the MFEP output are uncorrelated as long as they are not adjacent to each other. It also implies that even the adjacent samples are "weakly correlated" with correlation coefficient equal to 0.25, 0.07386, and 0.01965 for the rectangular, half-sine, and raised cosine pulse shapes, respectively. Motivated further by analytical tractability, the MFEP output samples are modeled to be uncorrelated. Therefore, they are independent since $y_M(t)$ is a complex Gaussian process.[10]

Denoting $\boldsymbol{\gamma}_{(N_r)} \triangleq (\gamma_{(1)}, \gamma_{(2)}, \ldots, \gamma_{(N_r)})$, we can derive the joint pdf of $\gamma_{(1)}, \gamma_{(2)}, \ldots, \gamma_{(N)}$ using the theory of "order statistics" [65] as

$$f_{\boldsymbol{\gamma}_{(N_r)}}\left(\{\gamma_{(i)}\}_{i=1}^{N_r}\right) = \begin{cases} N!\,f(\gamma_{(1)})f(\gamma_{(2)})\ldots f(\gamma_{(N)}), & \gamma_{(1)} > \gamma_{(2)} > \ldots > \gamma_{(N)} \\ 0, & \text{otherwise}. \end{cases} \tag{29.3.11}$$

[10] This justifies the standard assumption of "independent MFEP output samples" made in typical studies concerning Rake receivers (e.g., [9, 35]).

Substituting (29.3.4) into (29.3.11) together with the equal PDP assumption, the joint pdf of $\gamma_{(1)}, \gamma_{(2)}, \cdots, \gamma_{(N_r)}$ becomes

$$f_{\gamma_{(N_r)}}\left(\{\gamma_{(i)}\}_{i=1}^{N_r}\right) = \begin{cases} \frac{N_r!}{\Gamma^{N_r}} \exp\left(-\frac{1}{\Gamma}\sum_{i=1}^{N_r}\gamma_{(i)}\right), & \gamma_{(1)} > \gamma_{(2)} > \cdots > \gamma_{(N_r)} > 0 \\ 0, & \text{otherwise}. \end{cases} \tag{29.3.12}$$

The parameter Γ is the mean SNR of the MFEP output given by

$$\Gamma = \frac{R_y(t_1, t_1)}{R_n(t_1, t_1)} = \beta \frac{T_c}{T_d}\left(\frac{E_s}{N_0}\right), \tag{29.3.13}$$

where $R_n(t_1, t_2)$ is given by (29.2.19). It is important to note that the $\gamma_{(i)}$s are *no* longer independent, even though the underlying γ_is are independent.

29.4 SELECTIVE RAKE RECEIVER PERFORMANCE ANALYSIS

In this section, the theory developed in [41, 42, 43, 44, 45, 46, 47, 48, 49] for a more general diversity setting is applied to the study of SRake receivers with an arbitrary number of combined paths in a frequency-selective dense multipath channel.

29.4.1 Virtual path technique: The key idea

In general, the analysis of SRake receivers based on a chosen ordering of the paths at first appears to be complicated, since the SNR statistics of the ordered paths are necessarily dependent. Even the calculation of the *average* combiner output SNR alone can require a lengthy derivation as seen in [40]. Here, we alleviate this problem. The key idea is to transform the physical Rake receiver with ordered path variables into the virtual Rake receiver with a set of i.i.d. *virtual fingers* and to express the ordered path SNR variables as a linear function of i.i.d. virtual path SNR variables. The key advantage of this formulation is that it allows the SRake receiver output SNR to be expressed in terms of the i.i.d. virtual path SNR variables.

In this virtual Rake receiver framework, the *average* SNR of the combined output is obtained in a more concise manner than the derivation given in [40]. Furthermore, the extension to the derivation of the combiner output SNR variance as well as the evaluation of the SEP can be made in the virtual path domain. In this light, the derivation of the moments of the SRake receiver output SNR is essentially reduced to the calculation of the moments of the linear combination of i.i.d. random variables. Similarly, the evaluation of the SEP involving nested N_r-fold integrals essentially reduces to the evaluation of a single integral. The effectiveness of the virtual Rake receiver approach is apparent in the simplicity of the derivations.

29.4.2 Virtual path domain transformation

Consider the transformation of ordered multipath instantaneous SNRs into a new set of *virtual path* instantaneous SNRs, V_ns, using the relationship:

$$\gamma_{(i)} = \left[\beta \frac{T_{\mathrm{c}}}{T_{\mathrm{d}}} \left(\frac{E_{\mathrm{s}}}{N_0}\right)\right] \sum_{n=i}^{N_{\mathrm{r}}} \frac{V_n}{n} . \tag{29.4.1}$$

By direct substitution, we show in the following that the instantaneous SNRs of the virtual paths are i.i.d. normalized exponential random variables.

In view of the transformation matrix, (29.4.1) can be written as

$$\begin{bmatrix} \gamma_{(1)} \\ \gamma_{(2)} \\ \vdots \\ \gamma_{(N_{\mathrm{r}})} \end{bmatrix} = \underbrace{\begin{bmatrix} \frac{\Gamma}{1} & \frac{\Gamma}{2} & \cdots & \frac{\Gamma}{N_{\mathrm{r}}} \\ & \frac{\Gamma}{2} & \cdots & \frac{\Gamma}{N_{\mathrm{r}}} \\ & & \ddots & \vdots \\ & & & \frac{\Gamma}{N_{\mathrm{r}}} \end{bmatrix}}_{\triangleq \boldsymbol{T}} \begin{bmatrix} V_1 \\ V_2 \\ \vdots \\ V_{N_{\mathrm{r}}} \end{bmatrix} , \tag{29.4.2}$$

where $\Gamma = \left[\beta \frac{T_{\mathrm{c}}}{T_{\mathrm{d}}} \left(\frac{E_{\mathrm{s}}}{N_0}\right)\right]$, $\boldsymbol{T}$ is the "virtual path transformation matrix," and the entries of $\boldsymbol{T}$ below the main diagonal are zero. Denoting $\boldsymbol{V}_{N_{\mathrm{r}}} \triangleq [V_1, V_2, \dots, V_{N_{\mathrm{r}}}]^t$, we can then obtain the joint pdf of $V_1, V_2, \dots, V_{N_{\mathrm{r}}}$ using the distribution theory for transformations of random vectors [65], as

$$f_{\boldsymbol{V}_{N_{\mathrm{r}}}}(\{v_n\}_{n=1}^{N_{\mathrm{r}}}) = |\boldsymbol{T}| \, f_{\boldsymbol{\gamma}_{(N_{\mathrm{r}})}} \left(\{\gamma_{(i)}\}_{i=1}^{N_{\mathrm{r}}}\right) \Bigg|_{\boldsymbol{\gamma}_{(N_{\mathrm{r}})} = \boldsymbol{T} \boldsymbol{v}_{N_{\mathrm{r}}}} , \tag{29.4.3}$$

where $|\boldsymbol{T}|$ is the Jacobian of the transformation with $|\cdot|$ denoting the determinant and $\boldsymbol{v}_{N_{\mathrm{r}}} \triangleq [v_1, v_2, \dots, v_{N_{\mathrm{r}}}]^t$. Since $\boldsymbol{T}$ is an upper triangular matrix,

$$|\boldsymbol{T}| = \prod_{n=1}^{N_{\mathrm{r}}} \frac{\Gamma}{n} = \frac{1}{N_{\mathrm{r}}!} \left[\beta \frac{T_{\mathrm{c}}}{T_{\mathrm{d}}} \left(\frac{E_{\mathrm{s}}}{N_0}\right)\right]^{N_{\mathrm{r}}} . \tag{29.4.4}$$

The quantity in the exponent of (29.3.12), can be written in terms of the V_ns as

$$\begin{aligned} \frac{1}{\Gamma} \sum_{i=1}^{N_{\mathrm{r}}} \gamma_{(i)} &= \frac{1}{\left[\beta \frac{T_{\mathrm{c}}}{T_{\mathrm{d}}} \left(\frac{E_{\mathrm{s}}}{N_0}\right)\right]} \sum_{i=1}^{N_{\mathrm{r}}} \left[\beta \frac{T_{\mathrm{c}}}{T_{\mathrm{d}}} \left(\frac{E_{\mathrm{s}}}{N_0}\right)\right] \sum_{n=i}^{N_{\mathrm{r}}} \frac{V_n}{n} \\ &= \sum_{n=1}^{N_{\mathrm{r}}} \sum_{i=1}^{n} \frac{V_n}{n} \\ &= \sum_{n=1}^{N_{\mathrm{r}}} V_n . \end{aligned} \tag{29.4.5}$$

Note also that

$$\gamma_{(i)} = \left[\beta \frac{T_{\mathrm{c}}}{T_{\mathrm{d}}} \left(\frac{E_{\mathrm{s}}}{N_0}\right)\right] \sum_{n=i}^{N_{\mathrm{r}}} \frac{V_n}{n}$$
$$= \underbrace{\left[\beta \frac{T_{\mathrm{c}}}{T_{\mathrm{d}}} \left(\frac{E_{\mathrm{s}}}{N_0}\right)\right] \sum_{n=i+1}^{N_{\mathrm{r}}} \frac{V_n}{n}}_{=\gamma_{(i+1)}} + \left[\beta \frac{T_{\mathrm{c}}}{T_{\mathrm{d}}} \left(\frac{E_{\mathrm{s}}}{N_0}\right)\right] \frac{V_i}{i}, \tag{29.4.6}$$

and we arrive at the recursion formula for the virtual path transformation as

$$\gamma_{(i)} = \gamma_{(i+1)} + \left[\beta \frac{T_{\mathrm{c}}}{T_{\mathrm{d}}} \left(\frac{E_{\mathrm{s}}}{N_0}\right)\right] \frac{V_i}{i}, \qquad i = 1, \ldots, N_{\mathrm{r}}, \tag{29.4.7}$$

where $\gamma_{(N_{\mathrm{r}}+1)} = 0$. The quantity $\left[\beta \frac{T_{\mathrm{c}}}{T_{\mathrm{d}}} \left(\frac{E_{\mathrm{s}}}{N_0}\right)\right] \frac{V_i}{i}$ can be interpreted as the "difference between the adjacent ordered instantaneous SNRs," which implies that the virtual path transformation decouples the range of V_ns, and $0 < V_n < \infty$. When we use this together with (29.3.12), (29.4.4) and (29.4.5), (29.4.3) becomes

$$f_{\mathbf{V}_{N_{\mathrm{r}}}}(\{v_n\}_{n=1}^{N_{\mathrm{r}}}) = \begin{cases} e^{-\sum_{n=1}^{N_{\mathrm{r}}} v_n}, & 0 \leq v_n < \infty \\ 0, & \text{otherwise}. \end{cases} \tag{29.4.8}$$

Therefore,

$$f_{\mathbf{V}_{N_{\mathrm{r}}}}(\{v_n\}_{n=1}^{N_{\mathrm{r}}}) = \prod_{n=1}^{N_{\mathrm{r}}} f_{V_n}(v_n), \tag{29.4.9}$$

where $f_{V_n}(\cdot)$ is the pdf of V_n given by

$$f_{V_n}(v) = \begin{cases} e^{-v}, & 0 < v < \infty \\ 0, & \text{otherwise}. \end{cases} \tag{29.4.10}$$

Hence, the instantaneous SNRs of virtual paths are i.i.d. normalized exponential random variables.

The instantaneous SNR of the combiner output can now be expressed in terms of the instantaneous SNRs of the virtual paths as

$$\gamma_{\mathrm{SRake}} = \sum_{n=1}^{N_{\mathrm{r}}} b_n V_n, \tag{29.4.11}$$

where the coefficients b_n are given by

$$b_n = \begin{cases} \left[\beta \frac{T_{\mathrm{c}}}{T_{\mathrm{d}}} \left(\frac{E_{\mathrm{s}}}{N_0}\right)\right], & n \leq L \\ \left[\beta \frac{T_{\mathrm{c}}}{T_{\mathrm{d}}} \left(\frac{E_{\mathrm{s}}}{N_0}\right)\right] \frac{L}{n}, & \text{otherwise}. \end{cases} \tag{29.4.12}$$

Note that γ_{SRake} in (29.4.11) is now expressed in terms of i.i.d. virtual path variables as opposed to (29.3.3) where it was written in terms of dependent physical path variables.

29.4.3 Mean and variance of the combiner output SNR

The mean SNR at the output of the SRake combiner is a commonly accepted performance measure and is given by

$$\Gamma_{\text{SRake}} = \mathbb{E}\{\gamma_{\text{SRake}}\} \,. \tag{29.4.13}$$

This is calculated in [40] by substitution of the expression for γ_{SRake} in terms of the physical path variables given in (29.3.3) directly into (29.4.13) as

$$\Gamma_{\text{SRake}} = \mathbb{E}_{\{\gamma_{(i)}\}}\left\{\sum_{i=1}^{L}\gamma_{(i)}\right\} \tag{29.4.14}$$

$$= \int_0^\infty\int_0^{\gamma_{(1)}}\cdots\int_0^{\gamma_{(N_r-1)}}\left[\sum_{i=1}^{L}\gamma_{(i)}\right] \times \; f_{\gamma_{(N_r)}}\left(\{\gamma_{(i)}\}_{i=1}^{N_r}\right)\, d\gamma_{(N_r)}\ldots d\gamma_{(2)}d\gamma_{(1)}\,. \tag{29.4.15}$$

Since the statistics of the ordered paths are *no* longer independent, the evaluation of the average combiner output SNR in (29.4.15) involves nested integrals, which are in general cumbersome and can require a lengthy derivation as seen in [40]. We alleviate this situation by substituting the expression for γ_{SRake} in terms of the virtual path variables given in (29.4.11) into (29.4.13) as

$$\begin{aligned}
\Gamma_{\text{SRake}} &= \mathbb{E}_{\{V_n\}}\left\{\sum_{n=1}^{N_r} b_n V_n\right\} \\
&= \mathbb{E}_{\{V_n\}}\left\{\left(\sum_{n=1}^{L} V_n + \sum_{n=L+1}^{N_r}\frac{L}{n}V_n\right)\left[\beta\frac{T_c}{T_d}\left(\frac{E_s}{N_0}\right)\right]\right\} \\
&= L\left(1+\sum_{n=L+1}^{N_r}\frac{1}{n}\right)\left[\beta\frac{T_c}{T_d}\left(\frac{E_s}{N_0}\right)\right],
\end{aligned} \tag{29.4.16}$$

where we use the unity mean of the normalized exponential random variable ($\mathbb{E}\{V_n\} = 1$).

Next, we derive the variance of the combiner output SNR, σ^2_{SRake} to assess the effectiveness of the SRake receiver in a multipath environment. The calculation of σ^2_{SRake} in terms of the physical path variables would be even more involved than (29.4.15) since it involves averaging the products of dependent random variables (cross-terms) as

$$\sigma^2_{\text{SRake}} = Var\left\{\sum_{i=1}^{L} \gamma_{(i)}\right\} \tag{29.4.17}$$

$$= \int_0^{\infty}\int_0^{\gamma_{(1)}}\cdots\int_0^{\gamma_{(N_\text{r}-1)}} \left[\sum_{i=1}^{L} \gamma_{(i)} - \Gamma_{\text{SRake}}\right]^2$$
$$\times \; f_{\gamma_{(N_\text{r})}}\left(\{\gamma_{(i)}\}_{i=1}^{N_\text{r}}\right) d\gamma_{(N_\text{r})}\ldots d\gamma_{(2)}d\gamma_{(1)}\,. \tag{29.4.18}$$

This again is alleviated by the virtual path techniques as

$$\begin{aligned}
\sigma^2_{\text{SRake}} &= Var\left\{\sum_{n=1}^{N_\text{r}} b_n V_n\right\} \\
&= Var\left\{\left(\sum_{n=1}^{L} V_n + \sum_{n=L+1}^{N_\text{r}} \frac{L}{n} V_n\right)\left[\beta \frac{T_\text{c}}{T_\text{d}}\left(\frac{E_\text{s}}{N_0}\right)\right]\right\} \\
&= L\left(1 + L\sum_{n=L+1}^{N_\text{r}} \frac{1}{n^2}\right)\left[\beta \frac{T_\text{c}}{T_\text{d}}\left(\frac{E_\text{s}}{N_0}\right)\right]^2 ,
\end{aligned} \tag{29.4.19}$$

where we have used the fact that the virtual path variables V_ns are independent and the normalized exponential random variables have unit variance ($Var\{V_n\} = 1$). Note that the independence of the virtual path variables plays a key role in simplifying the derivations of (29.4.16) and (29.4.19).

We also define the normalized standard deviation of the combiner output SNR as

$$\begin{aligned}
\sigma_{\text{n,SRake}} &\triangleq 10\log_{10}\left\{\frac{\sqrt{\sigma^2_{\text{SRake}}}}{\Gamma_{\text{SRake}}}\right\} \\
&= 10\log_{10}\left\{\frac{\sqrt{\left(1 + L\sum_{n=L+1}^{N_\text{r}} \frac{1}{n^2}\right)}}{\sqrt{L}\left(1 + \sum_{n=L+1}^{N_\text{r}} \frac{1}{n}\right)}\right\} .
\end{aligned} \tag{29.4.20}$$

29.4.4 Error probability analysis in virtual Rake framework

We now turn our attention to the evaluation of the SEP, which depends on the statistics of the combiner output SNR. We obtain the SEP for an SRake receiver in a multipath-fading environment by averaging the conditional SEP over the channel ensemble, that is, by averaging $\mathbb{P}\{e \mid \gamma_{\text{SRake}}\}$ over the pdf of γ_{SRake} as

$$\begin{aligned}
P_{e,\text{SRake}} &= \mathbb{E}_{\gamma_{\text{SRake}}}\{\mathbb{P}\{e \mid \gamma_{\text{SRake}}\}\} \\
&= \int_0^{\infty} \mathbb{P}\{e \mid \gamma\} f_{\gamma_{\text{SRake}}}(\gamma) d\gamma\,,
\end{aligned} \tag{29.4.21}$$

where $\mathbb{P}\left\{e \mid \gamma_{\text{SRake}}\right\}$ is the *conditional* SEP, conditioned on the random quantity γ_{SRake} and $f_{\gamma_{\text{SRake}}}(\cdot)$ is the pdf of the combiner output SNR [66, 67, 68]. For coherent detection of M-ary phase-shift keying (MPSK), a representation for $\mathbb{P}\left\{e \mid \gamma_{\text{SRake}}\right\}$ involving a definite integral with *finite* limits, is given by [69, 66] as

$$\mathbb{P}\left\{e_{\text{MPSK}} \mid \gamma_{\text{SRake}}\right\} = \frac{1}{\pi}\int_0^{\Theta} e^{-\frac{\delta_{\text{MPSK}}}{\sin^2\theta}\gamma_{\text{SRake}}}\, d\theta\,, \tag{29.4.22}$$

where $\delta_{\text{MPSK}} = \sin^2(\pi/M)$ and $\Theta = \pi(M-1)/M$. Of course, for BPSK, $\delta_{\text{MPSK}} = 1$ and $\Theta = \pi/2$. If we substitute (29.4.22) into (29.4.21), the SEP of the SRake receiver becomes

$$P_{e,\text{SRake}} = \frac{1}{\pi}\int_0^{\Theta} \mathbb{E}_{\gamma_{\text{SRake}}}\left\{e^{-\frac{\delta_{\text{MPSK}}}{\sin^2\theta}\gamma_{\text{SRake}}}\right\}\, d\theta \tag{29.4.23}$$

$$= \frac{1}{\pi}\int_0^{\Theta}\int_0^{\infty} e^{-\frac{\delta_{\text{MPSK}}}{\sin^2\theta}\gamma} f_{\gamma_{\text{SRake}}}(\gamma)d\gamma\, d\theta\,. \tag{29.4.24}$$

Evaluation of (29.4.24) requires the knowledge of the pdf of the combiner output SNR γ_{SRake}, which can be quite difficult to evaluate. Alternatively, we can average over the channel ensemble with the technique of [70, 71], by substituting the expression for γ_{SRake} directly in terms of the physical path variables given in (29.3.3) as

$$P_{e,\text{SRake}} = \frac{1}{\pi}\int_0^{\Theta} \mathbb{E}_{\{\gamma_{(i)}\}}\left\{e^{-\frac{\delta_{\text{MPSK}}}{\sin^2\theta}\sum_{i=1}^{L}\gamma_{(i)}}\right\}\, d\theta \tag{29.4.25}$$

$$= \frac{1}{\pi}\int_0^{\Theta}\int_0^{\infty}\int_0^{\gamma_{(1)}}\cdots\int_0^{\gamma_{(N_r-1)}} e^{-\frac{\delta_{\text{MPSK}}}{\sin^2\theta}\sum_{i=1}^{L}\gamma_{(i)}}$$

$$\times\; f_{\gamma_{(N_r)}}\left(\{\gamma_{(i)}\}_{i=1}^{N_r}\right)\, d\gamma_{(N_r)}\cdots d\gamma_{(2)}d\gamma_{(1)}\, d\theta\,. \tag{29.4.26}$$

As in Section 29.4.3, the evaluation of (29.4.26) involves N_r-fold nested integrals, which are in general cumbersome and complicated to evaluate since the statistics of the ordered paths are *no* longer independent.

However, we can reformulate the SEP for the SRake receiver in terms of i.i.d. virtual path variables by substituting (29.4.11) into (29.4.23) as

$$P_{e,\text{SRake}} = \frac{1}{\pi}\int_0^{\Theta} \mathbb{E}_{\{V_n\}}\left\{e^{-\frac{\delta_{\text{MPSK}}}{\sin^2\theta}\sum_{n=1}^{N_r} b_n V_n}\right\}\, d\theta$$

$$= \frac{1}{\pi}\int_0^{\Theta}\int_0^{\infty}\int_0^{\infty}\cdots\int_0^{\infty} e^{-\frac{\delta_{\text{MPSK}}}{\sin^2\theta}\sum_{n=1}^{N_r} b_n v_n}\prod_{n=1}^{N_r} f_{V_n}(v_n)dv_n\, d\theta\,. \tag{29.4.27}$$

Since the V_ns are independent, (29.4.27) becomes

$$P_{e,\text{SRake}} = \frac{1}{\pi}\int_0^{\Theta} \prod_{n=1}^{N_r} \mathbb{E}_{V_n}\left\{e^{-\frac{\delta_{\text{MPSK}} b_n}{\sin^2\theta} V_n}\right\} d\theta$$
$$= \frac{1}{\pi}\int_0^{\Theta} \prod_{n=1}^{N_r} \psi_{V_n}\left(-\frac{\delta_{\text{MPSK}} b_n}{\sin^2\theta}\right) d\theta\,, \qquad (29.4.28)$$

where $\psi_{V_n}(\cdot)$ is the characteristic function (c.f.) of V_n and is given by

$$\psi_{V_n}(j\nu) \triangleq \mathbb{E}\left\{e^{+j\nu V_n}\right\} = \frac{1}{1-j\nu}\,. \qquad (29.4.29)$$

The effectiveness of the virtual path technique is apparent as we observe that the expectation operation in (29.4.25) no longer requires an N_r-fold nested integration.

Substituting (29.4.29) into (29.4.28), the SEP for the SRake receiver becomes

$$P_{e,\text{SRake}} = \frac{1}{\pi}\int_0^{\Theta}\left[\frac{\sin^2\theta}{\delta_{\text{MPSK}}\left[\beta\frac{T_c}{T_d}\left(\frac{E_s}{N_0}\right)\right] + \sin^2\theta}\right]^L$$
$$\times \prod_{n=L+1}^{N_r}\left[\frac{\sin^2\theta}{\delta_{\text{MPSK}}\left[\beta\frac{T_c}{T_d}\left(\frac{E_s}{N_0}\right)\right]\frac{L}{n} + \sin^2\theta}\right] d\theta\,. \qquad (29.4.30)$$

Thus, the derivation of the SEP for the SRake receiver involving the N_r-fold nested integrals in (29.4.26) essentially reduces to a single integral over θ involving trigonometric functions with finite limits. Note again that the independence of the virtual path variables plays a key role in simplifying the derivation of (29.4.30).

29.5 SPECIAL CASES

29.5.1 The single path receiver

The SP receiver, also known as the SD receiver, is the simplest form of a diversity system whereby the received signal is selected from *one* of N_r diversity paths [21]. The output SNR of the SP receiver is

$$\gamma_{\text{SP}} \triangleq \max_i \{\gamma_i\} = \gamma_{(1)}\,. \qquad (29.5.1)$$

This is a special case of the SRake receiver with $L = 1$. If we substitute $L = 1$ into (29.4.16), (29.4.19), and (29.4.20), the mean, the variance, and the normalized standard deviation of the SP receiver output SNR become

$$\Gamma_{\mathrm{SP}} = \left[\beta \frac{T_{\mathrm{c}}}{T_{\mathrm{d}}} \left(\frac{E_{\mathrm{s}}}{N_0}\right)\right] \sum_{n=1}^{N_{\mathrm{r}}} \frac{1}{n}, \tag{29.5.2}$$

$$\sigma^2_{\mathrm{SP}} = \left[\beta \frac{T_{\mathrm{c}}}{T_{\mathrm{d}}} \left(\frac{E_{\mathrm{s}}}{N_0}\right)\right]^2 \sum_{n=1}^{N_{\mathrm{r}}} \frac{1}{n^2}, \quad \text{and} \tag{29.5.3}$$

$$\sigma_{\mathrm{n,SP}} = 10 \log_{10} \left\{ \frac{\sqrt{\sum_{n=1}^{N_{\mathrm{r}}} \frac{1}{n^2}}}{\sum_{n=1}^{N_{\mathrm{r}}} \frac{1}{n}} \right\}, \tag{29.5.4}$$

respectively. Note that the special case given in (29.5.2) agrees with the well-known result for the mean SNR of the SD receiver given by [21, (5.2.8) page 316]. Similarly, setting $L = 1$ in (29.4.30), the SEP for the SP receiver becomes

$$P_{e,\mathrm{SP}} = \frac{1}{\pi} \int_0^{\Theta} \prod_{n=1}^{N_{\mathrm{r}}} \left[\frac{\sin^2\theta}{\delta_{\mathrm{MPSK}} \left[\beta \frac{T_{\mathrm{c}}}{T_{\mathrm{d}}} \left(\frac{E_{\mathrm{s}}}{N_0}\right)\right] \frac{1}{n} + \sin^2\theta} \right] d\theta \,. \tag{29.5.5}$$

29.5.2 The all-Rake receiver

For the ARake receiver, the signals from *all* Rake paths are weighted and combined to maximize the SNR at the combiner output. The output SNR of the ARake receiver is given by

$$\gamma_{\mathrm{ARake}} \triangleq \sum_{i=1}^{N_{\mathrm{r}}} \gamma_i = \sum_{i=1}^{N_{\mathrm{r}}} \gamma_{(i)} \,. \tag{29.5.6}$$

Note again that we can obtain the result for the ARake receiver from the SRake results given in (29.4.16), (29.4.19), (29.4.20), and (29.4.30) by setting $L = N_{\mathrm{r}}$, since ARake is a special case of SRake with $L = N_{\mathrm{r}}$. Therefore, the mean, the variance, and the normalized standard deviation of the ARake receiver output SNR become

$$\Gamma_{\mathrm{ARake}} = N_{\mathrm{r}} \left[\beta \frac{T_{\mathrm{c}}}{T_{\mathrm{d}}} \left(\frac{E_{\mathrm{s}}}{N_0}\right)\right], \tag{29.5.7}$$

$$\sigma^2_{\mathrm{ARake}} = N_{\mathrm{r}} \left[\beta \frac{T_{\mathrm{c}}}{T_{\mathrm{d}}} \left(\frac{E_{\mathrm{s}}}{N_0}\right)\right]^2, \quad \text{and} \tag{29.5.8}$$

$$\sigma_{\mathrm{n,ARake}} = 10 \log_{10} \left\{ \frac{1}{\sqrt{N_{\mathrm{r}}}} \right\}, \tag{29.5.9}$$

respectively. Note that the mean SNR for the ARake receiver given in (29.5.7) agrees with the well-known MRC receiver result given by [21, (5.2.16) page 319].

The SEP for the ARake receiver becomes

$$P_{e,\text{ARake}} = \frac{1}{\pi}\int_0^{\Theta}\left[\frac{\sin^2\theta}{\delta_{\text{MPSK}}\left[\beta\frac{T_\text{c}}{T_\text{d}}\left(\frac{E_\text{s}}{N_0}\right)\right]+\sin^2\theta}\right]^{N_\text{r}} d\theta\,. \tag{29.5.10}$$

29.6 NUMERICAL EXAMPLES

29.6.1 Mean and variance

In this section, the results derived in the previous sections for the mean and the variance of the SRake combiner output SNR are illustrated.

Figure 29.1 shows the average SNR of the SRake receiver (normalized by E_s/N_0) as a function of L for $T_\text{d} = 2$ μs and various chip rates, R_c. The data points denoted by the squares represent the average SNR of the SP receiver. The data points denoted by the stars represent the average SNR of the ARake receiver and serve as a upper bound for the average SNR of the SRake receiver.

Figure 29.2 shows the average SNR of the SRake receiver (normalized by E_s/N_0) versus R_c for $T_\text{d} = 2$ μs and various L. The data points denoted by the squares and the stars represent the average SNR of the SP receiver and the ARake receiver, respectively. Note that the curves for the SP receiver and the ARake receiver serve, respectively, as a lower and upper bound for the average SNR of the SRake receiver. The average SNR is improved with an increase in the number of fingers for a fixed BW as illustrated in Figures 29.1 and 29.2. For example, with a 5 MHz BW, a five-finger Rake structure results in a non-negligible improvement of 1.2 dB when compared to the three-finger Rake. On the other hand, for a fixed number of fingers, higher BW signals capture less total energy.

Although the average SNR is a commonly accepted performance measure [21], it does not convey complete information regarding the effect of multipath fading on an SRake receiver. To assess the effectiveness of the SRake receiver in the presence of multipath, we plot the normalized standard deviation of the combiner output SNR as a function of L for $T_\text{d} = 2$ μs and various R_c in Figure 29.3. The normalized standard deviation of the combiner output SNR for the SP receiver and the ARake receiver can be seen as the two limiting cases. Observe that the reduction in the SNR variation tapers off quickly with an increase in the number of Rake fingers.

Figure 29.4 shows the normalized standard deviation of the SRake receiver output SNR versus R_c for $T_\text{d} = 2$ μs and various L. Again, it can be clearly seen that the curves for the SP receiver and the ARake receiver upper and lower bound, respectively, the normalized standard deviation of the SRake receiver output SNR. Figure 29.4 illustrates the reduction in the SNR variation with an increase in BW for a fixed number of fingers. For example, 5 MHz signals achieve about 2.5 to 3.5 dB less variation compared to 1 MHz signals, whereas increasing the BW from 5 MHz to 10 MHz brings an improvement of less than 1.5 dB.

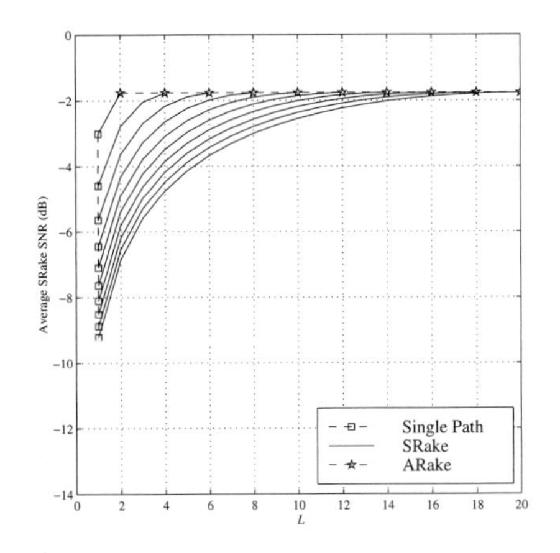

Figure 29.1. The average SNR gain of the SRake receiver (normalized by E_s/N_0) as a function of L for $T_d = 2\ \mu$s and various R_c. The solid curves are parameterized by different R_c starting from the uppermost curve with $R_c = 1$ MHz and decrease monotonically to the lowermost curve with $R_c = 10$ MHz. Reprinted from [15], copyright IEEE.

29.6.2 Symbol error probability

In this section, the results derived for SEP of the SRake receiver are illustrated with specific examples.

Figures 29.5–29.8 show the SEP versus the SNR, E_s/N_0, for various spreading BW's, or equivalently for various chip rates $R_c = 1/T_c$. We depict $R_c = 1$, 5, and 10 MHz with maximum excess channel delay $T_d = 2\ \mu$s for BPSK modulation with rectangular chip pulse waveforms.[11]

[11]The SEP with other chip pulse waveforms, such as the half-sine and the raised cosine, is not significantly different from that obtained with the rectangular pulse shape. The difference is in the attenuation of the mean SNR, and for comparison purposes, the rectangular pulse suffices.

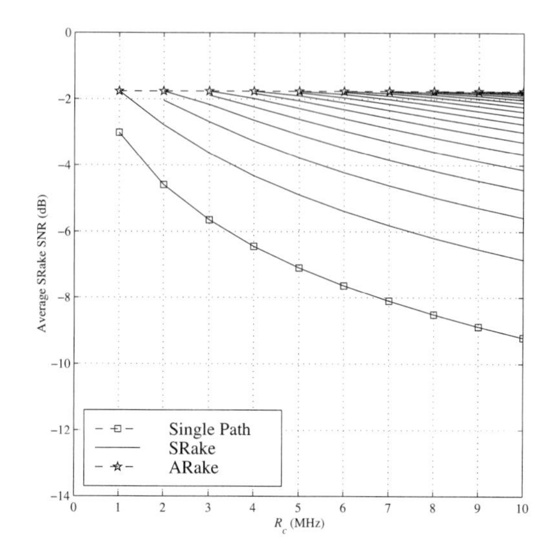

Figure 29.2. The average SNR gain of the SRake receiver (normalized by E_s/N_0) as a function of R_c for $T_d = 2\ \mu s$ and various L. The lowermost solid curve represents $L = 1$ and increases monotonically to the uppermost solid curve with $L = 20$. Reprinted from [15], copyright IEEE.

In Figures 29.5 and 29.6, the curves are parameterized by the number of combined paths L with the highest curve representing $L = 1$, or the single path (SP) receiver, in each graph. Each successively lower curve corresponds to an increasing L up to the maximum L denoted by N_r, or the ARake receiver performance. We see that an increase in the number of combined paths results in a lower SEP, but the improvement diminishes as the number of combined paths approaches N_r. For example with 5 MHz spreading BW, Figure 29.5 suggests that at a SEP of 10^{-4}, there is a 2.5 dB gain in using $L = 2$ paths from $L = 1$, but only over a 1 dB gain from $L = 2$ to 3. The gain in dB diminishes further when L is increased from 3 to 4 and beyond. This trend is also observed in Figure 29.6 for the spreading BW of 10 MHz. The diminishing returns suggest that only a *subset* of paths needs to be combined to give an acceptable level of performance.

Figures 29.7–29.8 compare the SEP of systems with $R_c = 1$ MHz and the case with $R_c = 5$ and 10 MHz. The dotted lines for $R_c = 1$ MHz are shown for $L = 1$ through 2. The solid lines depict the larger BW signals with the number of combined paths L increasing in powers of 2 up to N_r, i.e., $L = 1$, 2, 4, 8, 16, and 20 in Figure 29.8. The effects of the diminishing returns and the wide range in SEP are evident, as discussed above. It is also interesting to note the crossover in the curves between the two sets of R_c's. For example, it can be seen in Figure 29.8 that at SNRs above

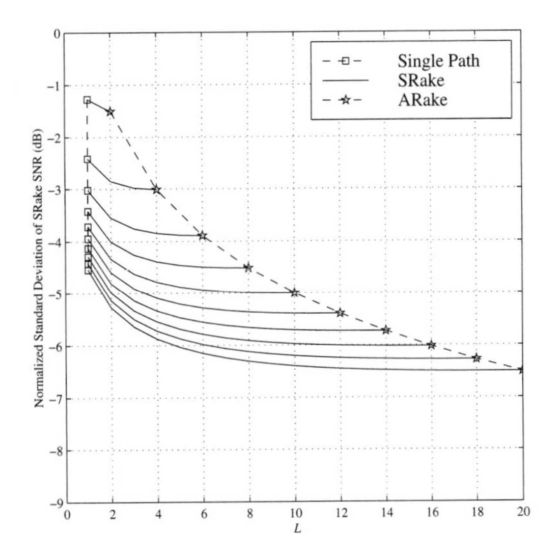

Figure 29.3. The normalized standard deviation of the SRake receiver output SNR as a function of L for $T_d = 2$ μs and various R_c. The solid curves are parameterized by different R_c starting from the uppermost curve with $R_c = 1$ MHz and decrease monotonically to the lowermost curve with $R_c = 10$ MHz. Reprinted from [15], copyright IEEE.

8.5 dB, the SEP for $R_c = 10$ MHz with $L = 4$ is lower than the $R_c = 1$ MHz with $L = 2$ curve. But at SNRs below 8.5 dB, the $R_c = 10$ MHz with $L = 4$ curve gives a higher SEP than $R_c = 1$ MHz with $L = 2$. We see that at low SNRs, more paths need to be combined in the SRake receiver with larger BWs in order to achieve better performance than the SRake receiver with smaller BWs. For instance, L would have to be increased to about 8 in Figure 29.8 for the $R_c = 10$ MHz case to achieve comparable performance with the $R_c = 1$ MHz with $L = 2$ case for low SNRs.

We can also see that the larger the chip rate R_c, the lower the achievable SEP. With $R_c = 1$ MHz, the ARake receiver can only achieve a SEP of about 10^{-2} at an SNR of 10 dB. For $R_c = 5$ MHz, the ARake receiver can achieve an SEP of better than 10^{-3} at the same SNR. Alternatively, the ARake receiver with 10 MHz can achieve 4×10^{-4} SEP at about 10 dB, or a gain of 8 dB over the system with 1 MHz. The drawback in using a larger spreading BW is that a greater number of paths need to be combined to get better performance. If only one path is selected ($L = 1$), then the system with a smaller BW has a lower SEP over some range of SNR. This can be seen in Figure 29.8 at 14 dB, for example.

It is apparent that a specified SEP can be achieved, in principle, with different

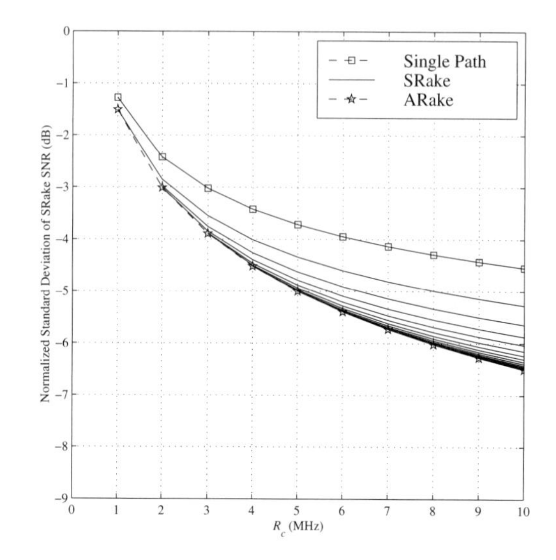

Figure 29.4. The normalized standard deviation of the SRake receiver output SNR as a function of R_{c} for $T_{\mathrm{d}} = 2\ \mu$s and various L. The uppermost solid curve represents $L = 1$ and decreases monotonically to the lowermost solid curve with $L = 20$. Reprinted from [15], copyright IEEE.

combinations of receiver complexity (the number of combined paths L), spreading BW of the signals (R_{c}), and the transmitted power (SNR). For example, Figure 29.7 shows that a SEP of 10^{-3} can be achieved with the combination of $R_{\mathrm{c}} = 1$ MHz and $L = 2$, or with $R_{\mathrm{c}} = 5$ MHz and $L = 8$. The performance difference between these cases is about 6 dB. Other combinations are also evident from the graphs. In general, larger spreading BWs reduce the power requirements as long as a sufficient number of paths are combined.

29.7 CONCLUSIONS AND COMMENTS

We derived exact SEP expressions as well as the average SNR and the variance of the combiner output SNR for an SRake receiver in a multipath fading environment. In particular, we considered frequency-selective WSSUS Gaussian channels with

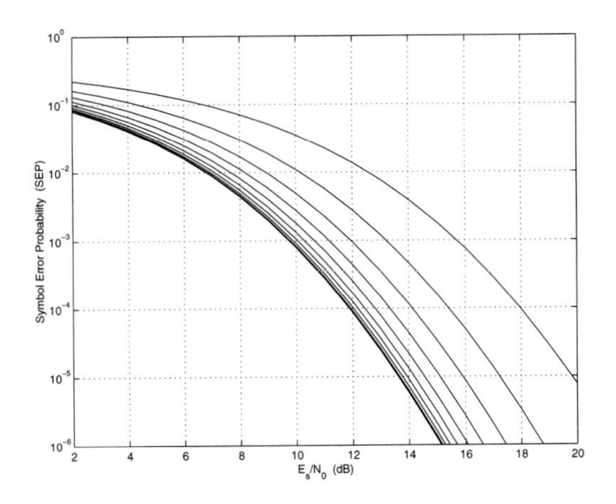

Figure 29.5. The symbol error probability of the SRake receiver vs. the SNR E_s/N_0 as a function of L for $R_c = 5$ MHz and $T_d = 2$ μs. The upper curve is for $L = 1$ and the successively lower curves are for increasing L up to $L = N_r = 10$. Reprinted from [17], copyright IEEE.

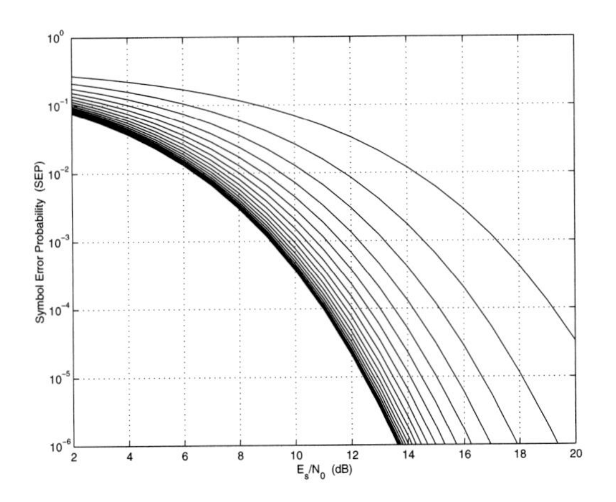

Figure 29.6. The symbol error probability of the SRake receiver vs. the SNR E_s/N_0 as a function of L for $R_c = 10$ MHz and $T_d = 2$ μs. The upper curve is for $L = 1$ and the successively lower curves are for increasing L up to $L = N_r = 20$. Reprinted from [17], copyright IEEE.

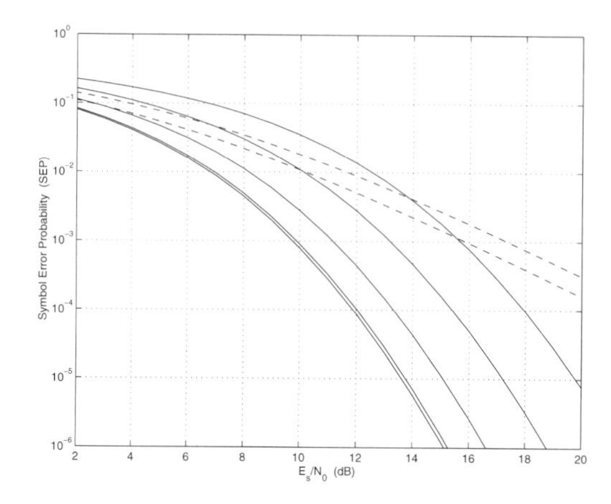

Figure 29.7. The symbol error probability of the SRake receiver vs. the SNR E_s/N_0 as a function of L with $T_d = 2$ μs. The dashed curves are for $R_c = 1$ MHz and the solid curves are for $R_c = 5$ MHz. The solid curves depict $L = 1$, 2, 4, 8, and 10 in successively lower positions. The dashed curves show $L = 1$ to $L = N_r = 2$.Reprinted from [17], copyright IEEE.

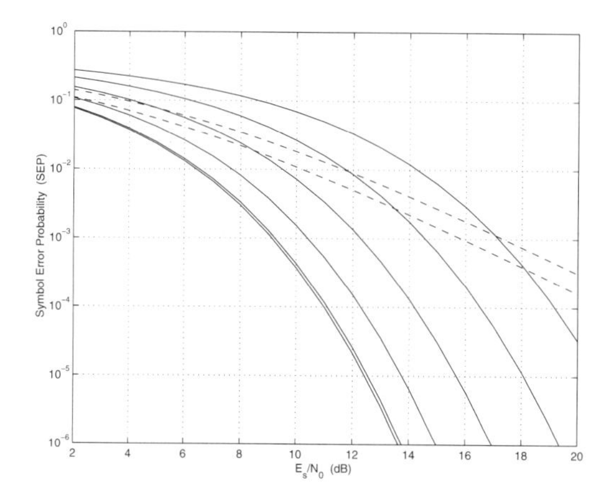

Figure 29.8. The symbol error probability of the SRake receiver vs. the SNR E_s/N_0 as a function of L with $T_d = 2$ μs. The dashed curves are for $R_c = 1$ MHz and the solid curves are for $R_c = 10$ MHz. The solid curves depict $L = 1$, 2, 4, 8, 16, and 20 in successively lower positions. The dashed curves show $L = 1$ to $L = N_r = 2$.Reprinted from [17], copyright IEEE.

constant power delay profile. We analyzed this system in the virtual Rake receiver domain, resulting in a simple derivation and formula of the mean, the variance, and the SEP for a given spreading BW and an *arbitrary* number of combined paths. The key idea was to transform the dependent ordered-path variables into a new set of i.i.d. *virtual paths* and to express the combiner output SNR as a linear combination of the i.i.d. virtual path SNR variables. In this framework, the derivation of the moments of the combiner output SNR and the SEP, which involves the evaluation of nested N_r-fold integrals, essentially reduces to the evaluation of a single integral.

For a fixed BW, the average SNR gain increases with an increase in the number of Rake fingers. On the other hand, for a fixed number of fingers, higher BW signals capture less total energy. Furthermore, we observed the reduction in the SNR variation with an increase in BW as well as the number of Rake fingers. We note that the SEP depends on the statistics of the SNR. As such, for a fixed BW, the SEP decreases with an increase in the number of combined paths in the SRake receiver. The decrease in the achievable SEP is much greater for larger BW signals at the expense of the receiver complexity. As the number of combined paths increases, the incremental gain in performance diminishes, suggesting that only a *subset* of paths needs to be combined to give an acceptable level of performance. In general, larger spreading BWs reduce the required power as long as a sufficient number of paths are combined. The results of this chapter enable trade-offs to be made between power, complexity, and BW to provide a specified SEP performance.

Bibliography

[1] R. A. Scholtz, "The spread-spectrum concept," *IEEE Trans. Commun.*, vol. COM-25, pp. 748–755, Aug. 1977.

[2] R. L. Pickholtz, D. L. Schilling, and L. B. Milstein, "Theory of spread-spectrum communications – A tutorial," *IEEE Trans. Commun.*, vol. COM-30, pp. 855–884, May 1982.

[3] M. K. Simon, J. K. Omura, R. A. Scholtz, and B. K. Levitt, *Spread Spectrum Communications Handbook.* New York, NY: McGraw-Hill, Inc., revised ed., 1994.

[4] R. L. Peterson, R. E. Ziemer, and D. E. Borth, *Introduction to Spread Spectrum Communications.* Englewood Cliffs, New Jersey 07632: Prentice Hall, first ed., 1995.

[5] F. Adachi, M. Sawahashi, and H. Suda, "Wideband DS-CDMA for next-generation mobile communications systems," *IEEE Commun. Mag.*, vol. 36, pp. 56–69, Sept. 1998.

[6] R. Price and P. E. Green, Jr., "A communication technique for multipath channels," *Proc. IRE*, vol. 46, pp. 555–570, Mar. 1958.

[7] G. Turin, "Introduction to spread-spectrum antimultipath techniques and their application to urban digital radio," *Proc. IEEE*, vol. 68, pp. 328–353, Mar. 1980.

[8] J. S. Lehnert and M. B. Pursley, "Multipath diversity reception of spread spectrum multiple-access communications," *IEEE Trans. Commun.*, vol. COM-35, pp. 1189–1198, Nov. 1987.

[9] J. G. Proakis, *Digital Communications.* New York, NY: McGraw-Hill, Inc., third ed., 1995.

[10] M. Z. Win and R. A. Scholtz, "Energy capture vs. correlator resources in ultra-wide bandwidth indoor wireless communications channels," in *Proc. Military Comm. Conf.*, vol. 3, pp. 1277–1281, Nov. 1997. Monterey, CA, **Invited Paper**.

[11] M. Z. Win and R. A. Scholtz, "On the energy capture of ultra-wide bandwidth signals in dense multipath environments," *IEEE Commun. Lett.*, vol. 2, pp. 245–247, Sept. 1998.

[12] M. Z. Win and R. A. Scholtz, "Infinite Rake and selective Rake receivers for ultra-wide bandwidth transmissions in multipath environments," *IEEE Commun. Lett.*, 2000. to appear.

[13] M. Z. Win and Z. A. Kostić, "Impact of spreading bandwidth on Rake reception in dense multipath channels," in *Proc. 8th Comm. Theory Mini Conf.*, pp. 78–82, June 1999. Vancouver, Canada.

[14] M. Z. Win and Z. A. Kostić, "Virtual path analysis of selective Rake receiver in dense multipath channels," *IEEE Commun. Lett.*, vol. 3, pp. 308–310, Nov. 1999.

[15] M. Z. Win and Z. A. Kostić, "Impact of spreading bandwidth on Rake reception in dense multipath channels," *IEEE J. Select. Areas Commun.*, vol. 17, pp. 1794–1806, Oct. 1999.

[16] M. Z. Win, G. Chrisikos, and N. R. Sollenberger, "Impact of spreading bandwidth and diversity order on the error probability performance of Rake reception in dense multipath channels," in *Proc. IEEE Wireless Commun. and Networking Conf.*, vol. 3, pp. 1558–1562, Sept. 1999. New Orleans, LA, **Invited Paper**.

[17] M. Z. Win, G. Chrisikos, and N. R. Sollenberger, "Effects of chip rate on selective rake combining," *IEEE Commun. Lett.*, vol. 4, pp. 233–235, 2000.

[18] M. Z. Win, G. Chrisikos, and N. R. Sollenberger, "Performance of Rake reception in dense multipath channels: Implications of spreading bandwidth and selection diversity order," *IEEE J. Select. Areas Commun.*, vol. 18, pp. 1516–1525, 2000.

[19] G.-T. Chyi, J. G. Proakis, and K. M. Keller, "On the symbol error probability of maximum-selection diversity reception schemes over a Rayleigh fading channel," *IEEE Trans. Commun.*, vol. 37, pp. 79–83, Jan. 1989.

[20] E. A. Neasmith and N. C. Beaulieu, "New results on selection diversity," *IEEE Trans. Commun.*, vol. 46, pp. 695–704, May 1998.

[21] W. C. Jakes, Ed., *Microwave Mobile Communications.* Piscataway, New Jersey, 08855-1331: IEEE Press, IEEE press classic reissue ed., 1995.

[22] N. C. Beaulieu and A. A. Abu-Dayya, "Analysis of equal gain diversity on Nakagami fading channels," *IEEE Trans. Commun.*, vol. 39, pp. 225–234, Feb. 1991.

[23] F. Adachi, K. Ohno, A. Higashi, T. Dohi, and Y. Okumura, "Coherent multicode DS-CDMA mobile radio access," *IEICE Transactions on Communications*, vol. E79-B, pp. 1316–1324, Sept. 1996.

[24] S. Abbas and A. Sheikh, "BER performance of the Rake receiver in a realistic mobile radio channel," in *Proc. 48th Veh. Technol. Conf*, May 1998.

[25] L. F. Chang, "Dispersive fading effects in CDMA radio systems," in *Proc. IEEE Int. Conf. on Universal Personal Comm.*, pp. 185–189, Oct. 1992. Dallas, TX.

[26] R. Herzog, J. Hagenauer, and A. Schmidbauer, "Soft-in/soft-out Hadamard despreader for iterative decoding in the IS-95(A) system," in *Proc. 47th Annual Int. Veh. Technol. Conf.*, pp. 1219–1222, May 1997. Phoenix, AZ.

[27] K. M. et al, "CDMA/TDD cellular systems for the 3rd generation mobile communication," in *Proc. 47th Annual Int. Veh. Technol. Conf.*, pp. 820–824, May 1997. Phoenix, AZ.

[28] J. M. Holtzman and L. M. Jalloul, "Rayleigh fading effect reduction with wideband DS/CDMA signals," *IEEE Trans. Commun.*, vol. 42, pp. 1012–1016, Feb./Mar./Apr. 1994.

[29] F. Amoroso and W. W. Jones, "Geometric model for DSPN satellite reception in the dense scatterer mobile environment," *IEEE Trans. Commun.*, vol. 41, pp. 450–453, Mar. 1993.

[30] F. Amoroso, "Effective bandwidth of DSPN signalling for mitigation of fading in the dense scatterers," *Electron. Lett.*, vol. 29, pp. 661–662, Apr. 1993.

[31] F. Amoroso, "Improved method for calculating mitigation bandwidth of DSPN signals," *Electron. Lett.*, vol. 29, pp. 1743–1745, Sept. 1993.

[32] L. M. Jalloul and J. M. Holtzman, "Multipath fading effect on wide-band DS/CDMA signals: Analysis, simulation, and measurements," *IEEE Trans. on Vehicul. Technol.*, vol. 43, pp. 801–807, Aug. 1994.

[33] D. L. Noneaker and M. B. Pursley, "On the chip rate of CDMA systems with doubly selective fading reception," *IEEE J. Select. Areas Commun.*, vol. 12, pp. 853–861, June 1994.

[34] D. Goeckel and W. Stark, "Performance of coded direct-sequence systems with Rake reception in a multipath fading environment," *European Trans. on Telecommun., Special Issue on Spread Spectrum Techniques*, vol. 6, pp. 41–52, Jan.-Feb. 1995.

[35] T. Eng and L. B. Milstein, "Coherent DS-CDMA performance in Nakagami multipath fading," *IEEE Trans. Commun.*, vol. 43, pp. 1134–1143, Feb./Mar./Apr. 1995.

[36] L. M. Jalloul and J. M. Holtzman, "Performance analysis of DS/CDMA with noncoherent M-ary orthogonal modulation in multipath fading channels," *IEEE J. Select. Areas Commun.*, vol. 12, pp. 862–870, June 1994.

[37] R. A. Scholtz and M. Z. Win, "Impulse radio," in *Wireless Communications* (S. G. Glisic and P. A. Leppänen, eds.), Kluwer Academic Publishers, 1997. **Invited Chapter**.

[38] T. Eng, N. Kong, and L. B. Milstein, "Comparison of diversity combining techniques for Rayleigh-fading channels," *IEEE Trans. Commun.*, vol. 44, pp. 1117–1129, Sept. 1996.

[39] H. Erben, S. Zeisberg, and H. Nuszkowski, "BER performance of a hybrid SC/MRC 2DPSK RAKE receiver in realistic mobile channels," in *Proc. 44th Annual Int. Veh. Technol. Conf.*, vol. 2, pp. 738–741, June 1994. Stockholm, Sweden.

[40] N. Kong and L. B. Milstein, "Combined average SNR of a generalized diversity selection combining scheme," in *Proc. IEEE Int. Conf. on Commun.*, vol. 3, pp. 1556–1560, June 1998. Atlanta, GA.

[41] M. Z. Win and J. H. Winters, "Analysis of hybrid selection/maximal-ratio combining of diversity branches with unequal SNR in Rayleigh fading," in *Proc. 49th Annual Int. Veh. Technol. Conf.*, vol. 1, pp. 215–220, May 1999. Houston, TX.

[42] M. Z. Win and J. H. Winters, "Analysis of hybrid selection/maximal-ratio combining in Rayleigh fading," in *Proc. IEEE Int. Conf. on Commun.*, vol. 1, pp. 6–10, June 1999. Vancouver, Canada.

[43] M.-S. Alouini and M. K. Simon, "Performance analysis of generalized selective combining over rayleigh fading channels," in *Proc. 8th Comm. Theory Mini Conf.*, pp. 110–114, June 1999. Vancouver, Canada.

[44] M. Z. Win and J. H. Winters, "Analysis of hybrid selection/maximal-ratio combining in Rayleigh fading," *IEEE Trans. Commun.*, vol. 47, pp. 1773–1776, Dec. 1999.

[45] M. Z. Win, R. K. Mallik, G. Chrisikos, and J. H. Winters, "Canonical expressions for the error probability performance for M-ary modulation with hybrid selection/maximal-ratio combining in Rayleigh fading," in *Proc. IEEE Wireless*

Commun. and Networking Conf., vol. 1, pp. 266–270, Sept. 1999. New Orleans, LA, **Invited Paper**.

[46] M. Z. Win, G. Chrisikos, and J. H. Winters, "Error probability for M-ary modulation using hybrid selection/maximal-ratio combining in Rayleigh fading," in *Proc. Military Comm. Conf.*, vol. 2, pp. 944 –948, Nov. 1999. Atlantic City, NJ.

[47] M. Z. Win and J. H. Winters, "Exact error probability expressions for H-S/MRC in Rayleigh fading: A virtual branch technique," in *Proc. IEEE Global Telecomm. Conf.*, vol. 1, pp. 537–542, Dec. 1999. Rio de Janeiro, Brazil.

[48] M. Win, N. Beaulieu, L. Shepp, B. Logan, and J. Winters, "Tight simple bounds on the error probability of hybrid diversity combining," in *Proc. IEEE International Symposium on Information Theory and Its Applications*, Nov. 2000.

[49] M. Z. Win and J. H. Winters, "Virtual branch analysis of symbol error probability for hybrid selection/maximal-ratio combining in Rayleigh fading," *IEEE Trans. Commun.*, 2000. accepted pending revision.

[50] Y. C. Yoon, R. Kohno, and H. Imai, "A spread-spectrum multiaccess system with cochannel interference cancellation for multipath fading channels," *IEEE J. Select. Areas Commun.*, vol. 11, pp. 1067–1075, Sept. 1993.

[51] B. R. Vojčić and R. L. Pickholtz, "Performance of coded direct sequence spread spectrum in a fading dispersive channel with pulsed jamming," *IEEE J. Select. Areas Commun.*, vol. 8, pp. 934–942, Sept. 1990.

[52] G. L. Turin, "The effects of multipath and fading on the performance of direct-sequence CDMA systems," *IEEE J. Select. Areas Commun.*, vol. SAC-2, pp. 597–603, July 1993.

[53] G. Turin, F. D. Clapp, T. L. Johnston, S. B. Fine, and D. Lavry, "A statistical model of urban multipath propagation," *IEEE Trans. on Vehicul. Technol.*, vol. VT-21, pp. 1–9, Feb. 1972.

[54] H. Hashemi, "Simulation of the urban radio propagation channel," *IEEE Trans. on Vehicul. Technol.*, vol. VT-28, pp. 213–225, Aug. 1979.

[55] D. C. Cox, "Delay doppler characteristics of multipath propagation at 910 MHz in a suburban mobile radio environment," *IEEE Trans. Antennas Propagat.*, vol. AP-20, pp. 625–635, Sept. 1972.

[56] D. C. Cox, "Time- and frequency-domain characterizations of multipath propagation at 910 MHz in a suburban mobile-radio environment," *Radio Science*, pp. 1069–1077, Dec. 1972.

[57] D. C. Cox, "910 MHz urban mobile radio propagation: Multipath characteristics in New York city," *IEEE Trans. Commun.*, vol. COM-21, pp. 1188–1194, Nov. 1973.

[58] D. C. Cox and R. P. Leck, "Distributions of multipath delay spread and average excess delay for 910-MHz urban mobile radio paths," *IEEE Trans. Antennas Propagat.*, vol. AP-23, pp. 206–213, Mar. 1975.

[59] P. F. Driessen, "Prediction of multipath delay profiles in mountainous terrain," *IEEE J. Select. Areas Commun.*, vol. 18, pp. 336–346, Mar. 2000.

[60] P. A. Bello, "Characterization of randomly time-variant linear channels," *IEEE Trans. Commun. Sys.*, vol. CS-11, pp. 360–393, Dec. 1963.

[61] J. D. Parsons, *The Mobile Radio Propagation Channel.* New York, NY: John Wiley & Sons, Inc., first ed., 1992.

[62] S. W. Golomb, *Shift Register Sequences.* Laguna Hills, CA: Aegean Park Press, revised ed., 1982.

[63] A. N. Shiryaev, *Probability.* New York: Springer-Verlag, second ed., 1995.

[64] T. F. Wong, T. M. Lok, and J. S. Lehnert, "Asynchronous multiple-access interference suppression and chip waveform selection with aperiodic random sequences," *IEEE Trans. Commun.*, vol. 47, pp. 103–114, Jan. 1999.

[65] P. J. Bickel and K. Doksum, *Mathematical Statistics: Basic Ideas and Selected Topics.* Oakland, California: Holden-Day, Inc., first ed., 1977.

[66] C. Tellambura, A. J. Mueller, and V. K. Bhargava, "Analysis of M-ary phase-shift keying with diversity reception for land-mobile satellite channels," *IEEE Trans. on Vehicul. Technol.*, vol. 46, pp. 910–922, Nov. 1997.

[67] A. Annamalai, C. Tellambura, and V. K. Bhargava, "A unified approach to performance evaluation of diversity systems on fading channels," in *Wireless Multimedia Network Technologies* (R. Ganesh and Z. Zvonar, eds.), Kluwer Academic Publishers, 1999.

[68] A. Annamalai, C. Tellambura, and V. K. Bhargava, "Exact evaluation of maximal-ratio and equal-gain diversity receivers for M-ary QAM on Nakagami fading channels," *IEEE Trans. Commun.*, vol. 47, pp. 1335–1344, Sept. 1999.

[69] J. W. Craig, "A new, simple and exact result for calculating the probability of error for two-dimensional signal constellations," in *Proc. Military Comm. Conf.*, pp. 25.5.1–25.5.5, 1991. Boston, MA.

[70] M. K. Simon and D. Divsalar, "Some new twists to problems involving the Gaussian probability integral," *IEEE Trans. Commun.*, vol. 46, pp. 200–210, Feb. 1998.

[71] M.-S. Alouini and A. Goldsmith, "A unified approach for calculating error rates of linearly modulated signals over generalized fading channels," in *Proc. IEEE Int. Conf. on Commun.*, vol. 1, pp. 459–463, June 1998. Atlanta, GA.

Chapter 30

CODE ACQUISITION

Code synchronization consists of two steps: acquisition and tracking. The code tracking is part of the code synchronization problem in which the relative delay τ, between the input and locally generated code, is within the tracking range. This range is most of the time defined as $|\tau| \leq T_c$, where T_c is the code chip interval. However, the initial delay τ between the two codes is much larger, and separate, most of the time completely different, algorithms should be used to reduce this delay to the tracking range. These algorithms are the subject of this section.

30.1 INTRODUCTION

A discussion of these algorithms starts with the maximum likelihood approach to provide insight into our best possible options [1]. Unfortunately, the best performance means the highest complexity. So, to compromise, we present a series of approximations.

For an initial insight into the problem, we first assume a biphase modulated direct-sequence (DS), spread-spectrum receiver, whose input waveform $r(t)$ is given by

$$r(t) = Ab(t)c\left(t + \zeta T_c\right)\cos\left(\omega t + \omega_D t + \theta\right) + n(t) \tag{30.1.1}$$

where the parameters A, ω, and θ stand for the waveform's amplitude, carrier radian frequency, and phase, respectively. Parameter ω_D is a radian frequency shift between transmitter and receiver, $c(t)$ is $a \pm 1$-valued spreading code with chip rate $R_c = T_c^{-1}$ chips/sec, delayed by ζT_c sec with respect to an arbitrary time reference, $d(t)$ is a binary data sequence (which might or might not be present during the acquisition mode), and $n(t)$ is additive noise, or a different type of interference or both. *The synchronizer's task consists of providing the receiver with reliable estimates $\hat{\zeta}, \hat{\omega}_D$, and $\hat{\theta}_c$ of the corresponding unknown quantities so that despreading and demodulation can follow.* For the code synchronization part, in particular, the equivalent problem is to align the unknown phase ζT_c of the incoming code as closely as possible with the known (and controllable) phase $\tilde{\zeta} T_c$ of a locally generated (at the receiver) identical despreading code $c\left(t + \tau T_c\right)$.

For the signal given by (30.1.1), the likelihood function is defined as [2]

$$\lambda\left(\tilde{\zeta}\right) = \int_0^T r(t)c\left(t;\tilde{\zeta}\right) dt \tag{30.1.2}$$

which is the time-domain (coherent) correlation operation between the observed waveform and the local code, positioned at the candidate offset $\tilde{\zeta}T_c$.

The ML synchronizing receiver should, in principle, create all possible time-offset versions of the known code waveform, correlate all of them with the received signal, and choose the ζ corresponding to the largest correlation as its estimate, $\hat{\zeta}_{ML}$. In view of the continuous range of values of ζ (and ω_D, if also required), this principle is impossible to realize and some type of range quantization is necessary, which is indeed what is done in practice. The resulting candidate values are called *bins* or *cells*. Instead of simultaneously generating $\lambda\left(\tilde{\zeta}\right)$ for all cells and choosing $\hat{\zeta} = \tilde{\zeta}$ which gives the largest $\lambda\left(\tilde{\zeta}\right)$, we can implement range quantization in a serial way by creating $\lambda\left(\tilde{\zeta}\right)$ for each cell one at a time, memorizing all of them, and then choosing the cell with the largest $\lambda\left(\tilde{\zeta}\right)$. For q cells, this means replacing q correlators with one correlator and a memory of size q. We obtain further reduction if we compare each new generated $\lambda\left(\tilde{\zeta}\right)$ to a given threshold and then reject or accept the cell as a synchro cell depending on whether or not $\lambda\left(\tilde{\zeta}\right)$ is larger than the threshold. Be aware that reduction in complexity as mentioned above will increase the average acquisition time. We discuss that later in this section. This is exactly the *coarse code synchronization* or *code acquisition* problem, the result of which is to resolve the code phase (or "epoch") ambiguity within the size of the cell. Since this remaining error is typically larger than desired, further operations are required to reduce it to acceptable levels. This remaining part of the synchronization task, namely, that of **fine** *synchronization* or *code tracking*, is performed by one of the available code-tracking loops, which we discuss in the next section.

For the two-dimensional problem mentioned previously (time plus frequency uncertainty), the corresponding cells will also be two-dimensional. Their total number will equal $q = (\Delta t/\delta t)(\Delta f/\delta f)$, where δt and δf stand for the cell dimensions in the respective directions.

In current practice, total parallelism is not feasible when the number of cells is very large (although it appears feasible for smaller uncertainty regions [3], [4]) and simpler solutions are necessary. One of the most familiar of such approaches is the simple technique of serial search, namely, one where the search starts from a specific cell and serially examines the remaining cells in some direction and in a prespecified order until the correct cell is found [5], [6]. Hence, serial search techniques do not account for any additional information gathered during the past search time. This could conceivably be used to alter the direction of search toward cells that show increased posterior likelihood of being the correct ones [7] (see also [8], [9] for related concepts in the radar target-search problem).

A serial search starts from a cell, which could be chosen totally arbitrarily (no prior information) or by some prior knowledge about a likely cell, and proceeds in a simple and easily implementable predirected manner. When the uncertainty space is two-dimensional and searching all possible cells serially appears to be time consuming, then a speed-up can be achieved by use of a bank of filters, each matched to a possible Doppler offset [10]. The same idea can be applied to the one-dimensional case (no frequency uncertainty). Now, a bank of correlators can be employed, each starting from a different point of the uncertainty region. This approach effectively amounts to dividing the search into many parallel subsearches, thereby reducing the total search time by a proportional amount.

Be aware that although it holds true that only one cell contains the exact delay and Doppler offsets of the incoming code, the set of desirable cells acceptable to the receiver includes a number of cells adjacent to the exact one. All these desirable cells are collectively called *hypothesis* H_1, and the remaining undesirable "out of sync" cells consitute *hypothesis* H_0.

The generic code acquisition model discussed so far can be summarized as follows. *The two-dimensional time/frequency code offset uncertainty within the noisy received waveform is quantized into a number of cells, which are typically searched in a serial fashion by a correlation receiver, although parallel multiple branches are also possible. As directed by an ML argument, the receiver creates a cross-correlation between the incoming waveform and the local code at a specific offset, whose output is used to decide whether the currently examined cell is a desirable* (H_1) *one. The process continues until one such cell is correctly identified. At that point, acquisition is terminated and tracking is initiated.*

30.2 ANALYSIS TOOLS: SIGNAL FLOW GRAPH METHOD

A significant amount of literature and effort has been devoted to analyzing the performance of both code acquisition and code tracking systems. The goal of the analysis is the maximization of some performance criteria discussed above (such as minimum acquisition time, maximum tracking time, minimum tracking error variance) by means of appropriate parameter selection such as threshold values in acquisition detectors or gain values in closed-loop code trackers. The analysis is necessarily specific to the structure under consideration, but there exist a few key tools and general concepts that recur in all analytical efforts.

As we identified previously, the two building blocks of any acquisition receiver are the detector and the search strategy. Accordingly, the purpose of analysis is to combine these two elements to derive the statistics of the acquisition time or the overall probability of detection. In fact, the impact of the specific detector structure on the performance measure can typically be summarized by a few important parameters *per cell*, as we illustrate with a simple example below. This idea implies that optimization of the detector structure and the choice of the search strategy can proceed independently.

There exist two general analytical directions for this type of analysis: the time-domain combinatorial technique and the transform-domain (or circular-state diagram) technique. The former proceeds along total-probability arguments and is occasionally faster and more insightful, whereas the latter seems more systematic and able to handle complicated search techniques.

We can represent the code acquisition decision process by using signal flow graph techniques. Each cell is represented by a node of a graph, and transitions between nodes depend on the outcome of the decision in a given cell. These transitions are characterized by branches connecting the nodes and are usually referred to as transmission functions. If $p_{ij}(n)$ is the probability for the process to move from node i to node j in exactly n steps, then

$$P_{i,j}(Z) = \sum_{n=0}^{\infty} z^n P_{i,j}(n) \tag{30.2.1}$$

is called the *generating function* or the geometric transform. In the following analysis we need a few relations derived from this definition. First, we can represent the first and the second derivative of this function as

$$\frac{\partial}{\partial Z} = P_{ij}(Z) = \sum_{n=0}^{\infty} n p_{ij}(n) z^{n-1} \tag{30.2.2}$$

$$\frac{\partial^2}{\partial Z^2} = P_{ij}(Z) = \sum_{n=0}^{\infty} n\,(n-1)\, p_{ij}(n) z^{n-2}. \tag{30.2.3}$$

The average number of steps to move from node i to node j is

$$\overline{n} = \sum_{n=0}^{\infty} n p_{ij}(n) \frac{\partial}{\partial Z} P_{ij}(Z)\bigg|_{Z=1}, \tag{30.2.4}$$

and the average time to do it can be represented as

$$\bar{t}_{ij} = T_{ij} = \overline{n}T = \left(\frac{\partial}{\partial Z} P_{ij}^{(Z)} \bigg|_{Z=1} \right) \cdot T \tag{30.2.5}$$

where T is the cell observation time needed to create the decision variable. For the variance, we start with the definition

$$\sigma_T^2 = \left(\overline{n^2} - \overline{n}^2 \right) T^2. \tag{30.2.6}$$

Bearing in mind that the second derivative of the generating function can be represented as

$$\frac{\partial^2}{\partial Z^2} P_{ij}(Z)\bigg|_{Z=1} = \sum_{n=0}^{\infty} n^2 p_{ij}(n) - \sum_{n=0}^{\infty} n p_{ij}(n) \overline{n^2} - \overline{n}, \tag{30.2.7}$$

the variance of time t_{ij} can be expressed in the following form:

$$\sigma_T^2 = \left[\frac{\partial^2 P_{ij}(Z)}{\partial Z^2} + \frac{\partial P_{ij}(Z)}{\partial Z} - \left(\frac{\partial P_{ij}(Z)}{\partial Z} \right)^2 \right]_{Z=1} T^2. \tag{30.2.8}$$

In what follows we use these few relations to analyze serial search code acquisition. To get an initial insight into this method, we start with code acquisition in the absence of multiple-access interference. We assume that there are q cells to be searched. Now, q may be equal to the length of the PN code to be searched or some multiple of it (for example, if the update size is one-half chip, q will be twice the code length to be searched). Further, assume that if a "hit" (output is above threshold) is detected by the threshold detector, then the system goes into a verification mode that can include both an extended duration dwell time and an entry into a code-loop tracking mode. In any event, we model the penalty of obtaining a false alarm as $K_{\tau D}$ sec, and the dwell time itself as τ_D sec. If a true hit is observed, the system has acquired the signal and the search is completed. Assume the false alarm probability P_{FA} and probability of detection P_D are given. We also assume that only one cell represents the synchro position. Let each cell be numbered from left to right so that the kth cell has an a priori probability of having the signal present, given that it was not present in cells 1 through $k-1$, of

$$P_k = \frac{1}{q+1-k}. \tag{30.2.9}$$

The generating function flow diagram given in Figure 30.1 uses the rule that at each node the sum of the probability emanating from the node equals unity. The unit time represents τ_b seconds, and $k\tau_D$ seconds are represented in z-transform by z. Consider node 1. The probability a priori of having the signal present is $P_1 = 1/q$, and the probability of it not being present in the cell is $1 - P_1$. Suppose the signal were not present. Then, we advance to the next node (node $1a$); since it corresponds to a probabilistic decision and not a unit time delay, no z multiplies the branch going to it. At node $1a$, one of two things happens. 1) A false alarm occurs, with probability $P_{FA} = \alpha$, which requires one unit of time to determine ($\tau_D sec$); then K units of time ($K\tau_D sec$) are needed to determine that there is no false alarm. Or 2) there is no false alarm with probability (1- α), which takes dwell time to determine, which requires the $(1-\alpha)\,z$ branch going to node 2.

Now, consider the situation at node 1 when the signal does occur there. If a hit occurs, then acquisition, as we have defined it, occurs and the process is terminated,

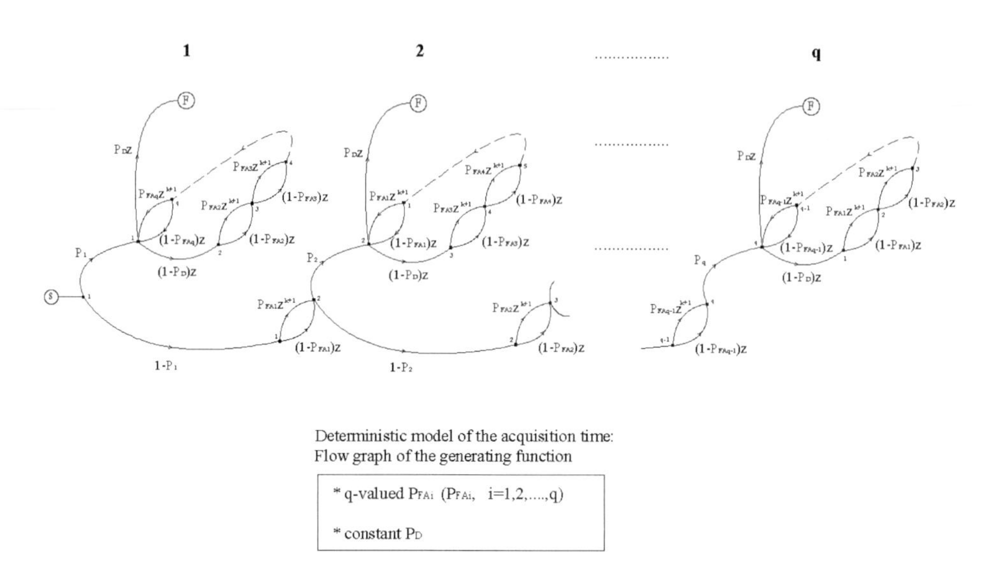

Figure 30.1. Decision-process flow diagram

hence the node F denoting "finish". If there was no hit at node 1 (the integrator output was below the threshold), which occurs with probability $1 - P_D$, then one unit of time would be consumed. This case is represented by the branch $(1 - P_D)\,z$, leading to node 2. At node 2 in the upper-left part of the diagram, either a false alarm occurs with probability α and delay $(K + 1)$ or a false alarm does not occur with a delay of 1 unit. The remaining portion of the generating function flow graph is a repetition of the portion just discussed with the appropriate node changes. We assume for the moment that there is only one spread spectrum signal in a Gaussian channel, so that P_{FA} and P_D are constant for each cell.

By using standard signal-flow, graph-reduction techniques, we can show that the overall transfer function between nodes s and F can be represented as

$$U(z) = \frac{(1-\beta)}{1-\beta z H^{q-1}} \frac{1}{q} \left[\sum_{l=0}^{q-1} H^1(z) \right] \tag{30.2.10}$$

moment generating function where

$$H(z) = \alpha z^{K+1} + (1-\alpha)z \quad \text{and} \quad \beta = 1 - P_D. \tag{30.2.11}$$

By using (30.2.5), we can give the mean acquisition time (after some algebra) as

$$\overline{T} = \frac{2 + (2 - P_D)\,(q-1)\,(1 + K P_{FA})}{2P_D} \tau_d \tag{30.2.12}$$

with τ_D being included in the formula to translate from our unit time scale. For the usual case when $q \gg 1$, $\overline{T}$ is given by

$$\overline{T} = \frac{(2 - P_D)(1 + KP_{FA})}{2P_D}(q\tau_d). \tag{30.2.13}$$

The variance of the acquisition time is given by (30.2.8). It can be shown that the expression for σ^2 is

$$\begin{aligned}\sigma^2 = \tau_D^2 \Big\{ &(1 + KP_{FA})^2 q^2 \left(\frac{1}{12} - \frac{1}{P_D} + \frac{1}{P_D^2} \right) \\ &+ 6q \left[K(K+1) P_{FA} \left(2P_D - P_D^2 \right) \right. \\ &+ \left. (1 + P_{FA}K)\left(4 - 2P_D - P_D^2\right) \right] + \frac{1 - P_D}{P_D^2} \Big\}.\end{aligned} \tag{30.2.14}$$

In addition, when $K(1 + KP_{FA}) \ll q$, then

$$\sigma^2 = \tau_D^2 (1 + KP_{FA})^2 q^2 \left(\frac{1}{12} - \frac{1}{P_D} + \frac{1}{P_d^2} \right). \tag{30.2.15}$$

As a partial check on the variance result, let $P_{FA} \to 0$ and $P_D \to 1$. Then, we have

$$\sigma^2 = \frac{(q\tau_D)^2}{12} \tag{30.2.16}$$

which is the variance of a uniformly distributed random variable, as we would expect for the limiting case. The above results provide a useful estimate of acquisition time for an idealized PN type system.

30.3 CODE ACQUISITION IN CDMA: QUASI-SYNCHRONOUS SYSTEMS

The previous section is limited to the case of the spread spectrum signal in a Gaussian channel. In that case, the probability of false alarm in all nonsynchro cells is the same. In a communication radio network, the interfering signal is a sum of Gaussian noise and overall multiple-access interference (MAI). In each cell, i, MAI has different value so that $P_{FAi} \neq P_{FAj}$ for each $i \neq j$. For such a case, under the assumption of a static channel, the serial acquisition process can be modeled again by the graph from Figure 30.1 with P_{FA} being different for each cell. We first deal with a simpler problem where probability of signal detection P_D does not depend on MAI. Besides being simpler, this model is still valid for an important class of these systems called quasi-synchronous CDMA networks. In these networks, all users are synchronized within the range between zero delay and the position of the first significant cross-correlation peak. Examples of such systems are described for both satellite and land mobile CDMA communication systems [11].

We obtain the average acquisition time by using the same steps as in the previous section. The details are presented in [12]. The result, after a cumbersome manipulation of very long equation, can be expressed as

$$\boxed{\overline{T}_{acq} = \left[2 + (q-1)\left(1 + k\overline{P}_{FA}\right)(2 - \alpha P_D)\right]\frac{\tau_d}{2P_D}}, \tag{30.3.1}$$

where

$$\alpha = \frac{1 + k\rho}{1 + k\overline{P}_{FA}} \tag{30.3.2}$$

with

$$\rho = \frac{2}{q(q-1)}\left(\sum_{i=1}^{q}(i-1)P_{FAi}\right) \tag{30.3.3}$$

and

$$\overline{P}_{FA} = \frac{1}{q}\sum_{i=1}^{q}P_{FAi}. \tag{30.3.4}$$

By inspection we can see from (30.3.1) that the minimum average acquisition time is obtained for large values of parameter α. Besides $\overline{P}_{FA}$, this parameter also depends on the position of the cells with high P_{FA_i} within the code delay uncertainty region. The set of P_{FA_i}, representing the probability distribution function of P_{FA}, will be called MAI pattern or MAI profile. From (30.3.2) and (30.3.3), we can see that for a large α, the products iP_{FA_i} should be large, which means larger P_{FA_i} for larger i. That let us hope the synchronization will be acquired before we get to the region with high P_{FA} or, in the case of the multiple sweep of the uncertainty region, we will have a smaller number of sweeps of the region.

Exact modeling of a CDMA channel with near-far problems is beyond the scope of this text, and a separate effort should be made to get an appropriate distribution of P_{FA_i} in different environments.

An example where the distribution of P_{Fai} is available is a cellular network. For a given set of codes used (e.g., Gold sequences), the base station knows exactly the sum of cross-correlation functions of the mutually interfering signals. So, for a given signal-to-noise ratio, which is controlled by a power control loop, the complete set of P_{Fai} is known. In this section we choose several examples to illustrate how the acquisition parameters depend on different patterns of P_{FAi}. The results for average acquisition time are summarized in Figure 30.2.

To facilitate the comparison of the results, we denote the simplified expression for the mean acquisition time with uniform $P_{FAi} = P$ as T_u. If the distribution P_{FAi} is linearly increasing, with notation $P_{FAq} = P_{FA\,max} = P$ and $P_{FA1} = P_{FA\,min} \ll P_{FA\,max}$ (network with severe near-far effect), the mean acquisition time is shorter than T_u. The same occurs with linearly decreasing distributions with $P_{FA1} = P_{FA\,max} = P$ and $P_{FAq} = P_{FA\,min} \ll P_{FA\,max}$, though in this case the difference is half of that obtained for linearly increasing distributions. In the former case, at the beginning of the acquisition process, P_{FAi} is low and gets higher later, while in the latter case, the acquisition process is more likely to make a wrong decision at the beginning of the process when P_{FAi} is higher than it is later on. That is why the mean time to acquisition for a linearly increasing distribution, as shown in Figure 30.2, is shorter than that for a linearly increasing distribution. The fact that in both cases $P_{FAi} \leq P$ explains why the mean acquisition times are shorter than the corresponding parameter for the uniform P_{FA} distribution.

The mean acquisition time for an alternating P_{FA1}, P_{FA2} distribution, where $P_{FA1} = P \gg P_{FA2}$, is also found to be shorter than T_u. This can be easily explained by the fact that in every second cell the probability of false alarm is low; therefore, compared with the uniform distribution, in those cells the process has less chance of giving a wrong indication of acquisition.

In the case of an inverse distribution, P_{FAi} decreases rapidly from the initial value P, and the corresponding mean acquisition time is not only lower than T_u but also lower than the mean acquisition time of the linearly decreasing distribution.

The quasi-uniform distribution can be seen as a case of uniform distribution where $P_{FAi} = P, \forall i$ with the exception of the jth cell, where $P_{FAi} = (1 + \mu)\, P$. If $\mu > 0$, and depending on the position of the jth cell, we expect to have a mean acquisition time higher than T_u. In fact, the lower the index of the irregular cell, the higher the probability of considering a wrong cell prior to acquisition and hence, the longer the acquisition time, and vice versa. This fully agrees with the result shown in Figure 30.2. Finally, the exponential distribution also has a shorter acquisition time than T_u, and when $q \gg 1$, the result is almost identical to that obtained for an inverse distribution.

Distribution of P_{FAi}, $i = 1,2,\ldots,q$	Mean acquisition time $\overline{T}_{acq}$ Exact expression $q >> 1$	Simplified expression ($q >> 1$ and $P_D \to 1$)	P_{FAi} pattern
1 Constant P_{FAi} $P_{FAi} = P,\ i = 1,2,\ldots,q$	$\frac{\tau_d}{2P_D}[2+(q-1)(1+kP)(2-P_D)]$	$T_u = \frac{\tau_d}{P_D}\left[1+\frac{q(1+kP)}{2}\right]$	
2 Linearly increasing P_{FAi} $P_{FA1} = P_{FA\,min},\ P_{FAq} = P_{FA\,max}$ $P = P_{FA\,max} >> P_{FAmin}$	$\frac{\tau_d}{2P_D}\left[2+\frac{(q-1)}{3}(3(2-P_D)+kP(3-2P_D))\right]$	$T_u - q\frac{kP}{3}\tau_d$	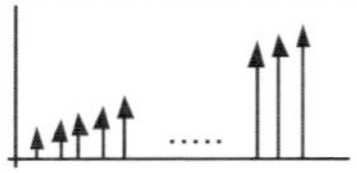
3 Linearly decreasing P_{FAi} $P_{FA1} = P_{FA\,max},\ P_{FAq} = P_{FA\,min}$ $P = P_{FA\,max} >> P_{FAmin}$	$\frac{\tau_d}{2P_D}\left[2+\frac{(q-1)}{3}(3(2-P_D)+kP(3-P_D))\right]$	$T_u - q\frac{kP}{6}\tau_d$	
4 Alternating P_{FA1} and P_{FA2} (even q) $P_{FAi} = \begin{cases} P_{FA1}, & i = 1,3,\ldots,q-1 \\ P_{FA2}, & i = 2,4,\ldots,q \end{cases}$ $P = P_{FA1} >> P_{FA2}$	$\frac{\tau_d}{2P_D}\left[2+\frac{(q-1)(2+kP)(2-P_D)}{2}\right]$	$T_u - q\frac{kP}{4}\tau_d$	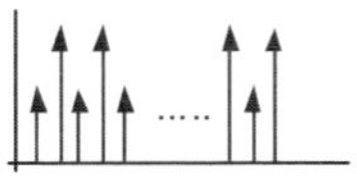
5 Inverse P_{FAi} distribution $P_{FAi} = P/i,\ i=1,2,\ldots,q;\quad s = \sum_{i=1}^{q} 1/i$	$\frac{\tau_d}{2P_D}\Big[2+(q-1)(2-P_D)-$ $-2k\frac{s}{q}P(1-P_D)+2kP(s-P_D)\Big]$	$T_u - (q-2s)\frac{kP}{2}\tau_d$	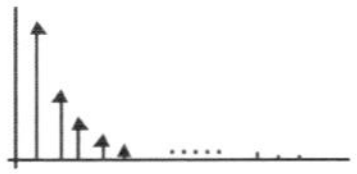
6 Quasi-uniform distribution $P_{FAi} = \begin{cases} P & \forall i, i \neq j \\ (1+\mu)P, & i = j \end{cases},\quad a = \frac{2}{q-1}(j-1)$ $\mu > 0$	$\frac{\tau_d}{2P_D}\Big[2+(q-1)[(2-P_D)(1+kP)+$ $+kP\frac{\mu}{q}(2-aP_D)]\Big]$	$T_u + (2-a)\mu\frac{kP}{2}\tau_d$	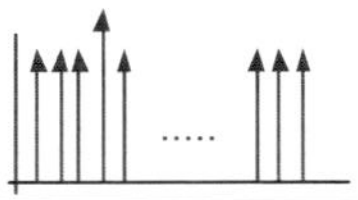
7 Exponential distribution $P_{FAi} = P\exp(\lambda i),\ i = 1,2,\ldots,q;\quad \lambda < 0$	$\frac{\tau_d}{2P_D}\left[2+q(2-P_D)+2kP\frac{\exp\lambda}{(1-\exp\lambda)}\right]$	$T_u - q\frac{kP}{2}\tau_d$	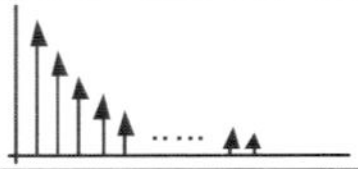

Figure 30.2. Exact and simplified expressions of mean acquisition time for different P_{FAi} distributions

30.4 CODE ACQUISITION IN CDMA: ASYNCHRONOUS SYSTEMS

In an asynchronous network MAI takes on different values in all cells, including the synchro cell, so in general P_D is different. For such a case, the average acquisition time becomes [13]

$$\overline{T}_{acq} = \frac{\tau_d}{2\tilde{P}_D}\left[2 + \left(1 + k\overline{P}_{FA}\right)(q-1)\left(2 - \alpha\tilde{P}_D\right) + 2k\left(\overline{P}_{FA} - \overline{P}_R\tilde{P}_D\right)\right] \quad (30.4.1)$$

where

$$\overline{P}_{FA} = \frac{1}{q}\sum_{i=1}^{q} P_{FAi}\,,\ \overline{P}_R = \frac{1}{q}\sum_{i=1}^{q}\frac{P_{FAi}}{P_{Di}},\ \tilde{P}_D = \left[\frac{1}{q}\sum_{i=1}^{q}\frac{1}{P_{Di}}\right]^{-1} \quad (30.4.2)$$

$$\alpha = \frac{1+k\rho}{1+k\overline{P}_{FA}} \text{ and } \rho = \frac{2}{q(q-1)}\left(\sum_{i=1}^{q}(i-1)P_{FAi}\right).$$

It is interesting to compare the expression for the mean acquisition time with previous results. Table 30.1 summarizes the results obtained for case #1, constant P_{FA} and P_D, case #2, q-valued P_{FA} and a constant P_D in quasi-synchronous networks, and case #3, q-valued P_{FA} and q-valued P_D in asynchronous networks.

The form of the three expressions provides insight into the major differences in average acquisition times for the three cases. In the expression for case #2, when compared with case #1, P_{FA} should be replaced by $\overline{P}_{FA}$, and P_D in the numerator should be modified by a factor α given by (30.4.2). The first factor takes into account the average P_{FA}, and the second modification takes into account the position of the initial search cell with respect to the distribution of P_{FAi}. In the expression for case #3, when compared with case #2, P_D should be replaced by $\tilde{P}_D$ in addition to a new term Δ that should be added to the numerator. This term can be expressed as $\Delta = 2k\left(\overline{P}_{FA} - \overline{P}_R\tilde{P}_D\right)$.

A first observation is that a sufficient condition for Δ to be zero is that P_{FA} or P_D or both have a constant distribution, that is, at least one of the following conditions is met: $P_{FAi} = P_{FA}, i = 1, 2, \ldots, q$ or $P_{Di} = P_D, i = 1, 2, \ldots, q$. A proof is straightforward from the definitions of $\overline{P}_{FA}$, $\overline{P}_R$ $\tilde{P}_D$ and Δ. Since $\overline{P}_{FA} \le \overline{P}_R$ and $\tilde{P}_D \le 1$, the sign of Δ cannot be determined without knowledge of the particular distributions of P_{FA} and P_D. From the definition of $\tilde{P}_D$, one can see that $\tilde{P}_D \to \overline{P}_D$ as long as $P_{Di} \approx 1, i = 1, 2, \ldots, q$. However, only one P_D needs to be small to cause a considerable reduction of the final value of $\tilde{P}_D$. The variation of $\tilde{P}_D$ also depends on the number of cells q.

Practical examples

To gain an additional insight into the system behavior, we present some approximations. For a CDMA network with K users, the received overall signal can be

Distribution of P_{FA} and P_D	Mean acquisition time $\overline{T}_{acq}$
case #1 $P_{FAi} = P_{FA}$, $\forall i$ fixed $P_{Di} = P_D$, $\forall i$ fixed	$\frac{\tau_d}{2P_D}\left[2 + (1 + kP_{FA})(q-1)(2 - P_D)\right]$
case #2 $P_{FAi} = \{P_{FA1}, P_{FA2}, \ldots, P_{FAq}\}\,(q - valued)$ $P_{Di} = P_D$, $\forall i$ fixed	$\frac{\tau_d}{2P_D}\left[2 + (1 + k\overline{P}_{FA})(q-1)(2 - \alpha P_D)\right]$
case #3 $P_{FAi} = \{P_{FA1}, P_{FA2}, \ldots, P_{FAq}\}\,(q - valued)$ $P_{Di} = \{P_{D1}, P_{D2}, \ldots, P_{Dq}\}\,(q - valued)$	$\frac{\tau_d}{2P_D}\left[2 + (1 + k\overline{P}_{FA})(q-1)\left(2 - \alpha\tilde{P}_D\right) + +2k\left(\overline{P}_{FA} - \overline{P}_R\tilde{P}_D\right)\right]$

Table 30.1. Mean acquisition time for different distributions of P_{FA} and P_D

represented as

$$r(t) = \sum_{i=1}^{K} b_i c_1 \cos(\omega t + \theta_i) + n(t) \tag{30.4.3}$$

where b_i, c_i, and θ_i are bit, code, and phase of the signal transmitted by user with index i.

For simplicity, a real signal is assumed. Extension to a complex signal (I&Q) is straightforward.

In the receiver with index k after frequency downconversion and correlation with code c_k, we have

$$\begin{aligned} y_k &= \sum_{i=1}^{K} b_i \rho_{ik}(\tau) \cos\epsilon_{ki} + n_k \qquad (30.4.4)\\ &= b_k \rho_{kk}(\tau) \cos\epsilon_{kk} + \sum_{\substack{i=1\\ i\neq k}}^{k} b_i \rho_{ik}(\tau) \cos\epsilon_{ki} + n_k \\ &= b_k \rho_{kk}(\tau) \cos\epsilon_{kk} + m_k(\tau) + n_k \end{aligned}$$

where $\rho_{ik}(\tau)$ is cross-correlation between codes with indices k and i and $\epsilon_{ki} = \theta_i - \hat{\theta}_k$. The multiple access interference term

$$m_k(\tau) = \sum_{\substack{i=1\\ i\neq k}}^{k} b_i \rho_{ik}(\tau) \cos\epsilon_{ki} \tag{30.4.5}$$

can be represented in each cell of the delay uncertainty region as a Gaussian zero-mean variable with variance

$$\sigma_{mk}^2(\tau) = \sum_{\substack{i=1\\ i\neq k}}^{k} \rho_{ik}^2(\tau). \tag{30.4.6}$$

To average out the impact of MAI on the code acquisition process, we approximate $\sigma_{mk}^2(\tau)$ with the average value $\sigma_m^2 = E_\tau\left\{\sigma_{mk}^2(\tau)\right\}$. Furthermore, we use discrete approximation of the correlation function $\rho(\tau) \to \rho(lT_c) \to \rho(l)$, where $l \in (1, N)$ and N is the code length. Therefore, the previous equation can be expressed as

$$\sigma_m^2 = E_\tau\left\{\sigma_{mk}^2(\tau)\right\} = \frac{K-1}{N}\sum_{l=0}^{N-1}\rho_{xy}^2(l) \tag{30.4.7}$$

where x and y are two arbitrary sequences from the same set of sequences. In the sequel, we evaluate this expression for different classes of codes that are considered for applications in UMTS systems. For m sequences we use the well-known relation [14], which is based on the Cauchy inequality

$$\sum_{l=0}^{N-1}\rho_{xy}^2(l) < \rho_x(0)\rho_y(0) + \sqrt{\sum_{l=0}^{N-1}\rho_x^2(l)} \times \sqrt{\sum_{l=0}^{N-1}\rho_y^2(l)} \tag{30.4.8}$$

and

$$\sum_{l=0}^{N-1}\rho_{xy}^2(l) > \rho_x(0)\rho_y(0) - \sqrt{\sum_{l=0}^{N-1}\rho_x^2(l)} \times \sqrt{\sum_{l=0}^{N-1}\rho_y^2(l)}. \tag{30.4.9}$$

For m sequence $\rho(0) = 1$ and $\rho(l) = -1/N$ for $l \neq 0$. Therefore, we have for the upper and lower bound on σ_m^2 the following expressions:

$$\sigma_m^{2\pm} = \frac{K-1}{N}\left\{1 \pm \frac{1}{N}\right\} \cong \frac{K-1}{N} \tag{30.4.10}$$

which for $N \gg 1$ becomes $(K-1)/N$. For short codes, this should be further elaborated in more detail. For a preferred pair of sequences [1], [14] obtained by decimation by factor $r = 2^k + 1$ and sequence length $N = 2^n - 1$, $e = gcd(2k, n)$, and e divides k, the three-valued cross-correlation is given as

$$N\rho_{xy}(l) = \begin{cases} -1; 2^n - 2^{n-e} - 1 \quad times \\ -1 + 2^{(n+e)/2}; 2^{n-e-1} + 2^{(n-e-2)/2} \quad times \\ -1 - 2^{(n+e)/2}; 2^{n-e-1} - 2^{(n-e-2)/2} \quad times. \end{cases} \tag{30.4.11}$$

For a specific choice of these parameters we have:
a) Gold sequences

$$r = 2^{\lfloor (n+2)/2 \rfloor} \tag{30.4.12}$$
$$\lfloor\ \rfloor int\ eger\ of$$
$$e = \begin{cases} 1;\ n\ odd \\ 2\ n\ even \end{cases}$$

b)Small set of Kassami sequences

$$n \quad even \\ r = 2^{n/2} + 1. \tag{30.4.13}$$

By using (30.4.11), we have the general case

$$\sigma_m^2 = \frac{K-1}{N^3} \left\{ 2^n - 2^{n-e} - 1 + \left(-1 + 2^{(n+e)/2}\right)^2 \left(2^{n-e-1} + 2^{(n-e-2)/2}\right) + \left(-1 - 2^{(n+e)/2}\right)^2 \left(2^{n-e-1} - 2^{(n-e-2)/2}\right) \right\}. \tag{30.4.14}$$

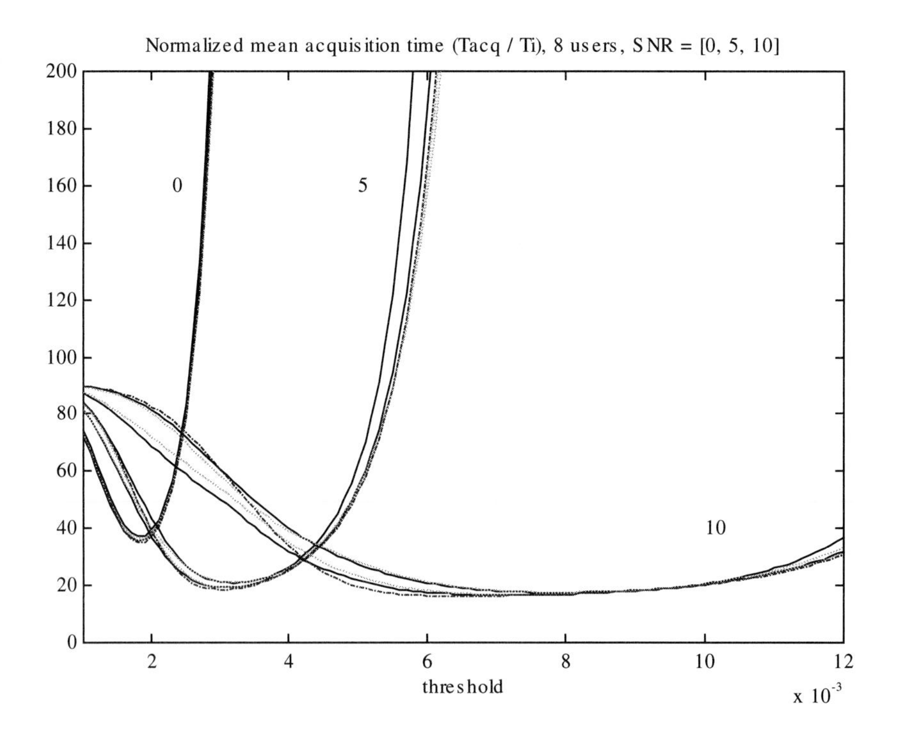

Figure 30.3. Upper and lower bounds of the mean acquisition time for 30 realizations of a random phase shift vector, $TiB = 10, K = 5$. Solid line: T_{acq1}; Dotted line: T_{acq2}; Dashdot line: T_{acq3}.

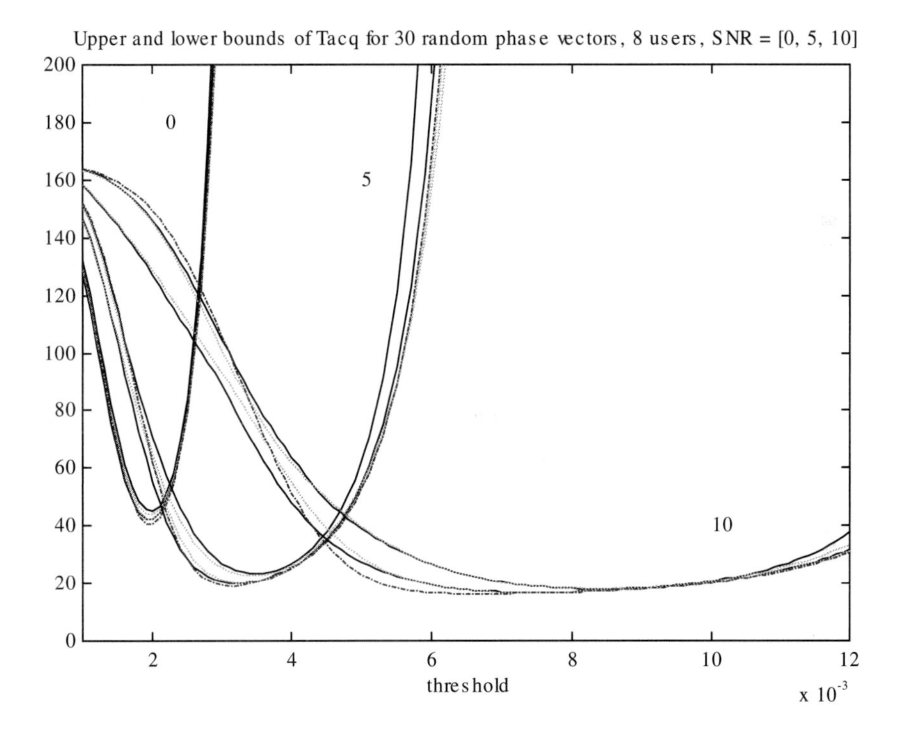

Figure 30.4. Upper and lower bounds of the mean acquisition time for 30 realizations of a random phase shift vector, $TiB = 10, K = 10$. Solid line: T_{acq1} ; Dotted line: T_{acq2} ; Dashdot line: T_{acq3}.

For illustration, we generate MAI by using eight Gold codes of length 31, generated by a preferred pair of polynomials $g(x) = 1 + x^2 + x^5$ and $g'(x) = 1 + x^2 + x^3 + x^4 + x^5$. Vector $\boldsymbol{\tau} = (\tau_1, \tau_2, \tau_3, \tau_4, \ldots, \tau_8)$ defines initial delays (in chips) of the sequences. By using known results for a square law detector followed by an integrator (integration interval T_i and signal bandwidth B), we can evaluate parameters $\overline{P}_{Di}, \overline{P}_{FAi}$. Figures 30.3 and 30.4 illustrate the preceding by showing three parallel results for the normalized, average acquisition time (T_{acq}/T_i).

T_{acq1} is obtained by using the exact result given by (31.3.1) (case #3 in Table 30.1), T_{acq2} is the approximation where the standard expression for T_{acq} is used (case #1 in Table 30.1) with $P_{FA} \Longrightarrow \overline{P}_{FA}$ and $P_D \Longrightarrow \overline{P}_D$, and T_{acq3} is the approximation where MAI is approximated by Gaussian noise with variance σ_m^2 given by (30.4.6) through (30.4.14).

In addition to the signal flow graph method discussed in this section, direct (time domain) and circular diagram methods are also used in this analysis. These methods are not discussed in this text; interested readers are referred to [15] and references therein.

In the case of multipath propagation, the previous analysis can be used with some approximations. Let us introduce a set of assumptions to simplify the analysis:

- Code period $NT_c \gg$ *delay spread* τ_{spread}
- The number of relevant signal components L
- Probability of detecting the l-th signal component P_{dl}
- Probability of detecting at least one component in a sweep of the delay uncertainty region is $P_{D_{total}} = 1 - \prod_l (1 - P_{dl})$

A good approximation for the average acquisition time in a multipath channel can be obtained by using results from the previous section and $P_D = P_{D_{total}}$.

Bibliography

[1] G. S. Glisic and B. Vucetic, *Spread Spectrum CDMA Systems for Wireless Commuications.* Artech House, 1997.

[2] S. Glisic and B. Vucetic, *Spread Spectrum CDMA Systems for Wireless Communications.* London, Artech House, 1997.

[3] L. D. Davisson and P. G. Flikkema, "Fast single-element PN acquisition for the TDRSS MA system," *IEEE Trans. Comm.*, vol. 36, pp. 1226–1235, 1988.

[4] E. A. Sourour and S. C. Gupta, "Direct-sequence spread-spectrum parallel acquisition in a fading mobile channel," *IEEE Trans. Comm.*, vol. 38, pp. 992–998, 1990.

[5] A. Polydoros and C. L. Weber, "A unified approach to serial search spread-spectrum code acquisition—part 1: General theory," *IEEE Trans on Communications*, vol. 32, pp. 542–549, 1984.

[6] G. F. Sage, "Serial synchronization of pseudonoise systems," *IEEE Trans. Comm.*, vol. 12, pp. 123–127, 1964.

[7] J. K. Holmes and K. T. Woo, "An optimum PN code search technique for a given a priori signal location density," in *NTC 78 Conf. Record*, (Birmingham, Alabama), pp. 18.6.1–18.6.5, 1978.

[8] C. Gumacos, "Analysis of an optimum sync search procedure," *IRE Trans. on Communications Systems*, vol. 11, pp. 89–99, 1963.

[9] E. C. Posner, "Optimal search procedures," *IEEE Trans. on Information Theory*, vol. IT-11, pp. 157–160, 1963.

[10] U. Cheng, W. Hurd, and J. Statman, "Spread spectrum code acquisition in the presence of Doppler shifts and data modulation," *IEEE Trans. Comm.*, vol. 38, pp. 241–250, 1990.

[11] J. W. Sheen and G. Stueber, "Effects of multipath fading on delay locked loops for spread spectrum systems," *IEEE Trans. Comm.*, vol. 42, pp. 1947–1956, 1994.

[12] M. Katz and S. Glisic, "Modeling of code acquisition process in cdma networks: Quasi-synchronous systems," *IEEE Trans. Communications*, vol. 46, pp. 1564–1569, 1998.

[13] M. Katz and S. Glisic, "Modeling of code acquisition process in cdma networks: Asynchronous systems," *IEEE J. Selected Areas Comm.*, vol. 18, pp. 73–86, 2000.

[14] R. Muirhead, *Aspects of Multivariate Statistical Theory*. New York: Wiley, 1982.

[15] A. Polydoros and S. Glisic, "Code synchronisation: A review of principles and techniques," in *Code Division Multiple Access* (S.Glisic and P.Leppanen, eds.), Kluwer Academic Publishers, 1995.

Chapter 31

CODE TRACKING

31.1 BASEBAND FULL-TIME EARLY-LATE TRACKING LOOP

Instead of searching for the maximum of a likelihood function, one can take a derivative of the function and use it as a control signal to change the local signal parameters for as long as it takes to get to the point where the derivative is zero. In general, schemes based on this approach are called *trackers*. Further, if the derivative is approximated as a difference, then the tracker becomes an *early-late tracker (ELT)* [1].

To gain an initial understanding of the code tracking loop operation, we consider first a baseband loop. A conceptual block diagram of the loop is shown in Figure 31.1. This is an ELT motivated scheme. In the code acquisition process, the normalized delay error between the input and local sequence $\delta = (\tau - \hat{\tau})/T_c$ is reduced to the range $\delta < 1$. The purpose of the code delay-tracking loop is to further reduce this error to zero and then to track any changes in τ. In the literature, this loop is known as a delay lock loop, or DLL [1]. The input signal is correlated with two locally generated, mutually delayed replicas of the PN code. After filtering, the useful component of the control signal $\epsilon(t)$ is used to control VCO output signal phase. For the analysis of the tracking error variance, results from standard phase lock loop theory can be used directly [1].

31.2 FULL-TIME EARLY-LATE NONCOHERENT TRACKING LOOP

Except in ranging, in most applications the input signal in DLL will be a complete DSSS signal. To get rid of information and unknown phase, we use a noncoherent structure, as shown in Figure 31.2. The tracking error variance can be expressed as [1]

$$\sigma_\delta^2 = \frac{\eta B_L}{S^2} \tag{31.2.1}$$

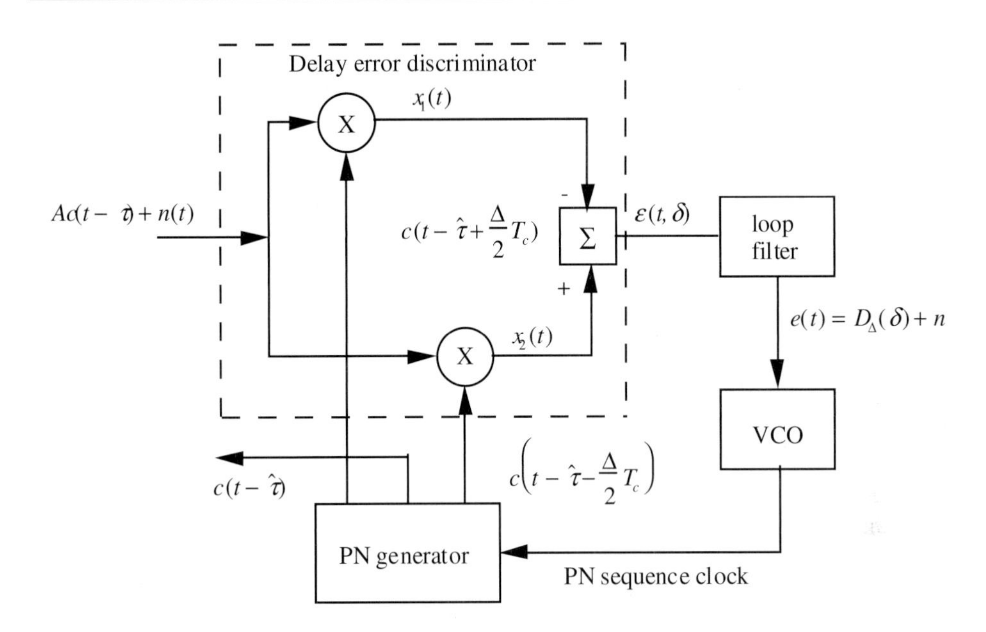

Figure 31.1. Baseband code tracking loop

where η is the equivalent output noise power density, B_L is the loop bandwidth, and S is the slope of the error detector S-curve. For the noncoherent early-late loop, this equation becomes [1]

$$\begin{aligned}\sigma_\delta^2 &= \frac{\eta B_L}{2A^4} \\ &= \frac{N_0 B_L}{2A^4}\left[N_0 B_N + \frac{A^2}{2}\right] \\ &= \frac{1}{2\rho_L}\left(1+\frac{2}{\rho_{if}}\right)\end{aligned} \tag{31.2.2}$$

where

$$\begin{aligned}\rho_L &= \frac{2A^2}{N_0 B_L} \\ \rho_{if} &= \frac{A^2}{N_0 B_N}.\end{aligned} \tag{31.2.3}$$

Parameter ρ_L is loop signal-to-noise power ratio and ρ_{if} is signal-to-noise power ratio at the output of IF (band-pass) filter. For a coherent DLL, σ_δ^2 is represented only by the first term of (31.2.2) [1]. The second term, which represents a degradation compared with the coherent loop, is due to additional modification of the derivative of the likelihood function, needed to get rid of data and signal phase.

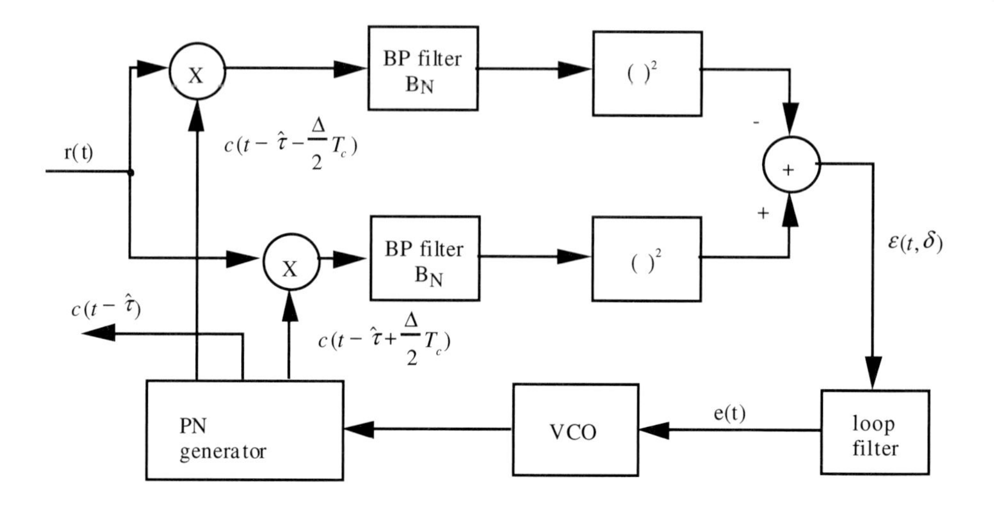

Figure 31.2. Full-time early-late noncoherent tracking loop

31.3 TAU-DITHER EARLY-LATE NONCOHERENT TRACKING LOOP

The loop operation is explained by reference to diagrams from Figure 31.3. The input signal plus noise is correlated with local PN sequence, and after BP filtering, the signal amplitude will be proportional to the correlation function $R_c(\delta)$. Suppose that δ is in region 1^0 shown in Figure 31.3. In periodic time intervals T_d, the phase of the local sequence is changed from $\hat{\tau} + \frac{\Delta}{2}T_c$ to $\hat{\tau} - \frac{\Delta}{2}T_c$, corresponding to the normalized delay error δ_{a1} and δ_{b1}, respectively. Whenever control signal $q(t) = \pm 1$ has a value $+1$, δ will be lower, and vice versa.

Envelope variations of the band-pass filter output signal for the two regions are shown in Figure 31.3. If these envelope variations are detected (square law device plus LP filter) and multiplied with the control signal, $q(t)$, $\epsilon(t, \delta)$, shown in the figure, is obtained. One can see that for the two regions, where δ is of the opposite sign, the DC components of $\epsilon(t, \delta)$ are also of the opposite sign. This provides a correct control signal for VCO. If $\delta = 0$, $q(t)$ will change δ from δ_{a3} to δ_{b3} and the envelope of $z(t)$ will be the same in both cases. This will result in a zero-value control signal. To use (31.2.1) for the tracking-error variance evaluation, we must evaluate the slope of the DC component at $\delta = 0$ and the power density function of the noise component of $e(t)$. For the case when $\Delta = 1,0$, $N \gg 1$ and $1/T_d = B_N/4$, the result is

$$\sigma_{\delta}^2 = \frac{1}{2\rho_L}\left(1.811 + \frac{3.261}{\rho_{if}}\right) \tag{31.3.1}$$

where ρ_L and ρ_{if} are given again by (31.2.3). The advantage of the Tau-dither loop

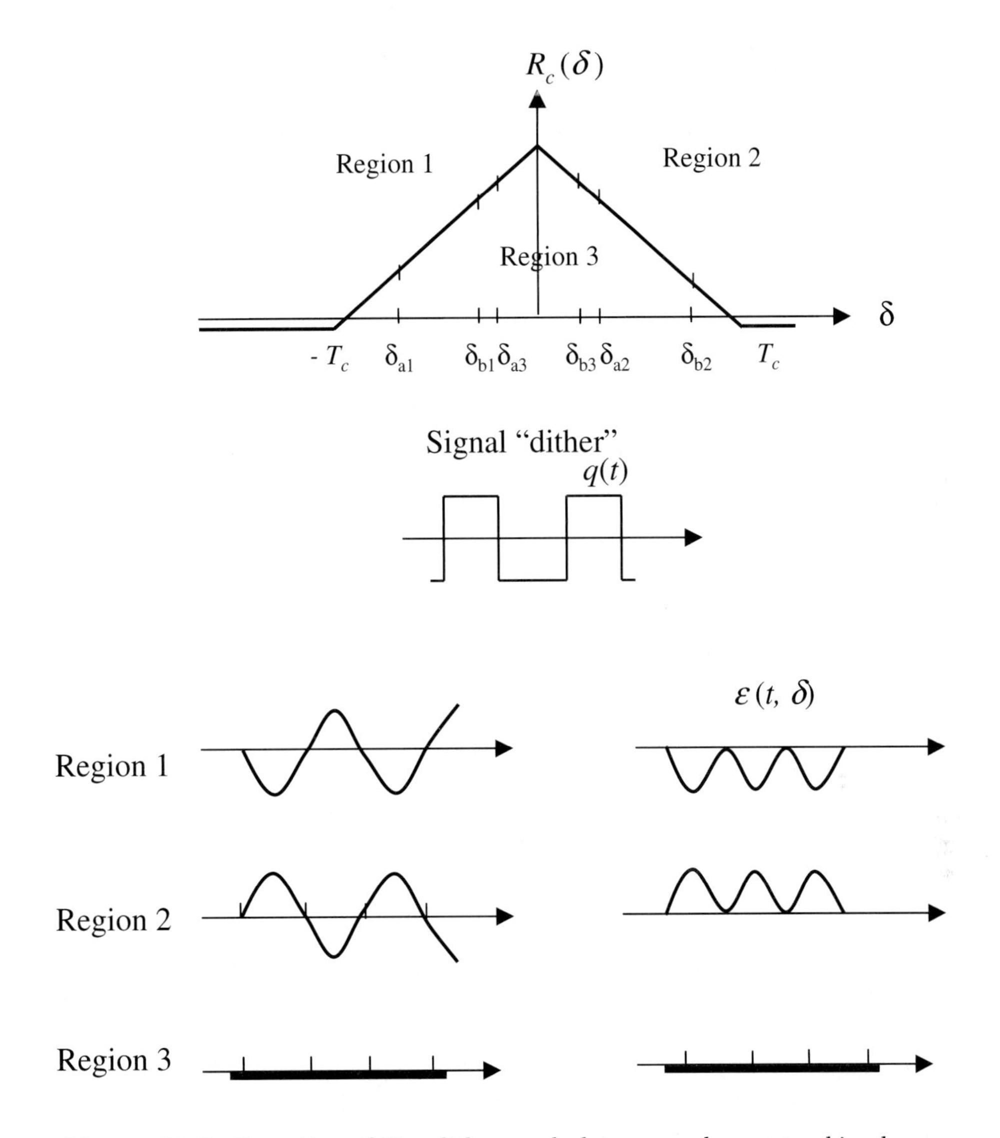

Figure 31.3. Operation of Tau-dither, early-late, noncoherent tracking loop

is low hardware complexity, but the tracking error is almost twice as large as in the case of the DLL. One should be aware that unbalanced circuitry in the DLL would result in a constant bias in delay tracking. One way around this problem is a double-dither, early-late, noncoherent tracking loop, described in [1]. In this loop, two arms are periodically switched, and the average value $\mathbf{D}(\delta)$ will be an effective control signal that will not be biased.

31.4 EFFECTS OF MULTIPATH FADING ON DELAY-LOCKED LOOPS

In this section, we discuss the effects of a specular multipath fading channel on the performance of a DLL. For this type of environment, the two-path channel model becomes

$$h(\tau) = \sqrt{2P}\left\{\delta(\tau-\tau_1)e^{j\theta_1} + h_2 e^{j\theta_2}\delta(\tau-\tau_1-\tau_d)\right\} \tag{31.4.1}$$

where θ_1 is a constant phase shift and h_2 and θ_2 are Rayleigh- and uniform-distributed random variables, respectively. When $\tau_d = 0$, the channel becomes the familiar frequency-nonselective Rician fading model.

In order to present some quantitative results, we need the following important system parameters: the power ratio of the main path to the second path, R; the bit SNR (SNR in data bandwidth) γ_d; the loop SNR for $\Delta = 1$, γL_0; and the ratio $\varsigma_0 = \gamma L_0/\gamma_d$. The parameters R, γ_d, and γL_0 are defined as follows:

$$R \underline{\underline{\Delta}} \frac{1}{E\left[h_2^2\right]} \tag{31.4.2}$$

$$\gamma_d \underline{\underline{\Delta}} \frac{PT_b}{N_0}$$

and

$$\gamma_{L_0} \underline{\underline{\Delta}} \frac{P}{N_0 B_L |\, \Delta = 1} \tag{31.4.3}$$

where T_b is the duration of an information bit and B_L is the closed-loop bandwidth for the case when $g_2 = 0$. That is,

$$B_L = \int_{-\infty}^{\infty} |H(f)|^2\, df \tag{31.4.4}$$

where $H(s)$ is the closed-loop transfer function.

In Figure 31.4, the DLL discriminator characteristic $S(\epsilon) = S_D(\epsilon) + S_1(\epsilon)$ is presented, where $S_D(\epsilon)$ is the desired discriminator characteristic and $S_I(\epsilon)$ is the interference that is caused by the effect of the second path. Figure 31.4 shows the overall discriminator characteristic $S(\epsilon) = S_D(\epsilon) + S_1(\epsilon)$ for the case of $\Delta = 1$, $h_1^2 = 1$, $h_2^2 = 0.5$, $\tau_d = 0.5T_c$, and $\theta_1 = \theta_2$.

By using the standard phase lock loop theory [2], we can evaluate the tracking-error variance for this case, with the results shown in Figure 31.5.

Effects of multipath fading on the normalized mean time to lose lock (MTLL) versus early-late discriminator offsets $\Delta/2$ are shown in Figure 31.6.

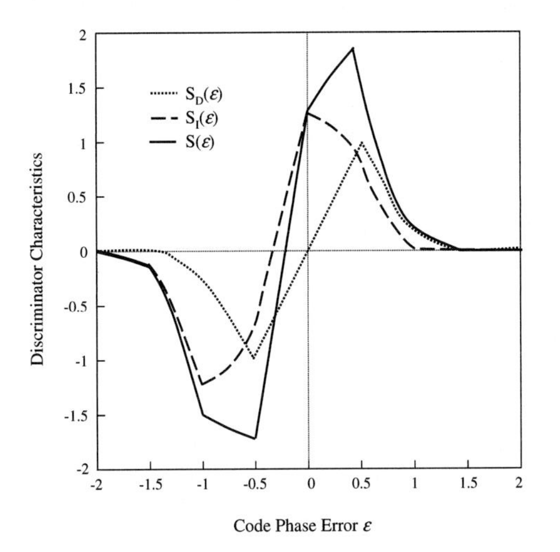

Figure 31.4. Early-late discriminator characteristics for the case of $\Delta = 0.5, h_1^2 = 1, h_2^2 = 0.5, \tau_d = 0.5T_c$, and $\theta_1 = \theta_2 \ \cdot S_D(\epsilon)$ is the desired discriminator characteristic, and $S_1(\epsilon)$ is the interference that is caused by the effect of the second path.

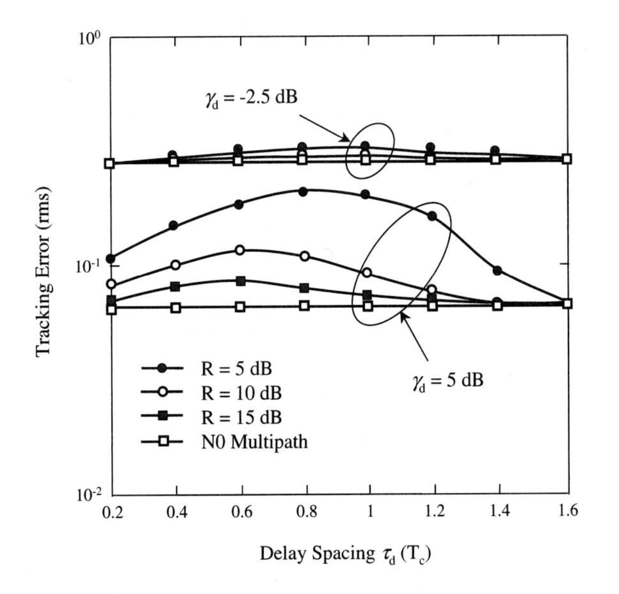

Figure 31.5. Effects of multipath fading on the tracking error performance with various delay spacings ($\Delta = 0.5, \zeta_0 = 100$)

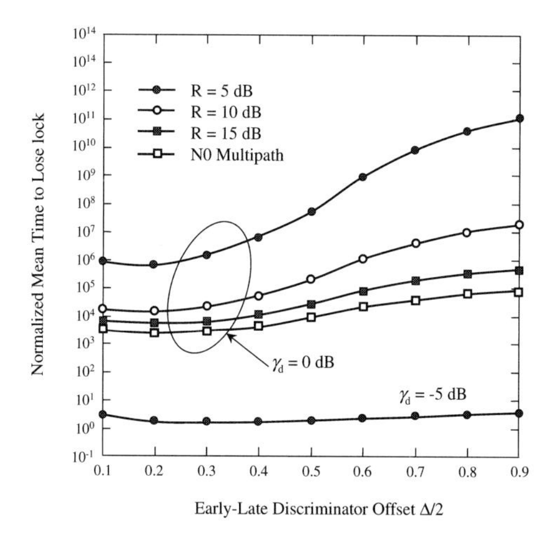

Figure 31.6. Effects of multipath fading on the MTLL performance with various early-late discriminator offsets ($\tau_d = 0.5, \zeta_0 = 100$)

Bibliography

[1] G. S. Glisic and B. Vucetic, *Spread Spectrum CDMA Systems for Wireless Commuications.* Artech House, 1997.

[2] J. W. Sheen and G. Stueber, "Effects of multipath fading on delay locked loops for spread spectrum systems," *IEEE Trans. Comm.*, vol. 42, pp. 1947–1956, 1994.

Chapter 32

CHANNEL ESTIMATION FOR CDMA SYSTEMS

32.1 IDENTIFICATION OF CHANNEL COEFFICIENTS

After code synchronization (acquisition and tracking), signal despreading can be performed. If the processing gain is large $T_b/T_c \geq 1$, after despreading the received low-pass equivalent discrete time signal is

$$y_k = x_k c_k + n_k \tag{32.1.1}$$

where h_k is channel gain. The residual fading is frequency nonselective because all multipath components are resolved in the despreading process in each finger of the Rake receiver. If x_k is a known training symbol and if the signal-to-noise ratio (SNR) is high, a good estimate of h_k can be easily computed as

$$h_k \approx y_k / x_k \underline{\underline{\Delta}} \tilde{c}_k \tag{32.1.2}$$

according to (32.2.17), where y_k is the received signal. However, most of the received symbols are not training symbols. In these cases, the available information for estimating h_k can be based upon prediction from the past-detected, data-bearing symbols $\overline{x}_i$ $(i < k)$.

Using a standard linear prediction approach, we formulate the predicted fading channel gain at time k as

$$\hat{c}_k = \sum_{i=1}^{N} b_i^* \tilde{c}_{k-i} \underline{\underline{\Delta}} \mathbf{b}(k)^H \tilde{\mathbf{c}}(k) \tag{32.1.3}$$

where

$$\tilde{\mathbf{h}}(k) = \left(\tilde{h}_{k-1}, \tilde{h}_{k-2}, \ldots, \tilde{h}_{k-N}\right)^T \tag{32.1.4}$$

is a vector of past corrected channel gain estimates and

$$\mathbf{b}(k) = \left(b_1, b_2, \ldots, b_N\right)^T \tag{32.1.5}$$

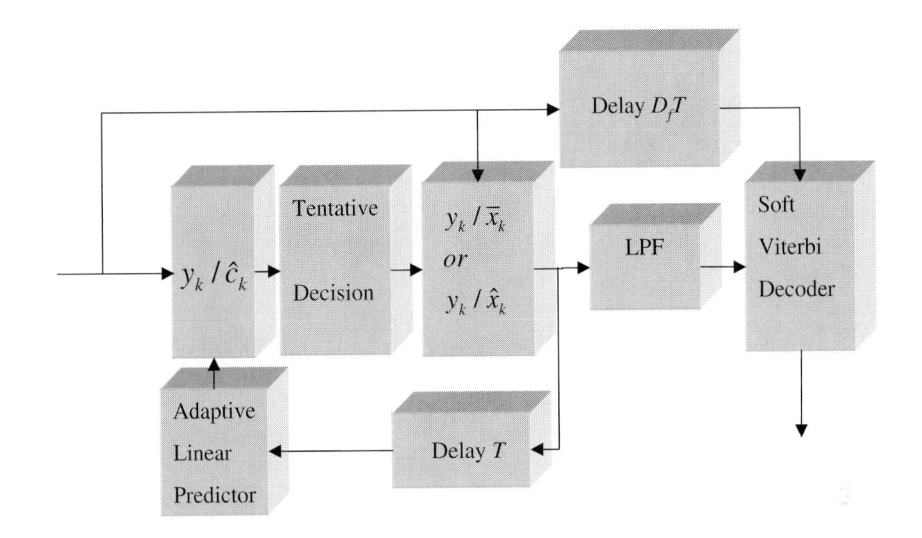

Figure 32.1. The DFALP algorithm for tracking phase and amplitude of frequency-nonselective fading channels

are the filter (linear predictor) coefficients at time k. The superscript T stands for transpose and H stands for Hermitian transpose. The constant N is the order of the linear predictor. The block diagram of the receiver is shown in Figure 32.1.

The updating process for the filter coefficients is defined as

$$\mathbf{b}(k+1) = \mathbf{b}(k) + \mu\left(\tilde{h}_k - \hat{h}_k\right)^{\star} \tilde{\boldsymbol{h}}(k). \tag{32.1.6}$$

Simulation results for BER versus predictor order N and the number of taps $2D_f + 1$ of the low pass filter are shown in Figures 32.2 and 32.3.

32.2 SUBSPACE-BASED METHOD

In this section we present a multiuser channel estimation problem through a signal subspace-based approach [1]. For this purpose, we present the received signal for K users as

$$r(t) = \sum_{k=1}^{K} r_k(t) + \eta_t \quad -\infty < t < \infty \tag{32.2.1}$$

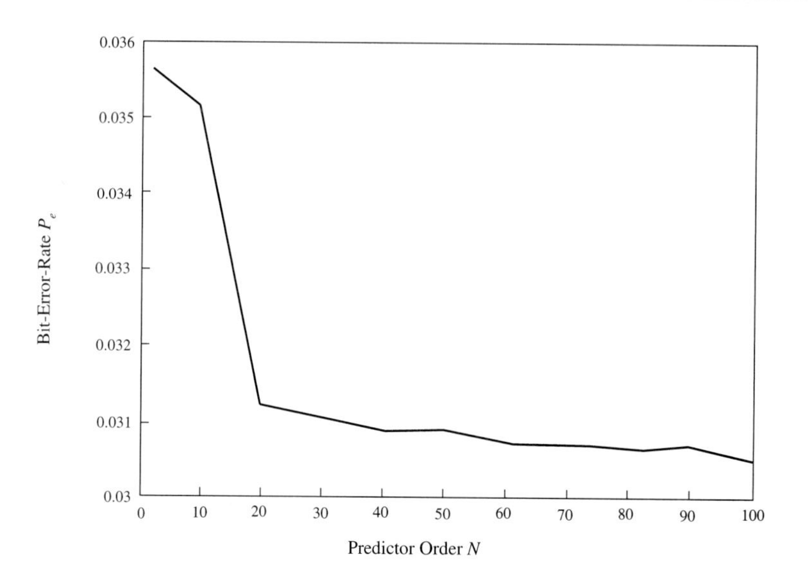

Figure 32.2. The BER of DFALP for different predictor order. $\nu = 60\ km/hour$; $f_s = 6000\ symbols/second$; $f = 5\ GHz$; $K = 4\ dB$; $f_mT = 0.0463, \gamma = 10\ dB/bit$; *and* $K_t = 5$.

$$\begin{aligned} r_k(t) &= h_k(t, \tau)^\star s_k(t) \\ &= \int_{-\infty}^{\infty} h_k(t, \alpha)\, s_k(\alpha)\, d\alpha\,. \end{aligned} \tag{32.2.2}$$

If phase-shift keying (PSK) is used to modulate the data, then the baseband complex envelope representation of the kth user's transmitted signal is given by

$$s_k(t) = A_k e^{j\phi_k} \sum_{1} e^{j(2\pi/M)m_k^{(i)}} c_k(t - iT) \tag{32.2.3}$$

where A_k is the transmitted signal amplitude, ϕ_k is the carrier phase relative to the local oscillator at the receiver, M is the size of the symbol alphabet, $m_k^{(i)} \in \{0, 1, \ldots, M-1\}$ is the transmitted symbol, $c_k(t)$ is the spreading waveform, and T is the symbol duration.

The spreading waveform is given by

$$c_k(t) = \sum_{n=0}^{N-1} p_{T_c}(t - nT_c)\, c_k^{(n)} \tag{32.2.4}$$

where $p_{T_c}(t)$ is a rectangular pulse, T_c is the chip duration ($T_c = T/N$), and $\left\{c_k^{(n)}\right\}$ for $n = 0, 1, \ldots, N-1$ is a signature sequence (possibly complex valued, since the

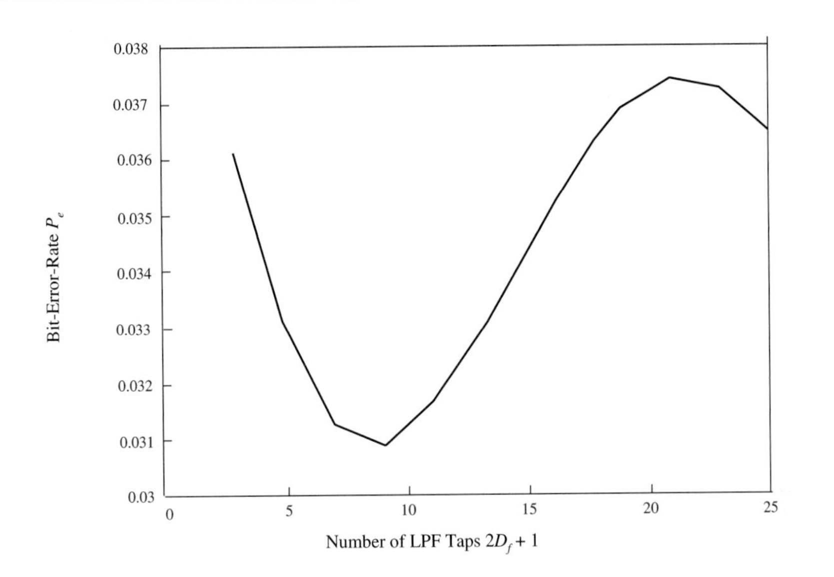

Figure 32.3. The BER of DFALP for different low-pass order. $\nu = 60\ km/hour$; $f_s = 6000\ symbols/second$; $f = 5\ GHz$; $K = 4\ dB$; $f_m T = 0.0463, \gamma = 10\ dB/bit$; *and* $K_t = 5$.

signature alphabet need not be binary). The chip-matched filter can be implemented as an integrate-and-dump circuit, and the discrete-time signal is given by

$$r[n] = \frac{1}{T_c} \int_{nt_c}^{(n+1)T_c} r(t)dt. \tag{32.2.5}$$

Thus, the received signal can be converted into a sequence of WSS random vectors by buffering $r[n]$ into blocks of length N

$$\mathbf{y}_i = [r\,[iN]\,r\,[1+iN]\cdots r\,[N-1+iN]]^T \in C^N \tag{32.2.6}$$

where the nth element of the ith observation vector is given by $y_{i,n} = r\,[n+iN]$.

Although each observation vector corresponds to one symbol interval, this buffering was done without regard to the actual symbol intervals of the users. Since the system is asynchronous, each observation vector will contain at least the end of the previous symbol (left) and the beginning of the current symbol (right) for each user. The factors due to the power, phase, and transmitted symbols of the kth user can be collected into a single complex constant factor $f_k^{(i)}$, e.g., some constant times $A_k e^{j\left[\phi_k + (2\pi/M)m_k^{(i)}\right]}$ and (32.2.6) becomes

$$\mathbf{y}_i = \sum_{k=1}^{K} \left[c_k^{(i-1)} \mathbf{u}_k^r + c_k^{(i)} \mathbf{u}_k^l\right) + \eta_i = \mathbf{A}\mathbf{c}_i + \eta_i \tag{32.2.7}$$

where $\eta_i = \left[\eta_{i,0}, \cdots, \eta_{i,N-1}\right]^T \in C^N$ is a Gaussian random vector. Its elements are zero mean with variance $\sigma^2 = N_0/2T_c$ and are mutually independent.

Vectors $\mathbf{u}_k^r$ and $\mathbf{u}_k^l$ are the right side of the kth user's code vector followed by zeroes and zeroes followed by the left side of the kth user's code vector, respectively. In addition, we have defined $\mathbf{f}_i = \left[f_1^{(i-1)} f_1^{(i)} \cdots f_K^{(i-1)} f_K^{(i)}\right]^T \in C^{2K}$ and the signal matrix $\mathbf{A} = \lfloor \mathbf{u}_1^r \ \mathbf{u}_1^l \ \cdots \ \mathbf{u}_K^r \ \mathbf{u}_K^l \rfloor \in C^{N \times 2K}$. We start with the assumption that each user's signal goes through a single propagation path with an associated attenuation factor and propagation delay. We assume that these parameters vary slowly with time, so that for sufficiently short intervals the channel is approximately a linear time-invariant system. The baseband channel impulse response can then be represented by a Dirac delta function as $h_k(t, \tau) = h_k(t) = h_k \delta(t - \tau_k)$, $\forall \tau$ where h_k is a complex valued attenuation weight and τ_k is the propagation delay. Since there is just a single path, we assume that h_k is incorporated into $f_k^{(i)}$ and concentrate solely on the delay.

If we define $\nu \in \{0, \cdots, N-1\}$ and $\gamma \in [0, 1]$ such that $(\tau_k/T_c) \bmod N = \nu + \gamma$ and if $\gamma = 0$, then the signal vectors become

$$\begin{aligned} \mathbf{u}_k^r &= \mathbf{c}_k^r(\nu) \\ &\equiv \left[c_k^{(N-\nu)} \cdots c_k^{(N-1)} 0 \cdots 0\right]^T \\ \mathbf{u}_k^l &= \mathbf{c}_k^l(\nu) \\ &\equiv \left[0 \cdots 0 \, c_k^{(0)} \cdots c_k^{(N-\nu-1)}\right]^T. \end{aligned} \tag{32.2.8}$$

Due to the nature of the chip-matched filter (an integrator), the samples for a non-zero γ can be represented as

$$\begin{aligned} \mathbf{u}_k^r &= (1-\gamma)\, \mathbf{c}_k^r(\nu) + \gamma \mathbf{c}_k^r(\nu+1) \\ \mathbf{u}_k^l &= (1-\gamma)\, \mathbf{c}_k^l(\nu) + \gamma \mathbf{c}_k^l(\nu+1). \end{aligned} \tag{32.2.9}$$

For a multipath transmission channel with L paths, the impulse response becomes

$$\begin{aligned} h_k(t, \tau) &= h_k(t) \\ &= \sum_{p=1}^{L} h_{k,p} \delta(t - \tau_{k,p}). \end{aligned} \tag{32.2.10}$$

The signal vectors become

$$\mathbf{u}_k^r = \sum_{p=1}^{L} h_{k,p} \left[(1-\gamma_{k,p})\, \mathbf{c}_k^r(\nu_{k,p}) + \gamma_{k,p} \mathbf{c}_k^r(\nu_{k,p}+1) \right], \tag{32.2.11}$$

$$\mathbf{u}_k^l = \sum_{p=1}^{L} h_{k,p} \left[(1-\gamma_{k,p})\, \mathbf{c}_k^l(\nu_{k,p}) + \gamma_{k,p} \mathbf{c}_k^l(\nu_{k,p}+1) \right].$$

By introducing the following notation

$$\mathbf{U}_k^r = \lfloor \mathbf{c}_k^r(0) \cdots \mathbf{c}_k^r(N-1) \rfloor \in C^{N \times N} \tag{32.2.12}$$
$$\mathbf{U}_k^l = \lfloor \mathbf{c}_k^l(0) \cdots \mathbf{c}_k^l(N-1) \rfloor \in C^{N \times N}$$

where the $\mathbf{c}_k$s are as defined in (32.2.8), we can express the signal vectors as a linear combination of the columns of these matrices

$$\mathbf{u}_k^r = \mathbf{U}_k^r \mathbf{h}_k^{'} \tag{32.2.13}$$
$$\mathbf{u}_k^l = \mathbf{U}_k^l \mathbf{h}_k^{'}$$

where $\mathbf{h}_k^{'}$ is the composite impulse response of the channel and the receiver front end, evaluated modulo the symbol period. Thus, the nth element of the impulse response is given by

$$h_{k,n}^{'} = \sum_{j=0}^{\infty} \frac{1}{T_c} \int_{jT+nT_c}^{jT+(n+1)T_c} h_k(t)^{\star} p_{T_c}(t) dt. \tag{32.2.14}$$

For delay spread $T_m < T/2$, at most two terms in the summation will be non-zero.

32.2.1 Estimating the signal subspace

The correlation matrix of the observation vectors is given by

$$\mathbf{R} = E\left[\mathbf{y}_i \mathbf{y}_i^{\dagger}\right] = \mathbf{A}\mathbf{F}\mathbf{A}^{\dagger} + \sigma^2 I \tag{32.2.15}$$

where $\mathbf{F} = E\left[\mathbf{f}_i \mathbf{f}_i^{\dagger}\right] \in C^{2K \times 2K}$ is diagonal. The correlation matrix can also be expressed in terms of its eigenvector decomposition

$$\mathbf{R} = \mathbf{V}\mathbf{D}\mathbf{V}^{\dagger} \tag{32.2.16}$$

where the columns of $\mathbf{V} \in C^{N\times N}$ are the eigenvectors of $\mathbf{R}$, and $\mathbf{D}$ is a diagonal matrix of the corresponding eigenvalues (λ_n). Furthermore,

$$\lambda_n = \begin{cases} d_n + \sigma^2, & if\ n \leq 2K \\ \sigma^2, & \text{otherwise} \end{cases} \tag{32.2.17}$$

where d_n is the variance of the signal vectors along the nth eigenvector and we assume that $2K < N$. Since the $2K$ largest eigenvalues of $\mathbf{R}$ correspond to the signal subspace, $\mathbf{V}$ can be partitioned as $\mathbf{V} = [\mathbf{V}_S \quad \mathbf{V}_N]$, where the columns of $\mathbf{V}_S = [\mathbf{v}_{S,1} \cdots \mathbf{v}_{S,2K}] \in C^{N\times 2K}$ form a basis for the signal subspace S_Y and $\mathbf{V}_N = [\mathbf{v}_{N,1} \cdots \mathbf{v}_{N,N-2K}] \in C^{N\times N-2K}$ spans the noise subspace N_Y. Readers less familiar with eigenvalue decomposition are referred to Appendix B of Part V. Since we would like to track slowly varying parameters, we form a moving average or a Bartlett estimate of the correlation matrix based on the J most recent observations

$$\hat{\mathbf{R}}_i = \frac{1}{J} \sum_{j=i-J+1}^{i} \mathbf{y}_j \mathbf{y}_j^{\dagger}. \tag{32.2.18}$$

It is well known [2] that the maximum likelihood estimate of the eigenvalues and associated eigenvectors of $\mathbf{R}$ is just the eigenvector decomposition of $\hat{\mathbf{R}}_i$. Thus, we perform an eigenvalue decomposition of $\hat{\mathbf{R}}_i$ and select the eigenvectors corresponding to the $2K$ largest eigenvalues as a basis for $\hat{S}_Y$.

32.2.2 Channel estimation

Consider the projection of a given user's signal vectors into the estimated noise subspace

$$\left.\begin{aligned} \mathbf{e}_k^r &= \left(\mathbf{u}_k^{r\dagger} \hat{\mathbf{V}}_N\right)^T \\ \mathbf{e}_k^l &= \left(\mathbf{u}_k^{l\dagger} \hat{\mathbf{V}}_N\right)^T \end{aligned}\right\} \in C^{N-2K}. \tag{32.2.19}$$

If $\mathbf{u}_k^r$ and $\mathbf{u}_k^l$ both lie in the signal subspace, then their sum $\mathbf{u}_k = \mathbf{u}_k^r + \mathbf{u}_k^l$ must also lie in $\mathbf{V}_S$. The projection of $\mathbf{u}_k$ into the estimated noise subspace

$$\tilde{\mathbf{e}}_k = \left(\mathbf{u}_k^{\dagger} \hat{\mathbf{V}}_N\right)^T \tag{32.2.20}$$

is a Gaussian random vector [3] and thus has probability density function

$$p_{\tilde{e}}\left(\tilde{\mathbf{e}}_k\right) = \frac{1}{det\left[\pi \mathbf{K}\right]} \exp\left\{-\tilde{\mathbf{e}}_k^{\dagger} \mathbf{K}^{-1} \tilde{\mathbf{e}}_k\right\}. \tag{32.2.21}$$

The covariance matrix $\mathbf{K}$ is a scalar multiple of the identity given by

$$\mathbf{K} = \frac{1}{J}\mathbf{u}_k^\dagger \mathbf{Q} \mathbf{u}_k \mathbf{I} \tag{32.2.22}$$

and

$$\mathbf{Q} = \sigma^2 \left[\sum_{k=1}^{2K} \frac{\lambda_k}{\left(\sigma^2 - \lambda_k\right)^2} \mathbf{v}_{S,k} \mathbf{v}_{S,k}^\dagger \right]. \tag{32.2.23}$$

Therefore, within an additive constant, the log-likelihood function of $\tilde{\mathbf{e}}_k$ is

$$\begin{aligned} \Lambda\left(\tilde{\mathbf{e}}_k\right) &= -\left(N - 2K\right) ln \left(\mathbf{u}_k^\dagger \mathbf{Q} \mathbf{u}_k\right) - J\frac{\tilde{\mathbf{e}}_k^\dagger \tilde{\mathbf{e}}_k}{\mathbf{u}_k^\dagger \mathbf{Q} \mathbf{u}_k} \\ &= -\left(N - 2K\right) ln \left(\mathbf{u}_k^\dagger \mathbf{Q} \mathbf{u}_k\right) - J\frac{\mathbf{u}_k^\dagger \mathbf{V}_N \mathbf{V}_N^\dagger \mathbf{u}_k}{\mathbf{u}_k^\dagger \mathbf{Q} \mathbf{u}_k}. \end{aligned} \tag{32.2.24}$$

The exact $\mathbf{V}_N$ and $\mathbf{Q}$ are unknown, but we can replace them with their estimates. The best estimates will minimize $\tilde{\mathbf{e}}_k$, resulting in the maximum of the likelihood function.

Unfortunately, maximizing this likelihood function is prohibitively complex for a general multipath channel, so we will consider only a single propagation path. In this case, the vector $\mathbf{u}_k$ is a function of only one unknown parameter: the delay τ_k. To form the timing estimate, we must solve

$$\hat{\tau}_k = \arg\max_{\tau_k \in [0,T)} \Lambda\left(\mathbf{u}_k\right). \tag{32.2.25}$$

Ideally, we would like to differentiate the log-likelihood function with respect to τ. However, the desired user's delay lies within an uncertainty region, $\tau_k \in [0, T]$, and $\mathbf{u}_k(\tau)$ is only piecewise continuous on this interval. To counter these problems, we divide the uncertainty region into N cells of width T_c and consider a single cell, $c_\nu \equiv [\nu T_c, (\nu + 1) T_c)$. We again define $\nu \in \{0, \cdots, N-1\}$ and $\gamma \in [0, 1)$ such that (τ/T_c) mod $N = \nu + \gamma$, and for $\tau \in c_v$ the desired user's signal vector becomes

$$\mathbf{u}_k\{\tau\} = (1 - \gamma)\,\mathbf{u}_k(\nu) + \gamma \mathbf{u}_k(\nu + 1) \tag{32.2.26}$$

and

$$\begin{aligned} \frac{d}{d\tau}\mathbf{u}_k(\tau) &= \mathbf{u}_k(\nu + 1) - \mathbf{u}_k(\nu) \\ &= a \text{ constant.} \end{aligned} \tag{32.2.27}$$

Thus, within a given cell, we can differentiate the log-likelihood function and solve for the maximum in closed form. We then choose whichever of the N solutions yields the largest value for (32.2.24). Details can be found in [1].

Under certain conditions, it may be possible to simplify this algorithm. Note that maximizing the log-likelihood function (32.2.24) is equivalent to maximizing

$$\Lambda\left(\tilde{\mathbf{e}}_k\right) = -\frac{N-2K}{J} ln\left(\mathbf{u}_k^{\dagger}\mathbf{Q}\mathbf{u}_k\right) - \frac{\mathbf{u}_k^{\dagger}\mathbf{V}_N\mathbf{V}_N^{\dagger}\mathbf{u}_k}{\mathbf{u}_k^{\dagger}\mathbf{Q}\mathbf{u}_k}. \qquad (32.2.28)$$

As $J \to \infty$, the leading term goes to zero; thus, for large observation windows, we can use the following approximation:

$$\Lambda\left(\tilde{\mathbf{e}}_k\right) \approx -\frac{\mathbf{u}_k^{\dagger}\mathbf{V}_N\mathbf{V}_N^{\dagger}\mathbf{u}_k}{\mathbf{u}_k^{\dagger}\mathbf{Q}\mathbf{u}_k}. \qquad (32.2.29)$$

This yields a much simpler expression for the stationary points [1]. The MUSIC algorithm is equivalent to (32.2.29) when one only maximizes the numerator and ignores the denominator, i.e., one assumes $\mathbf{u}_k^{\dagger}\mathbf{Q}\mathbf{u}_k$ is equal to 1 in (32.2.28) or (32.2.29). This yields an even simpler approximation for the log-likelihood function

$$\Lambda\left(\tilde{\mathbf{e}}_k\right) \approx -\mathbf{u}_k^{\dagger}\mathbf{V}_N\mathbf{V}_N^{\dagger}\mathbf{u}_k, \qquad (32.2.30)$$

which further simplifies solution for the stationary points [1].

For illustration, the simulation results for five users with length 31 Gold codes are presented in Figures 32.4 and 32.5.

A single desired user was acquired and tracked in the presence of strong multiple access interference (MAI). The power ratio between each of the four interfering users and the desired user is designated the MAI level.

The true log-likelihood estimate (32.2.24) is first compared with the large observation window approximation (32.2.29) and the MUSIC algorithm (32.2.30). This was done for a window size of 200 symbols and with a varying SNR. Figure 32.4 shows the probability of acquisition for each method, where acquisition is defined as $|\tau_k - \hat{\tau}_k| < \frac{1}{2}T_c$. Using the approximate log-likelihood function resulted in almost no drop in performance. Furthermore, when the SNR was poor, both probabilistic approaches considerably outperformed the MUSIC algorithm. In Figure 32.5, we compare the RMSE of the delay estimate once acquisition has occurred, i.e., after processing enough symbols to reach within half of one chip. The approximate log-likelihood function experiences a slight increase in error at low SNR, but again both probabilistic methods do better than MUSIC.

32.3 ADDITIONAL READINGS

In the previous section, we discussed an ML-based algorithm for MU channel estimation. The second-order statistics were used, so the information about the phase

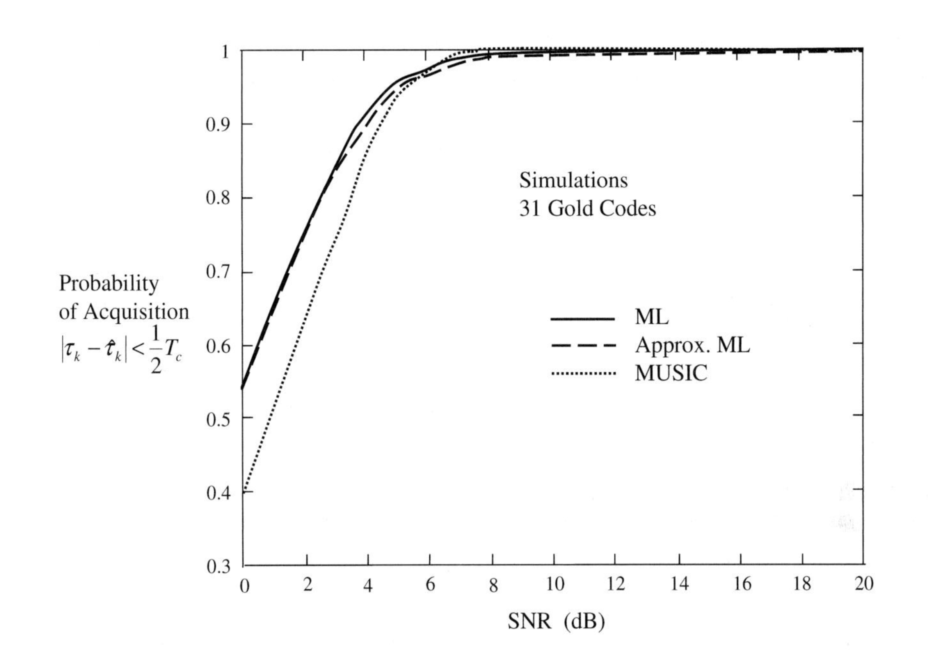

Figure 32.4. Probability of acquisition for the maximum likelihood (ML) estimator, the approximate ML, and the MUSIC algorithm $[K = 5, N = 31, J = 200, MAI = 20\ dB]$

could not be extracted. Due to complexity of the likelihood function, only the single-path propagation model has been considered. Further modification of the cost function should be introduced to make possible the multipath channel coefficient estimation. Some options, described in [1], remain still too complex to be considered for practical applications.

An algorithm for joint detection and estimation of amplitudes, but with known delays, was developed in [4] and [5]. In these papers, a tree-search method was used together with least-squares estimation.

An algorithm described in [6] is based on the idea of regenerating the MAI, subtracting it from the input signal, and then using the ML estimate algorithm separately for delays and amplitudes. The regeneration of MAI is based on a soft-decision estimation. A neural-network based algorithm [7] implicitly estimates delays and amplitudes, using the back-propagation algorithm. This kind of algorithm is rather slow and cannot be used for tracking fast-fading parameters.

A recursive signal processing using the Viterbi algorithm for joint tracking of amplitudes and delays was described in [8]. An extended Kalman estimator is used to update the amplitude and delay estimates for each survivor sequence in VA. The Nyquist samples are used as sufficient statistics. The resulting algorithm requires storage of the survivor sequences and is difficult to simplify. Different adaptive

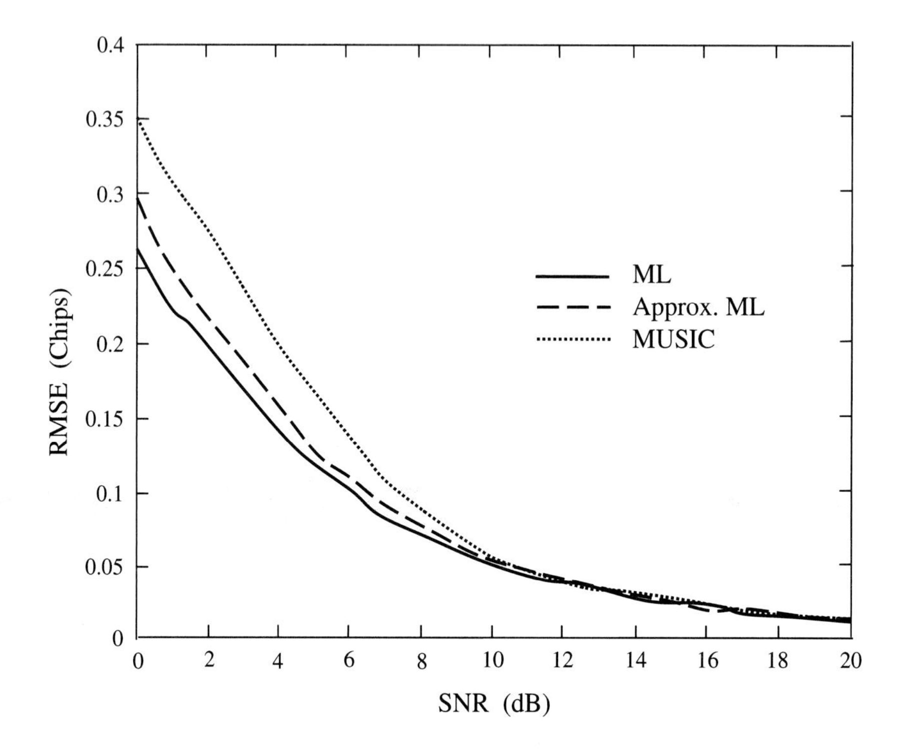

Figure 32.5. Root mean-squared error (RMSE) of the delay estimate in chips for the maximum likelihood (ML) estimator, the approximate ML, and the MUSIC algorithm $[K = 5, N = 31, J = 200, MAI = \; 20dB]$.

symbol-by-symbol multiuser detector (estimator) is presented in [9]. The algorithm starts from Bayesian recursion in the a posteriori composite user symbol probabilities. For updating, the likelihood of the received signal, given each composite symbol, is used. For known amplitudes and delays, this is a Gaussian density function. It is shown in [9] that for the case of unknown delays and amplitudes, the required likelihoods can be generated by use of EKF innovations. The K binary user symbols are viewed as a single composite symbol. For an asynchronous network, two adjacent symbols must be considered, so there are 2^{2K} data hypotheses. For direct implementation of the algorithm, an EKF update must be computed for each of the 2^{2K} hypotheses. The simplification presented in [9] is based on the fact that the 2^{2K} required Kalman filter innovations and measurement updates can be expressed in terms of estimated signal/received signal auto- and cross-correlation functions. These correlations should be computed in parallel, by special-purpose hardware with the correlator inputs running at the full CDMA bandwidth but with outputs updated only at the bit rate. It was also shown that there are only 2^K, rather than 2^{2K}, unique correlation functions. For further simplification, metric pruning was suggested. Only M composite data hypotheses, corresponding to the largest MAP metrics, should be used. Algorithms based on EKF for joint estimation of channel coefficients and delays for a single user are presented in [10], [11], [8].

In the case of advanced CDMA receivers employing any form of MAI cancellation, the maximum likelihood channel coefficients estimation problem can be approximated by an approach that extends the algorithm described in Figure 32.1. A comprehensive review of the literature on advanced CDMA receivers can be found in [12]. In this case, we could find the rough channel estimates by subtracting the estimated overall MAI from the matched filter outputs, removing the modulation by pilot or estimated symbols, and then additionally filtering the result of such preprocessing. Optimization of the predictor and smoother described in Figure 32.1 is additionally discussed in [13]. In systems with high processing gain, power control, and low cross-correlations between the users, the output of the matched filter of the kth user synchronized to the lth path can be considered separately as a digital phase-modulated signal in flat-fading and Gaussian noise.

For such channels, it was shown even in classical literature [14] that the optimum receiver results in a structure that follows Kailath's separation theorem [15]. In order words, the optimum receiver consists of an estimator that delivers MMSE estimates of the fading distortion and a detector that utilizes these estimates. In classic textbooks like Chapter 6 in [16], the detection of signals with random but time-invariant parameters is considered. Maximum likelihood estimates of the channel parameters are obtained from the signal received over the past several bit intervals. In [17], a pilot tone signaling to provide channel amplitude and phase information for the detection and the adaptive control of the transmitter is used.

In [18], [19], a finite-memory, minimum mean-squared error estimator with decision feedback is employed to estimate the fading distortion, and the likelihood ratio test of the detector is derived.

In [18], it was shown that the detector minimizes the probability of error if the estimator delivers MMSE estimates. Kam [20], [19] first recognized that the estimation of the complex fading coefficient equals the carrier recovery of the receiver.

To define the "maximum a posteriori" detector [14] showed that an adaptive part of the receiver can be identified as carrier recovery where the complex fading coefficient is estimated by use of Kalman filter.

In a system with relatively low processing gain (less than 20), the residual multipath component will still remain and the matched filter output will demonstrate frequency-selective fading resulting in intersymbol interference (ISI). Combining such signals in a Rake receiver would significantly reduce the overall combining gain. For that reason, pre-whitening (equalization) preprocessing would be needed prior to MRC. For this kind of operation, frequency-selective channel coefficients should be estimated.

A possible approach, as before, is to perform detection with an old channel estimate and then use the detected data to update the channel estimate in a decision-feedback manner. Such estimators are described for flat [14] or selective [21], [22] fading. They have the form of (LMS or Kalman type) recursive adaptive filters which, due to their averaging operation, exhibit a lag error. This effect often is aggravated by additional decision delays resulting from (whitened) matched filtering and equalization.

Channel estimator/predictor algorithms are based on optimization criteria; this approach minimizes the total error (lag plus noise) [23]. Unfortunately, the presence of an irreducible lag error limits robustness against faster fading channel conditions.

The situation is further complicated in the case of code transmission where final decisions are not readily available because of the deinterleaving devices and iterative (turbo) decoding algorithms. In this case, hard or tentative decisions may not be reliable since the large coding gains on fading channels [24] allow for the rather low link budget and the operating SNR is approaching the Shannon limit (-1.6 dB).

The main problem of using decision-aided channel estimators comes from the two stochastic processes—data and channel—being involved simultaneously.

The fading channel can be eliminated from detection by the introduction feed of forward synchronizer structures, which is an attractive means of achieving robust receiver adjustment, especially for all digital implementations.

For the flat-fading channels, entirely feed-forward estimator structures, which use periodically interleaved, known symbols (MMS pilot symbols), were described in [25], [26]. The fading channel samples are collected at the sampling rate above the Nyquist rate and smoothed; then, the final estimates are generated by interpolation. A similar solution is described also in [27], where interpolation of acquired samples of the channel impulse response is performed by a windowed sin(x)/x pulse and no further filtering is used.

In [28], the previous work is generalized and extended by a systematic derivation of the optimal feed-forward estimator. The result was a Wiener filter, also addressed in [29], [30].

Applying the Wiener theory to selective fading channel estimation is shown to result in a two-step procedure. First, a sequence of channel estimates (snapshots) is obtained through training segments multiplexed into the data stream. Using optimal training sequences leads to the taps of each estimate becoming statistically independent. Second, Wiener filtering is used to interpolate the acquired taps whereby the final estimate is generalized.

The above solutions are based on the use of a training sequence with or without additional exploitation of the feedback decisions. The existing UMTS standards proposals support such solutions by providing pilot symbols.

In general, in several situations the adoptive estimators may have to startup (or restart) without a training sequence. This problem has been extensively considered under the common title *blind estimator* (equalization).

In multipoint networks, whenever a channel from the master to one of the secondary stations goes down, it is clearly not feasible (or desirable) for the master to start sending a training sequence to reboot a particular line [31]. In digital communications over fading/multipath channels, a restart is required following a temporary path loss because of a severe fade [32].

Several blind equalizers that do not require a training sequence have been proposed in the literature [31], [33], [34]. Godard's equalizer [31] is now widely accepted and has been proposed for many applications, including the equalization of QAM (quadrature amplitude modulation) data signals. The objective of the Godard equalizer is to bring the receiver equalizer from a startup (or restart) to a point where it can reliably switch to a steady-state decision-directed algorithm (conventional equalizer). In general, a blind equalizer must be capable of getting close to the desired global solution ("open-eye" pattern) from an arbitrary initialization; once we have an open-eye pattern, a decision-directed equalizer can take over.

Given the channel estimate, we can easily design an equalizer, and vice versa (given an equalizer, we can calculate its inverse—the channel impulse response). Even so, the distinction between the two estimators is important from several points of view.

First, the structure (of the estimator) has a crucial effect on the performance of an equalizer. The decision feedback (nonlinear) equalizers do much better than the linear equalizers given the same information, where the performance criterion is the symbol-error rate.

Treating the channel estimator separately allows the use of analytically and computationally tractable and globally convergent, even possibly linear, techniques for channel estimation and then reverts to nonlinear structures for ISI equalization based on the channel impulse response estimate.

Second, as in the case of the Rake receiver, the channel estimate is sometimes of independent interest. Some additional examples, where an accurate estimate of the channel impulse response is needed without interruption of ongoing tasks, can be found in [35], [36].

Third, channel impulse response estimate is required to implement statistically efficient maximum likelihood and related symbol sequence estimators [37], [38]. These estimators do not yield a channel estimate as a by-product.

Some solutions are based on exploiting the higher-order cumulant statistics of the data, [39] - [40] and [41] - [42]. One of the advantages of the high-order cumulant statistics is that they are sensitive to the phase as well as magnitude of the transfer function of the underlying system. The second-order statistics cannot discriminate among systems that have identical, spectrally equivalent, minimum phase representation [43], [44], [45]. The transfer function of the overall (composite) channel is, in general, nonminimum phase, where the composite channel refers to the combined effect of the transmitting filter, the channel, the matched filter, the sampler, and the discrete-time noise whitening filter [37], [46]. In blind channel estimation problems, one wishes to fit an FIR filter to the data.

Hence, the papers dealing with (blind) FIR (or moving average: MA) system identification [47], [48], [41], [49] are relevant to the communications channel estimation problem (provided that appropriate but trivial modifications are made to account for complex data).

In the case of blind equalization (inverse system modeling), one fits a FIR inverse, hence, an autoregressive (AR), direct model to the data. Therefore, the papers dealing with (noncausal) AR system identification [50], [51] are also of interest for the study of the problem.

In [47] through [52], higher-order statistics have been explicitly used for channel estimation and for blind equalization. In [52], tricepstrum of the fourth-order cumulants is used to estimate the channel impulse response. In this paper, the coefficients of two (truncated) series expansion of the minimum phase and the maximum phase parts of the channel impulse response are estimated, instead of the direct estimation of the channel impulse response coefficients. The same approach in conjunction with a gradient LMS-type algorithm for recursive estimation of the channel impulse response is used in [47].

The result is then used for equalization. No convergency results have been presented in [47] and [52]. It is not clear how the approach of [47], [52] would perform for channels with deep spectral nulls, for which, strictly speaking, tricepstrum does not exist. In a Rake receiver, even one with processing gain as small as 4 (UMTS for 2 Mbits), this problem does not exist. In [48], an off-line iterative method is presented and analyzed for FIR model fitting, using the correlations and a partial set of fourth-order moments of the received signal. This work again does not discuss convergence results. The knowledge of the system order is assumed.

The drawbacks of the above approach are discussed in [49]. It has been shown that the approach presented in [48] is not necessarily consistent; therefore, it may fail to provide a reliable initial guess for the nonlinear approach of [48]. Also, identifiability issues for the nonlinear approach of [48] have been addressed. An interesting recursive (adaptive) approach to blind equalization has been presented in [53], [54]. The approach of [53] consists of estimating (updating) the coefficients of a (long) linear blind equalizer, using a novel criterion. Reference [54] combines a linear cumulant matching approach with the criterion in [53]; instead of estimating the equalizer parameter [54] directly, estimate the channel impulse response coefficients. This approach results in a nonlinear optimization problem that requires a good initial guess to converge. To this end, [53] proposes two new linear methods that are guaranteed to yield strongly consistent estimates. The linear methods, however, are not quite statistically efficient.

Bibliography

[1] J. S. Bensley and B. Aazhang, "Subspace-based channel estimation for code division multiple access communications and systems," *IEEE Trans. Comm.*, vol. 44, pp. 1009–1020, 1996.

[2] R. Muirhead, *Aspects of Multivariate Statistical Theory.* New York: Wiley, 1982.

[3] S. Kay, *Fundamentals of Statistical Signal Processing-Estimation Theory.* Englewood Cliffs, NJ: Prentice Hall, 1993.

[4] Z. Xie, C. Rushforth, R. Short, and T. Moon, "A tree-search algorithm for signal

detection and parameter estimation in multi-user communications," in *Proceedings of the IEEE Military Communications Conference*, (Monterey, CA), pp. 796–800, 1990.

[5] Z. Xie, C. Rushforth, R. Short, and T. Moon, "Joint signal detection and parameter estimation in multiuser communications," *IEEE Transactions on Communications*, vol. 41, pp. 1208–1216, 1993.

[6] D. Brady and J. Catipovic, "An adaptive, soft-decision multiuser receiver for underwater acoustical channels," in *Proceedings of the Asilomar Conference on Signals Systems and Computers*, (Pacific Grove, CA), pp. 1137–1141, 1992.

[7] B. Aazhang, B. Paris, and G. Orsak, "Neural networks for multiuser detection in code-division multiple-access communications," *IEEE Transactions on Communications*, vol. 40, pp. 1212–1222, 1992.

[8] R. lltis and A. Fuxjaeger, "A digital DS spread-spectrum receiver with joint channel and Doppler shift estimation," *IEEE Transactions on Communications*, vol. 39, pp. 1255–1267, 1991.

[9] R. lltis and Mailaender, "An adaptive multiuser detector with joint amplitude and delay estimation," *IEEE J. Selected Areas Comm.*, vol. 12, pp. 774–785, 1994.

[10] R. Iltis, "An EKF-based joint estimator for interference, multipath, and code delay in a DS spread-spectrum receiver," *IEEE Transactions on Communications*, vol. 42, pp. 1288–1299, 1994.

[11] R. lltis, "Joint estimation of PN code delay and multipath using the extended Kalman filter," *IEEE Transactions on Communications*, vol. 38, pp. 1677–1685, 1990.

[12] M. Juntti and S. Glisic, "Advanced CDMA for wireless communications," in *Wireless Communications: TDMA versus CDMA* (S. G. Glisic and P. A. Leppaenen, eds.), pp. 447–490, Kluwer, 1997.

[13] A. Maemmelae and V. P. Kaasila, "Smoothing and interpolation in a pilot symbol assisted diversity system," *International Journal of Wireless Information Networks*, vol. 4, pp. 205–214, 1997.

[14] R. Haeb and H. Meyr, "Systematic approach to carrier recovery and detection of digitally phase modulated signals on fading channels," *IEEE Transactions on Communications*, vol. 37, pp. 748–754, 1989.

[15] T. Kailath, "A general likelihood formula for random signals in Gaussian noise," *IEEE Trans. Information Theory*, vol. IT-15, 1969.

[16] J. C. Hancock and P. A. Wintz, *Signal Detection Theory.* New York: McGraw-Hill, 1966.

[17] R. Srinivasan, "Feedback communications over fading channels," *IEEE Trans. Communications*, vol. COM-29, pp. 50–57, 1981.

[18] P. K. Varshney and A. H. Haddad, "A receiver with memory for fading channels," *IEEE Trans. Communications*, vol. COM-26, pp. 278–283, 1978.

[19] P. Y. Kam and C. H. Teh, "An adaptive receiver with memory for slowly fading channels," *IEEE Trans. Communications*, vol. COM-32, pp. 654–659, 1984.

[20] P. Y. Kam and C. H. Teh, "Reception of psk signals over fading channels via quadrature amplitude estimation," *IEEE Trans. Communications*, vol. COM-31, pp. 1024–1027, 1983.

[21] A. P. Clark and S. Hariharan, "Efficient estimators for an HF radio link," *IEEE Trans. Communications*, vol. 38, pp. 1173–1180, 1990.

[22] S. McLaughlin, B. Mulgrew, and C. F. N. Cowan, "A performance study of the extended Kalman algorithm as an HF channel estimator," in *IEE 4th Int. Conf. HF Radio System, and Technique.*, pp. 335–338, 1988.

[23] S. Haykin, *Adaptive Filter Theory.* Englewood Cliffs, NJ: Prentice Hall, 1986.

[24] J. Hagenauer and E. Lutz, "Forward error correction coding for fading compensation in mobile satellite channels," *IEEE J. Sell Areas Commun.*, vol. SAC-5, pp. 215–225, 1987.

[25] A. Aghamohammadi and H. Meyr, "A new method for phase synchronization and automatic gain control of linearly-modulated signals on frequency-flat fading channels," *IEEE Trans. Communications*, vol. 39, pp. 25–29, 1991.

[26] M. L. Moher and J. H. Lodge, "Tcmp—a modulation and coding strategy for rician fading channels," *IEEE J. Sel. Areas Commun.*, vol. 7, pp. 1347–1355, 1989.

[27] N. W. K. Lo, D. D. Falconer, and A. U. H. Sheikh, "Adaptive equalization and diversity combining for mobile radio using interpolated channel estimates," *IEEE Trans. Vehicular Technol.*, vol. 40, pp. 636–645, 1991.

[28] S. A. Fechtel and H. Meyr, "Optimal parametric feedforward estimation of frequency selective fading radio channels," *IEEE Trans. Comm.*, vol. 42, pp. 1639–1650, 1994.

[29] J. K. Cavers, "An analysis of pilot symbol assisted modulation for Rayleigh fading channels," *IEEE Trans. Vehicular Technol.*, vol. 40, pp. 686–693, 1991.

[30] P. Hoeher, "TCM on frequency-selective land-mobile fading channels," in *Coded Modulation and Bandwidth Efficient Transmission*, (Tirrenia, Italy), pp. 317–328, 1991.

[31] D. N. Godard, "Self-recovering equalization and carrier tracking in two-dimensional data communication systems," *IEEE Trans. Communications*, vol. COM-28, pp. 1867–1875, 1980.

[32] N. K. Jablon, "Joint blind equalization, carrier recovery, and timing recovery for 64-QAM and 128-QAM sigad constellations," in *Proc. 1989 IEEE Intern. Conf. Commun.*, (Boston, MA), pp. 1043–1049, 1989.

[33] Y. Sato, "A method of self-recovering equalization for multilevel amplitude modulation," *IEEE Trans. Communications*, vol. COM-23, pp. 679–682, 1975.

[34] J. R. Treichler and M. G. Larimore, "New processing techniques based on the constant modulus adaptive algorithm," *IEEE Trans Acous. Speech Signal Proc.*, vol. ASSP-33, pp. 420–431, 1985.

[35] R. P. Gooch and J. C. Harp, "Blind channel identification using the constant modulus adaptive algorithm," in *Proc. 1988 IEEE Intern. Conf. Commun.*, (Philadelphia, PA), pp. 75–79, 1988.

[36] P. Balaban and R. Aviv, "Estimation of dispersive model parameters from time domain measurements on an operational FM channel," in *Proc. IEEE Glolal Telecommun. Conf.*, pp. 1500–1504, 1987.

[37] J. G. Proakis, *Digital Communications.* New York: McGraw-Hill, 2nd ed., 1989.

[38] E. M. Long and A. M. Bush, "Decision-directed sequential intersymbol interference sequence estimation for channels," in *Proc. 1989 IEEE Intern. Conf. Commun.*, (Boston, MA), pp. 841–845, 1989.

[39] J. K. Tugnait, "Adaptive IIR algorithms based on second- and higher-order statistics," in *Proc. 23rd Asilomar Conf. Signals Syst. Comp.*, (Pacific Grove, CA), pp. 571–575, 1989.

[40] C. L. Nikias and M. R. Raghuveer, "Bispectrum estimation: A digital signal processing framework," *Proc. IEEE*, vol. 75, pp. 869–891, 1987.

[41] in *Proc. Workshop on Higher-Order Spectral Analysis*, (Vail, Colorado), 1989.

[42] J. K. Tugnait, "Recovering the poles from fourth-order cumulanys of system output," in *Proc. 1988 American Control Conf.*, (Atlanta, GA), pp. 2090–2095, 1988.

[43] A. Benveniste, M. Goursat, and G. Ruget, "Robust identification of a nonminimum phase system: Blind adjustment of a linear equalizer in data communications," *IEEE Trans. Autom. Control*, vol. AC-25, pp. 385–399, 1980.

[44] M. Rosenblatt, *Stationary Sequences and Random Fields.* Boston: Birkhauser, 1985.

[45] J. K. Tugnait, "Identification of nonminimum phase linear stochastic systems," *Automatica*, vol. 22, pp. 457–464, 1986. Also an abridged version in *Proc. 23rd IEEE Decision Control Conf.*, pp. 342-347, Las Vegas, Dec. 1984.

[46] A. A. Giordano and F. M. Hsu, *Least Squares Estimation with Applications to Digital Signal Processing.* New York, NY: Wiley Interscience, 1985.

[47] D. Hatzinakos and C. L. Nikias, "Blind decision feedback equalization structures based on adaptive cumulant techniques," in *Proc. 1989 IEEE Intern. Conf. Commun.*, (Boston, MA), pp. 1278–1282, 1989.

[48] B. Porat and B. Friedlander, "Blind equalization of digital communication channels using high-order moments," *IEEE Trans. Acous. Speech Signal Proc.*, vol. ASSP-39, pp. 522–526, 1991.

[49] J. K. Tugnait, "Approaches to fir system identification with noisy data using higher-order statistics," *IEEE Trans. Acous. Speech Signal Proc.*, vol. ASSP-38, 1990. Also preliminary version in *Proc. Workshop Higher-Order Spectral Analysis'* pp. 13-18, Vail, CO, June 1989.

[50] J. K. Tugadt, "On selection of maximum cumulant lags for noncausal autoregressive model fitting," in *Proc. IEEE 1988 Intern. Conf. Acous. Speech Signal Proc.*, (New York, NY), pp. 3373–3375, 1988.

[51] J. K. Tugnait, "Consistent order selection for noncausal autoregressive models via higher-order statistics," *Automatica*, vol. 26, pp. 311–320, 1990.

[52] D. Hatzinakos and C. L. Nikias, "Estimation of multipath channel response in frequency selective channels," *IEEE J. Select. Areas Commun.*, vol. SAC-7, pp. 12–19, 1989.

[53] O. Shalvi and E. Weinstein, "New criteria for blind deconvolution of nonminimum phase systems (channels)," *IEEE Tran. Inform. Theory*, vol. IT-36, pp. 312–321, 1990.

[54] J. Tugnait, "Blind estimation of digital communication channel impulse response," *IEEE Transactions on Communications*, vol. 42, pp. 1606–1616, 1994.

Chapter 33

CAPACITY OF CDMA NETWORKS

33.1 PRELIMINARIES

We start with an overview of the existing work on the CDMA network capacity.

An overview of Code Division Multiple Access (CDMA) in general can be found in [1], [2], [3]. The CDMA capacity analysis covered in a number of papers [4], [5], Gilhousen et al. [4], motivated many researchers to simulate CDMA systems. Their analysis includes voice activity monitoring, sectorization (3 sectors), perfect uplink power control, and Gaussian approximation for multiple-access interference statistics.

The effect of imperfect power control on CDMA capacity has been studied in a number of papers [6], [7]. The conclusion of those references is that CDMA network capacity decreases rapidly as a function of power control error. Power control error is usually modeled as a lognormal random variable.

Proper choice of signal propagation parameters is important for studying CDMA network capacity. In general, the parameters can be divided into propagation losses, slow fading (shadowing), and fast fading. Path losses depend on the environment and cell type. Average attenuation grows as a power law of distance. A path loss exponent of 4 is mostly used in the literature. The shadowing effect is usually analyzed at the system level, and the standard model assumes lognormal distribution. Fast fading can be accounted for at the link level as the required signal-to-noise ratio for predefined quality of service. The influence of these propagation parameters is discussed, e.g., in [7]. The effects of user mobility on the CDMA capacity is studied in [8]. In that work, the results are based on simulations that take into account shadowing, call statistics, voice activity, cell sectorization, and user mobility. Perfect power control and ideal antenna directivity are assumed. The presented concepts are general, and they can be applied for any asynchronous CDMA cellular networks.

Discussion of Erlang capacity in a power-controlled CDMA system is presented in [9]. Similar capacity gains, as reported in [4], are expected in Erlang capacity as well. Reverse-link Erlang capacity under nonuniform cell loading is reported in [10].

The effects of adaptive base station antenna arrays on CDMA capacity are studied e.g., in [11], [12], and [13]. The results show that significant capacity gains can be achieved with quite simple techniques. Additional analysis of the outage probability in cellular CDMA is presented in [14] and [15]. Both Nakagami and Rician fading coupled with lognormal shadowing are taken into account in the propagation model. Additionally, voice activity factor is included in [14].

The effect of more sophisticated receiver structures (like multiuser detectors, MUD, or joint detectors) on CDMA or hybrid systems capacity are examined in [16], [17]. Results of [16] show a roughly twofold increase in capacity with MUD efficiency of 65% compared to conventional receivers. The effect of the fractional cell load on the coverage of the system is presented in [18]. The coverage of the MUD-CDMA uplink was less affected by the variation in cell loading than in conventional systems. References [19] and [20] describe a CDMA system where joint detection data estimation is used with coherent receiver antenna diversity. This system can be used as a hybrid multiple-access scheme with TDMA and FDMA components. In [17], significant capacity gains are reported when zero-forcing multiuser detectors are used instead of conventional single-user receivers.

In most of the references, it has been assumed that the service of interest is low-rate speech. In next generation systems, however, mixed services including high-rate data must be taken into account. This is done in [21], where the performance of an integrated voice/data system is presented.

From the above literature survey, it can be seen that generally, capacity evaluation of CDMA networks and comparison of CDMA and TDMA systems have been an important and controversial issue. One of the reasons for such a situation is a lack of a systematic, easy-to-follow mathematical framework for this evaluation. The situation is complicated by the fact that a lot of parameters are involved, and some of the system components are rather complex, resulting in an imperfect operation. The analysis of an advanced CDMA network should, in general, take all these elements into account, including their imperfections, and formulate an expression for the system capacity in a form that can be used in practice.

33.2 SYSTEM MODEL

The complex envelope of the signal transmitted by user $k \in \{1, 2, \ldots, K\}$ in the nth symbol interval $t \in [nt, (n+1)T]$ is

$$s_k = A_k s_k^{(n)} (t - \tau_k) \tag{33.2.1a}$$

$$s_k^{(n)} = S_k^{(n)} e^{j\phi_{k0}}, \tag{33.2.1b}$$

$$\begin{aligned} S_k^{(n)} &= b_{ik}^{(n)} \varphi(t - nT) \sum_{m=0}^{N_c-1} c_{ikm}^{(n)} \Psi(t - mT_c) \\ &\quad + j b_{qk}^{(n)} \varphi(t - nT - \epsilon_b T) \sum_{m=0}^{N_c-1} c_{qkm}^{(n)} \Psi(t - mT_c - \epsilon_c T_c) \\ &= b_{ik}^{(n)} \varphi(t - nT) c_{ik}(t) + j b_{qk}^{(n)} \varphi(t - nT - \epsilon_b T) c_{qk}(t) \end{aligned} \tag{33.2.1c}$$

where $A_k = \sqrt{E_k}$ is the transmitted amplitude of user k, E_k is the energy per symbol, τ_k is the signal delay, ϕ_{k0} is the transmitted signal carrier phase, b_{ik} and b_{qk} are two information bits in I- and Q-channel, respectively, $\varphi(t)$ is the bit pulse shape, ϵ_b is the bit offset in Q-channel, $c_{ikm}^{(n)}$ and $c_{qkm}^{(n)}$ are the mth chips of the kth user PN codes in I- and Q-channel, respectively, $\Psi(t)$ is the chip shape, and ϵ_c is the chip offset. In practical applications, ϵ_b and ϵ_c will have values either 0 or $\frac{1}{2}$. Equation (33.2.1) is general, and different combinations of the signal parameters cover most of the signal formats of practical interest. To simplify notation, we represent $S_k^{(n)}$ in (33.2.1) as

$$S_k^{(n)} = S_k = S_{ik} + jS_{qk} = d_{ik}c_{ik} + jd_{qk}c_{qk} \tag{33.2.2a}$$

$$d_{ik} = b_{ik}^{(n)} \varphi(t - nT) \tag{33.2.2b}$$

$$c_{ik} = \sum_{m=0}^{N_c-1} c_{ikm}^{(n)} \Psi(t - mT_c) \tag{33.2.2c}$$

$$d_{qk} = b_{qk}^{(n)} \varphi(t - nT - \epsilon_b T) \tag{33.2.2d}$$

$$c_{qk} = \sum_{m=0}^{N_c-1} c_{qkm}^{(n)} \Psi(t - mT_c - \epsilon_c T_c). \tag{33.2.2e}$$

The channel impulse responses consist of discrete multipath components represented as

$$h_k^{(n)} = \sum_{l=1}^{L} h_{kl}^{(n)} \delta\left(t - \tau_{kl}^{(n)}\right) = \sum_{l=1}^{L} H_{kl}^{(n)} e^{j\phi_{kl}} \delta\left(t - \tau_{kl}^{(n)}\right) \tag{33.2.3a}$$

$$h_{kl}^{(n)} = H_{kl}^{(n)} e^{j\phi_{kl}} \tag{33.2.3b}$$

where L is the number of multipath components of the channel, $h_{kl}^{(n)}$ is the complex coefficient (gain) of the lth path of user k at symbol interval with index n, $\tau_{kl}^{(n)} \in [0, T_m)$ is delay of the lth path component of user k in symbol interval n, and $\delta(t)$ is the Dirac delta function. We assume that T_m is the delay spread of the channel. In what follows, we drop index n whenever doing so does not produce ambiguity. It is also assumed that $T_m < T$. The overall received signal during N_b symbol intervals can be represented as

$$r(t) = Re\left\{ e^{j\omega_0 t} \sum_{n=0}^{N_b-1} \sum_{k=1}^{K} s_k^{(n)}(t) \star h_k^{(n)}(t) \right\} + z(t) \tag{33.2.4}$$
$$= Re\left\{ e^{j\omega_0 t} \sum_{n=0}^{N_b-1} \sum_k \sum_l a_{kl} S_k^{(n)}(t - nT - \tau_k - \tau_{kl}) \right\} + z(t)$$

where $a_{kl} = A_k H_{kl}^{(n)} e^{j\Phi_{kl}}$, $\Phi_{kl} = \phi_0 + \phi_{k0} - \phi_{kl}$, ϕ_0 is frequency downconversion phase error, $z(t)$ is a complex zero-mean additive white Gaussian noise process with two-sided power spectral density σ^2, and ω_0 is the carrier frequency. The complex matched filter of user k will create two correlation functions for each path

$$y_{ikl}^{(n)} = \int_{nT+\tau_k+\tau_{kl}}^{(n+1)T+\tau_k+\tau_{kl}} r(t) c_{ik}(t - nT - \tau_k + \tau_{kl}) \cos\left(\omega_0 t + \tilde{\phi}_{kl}\right) dt = \tag{33.2.5}$$
$$= \sum_{k'} \sum_{l'} A_{k'l'} \left[d_{ik'} \rho_{ik'l',ikl} \cos \epsilon_{k'l',kl} + d_{qk'} \rho_{qk'l',ikl} \sin \epsilon_{k'l',kl} \right]$$
$$= \sum_{k'} \sum_{l'} y_{ikl}(k'l')$$

where $\tilde{\Phi}_{kl}$ is the estimate of Φ_{kl} and

$$y_{ikl}(k'l') = \gamma_{iikl}(k'l') + y_{iqkl}(k'l') = \tag{33.2.6}$$
$$= A_{k'l'} \left[d_{ik'} \rho_{ik'l',ikl} \cos \epsilon_{k'l',kl} + d_{qk'} \rho_{qk'l',ikl} \sin \epsilon_{k'l',kl} \right]$$

and

$$y_{qkl}^{(n)} = \int_{nT+\tau_k+\tau_{kl}}^{(n+1)T+\tau_k+\tau_{kl}} r(t) c_{qk}(t - nT - \tau_k + \tau_{kl}) \sin\left(\omega_0 t + \tilde{\Phi}_{kl}\right) dt \tag{33.2.7a}$$
$$= \sum_{k'} \sum_{l'} A_{k'l'} \left[d_{qk'} \rho_{qk'l',qkl} \cos \epsilon_{k'l',kl} + d_{ik'} \rho_{ik'l',qkl} \sin \epsilon_{k'l',kl} \right]$$
$$= \sum_{k'} \sum_{l'} y_{qkl}(k'l')$$
$$y_{qkl}(k'l') = y_{qqkl}(k'l') + y_{qikl}(k'l')$$
$$= A_{k'l'} \left[d_{qk'} \rho_{qk'l',qkl} \cos \epsilon_{k'l,'kl} - d_{ik'} \rho_{ik'l',qkl} \sin \epsilon_{k'l',kl} \right], \tag{33.2.7b}$$

where $\rho_{x,y}$ are cross-correlation functions between the corresponding code components x and y. Each of these components is defined with three indices. Parameter $\epsilon_{a,b} = \Phi_a - \tilde{\Phi}_b$, where a and b are defined with two indices. Let the vectors $\Im(\,)$ of MF output samples for the nth symbol interval be defined as

$$\mathbf{y}_{ik}^{(n)} = \Im^L\left(y_{ikl}^{(n)}\right) = \left(y_{ik1}^{(n)}, y_{ik2}^{(n)}, \cdots, y_{ikL}^{(n)}\right)^T, \in C^L, \tag{33.2.8a}$$

$$\mathbf{y}_{qk}^{(n)} = \Im^L\left(y_{qkl}^{(n)}\right) = \left(y_{qk1}^{(n)}, y_{qk2}^{(n)}, \cdots, y_{qkL}^{(n)}\right)^T, \in C^L, \tag{33.2.8b}$$

$$\mathbf{y}_{k}^{(n)} = \mathbf{y}_{ik}^{(n)} j\mathbf{y}_{qk}^{(n)}, \tag{33.2.8c}$$

$$\mathbf{y}^{(n)} = \Im^k\left(\mathbf{y}_k^{T(n)}\right) \in C^{KL}, \tag{33.2.8d}$$

$$\mathbf{y} = \Im^{N_b}\left(\mathbf{y}^{T(n)}\right), \in C^{N_bKL}. \tag{33.2.8e}$$

Let, in general, $\mathbf{R}^{(n)}(i) \in [-1,1]^{KL\times KL}$ be a cross-correlation matrix with partition

$$\mathbf{R}^{(n)}(i) = \begin{pmatrix} \mathbf{R}_{1,1}^{(n)}(i) & \mathbf{R}_{1,2}^{(n)}(i) & \ldots & \mathbf{R}_{1,K}^{(n)}(i) \\ \mathbf{R}_{2,1}^{(n)}(i) & \mathbf{R}_{2,2}^{(n)}(i) & \ldots & \mathbf{R}_{2,K}^{(n)}(i) \\ \vdots & \vdots & \ddots & \vdots \\ \mathbf{R}_{K,1}^{(n)}(i) & \mathbf{R}_{K,2}^{(n)}(i) & \ldots & \mathbf{R}_{K,K}^{(n)}(i) \end{pmatrix} \in R^{KL\times KL}$$
$$= \Im\Im^K\left(\mathbf{R}_{k,k'}^{(n)}, (i)\right). \tag{33.2.9}$$

We now define four specific matrices of the form given by (33.2.9) with the following notation

$$\mathbf{R}^{ii(n)}(i) = \Im\Im^K\left(\mathbf{R}_{k,k'}^{ii(n)}(i)\right) \tag{33.2.10a}$$

$$\mathbf{R}^{qi(n)}(i) = \Im\Im^K\left(\mathbf{R}_{k,k'}^{qi(n)}(i)\right) \tag{33.2.10b}$$

$$\mathbf{R}^{iq(n)}(i) = \Im\Im^K\left(\mathbf{R}_{k,k'}^{iq(n)}(i)\right) \tag{33.2.10c}$$

$$\mathbf{R}^{qq(n)}(i) = \Im\Im^K\left(\mathbf{R}_{k,k'}^{qq(n)}(i)\right) \tag{33.2.10d}$$

where matrices $\mathbf{R}_{k,k'}^{ab(n)}(i) \in R^{L\times L}, \forall k,k' \in \{1,2,\ldots,K\}$ have elements

$$\left(R_{k,k'}^{ii(n)}(i)\right)_{l,l'} = \cos\epsilon_{k'l',kl} \times \int_{-\infty}^{\infty} c_{ik}^{(n)}(t-\tau_k-\tau_{kl}) \tag{33.2.11a}$$
$$c_{ik}^{(n-i)}(t+iT-\tau_{k'}-\tau_{k'l'})\,dt,$$
$$\left(R_{k,k'}^{qi(n)}(i)\right)_{l,l'} = \sin\epsilon_{k'l',kl} \times \int_{-\infty}^{\infty} c_{qk}^{(n)}(t-\tau_k-\tau_{k,l}) \tag{33.2.11b}$$
$$c_{ik'}^{(n-i)}(t+iT-\tau_{k'}-\tau_{k'l'})\,dt,$$
$$\left(R_{k,k'}^{iq(n)}(i)\right)_{l,l'} = -\sin\epsilon_{k'l',kl} \times \int_{-\infty}^{\infty} c_{ik}^{(n)}(t-\tau_k-\tau_{kl}) \tag{33.2.11c}$$
$$c_{qk}^{(n-i)}(t+iT-\tau_{k'}-\tau_{k'l'})\,dt,$$
$$\left(R_{k,k'}^{qq(n)}(i)\right)_{l,l'} = \cos\epsilon_{k'l',kl} \times \int_{-\infty}^{\infty} c_{qk}^{(n)}(t-\tau_k-\tau_{kl}) \tag{33.2.11d}$$
$$c_{qk'}^{(n-i)}(t+iT-\tau_{k'}-\tau_{k'l'})\,dt,$$
$$\forall l,l' \in \{1,2,\ldots,K\}.$$

The vector 33.2.8ad can be expressed as [22]

$$\begin{aligned}\mathbf{y}^{(n)}(\mathbf{R},\mathbf{H},\mathbf{A},\mathbf{d}) &= \mathbf{R}^{(n)}(2)\,\mathbf{H}^{(n-2)}\mathbf{A}\mathbf{d}^{(n-2)} + \\ &\quad +\mathbf{R}^{(n)}(1)\,\mathbf{H}^{(n-1)}\mathbf{A}\mathbf{d}^{(n-1)} \\ &= +\mathbf{R}^{(n)}(0)\,\mathbf{H}^{(n)}\mathbf{A}\mathbf{d}^{(n)} + \\ &\quad +\mathbf{R}^{(n)}(-1)\,\mathbf{H}^{(n+1)}\mathbf{A}\mathbf{d}^{(n+1)} \\ &\quad +\mathbf{R}^{(n)}(-2)\,\mathbf{H}^{(n+2)}\mathbf{A}\mathbf{d}^{(n+2)} + \mathbf{w}^{(n)}\end{aligned} \tag{33.2.12}$$

where

$$\mathbf{A} = diag\,(A_1, A_2, \ldots, A_k) \in R^{K\times K} \tag{33.2.13}$$

is a diagonal matrix of transmitted amplitudes,

$$\mathbf{H}^{(n)} = diag\left(\mathbf{H}_1^{(n)}, \mathbf{H}_2^{(n)}, \ldots, \mathbf{H}_K^{(n)}\right) \in R^{KL\times KL} \tag{33.2.14}$$

is the matrix of channel coefficient vectors,

$$\mathbf{H}_k^{(n)} = \left(H_{k,l}^{(n)}, H_{k,2}^{(n)}, \ldots H_{k,L}^{(n)}\right)^T \in C^L, \tag{33.2.15a}$$
$$\mathbf{d}^{(n)} = \left(d_1^{(n)}, d_2^{(n)}, \ldots d_K^{(n)}\right)^T \in \Xi^K, \tag{33.2.15b}$$

is the vector of the transmitted data, and $\mathbf{w}^{(n)} \in C^{KL}$ is the output vector due to noise. It is easy to show that $\mathbf{R}^{(n)}(i) = \mathbf{0}_{KL}, \forall|i| > 2$ and $\mathbf{R}^{(n)}(-i) = \mathbf{R}^{T(n+1)}(i)$, where $\mathbf{0}_{KL}$ is an all-zero matrix of size $KL \times KL$. Thus, the concatenation vector of the matched filter outputs 33.2.8ae has the expression

$$\mathbf{y}(R, H, A, \mathbf{d}) = RHA\mathbf{d} + \mathbf{w} = RH\mathbf{h} + \mathbf{w} \tag{33.2.16}$$

where

$$R = \begin{pmatrix} \mathbf{R}^{(0)}(0) & \mathbf{R}^{T(1)}(1) & \mathbf{R}^{T(2)}(2) & \cdots & \mathbf{0}_{KL} \\ \mathbf{R}^{(1)}(1) & \mathbf{R}^{(1)}(0) & \mathbf{R}^{T(2)}(1) & \cdots & \mathbf{0}_{KL} \\ \mathbf{R}^{(2)}(2) & \mathbf{R}^{(2)}(1) & \mathbf{R}^{(2)}(0) & \cdots & \mathbf{0}_{KL} \\ \vdots & \vdots & \vdots & \ddots & \vdots \\ \mathbf{0}_{KL} & \mathbf{0}_{KL} & \mathbf{0}_{KL} & \cdots & \mathbf{R}^{(N_b-1)}(0) \end{pmatrix} \in R^{N_bKL \times N_bKL}, \tag{33.2.17a}$$

$$H = diag\left(\mathbf{H}^{(0)}, \mathbf{H}^{(1)}, \ldots \mathbf{H}^{(N_b-1)}\right) \in C^{N_bKL \times N_bKL}, \tag{33.2.17b}$$

$$A = diag\left(\mathbf{A}, \mathbf{A}, \ldots, \mathbf{A}\right) \in R^{N_bK \times N_bK}, \tag{33.2.17c}$$

$$\mathbf{d} = \left(\mathbf{d}^{T(0)}, \mathbf{d}^{T(1)}, \ldots, \mathbf{d}^{T(T_b-1)}\right)^T \in \Xi^{N_bK}, \tag{33.2.17d}$$

$\mathbf{h} = A\mathbf{d}$ is the data-amplitude product vector, and w is the Gaussian noise output vector with zero mean and covariance matrix $\sigma^2 R$. If we define

$$\mathbf{y}_{ii} = \mathbf{y}\left(R^{ii}, H, A, \mathbf{d}_i\right), \tag{33.2.18a}$$

$$\mathbf{y}_{qi} = \mathbf{y}\left(R^{qi}, H, A, \mathbf{d}_q\right), \tag{33.2.18b}$$

$$\mathbf{y}_{iq} = \mathbf{y}\left(R^{iq}, H, A, \mathbf{d}_i\right), \tag{33.2.18c}$$

$$\mathbf{y}_{qq} = \mathbf{y}\left(R^{qq}, H, A, \mathbf{d}_q\right), \tag{33.2.18d}$$

then we have

$$\mathbf{y}_i = \mathbf{y}_{ii} + \mathbf{y}_{qi}, \tag{33.2.19a}$$

$$\mathbf{y}_q = \mathbf{y}_{iq} + \mathbf{y}_{qq}. \tag{33.2.19b}$$

33.3 PERFORMANCE ANALYSIS: CDMA SYSTEM CAPACITY

The starting point in the evaluation of CDMA system capacity is parameter $y_m = E_{bm}/N_0$, the received signal energy per symbol per overall noise density in a given reference receiver with index m. For this analysis, we can represent this parameter in the general case as

$$y_m = \frac{E_{bm}}{N_0} = \frac{ST}{I_{oc} + I_{oic} + I_{oin} + \eta_{th}} \tag{33.3.1}$$

where I_{oc}, I_{oic}, and I_{oin} are power densities of intracell, intercell, and overlay-type internetwork interference, respectively, and η_{th} is thermal noise power density. Parameter S is the overall received power of the useful signal, and $T = 1/R_b$ is the information bit interval. Contributions of I_{oic} and I_{oin} to N_0 have been discussed in a number of papers [4], [5]. To minimize repetition in our analysis, we parameterize this contribution by introducing

$$\eta_0 = I_{oic} + I_{oin} + \eta_{th} \tag{33.3.2}$$

and concentrate on the analysis of the intracell interference in a CDMA network, based on advanced receivers using imperfect Rake and MAI cancellation. A general block diagram of the receiver is shown in Figure 33.1. An extension of the analysis to both intercell and internetwork interference is straightforward.

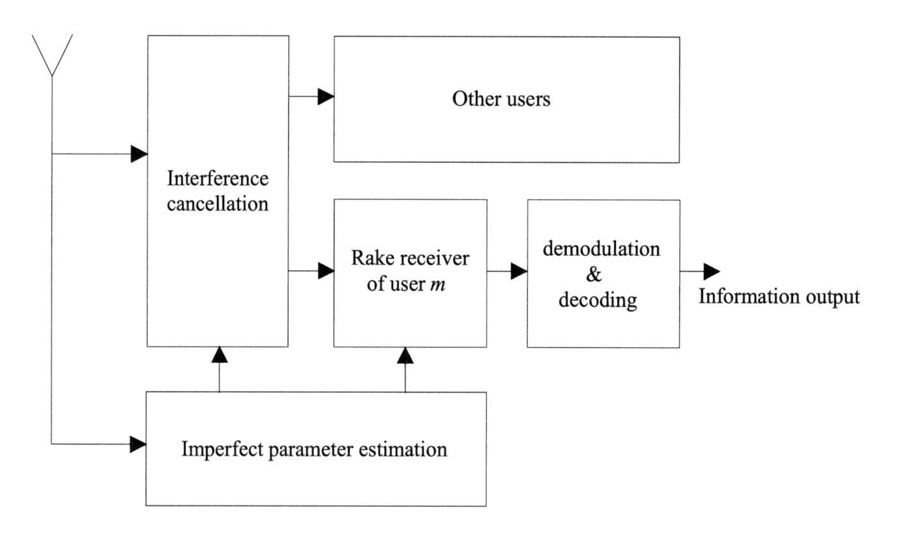

Figure 33.1. General receiver block diagram

33.4 MULTIPATH CHANNEL: NEAR-FAR EFFECT AND POWER CONTROL

We start with the rejection combiner that will choose the first multipath signal component and reject (suppress) the others. In this case, (33.3.1) for I-channel becomes

$$Y_{ibm} = \frac{\alpha_{iim1}(m1) S/R_b}{\left\{ \alpha_{iqm1}(m1) + \sum_{k'=1}^{K} \sum_{\substack{l'=1(k' \neq m) \\ l'=2(k'=m)}}^{L} \alpha_{im1}(k'l') \right\} S/R_b + \eta_0} \tag{33.4.1}$$

$$= \frac{\alpha_{iim1}(m1)}{\alpha_{iqm1}(m1) + \sum_{k'=1}^{K} \sum_{\substack{l'=1(k' \neq m) \\ l'=2(k'=m)}}^{L} \alpha_{im1}(k'l') + \eta_0 R_b + /S}$$

where $\alpha_x(z)$, (for $x = iim1, iqm1, im1$ and $z = m1, k'l'$) is the power coefficient defined as $\alpha_x(z) = E_\epsilon \left\{ |y_x(z)|^2 \right\} /S$, S is the normalized power level of the received signal, and parameters $y_x(z)$ are in general defined by (33.2.5) and (33.2.19). $E_\epsilon \{ \}$ stands for averaging with respect to corresponding phases $\epsilon_{a,b}$ defined by (33.2.7a). Based on this scheme, we have

$$\alpha_{im1}(k'l') = E_\epsilon \left\{ y_{im1}^2(k'l') \right\} = A_{k'l'}^2 \rho_{im1}^2(k'l')/2 \Longrightarrow A_{k'l'}^2/2 \tag{33.4.2}$$

where $\rho_{im1}^2(k'l') = \rho_{ik'l',im1}^2 + \rho_{qk'l',im1}^2, \rho^2 = E_\rho \left\{ \rho_{ik'l',im1}^2 + \rho_{qk'l',im1}^2 \right\}$, and normalization $A_{k'l'}^2 q^2/2 \Longrightarrow A_{k'l'}^2/2$.

Similar equations can be obtained for Q-channel, too. It has been assumed that all "interference per path" components are independent. In what follows, we simplify the notation by dropping all indices $im1$ so that $\alpha_{im1}(k'l') \Longrightarrow \alpha_{kl}$. With no power control (npc), α_{kl} depends only on the channel characteristics. In partial power control (ppc), only the first multipath component of the signal is measured and used in power control (open or closed) loop. Full power control (fpc) normalizes all components of the received signal, and Rake power control (rpc) normalizes only those components combined in the Rake receiver. These concepts for ideal operation are defined by the following equations:

$$npc \Longrightarrow \alpha_{kl} = \alpha_{kl}, \ \forall k, l, \tag{33.4.3a}$$

$$ppc \Longrightarrow \alpha_{kl} = 1, \ \forall k, \tag{33.4.3b}$$

$$fpc \Longrightarrow \sum_{l=1}^{L} \alpha_{kl} = 1, \ \forall k, \tag{33.4.3c}$$

$$rpc \Longrightarrow \sum_{l=1}^{L_0} \alpha_{kl} = 1, \ \forall k \tag{33.4.3d}$$

where L_0 is the number of fingers in the Rake receiver. The contemporary theory in this field does not recognize these options, thus causing misunderstanding and misconceptions in the interpretation of the power control problem in a CDMA network. Although fpc is not feasible, in practice, the analysis including fpc should provide the reference results for the comparison with other, less efficient, options.

Another problem in the interpretation of the results in the analysis of the power control imperfections is caused by the assumption that all users in the network have the same problem with power control. Hence, the imperfect power control is characterized with the same variance as the power control error. This is more than a pessimistic assumption and yet it has been used very often in analyses published so far. If we now introduce matrix $\boldsymbol{\alpha}_m$ with coefficients $||\alpha_{kl}||$, $\forall k, l$ except for $\alpha_{m1} = 0$ and use notation $\mathbf{1}$ for vector of all ones, (33.4.1) becomes

$$Y_{bm} = \frac{\alpha_{m1}}{\mathbf{1} \cdot \boldsymbol{\alpha}_\mathbf{m} \cdot \mathbf{1}^T + \eta_0 R_b / S}. \tag{33.4.4}$$

Compared with (33.4.1), index i is dropped to indicate that the same form of equation is valid for both, the I- and Q-channel defined by (33.2.19).

33.5 MULTIPATH: RAKE RECEIVER AND INTERFERENCE CANCELLING

If an L_0-fingers Rake receiver ($L_0 \leq L$) with combiner coefficients w_{mr} ($r = 1, 2, \ldots, L_0$) and interference canceller is used, the signal-to-noise ratio becomes

$$Y_{bm} = \frac{r_m^{(L_0)}}{f(m, \boldsymbol{\alpha}, \mathbf{c}, \mathbf{r}) K + \zeta_0 \eta R_b / S} \tag{33.5.1}$$

where

$$\zeta_0 = \sum_{r=1}^{L_0} w_{mr}^2 = \mathbf{w}_m \mathbf{w}_m^T; \; \mathbf{w}_m = (w_{m1}, w_{m2}, \ldots, w_{mn}) \tag{33.5.2}$$

is due to Gaussian noise processing in the Rake receiver, and noise density η_0 becomes η because of additional signal processing. Also, we have

$$\begin{aligned} f(m, \mathbf{a}, \mathbf{c}, \mathbf{r}) &= \frac{1}{K} \sum_{\substack{k=1 \\ k \neq m}}^{K} \sum_{r=1}^{L_0} \sum_{l=1}^{L} w_{mr}^2 \alpha_{kl} (1 - C_{kl}) \\ &+ \frac{1}{K} \sum_{r=1}^{L_0} \sum_{\substack{l=1 \\ l \neq r}}^{L} w_{mr}^2 \alpha_{ml} (1 - C_{ml}) \\ &= \frac{1}{K} \left\{ \mathbf{w}_m \left(\mathbf{1} \cdot \boldsymbol{\alpha}_{cmr} \cdot \mathbf{1}^T \right) \mathbf{w}_m^T \right\} \end{aligned} \tag{33.5.3}$$

with $\boldsymbol{\alpha}_{cmr}$ being a matrix of size $K \times L$ with coefficients $||\alpha_{kl}(1 - C_{kl})||$ except for $\alpha_{mr}(1 - C_{mr}) = 0$ and C_{ml} is efficiency of the canceller. Parameter $r_m^{(L_0)}$ in 33.5.1), called Rake receiver efficiency, is given as

$$r_m^{(L_0)} = \left(\sum_{r=1}^{L_0} w_{mr} \cos \epsilon_{\theta mmr} \sqrt{\alpha_{mr}}\right)^2 = \left(\mathbf{w}_m \cdot \boldsymbol{\alpha}_{mm\sqrt{}}\right)^2 \tag{33.5.4}$$

with $\boldsymbol{\alpha}_{mm\sqrt{}} = \left(\cos \epsilon_{\theta mm1} \sqrt{\alpha_{m1}}, \cos \epsilon_{\theta mm2} \sqrt{\alpha_{m2}}, \ldots\right)^T$. Parameter $\epsilon_{\theta mmr} = \theta_{mmr} - \hat{\theta}_{mmr}$ is the carrier-phase synchronization error in receiver m for the signal of user m in path r. We drop index mkl whenever doing so does not result in ambiguity. In the sequel, we use the following notation: $\alpha_{kl} = A_{kl}^2/2$, $\hat{A}_{mkl}$ is the estimation of A_{kl} by receiver m, $\epsilon_a = \Delta A_{mkl}/A_{kl} = \left(A_k l - \hat{A}_{mkl}\right)/A_{kl}$ is the relative amplitude estimation error, ϵ_m = BER = bit error rate, which can be represented as

$$\overline{m\hat{m}} = 1 - 2BER \tag{33.5.5}$$

ϵ_θ = carrier phase estimation error.

For the equal gain combiner (EGC), the combiner coefficients are given as $w_{mr} = 1$. Having in mind the notation used so far in the sequel, we drop index m for simplicity. For the maximal ratio combiner (MRC), the combiner coefficients are based on estimates as

$$\hat{w}_r = \frac{\cos \epsilon_{\theta r}}{\cos \epsilon_{\theta 1}} \cdot \frac{\hat{A}_r}{\hat{A}_1} \cong \frac{\left(1 - \epsilon_{\theta r}^2/2\right)}{\left(1 - \epsilon_{\theta 1}^2/2\right)} \cdot \frac{A_r\left(1 - \epsilon_{ar}\right)}{A_1\left(1 - \epsilon_{a1}\right)}, \tag{33.5.6a}$$

$$E\left\{\hat{w}_r\right\} = w_r\left(1 - \sigma_{\theta r}^2\right)\left(1 + \sigma_{\theta 1}^2\right)\left(1 - \epsilon_{ar}\right)\left(1 + \epsilon_{a1}\right), \tag{33.5.6b}$$

$$E\left\{\hat{w}_r^2\right\} = w_r^2\left(1 - 2\sigma_{\theta r}^2 + 3\sigma_{\theta r}^4\right)\left(1 + 2\sigma_{\theta 1}^2 - 3\sigma_{\theta 1}^4\right) \left(1 - \epsilon_{ar}\right)^2\left(1 + \epsilon_{a1}\right)^2. \tag{33.5.6c}$$

Averaging (33.5.4) gives for EGC

$$\begin{aligned} E\left\{r^{(L_0)}\right\} &= E\left\{\left(\sum_{r=1}^{L_0} \cos \epsilon_{\theta r} \sqrt{\alpha_r}\right)^2\right\} = \\ &= E\left\{\left(\sum_{r=1}^{L_0} \left(1 - \epsilon_{\theta r}^2/2\right) \sqrt{\alpha_r}\right)^2\right\} = \\ &= \sum_r \sum_{\substack{l \\ l \neq r}} \left(1 - \sigma_{\theta r}^2\right)\left(1 - \sigma_{\theta l}^2\right) \sqrt{\alpha_r \alpha_l} + \sum_r \alpha_r\left(1 - 2\sigma_{\theta r}^2 + 3\sigma_{\theta r}^4\right). \end{aligned} \tag{33.5.7}$$

For MRC, the same relation becomes

$$E\left\{r^{(L_0)}\right\} = E\left\{\left(\sum_{r=1}^{L_0} \frac{\alpha_r}{\sqrt{\alpha_1}} \frac{\left(1-\epsilon_{\theta r}^2/2\right)^2}{\left(1-\epsilon_{\theta 1}^2/2\right)} \frac{\left(1-\epsilon_{ar}\right)}{\left(1-\epsilon_{a1}\right)}\right)^2\right\} = \tag{33.5.8}$$

$$= E\left\{\sum_{r=1}^{L_0} \frac{\alpha_r^2}{\alpha_1} \frac{\left(1-\epsilon_{\theta r}^2/2\right)^4}{\left(1-\epsilon_{\theta 1}^2/2\right)^2} \frac{\left(1-\epsilon_{ar}\right)^2}{\left(1-\epsilon_{a1}\right)^2}\right\} +$$

$$+ \sum_r \sum_{\substack{l \\ l \neq r}} \frac{\alpha_r \alpha_l}{\alpha_1} \left(1 - 2\sigma_{\theta r}^2 + 3\sigma_{\theta r}^4\right)\left(1 - 2\sigma_{\theta l}^2 + 3\sigma_{\theta l}^4\right) \times$$

$$\times \left(1 + 2\sigma_{\theta 1}^2 - 3\sigma_{\theta 1}^4\right)\left(1-\epsilon_{ar}\right)\left(1-\epsilon_{al}\right)\left(1+\epsilon_{a1}\right)^2.$$

In order to evaluate the first term, we use limits. For the upper limit, we have

$$\epsilon_{\theta r}^2 \Longrightarrow \epsilon_{\theta 1}^2. \tag{33.5.9}$$

By using this, we have

$$\frac{\left(1-\epsilon_{\theta r}^2/2\right)^4}{\left(1-\epsilon_{\theta 1}^2/2\right)^2} \Longrightarrow \left(1-\epsilon_{\theta 1}^2/2\right)^2, \tag{33.5.10}$$

and the first term becomes [23]

$$\sum_{r=1}^{L_0} \frac{\alpha_r^2}{\alpha_1} \left(1 - 2\sigma_{\theta 1}^2 + 3\sigma_{\theta 1}^4\right) \frac{\left(1-\epsilon_{ar}\right)^2}{\left(1-\epsilon_{a1}\right)^2}. \tag{33.5.11}$$

For the lower limit, we use

$$\epsilon_{\theta 1}^2 \Longrightarrow \epsilon_{\theta r}^2, \tag{33.5.12}$$

and the first term becomes

$$\sum_{r=1}^{L_0} \frac{\alpha_r^2}{\alpha_1} \left(1 - 2\sigma_{\theta r}^2 + 3\sigma_{\theta r}^4\right) \frac{\left(1-\epsilon_{ar}\right)^2}{\left(1-\epsilon_{a1}\right)^2}. \tag{33.5.13}$$

For a signal with I- and Q-components parameter $\cos\epsilon_{\theta r}$ should be replaced by

$$\cos\epsilon_{\theta r} \Longrightarrow \cos\epsilon_{\theta r} + m\rho\sin\epsilon_{\theta r} \tag{33.5.14}$$

where m is the information in the interfering channel (I or Q), and ρ is the cross-correlation between the codes used in the I- and Q-channel. For small tracking errors, this term can be replaced as

$$\cos\epsilon_{\theta r} + m\rho\sin\epsilon_{\theta r} \approx 1 + m\rho\epsilon - \epsilon^2/2 \tag{33.5.15}$$

where we further simplify the notation dropping the subscript $(\,)_{\theta r}$. By using (33.5.15) in (33.5.6–33.5.14), we can derive similar expressions for the complex signal format.

33.6 INTERFERENCE CANCELLER: NONLINEAR MULTIUSER DETECTORS

For the system performance evaluation, we need a model for the canceller efficiency. Linear multiuser structures might not be of much interest in the next generation of the mobile communication systems, where use of long codes will be attractive. An alternative approach is nonlinear (multistage) multiuser detection; that would include channel estimation parameters, too. This approach would be based on interference estimation and cancellation schemes (OKI standard-IS-665/ITU recommendation M.1073 or UMTS recently defined by ETSI) [24], [25].

In general, if the estimates of (33.2.19) are denoted as $\hat{\mathbf{y}}_i$ and $\hat{\mathbf{y}}_q$ then the residual interference after cancellation can be expressed as

$$\begin{aligned} \Delta\mathbf{y}_i &= \mathbf{y}_i - \hat{\mathbf{y}}_i \\ \Delta\mathbf{y}_q &= \mathbf{y}_q - \hat{\mathbf{y}}_q \\ \Delta\mathbf{y} &= \Delta\mathbf{y}_i + j\Delta\mathbf{y}_q = Vec\{\Delta y_\varsigma\} \end{aligned} \tag{33.6.1}$$

where index $\varsigma \Longrightarrow k, l$ spans all combinations of k and l. By using (33.6.1), we can obtain each component $\alpha_{kl}(1 - C_{kl})$ in (33.5.3) as a corresponding entry of $Vec\left\{|\Delta y_\varsigma|^2\right\}$. To further elaborate these components, we use a simplified notation and analysis.

After frequency downconversion and despreading, the signal from user k, received through path l at receiver m, would have the form

$$\hat{S}_{mkl} = \hat{A}_{mkl}\hat{m}_k \cos\hat{\theta}_{mkl} = (A_{mkl} + \Delta A_{mkl})\,\hat{m}_k \cos(\theta_{mkl} + \epsilon_{\theta mkl}) \tag{33.6.2a}$$

for a single-signal component, and

$$\hat{S}^i_{mkl} = \hat{A}_{mkl}\hat{m}_{ki} \cos\theta_{mkl} + \hat{A}_{mkl} + \hat{m}_{kq} \sin\theta_{mkl}, \tag{33.6.2b}$$

$$\hat{S}^q_{mkl} = -\hat{A}_{mkl}\hat{m}_{ki} \sin\theta_{mkl} + \hat{A}_{mkl} + \hat{m}_{kq} \cos\theta_{mkl} \tag{33.6.2c}$$

for a complex (I & Q) signal structure. In a given receiver m, components $\hat{S}^i_{mkl}$ and $\hat{S}^q_{mkl}$ correspond to Δy_{ikl} and Δy_{qkl}. Parameter A_{mkl} includes both amplitude and correlation function. In (33.6.2a), ΔA_{mkl} and $\epsilon_{\theta mkl}$ are amplitude and phase estimation errors. The canceller would create $S_{mkl} - \hat{S}_{mkl} = \Delta S_{mkl}$, and the power of this residual error (with index m dropped for simlicity) would be

$$E_\theta\left\lfloor(\Delta S_{kl})^2\right\rfloor = E_\theta\left[A_{kl} m_k \cos\theta_{kl} - (A_{kl} + \Delta A_{kl})\,\hat{m}_k \cos(\theta_{kl} + \epsilon_{\theta k})\right]^2 \tag{33.6.3}$$

where $E_\theta[\,]$ stands for averaging with respect to θ_{kl} and m_k. Parameter $(\Delta S_{kl})^2$ corresponds to $|\Delta y_n|^2$. This can be represented as

$$E_\theta\left\lfloor(\Delta S_{kl})^2\right\rfloor = \alpha_{kl}\left\lfloor 1 + (1+\epsilon_a)^2 - 2(1+\epsilon_a)\cdot(1-2\epsilon_m)\cos\epsilon_\theta\right\rfloor. \tag{33.6.4}$$

From this equation we have

$$1 - C_{kl} = \frac{1}{\alpha_{kl}} E_\theta\left[(\Delta S_{kl})^2\right] \tag{33.6.5}$$

and

$$C_{kl} = 2(1+\epsilon_a)(1-2\epsilon_m)\cos\epsilon_\theta - (1+\epsilon_a)^2. \tag{33.6.6}$$

Expanding $\cos\epsilon_\theta$ as $1-\epsilon_\theta^2/2$ and averaging gives

$$C_{kl} = 2(1+\epsilon_a)(1-2\epsilon_m)\left(1-\sigma_\theta^2\right) - (1+\epsilon_a)^2. \tag{33.6.7}$$

For zero-mean ϵ_θ,

$$\sigma_\theta^2 = E\left[\epsilon_\theta^2/2\right] \tag{33.6.8}$$

is the carrier phase tracking error variance. For complex (I & Q) signal structures, cancellation efficiency in the I- and Q-channel can be represented as

$$C_{kl}^i = 4(1+\epsilon_a)(1-2\epsilon_m)\left(1+\sigma_\theta^2\right) - 2(1+\epsilon_a)^2 - 1, \tag{33.6.9a}$$

$$C_{kl}^q = 4(1+\epsilon_a)(1-2\epsilon_m)\left(1+\sigma_\theta^2\right) - 2(1+\epsilon_a)^2 - 1. \tag{33.6.9b}$$

So, in this case the canceller efficiency is expressed in terms of amplitude, phase, and data estimation errors. These results should now be used for analysis of the impact of a large scale of channel estimators on overall CDMA network sensitivity. The performance measure of any estimator is the parameter-estimation error variance that should be directly used in (33.6.9) for cancellation efficiency and (33.5.6—33.5.15) for a Rake receiver. If joint parameter estimation is used, based on ML criteria, then the Cramer Rao bound could be used for these purposes. For a Kalman-type estimator, the error covariance matrix is available for each iteration of estimation. If each parameter is estimated independently, then for carrier phase and code-delay estimation error, a simple relation $\sigma_{\theta,\tau}^2 = 1/SNR_L$ can be used, where SNR_L is the signal-to-noise-ratio in the tracking loop. For the evaluation of this SNR_L, the noise power is, in general, given as $N = B_L N_0$. For this case, the noise density N_0 is approximated as a ratio of the overall interference plus noise power divided by the signal bandwidth. The loop bandwidth will be proportional to f_D where f_D is the fading rate (Doppler). The higher the f_D, the higher the loop noise bandwith and the higher the equivalent noise power ($N_0 f_D$). If interference cancellation is performed prior to parameter estimation, N_0 is obtained from $f\,(\,)$ defined by (33.5.3). If parameter estimation is used without interference cancellation, the same $f\,(\,)$ is used with $C_{kl}=0$. In addition,

$$\epsilon_a \Longrightarrow \frac{A-\hat{A}(1-\epsilon_\tau)}{A}\frac{\Delta A+\hat{A}\epsilon_\tau}{A} \tag{33.6.10}$$

$$\epsilon_a \Longrightarrow \epsilon_A + \epsilon_\tau(1-\epsilon_A)$$

where ϵ_τ is the code delay estimation error and $\epsilon_A = \left(A - \hat{A}\right)/A = 1 - \hat{A}/A$. For noncoherent estimation we have [26]

$$\frac{\hat{A}}{A}\left(\frac{\tau}{4\rho}\right)^{1/2} \exp\left(-\frac{\rho}{2}\right)\left\{1 + \rho I_0\left(\frac{\rho}{2}\right) + \rho I_1\left(\frac{\rho}{2}\right)\right\} \tag{33.6.11}$$

where $I_0\,(\,)$ and $I_1\,(\,)$ are Bessel zero and first-order functions, respectively, and ρ is the signal-to-noise ratio.

33.7 APPROXIMATIONS

If we assume that the channel estimation is perfect ($\epsilon_a = \epsilon_\theta = 0$), parameter C_{mkl} becomes

$$C_{kl} = 2\left(1 - 2\epsilon_m\right) - 1 = 1 - 4\epsilon_m \tag{33.7.1}$$

for DPSK modulation $\epsilon_m = (1/2)\exp\left(-y/2\right)$ where y is the signal-to-noise ratio and for CPSK $\epsilon_m = (1/2)\,erfc\left(\sqrt{y}\right)$. So, we have

$$C_{kl} = 1 - 2e^{-y} \text{ for DPSK,} \tag{33.7.2}$$

$$C_{kl} = 1 - 2erfc\left(\sqrt{y}\right) \text{ for CPSK.} \tag{33.7.3}$$

For large y, $C_{kl} \Longrightarrow 1$, and for small y in the DPSK system, we have $e^{-y} \cong 1 - y$ and $C_{kl} \cong 2y - 1$. This can be presented as $C_{kl} = 2Y_b - 1$, where Y_b is given by (33.5.1). Bearing in mind that Y_b depends on C_{kl} we can solve the whole equation through an iterative procedure starting with an initial value of $C_{kl} = 0, \forall, m, k, l$. Similar approximations can be obtained for σ_θ^2 and ϵ_a. From a practical point of view, an attractive solution could be a scheme that would estimate and cancel only the strongest interference (e.g., successive interference cancellation schemes [27], [28]).

33.8 OUTAGE PROBABILITY

The previous section already completely defines the simulation scenario for the system performance analysis. For the numerical analysis, further assumptions and specifications are necessary. First of all, we need the channel model. Exponential multipath intensity profile (MIP) is a widely used analytical model realized as a tapped delay line [29]. It is very flexible in modeling different propagation scenarios. The decay of the profile and the number of taps in the model can vary. Averaged power coefficients in the multipath intensity profile are

$$\overline{\alpha_l} = \overline{\alpha_0}e^{-\lambda l} \quad l, \lambda \geq 0 \tag{33.8.1}$$

where λ is the decay parameter of the profile. Power coefficients should be normalized as

$$\sum_{l=0}^{L-1} \overline{\alpha_0} e^{-\lambda l} = 1. \tag{33.8.2}$$

For $\lambda = 0$, the profile will be flat. The number of resolvable paths depends on the channel chip rate, and this number must be taken into account. We start from (33.5.1) and look for the average system performance for $\rho^2 = 1/G$, where $G = W/R_b$ is the system processing gain and W is the system bandwidth (chip rate). The average signal to noise ratio is expressed as

$$\overline{Y_b} = \frac{r^{(L_0)} G}{f(\boldsymbol{\alpha}) K + \zeta_0 \eta' W/S}. \tag{33.8.3}$$

Now, if we accept some quality of transmission, BER $= 10^{-e}$, that can be achieved with given SNR $= Y_0$, then with the equivalent average interference density $\eta_0 = I_{oic} + I_{oin} + \eta_{th}$, the signal-to-noise ratio will be

$$Y_0 = \frac{r^{(L_0)} G}{\eta_0}. \tag{33.8.4}$$

To evaluate the outage probability P_{out}, we need to evaluate [4]

$$\begin{aligned} P_{out} &= Pr\left(BER > 10^{-e}\right) && (33.8.5) \\ &= Pr\left(MAI + \frac{\eta W}{S} > \eta_0\right) \\ &= Pr\left(MAI > \eta_0 - \frac{\eta W}{S}\right) \\ &= Pr\left(MAI > \delta\right) && (33.8.6) \end{aligned}$$

where δ is given as

$$\delta = \frac{r^{(L_0)} G}{Y_0} - \frac{\eta W}{S}. \tag{33.8.7}$$

It can be shown that this outage probability can be represented by the Gaussian integral

$$P_{out} = Q\left(\frac{\delta - m_g}{\sigma_g}\right). \tag{33.8.8}$$

From (33.8.3) we have for the system capacity K with ideal system components

$$K_{max} = \frac{r_0^{(L_0)} G}{Y_0 f_0(\boldsymbol{\alpha})} - \zeta_0 \eta W / S f_0(\boldsymbol{\alpha}) \tag{33.8.9}$$

Because of imperfections in the operation of the Rake receiver and interference canceller, this capacity is reduced to

$$K' = \frac{r^{(L_0)} G}{Y_0 f(\boldsymbol{\alpha})} - \zeta_0 \eta' W / S f(\boldsymbol{\alpha}). \tag{33.8.10}$$

where $r_o^{(L_0)}$ and $f_0(\boldsymbol{\alpha})$ are now replaced by real parameters $r^{(L_0)}$ and $f(\boldsymbol{\alpha})$ that take into account those imperfections. The system sensitivity function is defined as

$$\Re = \frac{K_{max} - K'}{K_{max}} = \frac{1}{K_{max}} \left\{ \frac{\Delta r^{(n)} G}{Y_0 f(\boldsymbol{\alpha}) f_0(\boldsymbol{\alpha})} - \frac{\zeta_0 \eta' W \Delta f(\boldsymbol{\alpha})}{S f(\boldsymbol{\alpha}) f_0(\boldsymbol{\alpha})} \right\} \tag{33.8.11}$$

where

$$\Delta r^{(L_0)} = r_0^{(L_o)} f(\boldsymbol{\alpha}) - r^{(L_0)} f_0(\boldsymbol{\alpha}), \tag{33.8.12}$$

$$\Delta f(\boldsymbol{\alpha}) = f(\boldsymbol{\alpha}) - f_0(\boldsymbol{\alpha}). \tag{33.8.13}$$

33.9 ILLUSTRATIONS

In this section, we present some numerical results for illustration. The results are obtained for a channel with exponential multipath delay profile with decay factor λ. The number of paths is $L = 4$. The system parameters are chosen from UMTS recommendations: processing gain $G = 256(24\ dB)$, the required signal-to-noise ratio has been set to $Y_0 = 2$ (3 dB), bit rate $R_b = 16\ kbit/s$, and chip rate $R_c = 4.096$ Mchips/s. BPSK modulation is assumed. For all illustrations, the maximum signal-to-noise ratio in the first Rake finger with perfect intracell interference cancellation is normalized to 3.2 (5 dB). Parameter estimation is made before interference cancellation (worst case). Cancelling efficiency for maximum capacity is calculated according to (33.7.3). When estimation errors are included, cancelling efficiencies follow (33.6.7). Carrier-phase tracking-error variance is assumed to be $\sigma_\theta^2 = 1/SNR_L$. For MRC, the amplitude estimation error is approximated from (33.6.11) to follow $\epsilon_a = 1/4SNR_L$.

In Figure 33.2 we present network capacity K' (33.8.10) versus the number of Rake fingers for equal gain combiner (EGC) and $\lambda = 0.25$. For the number of fingers $L_0 > 2$, we see that the capacity starts to decrease due to imperfections in the parameter estimation. The situation is more critical if the fading rate (f_D)

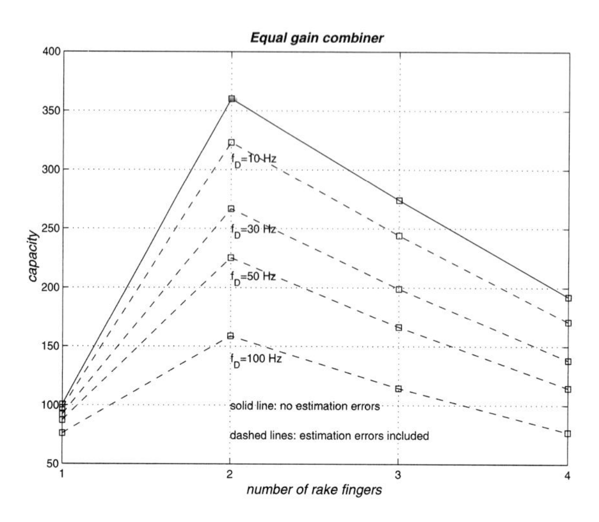

Figure 33.2. Capacity (K') versus the number of Rake fingers for EGC ($\lambda = 0.25$)

is higher. The same parameter for $\lambda = 0$ is presented in Figure 33.3. The signal component in each finger is now stronger, so that capacity will constantly increase if the number of fingers is increased. Once again, the capacity is lower if the fading rate is higher.

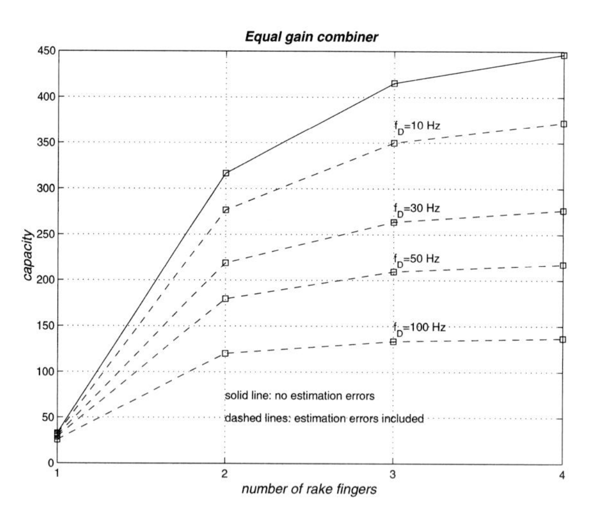

Figure 33.3. Capacity (K') versus the number of Rake fingers for EGC ($\lambda = 0$)

In the case of maximal ratio combiner (MRC), capacity versus the number of Rake fingers for $\lambda = 0.25$ is shown in Figure 33.4. As in Figure 33.2, the capacity decreases with n due to parameter estimation errors, but this is much less critical

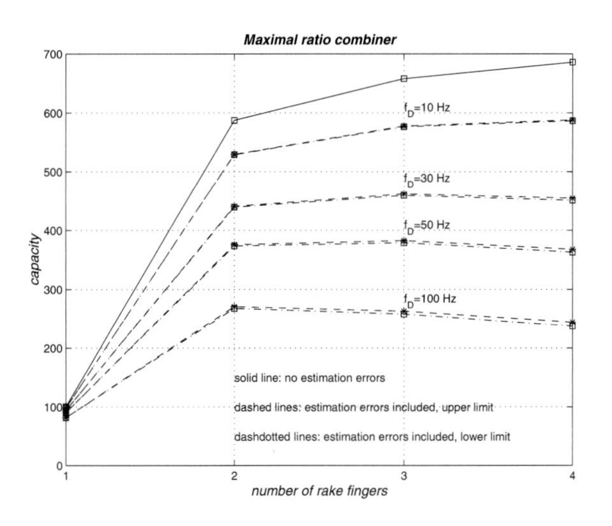

Figure 33.4. Capacity (K') versus the number of Rake fingers for MRC ($\lambda = 0.25$)

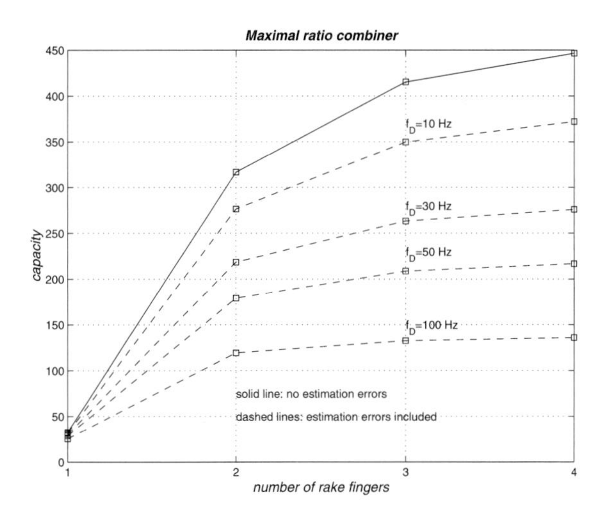

Figure 33.5. Capacity (K') versus the number of Rake fingers for MRC ($\lambda = 0$)

than in the case of EGC. The same parameter for $\lambda = 0$ is shown in Figure 33.5. The signal component in each finger is now stronger, so that the capacity constantly increases if the number of fingers is increased. Once again, the capacity is lower if the fading rate is higher.

In Figure 33.6 we present CDMA network sensitivity function versus the number of Rake fingers for EGC and $\lambda = 0$. For larger n, the sensitivity (percentage of lost capacity) is increased by imperfections in the parameter estimation process. The

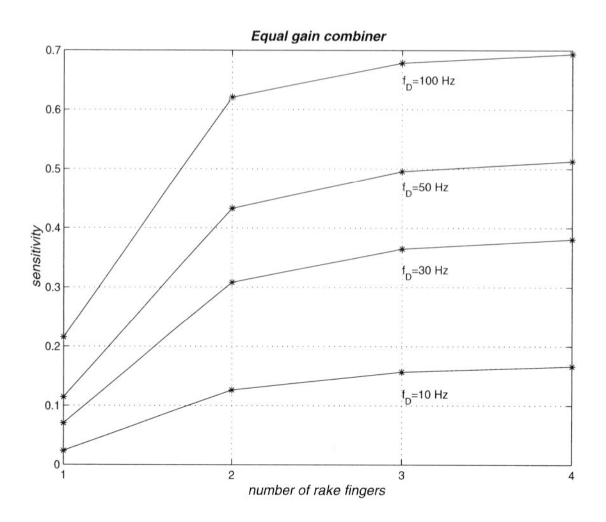

Figure 33.6. CDMA network sensitivity function versus the number of Rake fingers for EGC ($\lambda = 0$)

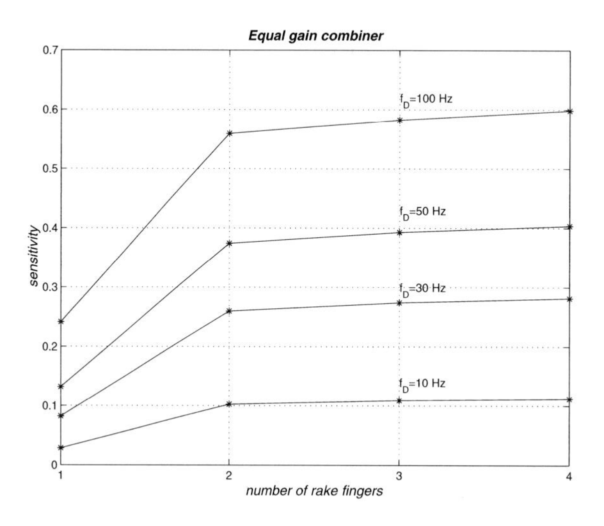

Figure 33.7. CDMA network sensitivity function versus the number of Rake fingers for EGC ($\lambda = 0.25$)

situation is more critical for faster fading. The same parameter for $\lambda = 0.25$ is presented in Figure 33.7. Because of the weaker signal components in Rake fingers with higher index, the network becomes now more sensitive when n increases.

In the case of MRC, CDMA network sensitivity function versus the number of Rake fingers for $\lambda = 0$ is shown in Figure 33.8. Very much the same behavior as

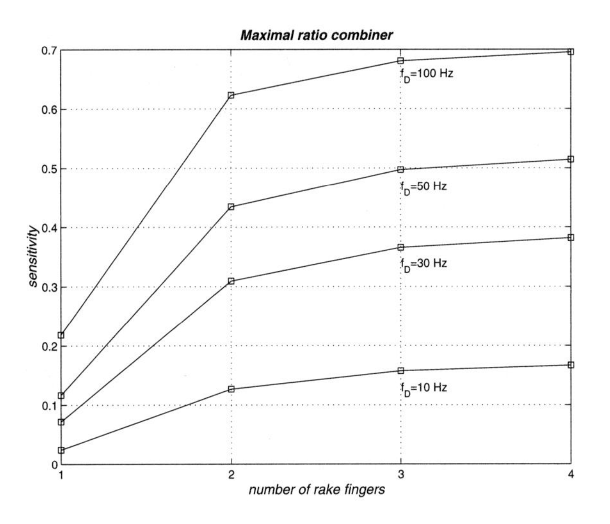

Figure 33.8. CDMA network sensitivity function versus the number of Rake fingers for MRC ($\lambda = 0$)

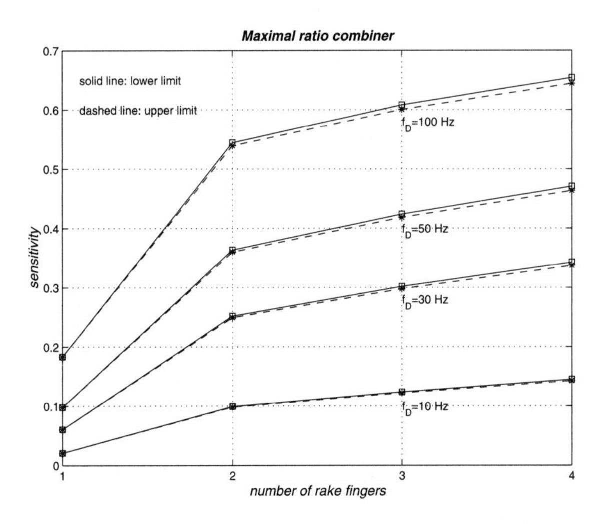

Figure 33.9. CDMA network sensitivity function versus the number of Rake fingers for MRC ($\lambda = 0.25$)

in Figure 33.6 is demonstrated except that the sensitivity is slightly higher due to additional errors in combiner coefficient estimation. Finally, if $\lambda = 0.25$, the same parameter is shown in Figure 33.9. The sensitivity function is now lower because the contribution of the weaker signal components is reduced proportionally.

As a general conclusion one can see that capacity for the EGC (Figure 33.2) is lower than for the MRC (Figure 33.4) due to the inferior combining scheme. On the other hand, the sensitivity for the MRC (Figure 33.9) is higher than for the EGC (Figure 33.7) because the former is using estimated (with errors) signal phases and amplitudes in the combiner. At the same time, EGC has all combining coefficients equal to 1 and only estimated parameters (with errors) are signal phases.

Bibliography

[1] R. L. Pickholtz, L. B. Milstein, and D. L. Schilling, "Spread Spectrum for Mobile Communications," *IEEE Trans. Vehicular Techn.*, vol. 40, pp. 313–321, 1991.

[2] R. Kohno and L. B. Milstein, "Spread spectrum access methods for wireless communications," *IEEE Commun. Mag.*, pp. 58–67, 1995.

[3] W. C. Y. Lee, "Overview of cellular CDMA," *IEEE Trans. Vehicular Techn.*, vol. 40, pp. 291–301, 1991.

[4] K. S. Gilhousen *et al.*, "On the capacity of a cellular cdma system," *IEEE Trans. Vehicular Techn.*, vol. 40, pp. 303–312, 1991.

[5] W. Granzow and W. Koch, "Potential capacity of TDMA and CDMA cellular telephone systems," in *Proc. IEEE ISSSTA'92*, pp. 243–246, 1992.

[6] C.-C. Lee and R. Steele, "Closed-loop power control in CDMA systems," in *IEE Proc.-Commun.*, vol. 143, pp. 231–239, 1996.

[7] G. Falciasecca *et al.*, "Influence of propagation parameters on cellular CDMA capacity and effects of imperfect power control," in *Proc. IEEE ISSSTA'92*, pp. 255–258, 1992.

[8] A. Baiocchi *et al.*, "Effects of user mobility on the capacity of a CDMA cellular network," *European Trans. Telecommunications*, vol. 7, pp. 305–314, 1996.

[9] A. M. Viterbi and A. J. Viterbi, "Erlang capacity of a power controlled CDMA system," *IEEE J. Select. Areas Commun.*, vol. 11, 1993.

[10] M. A. Landolsi, V. V. Veeravalli, and N. Jain, "New results on the reverse link capacity of CDMA cellular networks," in *Proc. IEEE VTC'96*, pp. 1462–1466, 1996.

[11] J. C. Liberti Jr. and T. S. Rappaport, "Analytical results for capacity improvements in cdma," *IEEE Trans. Vehicular Techn.*, vol. 43, pp. 680–690, 1994.

[12] A. F. Naguib *et al.*, "Capacity improvement with base-station antenna arrays in cellular CDMA," *IEEE Trans. Vehicular Techn.*, vol. 43, pp. 691–698, 1994.

[13] J. E. Miller and S. L. Miller, "DS-SS-CDMA uplink performance with imperfect power control and a base station antenna array," in *Proc. IEEE VTC'96*, pp. 66–70, 1996.

[14] L. Tomba, "Outage probability in CDMA cellular systems with discontinuous transmission," in *Proc. IEEE ISSSTA'96*, pp. 481–485, 1996.

[15] L. Tomba, "Computation of the outage probability in Rice fading radio channels," *European Trans. Telecomm.*, vol. 8, pp. 127–134, 1997.

[16] S. Haemaelaeinen, H. Holma, and A. Toskala, "Capacity evaluation of a cellular CDMA uplink with multiuser detection," in *Proc. IEEE ISSSTA'96*, pp. 339–343, 1996.

[17] S. Manji and N. B. Mandayam, "Outage probability for a zero forcing multiuser detector with random signature sequences," in *Proc. IEEE VTC'98*, pp. 174–178, 1998.

[18] H. Holma, A. Toskala, and T. Ojanperae, "Cellular coverage analysis of wideband MUD-CDMA system," in *Proc. IEEE PIMRC'97*, pp. 549–553, 1997.

[19] J. Blanz, A. Klein, M. Nabhan, and A. Steil, "Capacity of a cellular mobile radio system applying joint detection," in *COST 231 TD94 002*, 1994.

[20] J. Blanz, A. Klein, M. Nabhan, and A. Steil, "Performance of a cellular hybrid C/TDMA mobile radio system applying joint detection and coherent receiver antenna diversity," *IEEE J. Select. Areas Commun.*, vol. 12, pp. 568–579, 1994.

[21] W. Huang and V. K. Bhargava, "Performance evaluation of a DS/CDMA cellular system with voice and data services," in *Proc. IEEE PIMRC'96*, pp. 588–592, 1996.

[22] M. Juntti and S. Glisic, *Advanced CDMA for Wireless Communications*, pp. 447–490. Kluwer Academic Publishers.

[23] R. E. Ziemer and W. H. Tranter, *Principles of Communications, Systems, Modulation, and Noise.* John Wiley & Sons, 1995.

[24] A. Fukasawa *et al.*, "Wideband CDMA system for personal radio communications," *IEEE Communications Magazine*, pp. 116–123, 1996.

[25] F. Adachi and M. Sawahashi, "Wideband wireless access based on DS-CDMA," *IEICE Trans. Commun.*, vol. E81-B, pp. 1305–1316, 1998.

[26] H. Meyr and G. Ascheid, *Synchronization in Digital Communications—Vol. 1, Phase-, Frequency-Locked Loops, and Amplitude Control.* New York: John Wiley & Sons, 1990.

[27] P. Patel and J. Holtzman, "Analysis of a simple successive interference cancellation scheme in a DS/CDMA system," *IEEE J. Select. Areas Commun.*, vol. 12, pp. 796–807, 1994.

[28] S. Moshavi, "Multi-user detection for DS-CDMA communications," *IEEE Communications Magazine*, pp. 124–135, 1996.

[29] T. Eng and L. B. Milstein, "Comparison of hybrid FDMA/CDMA systems in frequency selective Rayleigh fading," *IEEE J. Select. Areas Commun.*, vol. 12, pp. 938–951, 1994.

Appendix A

CORRELATION FUNCTION OF THE MFEP OUTPUT

A.1 SLOWLY VARYING CHANNEL ASSUMPTION

For a slowly varying channel, i.e., $h(t, \tau)$ varying slowly over the time period on the order of a symbol duration, $R_h(t_1 - \alpha_1, t_2 - \alpha_2; \tau_1, \tau_2)$ does not change appreciably over the two-dimensional region $(\alpha_1, \alpha_2) \in [0, T_s] \times [0, T_s]$, and it can be approximated by

$$R_h(t_1 - \alpha_1, t_2 - \alpha_2; \tau_1, \tau_2) \approx R_h(t_1 - \frac{T_s}{2}, t_2 - \frac{T_s}{2}; \tau_1, \tau_2). \tag{A.1.1}$$

In this case, equation (29.2.15) becomes

$$R_y(t_1, t_2) = \int_{-\infty}^{+\infty} \int_{-\infty}^{+\infty} R_h(t_1 - \frac{T_s}{2}, t_2 - \frac{T_s}{2}; \tau_1, \tau_2) \widetilde{R}_s^*(t_1 - T_s - \tau_1)\widetilde{R}_s(t_2 - T_s - \tau_2)\, d\tau_1 d\tau_2 , \tag{A.1.2}$$

where

$$\widetilde{R}_s(t - T_s - \tau) \triangleq \int_0^{+\infty} s(t - \alpha - \tau) f_M(\alpha)\, d\alpha . \tag{A.1.3}$$

A.2 UNCORRELATED SCATTERING ASSUMPTION

For an uncorrelated scattering (US) channel, defined as one with channel values at different path delays being uncorrelated,

$$R_h(t_1, t_2; \tau_1, \tau_2) = R_h(t_1, t_2; \tau_1)\delta(\tau_2 - \tau_1). \tag{A.2.1}$$

Therefore, the correlation function of the MFEP output for a slowly varying US channel becomes

$$R_y(t_1, t_2) = \int_{-\infty}^{+\infty} R_h(t_1 - \frac{T_s}{2}, t_2 - \frac{T_s}{2}; \tau) \widetilde{R}_s^*(t_1 - T_s - \tau)\widetilde{R}_s(t_2 - T_s - \tau)\, d\tau . \tag{A.2.2}$$

A.3 WIDE-SENSE STATIONARY CHANNEL ASSUMPTION

Many wireless communications channels can be modeled to possess channel statistics that remain "stationary" over short time intervals (or over small spatial distances). To be precise, these channels are not necessarily stationary in a strict sense nor in the second-order. However under translations over short time intervals, their second-order statistics are invariant and can be approximated as being wide-sense stationary (WSS).[1] For WSS channels, the correlation function depends on t_1 and t_2 only through the difference $t_2 - t_1$, i.e.,

$$R_{\mathrm{h}}(t_1, t_2; \tau_1, \tau_2) = R_{\mathrm{h}}(t_2 - t_1; \tau_1, \tau_2) . \tag{A.3.1}$$

With (A.3.1), the correlation function of the MFEP output for a slowly varying WSSUS channel becomes

$$R_{\mathrm{y}}(t_1, t_2) = \int_{-\infty}^{+\infty} R_{\mathrm{h}}(t_2 - t_1; \tau) \widetilde{R}_{\mathrm{s}}^*(t_1 - T_{\mathrm{s}} - \tau) \widetilde{R}_{\mathrm{s}}(t_2 - T_{\mathrm{s}} - \tau)\, d\tau . \tag{A.3.2}$$

In SS parlance, $\widetilde{R}_{\mathrm{s}}(\tau)$ is the periodic time autocorrelation function of the baseband spread signature sequence. For a reasonable sequence design, $\widetilde{R}_{\mathrm{s}}(\tau)$ possesses small side lobes and a narrow peak over the interval $|\tau| < 1/B_{\mathrm{s}}$, where B_{s} denotes the spreading BW [2]. For DS-CDMA systems using pseudorandom sequences and rectangular chip pulse shape, the function $\widetilde{R}_{\mathrm{s}}(\tau)$ can be written as

$$\widetilde{R}_{\mathrm{s}}(\tau) = \begin{cases} E_{\mathrm{s}} \left[1 - \frac{|\tau|}{T_{\mathrm{c}}}(1 + \frac{1}{N})\right], & |\tau| < T_{\mathrm{c}} \\ -\frac{E_{\mathrm{s}}}{N}, & T_{\mathrm{c}} \le |\tau| \le NT_{\mathrm{c}} , \end{cases} \tag{A.3.3}$$

where E_{s} is the peak value of the autocorrelation function, $N = T_{\mathrm{s}}/T_{\mathrm{c}}$ is known as the processing gain, and T_{c} is the chip duration [2, 3, 4, 5, 6]. The spreading BW B_{s} is roughly equal to the chip rate R_{c} which is defined by $R_{\mathrm{c}} = 1/T_{\mathrm{c}}$. The sidelobes of $\widetilde{R}_{\mathrm{s}}(\tau)$ relative to the peak value are $-1/N$. For DS-CDMA systems with large processing gain, $-1/N$ is small and, from a practical viewpoint, (A.3.3) can be approximated by

$$\widetilde{R}_{\mathrm{s}}(\tau) = \begin{cases} E_{\mathrm{s}} \left[1 - \frac{|\tau|}{T_{\mathrm{c}}}\right], & |\tau| < T_{\mathrm{c}} \\ 0, & T_{\mathrm{c}} \le |\tau| \le NT_{\mathrm{c}} . \end{cases} \tag{A.3.4}$$

This implies that

$$\widetilde{R}_{\mathrm{s}}^*(t_1 - T_{\mathrm{s}} - \tau) \widetilde{R}_{\mathrm{s}}(t_2 - T_{\mathrm{s}} - \tau) \approx 0 , \qquad |t_2 - t_1| \ge \frac{2}{B_{\mathrm{s}}} . \tag{A.3.5}$$

[1] Definitions for different kinds of stationary can be found in [1].

Together with the slowly varying channel assumption, this gives

$$\begin{aligned} R_{\mathrm{h}}(t_2 - t_1; \tau) &\approx R_{\mathrm{h}}(0; \tau) \qquad |t_2 - t_1| < \frac{2}{B_{\mathrm{s}}} \\ &\triangleq P_{\mathrm{h}}(0; \tau) . \end{aligned} \tag{A.3.6}$$

The function $P_{\mathrm{h}}(0;\tau)$ is known as the power delay profile or multipath intensity profile. By use of (A.3.6), the correlation function of the MFEP output for a slowly varying WSSUS channel reduces to

$$R_{\mathrm{y}}(t_1, t_2) = \int_{-\infty}^{+\infty} P_{\mathrm{h}}(0, \tau_1) \underbrace{\widetilde{R}_{\mathrm{s}}^*(t_1 - T_{\mathrm{s}} - \tau)\widetilde{R}_{\mathrm{s}}(t_2 - T_{\mathrm{s}} - \tau)}_{=0\ \forall \quad |t_2 - t_1| \geq \frac{2}{B_{\mathrm{s}}}} d\tau . \tag{A.3.7}$$

Bibliography

[1] W. A. Gardner, *Introduction to Random Processes with Applications to Signals and Systems.* McGraw-Hill, Inc., second ed., 1990.

[2] S. W. Golomb, *Shift Register Sequences.* Laguna Hills, CA: Aegean Park Press, revised ed., 1982.

[3] R. A. Scholtz, "The spread-spectrum concept," *IEEE Trans. Commun.*, vol. COM-25, pp. 748–755, Aug. 1977.

[4] R. L. Pickholtz, D. L. Schilling, and L. B. Milstein, "Theory of spread-spectrum communications – A tutorial," *IEEE Trans. Commun.*, vol. COM-30, pp. 855–884, May 1982.

[5] M. K. Simon, J. K. Omura, R. A. Scholtz, and B. K. Levitt, *Spread Spectrum Communications Handbook.* New York, NY, 10020: McGraw-Hill, Inc., revised ed., 1994.

[6] R. L. Peterson, R. E. Ziemer, and D. E. Borth, *Introduction to Spread Spectrum Communications.* Englewood Cliffs, New Jersey 07632: Prentice Hall, first ed., 1995.

Appendix B

LINEAR AND MATRIX ALGEBRA

B.1 DEFINITIONS

Consider an $m \times n$ matrix $\mathbf{R}$ with elements r_{ij}, $i = 1, 2, \ldots, m; j = 1, 2, \ldots, n$. A shorthand notation for describing $\mathbf{R}$ is

$$[\mathbf{R}]_{ij} = r_{ij}.$$

The transpose of $\mathbf{R}$, which is denoted by bfR^T, is defined as the $n \times m$ matrix with elements $\mathbf{r}_{ji}$ or

$$\left[\mathbf{R^T}\right]_{ij} = r_{ij}.$$

A square matrix is one for which $m = n$. A square matrix is symmetric if $\mathbf{R}^T = \mathbf{R}$. The rank of a matrix is the number of linearly independent rows or columns, whichever is less. The inverse of a square $n \times n$ matrix is the square $n \times n$ matrix $\mathbf{R}^{-1}$ for which

$$\mathbf{R}^{-1}\mathbf{R} = \mathbf{R}\mathbf{R}^{-1} = \mathbf{I}$$

where $\mathbf{I}$ is the $n \times n$ identity matrix. The inverse will exist if and only if the rank of $\mathbf{R}$ is n. If the inverse does not exist, then $\mathbf{R}$ is singular. The determinant of a square $n \times n$ matrix is denoted by $det\,(\mathbf{R})$. It is computed as

$$det\,(\mathbf{R}) = \sum_{j=1}^{n} r_{ij} C_{ij}$$

where

$$C_{ij} = (-1)^{i+j} M_{ij}.$$

M_{ij} is the determinant of the submatrix of $\mathbf{R}$ obtained by deleting the ith row and jth colunm and is termed the minor of r_{ij}. C_{ij} is the cofactor of r_{ij}. Note that any choice of i for $i = 1, 2, \ldots, n$ will yield the same value for $det(\mathbf{R})$. A quadratic form Q is defined as

$$Q = \sum_{i=1}^{n} \sum_{j=1}^{n} r_{ij} x_i x_j.$$

In defining the quadratic form it is assumed that $r_{ji} = r_{ij}$. This entails no loss in generality since any quadratic function can be expressed in this manner. Q can also be expressed as

$$Q = \mathbf{x}^T \mathbf{R} \mathbf{x}$$

where $\mathbf{x} = [x_1 x_2 \ldots x_n]^T$ and $\mathbf{R}$ is a square $n \times n$ matrix with $r_{ji} = r_{ij}$ or $\mathbf{R}$ is a symmetric matrix.

A square $n \times n$ matrix $\mathbf{R}$ is positive semidefinite if $\mathbf{R}$ is symmetric and

$$\mathbf{x}^T \mathbf{R} \mathbf{x} \geq 0$$

for all $\mathbf{x} \neq \mathbf{0}$. If the quadratic form is strictly positive, then $\mathbf{R}$ is positive definite. When referring to a matrix as positive definite or positive semidefinite, one always assumes that the matrix is symmetric. The trace of a square $n \times n$ matrix is the sum of its diagonal elements or

$$tr(\mathbf{R}) = \sum_{i=1}^{n} r_{ii}.$$

A partitioned $m \times n$ matrix $\mathbf{R}$ is one that is expressed in terms of its submatrices. An example is the 2×2 partitioning

$$\mathbf{R} = \begin{bmatrix} \mathbf{R}_{11} & \mathbf{R}_{12} \\ \mathbf{R}_{21} & \mathbf{R}_{22} \end{bmatrix}.$$

Each "element" $\mathbf{R}_{ij}$ is a submatrix of $\mathbf{R}$. The dimensions of the partitions are given as

$$\begin{bmatrix} k \times l & k \times (n-l) \\ (m-k) \times l & (m-k) \times (n-l) \end{bmatrix}.$$

B.2 SPECIAL MATRICES

A diagonal matrix is a square $n \times n$ matrix with $r_{ij} = 0$ for $i \neq j$, or all elements off the principal diagonal are zero. A diagonal matrix appears as

$$\mathbf{R} = \begin{bmatrix} r_{11} & 0 & \cdots & 0 \\ 0 & r_{22} & \cdots & 0 \\ \vdots & \vdots & \ddots & \vdots \\ 0 & 0 & \cdots & r_{nn} \end{bmatrix}.$$

A diagonal matrix will sometimes be denoted by diag $(r_{11}, r_{22}, \ldots\ldots, r_{nn})$. The inverse of a diagonal matrix is found by simple inversion of each element on the principal diagonal. A generalization of the diagonal matrix is the square $n \times n$ block diagonal matrix

$$\mathbf{R} = \begin{bmatrix} R_{11} & 0 & \ldots\ldots\ldots & 0 \\ 0 & R_{22} & \ldots\ldots\ldots & 0 \\ . & & & \\ 0 & 0 & \ldots\ldots\ldots\ldots & R_{kk} \end{bmatrix}$$

in which all submatrices $\mathbf{R}_{ii}$ are square and the other submatrices are identically zero. The dimensions of the submatrices need not be identical. For instance, if $k = 2, \mathbf{R}_{11}$ might have dimension 2×2, while $\mathbf{R}_{22}$ might be a scalar. If all $\mathbf{R}_{ii}$ are nonsingular, then the inverse is easily found as

$$\mathbf{R}^{-1} = \begin{bmatrix} R_{11}^{-1} & 0 & \ldots\ldots\ldots & 0 \\ 0 & R_{22}^{-1} & \ldots\ldots\ldots & 0 \\ . & & & \\ 0 & 0 & \ldots\ldots\ldots\ldots & R_{kk}^{-1} \end{bmatrix}.$$

Also, the determinant is

$$det\,(\mathbf{R}) = \prod_{i=1}^{n} det\,(\mathbf{R}_{ii}).$$

A square $n \times n$ matrix is orthogonal if

$$\mathbf{R}^{-1} = \mathbf{R}^{T}.$$

For a matrix to be orthogonal, the columns (and rows) must be orthonormal, or if

$$\mathbf{R} = [\mathbf{r}_1 \; \mathbf{r}_2 \; \ldots \; \mathbf{r}_n]$$

where $\mathbf{r}_i$ denotes the ith column, the conditions

$$\mathbf{r}_i^T \mathbf{r}_j = \begin{cases} 0 \text{ for } i \neq j \\ 1 \text{ for } i = j \end{cases}$$

must be satisfied.

An idempotent matrix is a square $n \times n$ matrix that satisfies

$$\mathbf{R}^2 = \mathbf{R}.$$

This condition implies that $\mathbf{R}^l = \mathbf{R}$ for $l \geq 1$. An example is the projection matrix

$$\mathbf{R} = \mathbf{H}\left(\mathbf{H}^T\mathbf{H}\right)^{-1}\mathbf{H}^T$$

where $\mathbf{H}$ is an $m \times n$ full rank matrix with $m > n$.

A square $n \times n$ Toeplitz matrix is defined as

$$[\mathbf{R}]_{ij} = r_{i-j}$$

or

$$\mathbf{R} = \begin{bmatrix} r_0 & r_{-1} & r_{-2} & \cdots & r_{-(n-1)} \\ r_1 & r_0 & r_{-1} & \cdots & r_{-(n-2)} \\ \vdots & \vdots & \vdots & \vdots & \vdots \\ r_{n-1} & r_{n-2} & r_{n-3} & \cdots & r_0 \end{bmatrix}.$$

Each element along a northwest-southeast diagonal is the same. If in addition, $r_{-k} = r_k$, then $\mathbf{R}$ is symmetric Toeplitz.

B.3 MATRIX MANIPULATION AND FORMULAS

Some useful formulas for the algebraic manipulation of matrices are summarized in this section. For $n \times n$ matrices $\mathbf{R}$ and $\mathbf{P}$, the following relationships are useful.

$$\begin{aligned}
(\mathbf{RP})^T &= \mathbf{P}^T\mathbf{R}^T \\
\left(\mathbf{R}^T\right)^{-1} &= \left(\mathbf{R}^{-1}\right)^T \\
(\mathbf{RP})^{-1} &= \mathbf{P}^{-1}\mathbf{R}^{-1} \\
det\left(\mathbf{R}^T\right) &= det\left(\mathbf{R}\right) \\
det\left(c\mathbf{R}\right) = c^n det\left(\mathbf{R}\right)\ \ (c\ \text{ a scalar}) \\
det\left(\mathbf{RP}\right) &= det\left(\mathbf{R}\right) det\left(\mathbf{P}\right) \\
det\left(\mathbf{R}^{-1}\right) &= \frac{1}{det\left(\mathbf{R}\right)} \\
tr\left(\mathbf{RP}\right) &= tr\left(\mathbf{PR}\right) \\
tr\left(\mathbf{R}^T\mathbf{P}\right) &= \sum_{i=1}^{n}\sum_{j=1}^{n}\left[\mathbf{R}\right]_{ij}\left[\mathbf{P}\right]_{ij}
\end{aligned}$$

For vectors $\mathbf{x}$ and $\mathbf{y}$, we have

$$\mathbf{y}^T\mathbf{x} = tr\left(\mathbf{x}\mathbf{y}^T\right).$$

It is frequently necessary to determine the inverse of a matrix analytically. To do so, one can use the following formula. The inverse of a square $n \times n$ matrix is

$$\mathbf{R}^{-1} = \frac{\mathbf{C}^T}{det\left(\mathbf{R}\right)}$$

where $\mathbf{C}$ is the square $n \times n$ matrix of cofactors $\mathbf{R}$. The cofactor matrix is defined by

$$\left[\mathbf{C}\right]_{ij} = (-1)^{i+j} M_{ij}$$

where M_{ij} is the minor of r_{ij} obtained by deleting the ith row and jth column of $\mathbf{R}$. Another formula that is quite useful is the matrix inversion lemma

$$(\mathbf{R} + \mathbf{PCD})^{-1} = \mathbf{R}^{-1} - \mathbf{R}^{-1}\mathbf{P}\left(\mathbf{DR}^{-1}\mathbf{P} + \mathbf{C}^{-1}\right)^{-1}\mathbf{DR}^{-1}$$

where it is assumed that $\mathbf{R}$ is $n \times n$, $\mathbf{P}$ is $n \times m$, $\mathbf{C}$ is $m \times m$, and $\mathbf{D}$ is $m \times n$ and that the indicated inverses exist. A special case known as Woodbury's identity results for $\mathbf{P}$ and $n \times 1$ column vector $\mathbf{u}$, $\mathbf{C}$ a scalar of unity, and $\mathbf{D}$ a $1 \times n$ row vector $\mathbf{u}^T$. Then,

$$\left(\mathbf{R} + \mathbf{u}\mathbf{u}^T\right)^{-1} = \mathbf{R}^{-1} - \frac{\mathbf{R}^{-1}\mathbf{u}\mathbf{u}^T\mathbf{R}^{-1}}{1 + \mathbf{u}^T\mathbf{R}^{-1}\mathbf{u}}.$$

One can manipulate partitioned matrices according to the usual rules of matrix algebra by considering each submatrix as an element. For multiplication of partitioned matrices, the submatrices that are multiplied together must be conformable. As an illustration, for 2×2 partitioned matrices:

$$\begin{aligned}\mathbf{RP} &= \begin{bmatrix}\mathbf{R}_{11} & \mathbf{R}_{12}\\ \mathbf{R}_{21} & \mathbf{R}_{22}\end{bmatrix}\begin{bmatrix}\mathbf{P}_{11} & \mathbf{P}_{12}\\ \mathbf{P}_{21} & \mathbf{P}_{22}\end{bmatrix}\\ &= \begin{bmatrix}\mathbf{R}_{11}\mathbf{P}_{11}+\mathbf{R}_{12}\mathbf{P}_{21} & \mathbf{R}_{11}\mathbf{P}_{12}+\mathbf{R}_{12}\mathbf{P}_{22}\\ \mathbf{R}_{21}\mathbf{P}_{11}+\mathbf{R}_{22}\mathbf{P}_{21} & \mathbf{R}_{21}\mathbf{P}_{12}+\mathbf{R}_{22}\mathbf{P}_{22}\end{bmatrix}.\end{aligned}$$

The transposition of a partitioned matrix is formed by transposing the submatrices of the matrix and applying T to each submatrix. For a 2×2 partitioned matrix:

$$\begin{bmatrix}\mathbf{R}_{11} & \mathbf{R}_{12}\\ \mathbf{R}_{21} & \mathbf{R}_{22}\end{bmatrix}^T = \begin{bmatrix}\mathbf{R}_{11}^T & \mathbf{R}_{21}^T\\ \mathbf{R}_{12}^T & \mathbf{R}_{22}^T\end{bmatrix}$$

The extension of these properties to arbitrary partitioning is straightforward.

Determination of the inverses and determinants of partitioned matrices is facilitated by the following formulas. Let R be a square $n \times n$ matrix partitioned as

$$\mathbf{R} = \begin{bmatrix}\mathbf{R}_{11} & \mathbf{R}_{12}\\ \mathbf{R}_{21} & \mathbf{R}_{22}\end{bmatrix} = \begin{bmatrix} k \times k & k \times (n-k)\\ (n-k)\times k & (n-k)\times(n-k)\end{bmatrix}.$$

Then,

$$\mathbf{R}^{-1} = \begin{bmatrix}\left(\mathbf{R}_{11} - \mathbf{R}_{12}\mathbf{R}_{22}^{-1}\mathbf{R}_{21}\right)^{-1} & -\left(\mathbf{R}_{11} - \mathbf{R}_{12}\mathbf{R}_{22}^{-1}\mathbf{R}_{21}\right)^{-1}\mathbf{R}_{12}\mathbf{R}_{22}^{-1}\\ -\left(\mathbf{R}_{22} - \mathbf{R}_{21}\mathbf{R}_{11}^{-1}\mathbf{R}_{12}\right)^{-1}\mathbf{R}_{21}\mathbf{R}_{11}^{-1} & \left(\mathbf{R}_{22} - \mathbf{R}_{21}\mathbf{R}_{11}^{-1}\mathbf{R}_{12}\right)^{-1}\end{bmatrix}$$

$$\begin{aligned}det\left(\mathbf{R}\right) &= \left(det\mathbf{R}_{22}\right) det\left(\mathbf{R}_{11} - \mathbf{R}_{12}\mathbf{R}_{22}^{-1}\mathbf{R}_{21}\right)\\ &= det\left(\mathbf{R}_{11}\right) det\left(\mathbf{R}_{22} - \mathbf{R}_{21}\mathbf{R}_{11}^{-1}\mathbf{R}_{12}\right)\end{aligned}$$

where the inverses of $\mathbf{R}_{11}$ and $\mathbf{R}_{22}$ are assumed to exist.

B.4 THEOREMS

Some important theorems are summarized in this section.

1. A square $n \times n$ matrix $\mathbf{R}$ is invertible (nonsingular) if and only if its columns (or rows) are linearly independent or, equivalently, if its determinant is nonzero. In such a case, $\mathbf{R}$ is full rank.

   Otherwise, it is singular.

2. A square $n \times n$ matrix $\mathbf{R}$ is positive definite if and only if

   - it can be written as

     $$\mathbf{R} = \mathbf{C}\mathbf{C}^T$$

     where bfC is also $n \times n$ and is full rank and hence invertible, or

   - the principal minors are all positive. (The ith principal minor is the determinant of the submatrix formed by deleting all rows and columns with an index greater than i.) If $\mathbf{R}$ can be written as in the previous equation but $\mathbf{C}$ is not full rank or the principal minors are only nonnegative, then $\mathbf{R}$ is positive semidefinite.

3. If $\mathbf{R}$ is positive definite, then the inverse exists and may be found from the previous equation as

   $$\mathbf{R}^{-1} = \left(\mathbf{C}^{-1}\right)^T \left(\mathbf{C}^{-1}\right).$$

4. Let $\mathbf{R}$ be positive definite. If $\mathbf{P}$ is an $m \times n$ matrix of full rank with $m \leq n$, then $\mathbf{P}\mathbf{R}\mathbf{P}^T$ is also positive definite.

5. If $\mathbf{R}$ is positive definite (positive semidefinite), then

   - the diagonal elements are positive (nonnegative)
   - the determinant of $\mathbf{R}$, which is a principal minor, is positive (nonnegative).

B.5 EIGENDECOMPOSITION OF MATRICES

An eigenvector of a square $n \times n$ matrix $\mathbf{R}$ is an $n \times 1$ vector $\mathbf{v}$ satisfying

$$\mathbf{R}\mathbf{v} = \lambda\mathbf{v}$$

for some scalar λ, which may be complex. λ is the eigenvalue of $\mathbf{R}$ corresponding to the eigenvector $\mathbf{v}$. It is assumed that the eigenvector is normalized to have unit length or $\mathbf{v}^T\mathbf{v} = 1$. If $\mathbf{R}$ is symmetric, then one can always find n linearly independent eigenvectors, although they will not in general be unique. An example is the identity matrix for which any vector is an eigenvector with eigenvalue

1. If $\mathbf{R}$ is symmetric, then the eigenvectors corresponding to distinct eigenvalues are orthonormal or $\mathbf{v}_i^T\mathbf{v}_j = \delta_{ij}$ and the eigenvalues are real. If, furthermore, the matrix is positive definite (positive semidefinite), then the eigenvalues are positive (nonnegative). For a positive semidefinite matrix, the rank is equal to the number of non-zero eigenvalues.

The defining previous relation can also be written as

$$\mathbf{R}\left[\mathbf{v}_1\ \mathbf{v}_2\ \cdots \mathbf{v}_n\right] = \left[\lambda_1\mathbf{v}_1\ \lambda_2\mathbf{v}_2 \cdots \lambda_n\mathbf{v}_n\right]$$

or

$$\mathbf{RV} = \mathbf{V}\Lambda$$

where

$$\begin{aligned} \mathbf{V} &= \left[\mathbf{v}_1\ \mathbf{v}_2\ \cdots \mathbf{v}_n\right] \\ \Lambda &= diag\left(\lambda_1, \lambda_2, \cdots, \lambda_n\right). \end{aligned}$$

If $\mathbf{R}$ is symmetric so that the eigenvectors corresponding to distinct eigenvalues are orthonormal and the remaining eigenvectors are chosen to yield an orthonormal eigenvector set, then the $\mathbf{V}$ is an orthonormal matrix. As such, its inverse is $\mathbf{V}^T$, so that the previous equation becomes

$$\begin{aligned} \mathbf{RV} &= \mathbf{V}\Lambda\mathbf{V}^T \\ &= \sum_{i=1}^{n} \lambda_1 \mathbf{v}_i \mathbf{v}_i^T. \end{aligned}$$

Also, the inverse is easily determined as

$$\begin{aligned} \mathbf{R}^{-1} &= \mathbf{V}^{T-1}\Lambda^{-1}\mathbf{V}^{-1} \\ &= \mathbf{V}\Lambda^{-1}\mathbf{V}^T \\ &= \sum_{i=1}^{n} \frac{1}{\lambda_i}\mathbf{v}_i\mathbf{v}_i^T. \end{aligned}$$

A final useful relationship follows as

$$\begin{aligned} det\left(\mathbf{R}\right) &= det\left(\mathbf{V}\right) det\left(\Lambda\right) det\left(\mathbf{V}^{-1}\right) \\ &= det\left(\Lambda\right) \\ &= \prod_{i=1}^{n} \lambda_i. \end{aligned}$$

Appendix C

SYMBOLS

$\mathbf{A}^{\dagger}$	Pseudoinverse of $\mathbf{A}$.
$\mathbf{a}^{(k)}$	Distorted code vector, distorted user signature.
$\mathbf{C}^{(k)}$	Code vector delay matrix for code vector k.
L_s	Sequence length.

SUBJECT INDEX

E

Q

R

ABOUT THE EDITOR

Andreas F. Molisch is Associate Professor for mobile communications and head of the "mobile radio air interface" research group in the mobile radio department of the Institut fuer Nachrichtentechnik und Hochfrequenztechnik (INTHFT), Technische Universitaet Wien, Vienna, Austria.

He received the Dipl. Ing., Dr. techn., and Habilitation degrees in 1990, 1994, and 1999, respectively. Since 1991, he has been with the INTHFT in various capacities. From 1999 to 2000, he was on a part-time leave working at FTW Telecommunications Research Center Vienna. In summer 2000, he was a guest researcher at AT&T Laboratories, Red Bank, NJ.

He has done research in the areas of SAW filters, radiative transfer in atomic vapors, and atomic line filters. His current research interests are bit error probabilities of TDMA mobile communications systems, the measurement and modeling of mobile radio channels, smart antennas, MIMO systems, and OFDM.

Dr. Molisch is the author, co-author, or editor of two books, four book sections, more than forty journal papers, and numerous conference papers. He has participated in the European research initiatives COST 231 and COST 259 as Austrian representative. He received the 1991 GIT-Award of the Austrian Society of Electrical Engineering and the Kardinal-Innitzer Award 1999 for best engineering habilitation thesis, and was appointed Senior Member of the IEEE in 2000.